Computer Vision and Image Processing

Computer Vision and Image Processing:
A Practical Approach Using CVIPtools

Scott E Umbaugh, Ph.D.

To join a Prentice Hall PTR
mailing list, point to:
http://www.prenhall.com/mail_lists/

Prentice Hall PTR
Upper Saddle River, NJ 07458

Library of Congress Cataloging-in-Publication Data

Umbaugh, Scott E.
Computer vision and image processing : a practical approach using CVIPtools / Scott E Umbaugh.
p. cm.
Includes bibliographical references and index.
ISBN 0-13-264599-8 (hardcover)
1. Computer vision. 2. Image processing—Digital techniques. 3. Application software. I. Title.
TA1634.U48 1998
006.4'2—dc21 97-26438
CIP

Editorial/production supervision: *BooksCraft, Inc., Indianapolis, IN*
Cover design director: *Jerry Votta*
Original cover photo: *Mark Zuke*
Processed cover images: *CVIPtools and Scott Umbaugh*
Acquisitions editor: *Bernard Goodwin*
Manufacturing manager: *Alexis R. Heydt*
Marketing manager: *Miles Williams*

Reprinted with corrections June, 1999

Prentice Hall books are widely used by corporations and government agencies for training, marketing, and resale.

The publisher offers discounts on this book when ordered in bulk quantities. For more information, contact: Corporate Sales Department
Phone: 800-382-3419
Fax: 201-236-7141
E-mail: corpsales@prenhall.com.

Or write: Prentice Hall PTR
Corp. Sales Dept.
One Lake Street
Upper Saddle River, NJ 07458

Printed in the United States of America

10 9 8 7 6 5 4 3 2

ISBN: 0-13-264599-8

Prentice-Hall International (UK) Limited, *London*
Prentice-Hall of Australia Pty. Limited, *Sydney*
Prentice-Hall Canada Inc., *Toronto*
Prentice-Hall Hispanoamericana, S.A., *Mexico*
Prentice-Hall of India Private Limited, *New Delhi*
Prentice-Hall of Japan, Inc., *Tokyo*
Prentice-Hall Asia Pte. Ltd., *Singapore*
Editora Prentice-Hall do Brasil, Ltda., *Rio de Janeiro*

Table of Contents

Part II
CVIPtools

Part III
Appendices

To
LaDonna, Robin, and David

Preface

Computer vision deals with the processing of image data for use by a computer. The automatic classification of blood cells in medical images and the robotic control of an unmanned lunar rover are examples of computer vision applications. *Image processing* involves the manipulation of image data for viewing by people. Examples include special effects imaging for motion pictures and the restoration of satellite images distorted by a faulty lens. *Computer imaging* blends the techniques of both computer vision and image processing; consequently, it is a rapidly growing and exciting field to be involved in today. This book presents a unique approach to the practice of computer imaging and will be of interest both to those who want to learn more about the subject and to those who just want to use computer imaging techniques.

Why Write This Book?

This book takes an applications-oriented approach to computer vision and image processing (CVIP) and brings together these two separate but related fields with an engineering perspective. Although a number of good theory-based textbooks are available, they are primarily of two types: those that were developed from an electrical engineering, specifically a digital signal processing perspective (image processing) and those that have evolved from a computer science perspective (computer vision). In recent years a few books have been written that combined these fields, but they, too, have been primarily theoretical in nature. I felt that there was a need for an application-oriented book that would bring these two perspectives together, and this book fills that gap.

The book's development was initiated by my research experience while working toward a PhD in electrical engineering. I encountered two problems then. First, the research and development information available in the professional journals was often incomplete, too specific, or too esoteric. Second, I found that I spent a major portion of my time developing software, severely limiting time available for research and development. The first problem I hope to solve by writing this book—I have organized and clearly presented paradigms and methods for the development of CVIP applications. I attacked the second problem by developing a set of reusable software tools for CVIP applications and research and, in the process, developed a valuable environment for learning about computer imaging.

Who Will Use This Book?

Computer Vision and Image Processing: A Practical Approach Using CVIPtools is intended for use by working professionals in research and development, the commercial sectors, and the academic community. This includes practicing engineers, consultants, and programmers, as well as graphics artists, medical imaging professionals, and multimedia specialists, among others. The book can be used for self study and is of interest to anyone involved with developing computer imaging applications, whether they are engineers, geographers, biologists, oceanographers, or astronomers. At the university it can be used in any computer imaging applications-oriented course, typically in senior-level or graduate courses, or as a laboratory text in a standard computer vision and/or image processing course. It is essentially conceptual in nature, with only as much theory as is necessary to understand the use of the algorithms.

The prerequisites for the book are an interest in the field, a basic background in computers, and a basic math background (algebra/trigonometry). Knowledge of the C programming language will be necessary for those intending to develop algorithms at the programming level. Some background in signal and system theory is required for those intending to gain a deep understanding of the sections on transforms and compression. However, the book is written so that those individuals who do not have this background can learn to use the tools and achieve a conceptual understanding of the material.

Approach

Throughout the book I have opted to use what I call a just-in-time approach to learning. Instead of presenting techniques or mathematical tools when they fit into a nice, neat *theoretical framework*, topics are presented as they become necessary for *practical applications*. For example, the mathematical process of convolution is introduced when it is needed for an image zoom algorithm, and morphological operations are introduced when morphological filtering operations are needed after image segmentation. This approach provides you with the motivation to learn and use the tools and topics because you will see an immediate need for them. It also makes the book more useful to working professionals who may not work through the book sequentially but will refer to a specific section as the need arises.

Organization of the Book

The book is divided into two major parts: I. Computer Vision and Image Processing Fundamentals, and II. CVIPtools. The first part of the book contains all the basic concepts, definitions, models, and algorithms necessary to understand computer imaging. Each chapter includes numerous references and examples for the material presented. The material is presented in a conceptual and application-oriented manner, so that you will immediately understand how each topic fits into the overall framework of CVIP applications development.

Chapter 1 provides an overview of the computer imaging field as well as a basic background in computer imaging systems, human visual perception, and image representation. Chapters 2–5 correspond to the major application areas of CVIP: image analysis, image restoration, image enhancement, and image compression. A system model is presented in the introduction to each of these chapters, which is used to help you understand how each specific algorithm relates to the overall process under discussion. The book follows a logical approach meant to model the processes an experienced CVIP applications developer follows.

The second part of the book, CVIPtools, is all about the software. Chapter 6 describes how to use the software, with examples, and Chapter 7 describes applications that have been developed with CVIPtools. The inclusion of applications that have been developed using the software provides insight into the applications development process. Chapter 8 describes the software development environment provided with CVIPtools and includes programming and tutorial exercises. Including these exercises in the book should encourage you to become actively involved in the learning process and will provide some immediate practical experience. Chapter 9 contains brief library descriptions and function prototypes to allow software developers to use all the CVIPtools functions, and thus develop their own applications. More extensive documentation on the functions, including examples, descriptions, and complete UNIX man (manual) pages are on the CD-ROM in standard UNIX format as well as HTML format.

THE CVIPTOOLS SOFTWARE DEVELOPMENT ENVIRONMENT

The software development environment includes an extensive set of C libraries, a skeleton program for using the libraries called CVIPlab, a Tcl shell called CVIPtcl, and a GUI-based program for the exploration of computer imaging called CVIPtools. The CVIPlab program and all the libraries are platform independent and all are ANSI-C compatible. The Tcl shell contains all the high-level CVIPtools functions, and Tcl scripts can be written for rapid prototyping. Initially developed for UNIX platforms, CVIPtools has recently been ported to Windows NT/95. The only difference between the Windows and UNIX versions is that the CVIPtools RAM-based image viewer requires X-windows and is thus UNIX specific. The CVIPtools software, the libraries, the CVIPlab program, the image files, and the associated documentation are included on the CD-ROM.

The CVIPtools software has been used in projects funded by the National Institutes of Health and the U.S. Department of Defense. Because it is a university-sponsored project, it is continually being upgraded and expanded, and updates are available via the information superhighway (World Wide Web). This software allows you to learn about CVIP topics in an interactive and exploratory manner and to develop your own programming expertise with the CVIPlab program and the associated laboratory exercises. With the CVIPlab program you can link any of the already defined CVIPtools functions, ranging from general-purpose input/output and matrix functions to more advanced transform functions and complex CVIP algorithms. Some of these functions are state-of-the-art algorithms because CVIPtools is constantly being improved by the Computer Vision and Image Processing Laboratory at Southern Illinois University at Edwardsville (SIUE).

Acknowledgments

I thank Southern Illinois University at Edwardsville, specifically the School of Engineering and the Electrical Engineering Department, for their support of this endeavor. I also thank all the students who have taken my computer imaging courses and provided valuable feedback regarding the learning and teaching of the topics. Additionally, in the lab, the students' feedback has been very helpful with the CVIPtools software development. I thank the reviewers of the book for their valuable comments and suggestions, as well as their encouragement: A. Murat Tekalp, University of Rochester; Harley R. Myler, University of Central Florida; Michael Zanger, Maze Systems; Dr. William V. Stoecker, Stoecker and Associates; Dr. Randy Moss, University of Missouri-Rolla; and Dr. Raghu Bollini, Southern Illinois University at Edwardsville.

I also thank the publisher, Prentice Hall PTR, for having the insight, foresight, and good taste to publish the book. Bernard Goodwin and Sophie Papanikolaou and their staff have been very helpful throughout the project, and a special thanks goes to them. I thank BooksCraft, Inc., for all their good work producing the book. Sara Black, who wielded the editor's pen with great skill, and Don MacLaren, BooksCraft's president, deserve a thank you for their excellent work. Thanks to New York University Department of Dermatology for permission to use the skin tumor images, the U.S. Army for permission to use the helicopter images, Stealth Technologies and Washington University for the deformable template MRI images, MIT for permission to use the classic cameraman image, George Dean for the clown photo, and Mark Zuke for the many original photographs he supplied. I also thank the people who appear in the images including, Kim Jackson, Scott Smith, Christopher Umbaugh, Chad Umbaugh, Robin Umbaugh, David Umbaugh, and Jeffrey Zuke.

The CVIPtools software has been developed primarily by myself and a few graduate students: Gregory Hance, Arve Kjoelen, Kun Luo, Mark Zuke, and Yansheng Wei; without their hard work and dedication, CVIPtools would not be as comprehensive and robust as it is. Other students who have contributed directly to its development include Zhen Li, Wenxing Li, Wen Zheng, and Sreenivas Makam. Additional students who have contributed include Kui Cai, Mark Heffron, Asif Haswarey, Melvin Johnson, Ambal Ramachandran, Sridhar Ramalingam, Ramesh Reddy, John Creighton, Brad Walker, Jiaxin Tan, Hong Niu, Dave Lyons, Simon Low, Marc Thompson, Muthu Sankarasubbu, Srinivas Madiraju, Frank Smith, Steve Costello, Joseph Tsai, Ivan Lambov, Brad Noble, Robert McClean, Lance Kendrick, Chandra Swaminathan, Praveen Chandra, Jihong Zhou, and Bob Washington.

I also thank those whose public domain software we have used in various portions of CVIPtools: Jef Pokanzer's pbmplus, Sam Leffler's TIFF library, and Paul Heckbert's Graphics Gems. Thanks to Sven Delmas, the creator of XF, which was used extensively in our GUI development.

Additionally, I thank Arve Kjoelen, who, in addition to being one of the primary CVIPtools developers, was instrumental in setting up our World Wide Web site, as well as providing necessary system support for this project. Arve also deserves thanks for his system administration support of the Computer Vision and Image Processing Laboratory here at SIUE. Additionally, his master's thesis serves as the basis for one of the applications described in Section 7.4.

Gregory Hance deserves credit for the first version of CVIPtools and for helping us get the graphical user interface off the ground. His master's thesis serves as the basis for Section 7.2. In that section I describe an application that was partially funded by the National Institutes of Health, thanks to the efforts of Dr. Randy Moss and Dr. William V. (Van) Stoecker at the University of Missouri-Rolla. Greg, Randy, and Van also contributed to this section via a paper that was published in the *IEEE Engineering in Medicine and Biology Society Magazine*.

I thank Melvin Johnson with the U.S. Army for working hard to help us get support for CVIPtools-related projects. These projects were funded through Systems Dynamics International, Stealth Technologies, and Camber Corporation. These applications are described in Sections 7.3, 7.5, and 7.6. Lance Kendrick's work is the basis for the application described in Section 7.5, and I thank him and Dr. Kurt Smith, President of Surgical Navigation Technologies, for their permission to discuss the Deformable Templates methods.

Kun Luo and Yansheng Wei deserve recognition for our X-windows image viewer, with a thanks to John Bradley, who wrote the XV image viewer, which served as the inspiration for our viewer. Kun Luo and Zhen Li deserve recognition for the implementation of the first version of CVIPtcl, a Tcl shell upon which our GUI is based. Kun Luo dedicated a major amount of his time to the development of the CVIPtools software and helped us greatly in solving many problems. I thank Yansheng Wei for his hard work and dedication on the GUI implementation and for all the related software work he has done. Yansheng Wei made important contributions to the organization of the software, making the software portable, and has provided invaluable service to the project. Both Kun Luo and Yansheng Wei deserve a special thanks for their extraordinary efforts and contributions to CVIPtools.

A graduate student who has worked with me for the past two years, Mark Zuke, deserves credit for helping extensively with the development of the software and the book itself. He is responsible for the GUI design for CVIPtools, the Help pages, the organization of the UNIX man pages, and assistance with the organization of the project in general. For the book, he helped me create all the figures, as well as provided many of the original photographs. Mark has also been very helpful in providing feedback on the book as it was being written. He was also involved with the software development and documentation for the applications described in Sections 7.3 and 7.6. Overall, Mark's contributions to this project have been substantial, and his extra efforts deserve special recognition.

Finally, I thank my family for all their contributions; without them this book would not have been possible. I thank my mom, who instilled in me a love of learning and a sense of wonder about the world around me; my dad, who taught me how to think like an engineer and the importance of self-discipline and hard work; and my brothers, who were there when I was growing up (need I say more?). And my own family, my lovely wife and beautiful children, who made all the sacrifices during the writing of this book, deserve my heartfelt thanks. They endured the late nights and many working weekends, and I thank them for their patience and encouragement during this process.

I

Computer Vision and Image Processing Fundamentals

CHAPTER 1

Introduction to Computer Vision and Image Processing

1.1 Overview: Computer Imaging

Computer imaging is a fascinating and exciting area to be involved in today. The advent of the information superhighway, with its ease of use via the World Wide Web, combined with the advances in computer power have brought the world into our offices and into our homes. One of the most interesting aspects of this information revolution is the ability to send and receive complex data that transcend ordinary written text. Visual information, transmitted in the form of digital images, is becoming a major method of communication in the modern age. *Computer imaging* can be defined as the acquisition and processing of visual information by computer. The importance of computer imaging is derived from the fact that our primary sense is our visual sense, and the information that can be conveyed in images has been known throughout the centuries to be extraordinary—one picture *is* worth a thousand words.

Fortunately, this is the case because the computer representation of an image requires the equivalent of many thousands of words of data, and without a corresponding amount of information the medium would be prohibitively inefficient. The massive amount of data required for images is a primary reason for the development of many subareas within the field of computer imaging, such as image compression and segmentation. Another important aspect of computer imaging involves the ultimate "receiver" of the visual information—in some cases the human visual system and in others the computer itself.

This distinction allows us to separate the field of computer imaging into two primary categories: 1) computer vision and 2) image processing. In *computer vision* applications the processed (output) images are for use by a computer, whereas in *image processing* applications the output images are for human consumption. The human visual system and the computer as a vision system have varying limitations and strengths, and the computer imaging specialist needs to be aware of the functionality of these two very different systems.

These two categories are not totally separate and distinct. The boundaries that separate the two are fuzzy, but this definition allows us to explore the differences between the two and to understand how they fit together (Figure 1.1-1). Historically, the field of image processing grew from electrical engineering as an extension of the signal processing branch, whereas the computer science discipline was largely responsible for developments in computer vision. Recently, these two primary groups have come together to create the modern field of computer imaging. At many universities these two are still separate and distinct, but the commonalities and the perceived needs of each are rapidly bringing the two together.

Figure 1.1-1 Computer Imaging

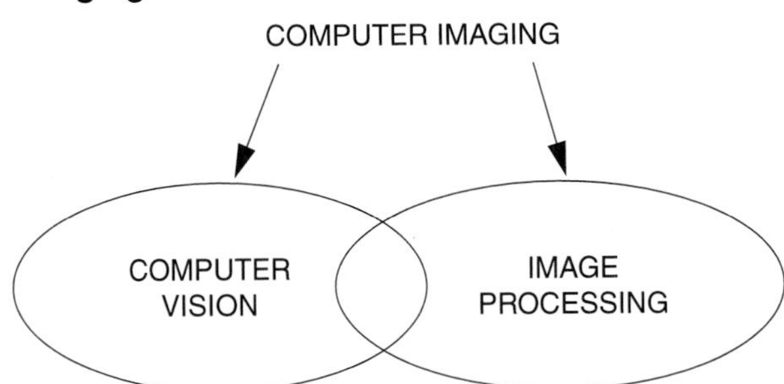

Computer imaging can be separated into two different but overlapping areas.

1.2 Computer Vision

Computer vision is computer imaging where the application does not involve a human being in the visual loop. In other words, the images are examined and acted upon by a computer. Although people are involved in the development of the system, the final application requires a computer to use the visual information directly. One of the major topics within the field of computer vision is image analysis.

Image analysis involves the examination of the image data to facilitate solving a vision problem. The image analysis process involves two other topics: feature extraction and pattern classification. *Feature extraction* is the process of acquiring higher-level image information, such as shape or color information, and *pattern classification* is the act of taking this higher-level information and identifying objects within the image.

The field of computer vision may be best understood by considering different types of applications. Many of these applications involve tasks that either are tedious for people to perform, require work in a hostile environment, require a high rate of processing, or require access and use of a large database of information. Computer vision systems are used in many and various types of environments—from manufacturing plants to hospital surgical suites to the surface of Mars. For example, in manufacturing systems, computer vision is often used for quality control. There, the computer vision system will scan manufactured items for defects and provide control signals to a robotic manipulator to remove defective parts automatically. One interesting example of this type of system involves the creation of one of the first real draft beverages in a can.

The main difference between a draft beverage and a canned beverage is that the draft drink is not carbonated until it is to be consumed, so the problem is to not carbonate the beverage until the can is opened. To do this, the manufacturer inserted a small device that contains the gas used to carbonate the beverage into the can. One of the difficulties they encountered was that if the size of the device varied by a very, very small amount (±01. mm) it would not operate correctly—too big and it deforms the can, too small and it will float to the top of the can. A computer vision system was developed to measure the devices and reject those not within the specified tolerance.

Computer vision systems are used in many different areas within the medical community, with the only certainty being that the types of applications will continue to grow. Current examples of medical systems being developed include: systems to diagnose skin tumors automatically, systems to aid neurosurgeons during brain surgery, and systems to perform clinical tests automatically. Systems that automate the diagnostic process are being developed primarily to be used as tools by medical professionals where specialists are unavailable or to act as consultants to the primary care givers, but they may serve their most useful purpose in the training of medical professionals. Many of these types of systems are highly experimental, and it may be a long time before we actually see computers playing doctor like the holographic doctor in the Star Trek series. Computer vision systems that are being used in surgical suites have already been used to improve the surgeon's ability to "see" what is happening in the body during surgery and consequently improve the quality of medical care available. Systems are also currently being used for tissue and cell analysis. For example, they are being used to automate applications that require the identification and counting of certain types of cells.

The field of law enforcement and security is an active area for computer vision system development, with applications ranging from automatic identification of fingerprints to DNA analysis. Security systems to identify people by retinal scans, facial scans, and the veins in the hand have been developed. Infrared imaging to count bugs has been used at Disney World to help keep their greenery green. Currently, systems are in place to check our highways for speeders automatically, and in the future, computer vision systems may be used to automate our transportation systems fully to make travel safer. The U.S. space program and the Defense Department, with their needs for robots with visual capabilities, are actively involved in research and development of computer vision systems. Applications range from autonomous vehicles to target tracking and identification. Satellites orbiting the earth collect massive amounts of image data every day, and these images are automatically scanned to aid in making maps, predicting the weather, and helping us to understand the changes taking place on our home planet.

1.3 IMAGE PROCESSING

Image processing is computer imaging where the application involves a human being in the visual loop. In other words, the images are to be examined and acted upon by people. For these types of applications, we require some understanding of how the human visual system operates. The major topics within the field of image processing include image restoration, image enhancement, and image compression. As was previously mentioned, image analysis is often used as preliminary work in the development of image processing algorithms, but the primary distinction between computer vision and image processing is that the output image is to be used by a human being.

Image restoration is the process of taking an image with some known, or estimated, degradation, and restoring it to its original appearance. Image restoration is often used in the field of photography or publishing where an image was somehow degraded but needs to be improved before it can be printed. For this type of application, we need to know something about the degradation process in order to develop a model for the distortion. When we have a model for the degradation process, we can apply the inverse process to the image to restore it to its original form. This type of image restoration is often used in space exploration—for example, to eliminate artifacts generated by mechanical jitter in a spacecraft (Figure 1.3-1) or to compensate for distortion in the optical system of a telescope.

Image enhancement involves taking an image and improving it visually, typically by taking advantage of the human visual system's response. One of the simplest and often most dramatic enhancement techniques is to simply stretch the contrast of an image (Figure 1.3-2). Enhancement methods tend to be problem specific. For example, a method that is used to enhance satellite images may not be suitable for enhancing medical images. Although enhancement and restoration are similar in aim, to make an image look better, they differ in how they approach the problem. Restoration methods attempt to model the distortion to the image and reverse this degradation, whereas enhancement methods use knowledge of the human visual system's response to improve an image visually.

Image compression involves reducing the typically massive amount of data needed to represent an image. This is done by eliminating data that are visually unnecessary and by taking advantage of the redundancy that is inherent in most

Figure 1.3-1 Image Restoration

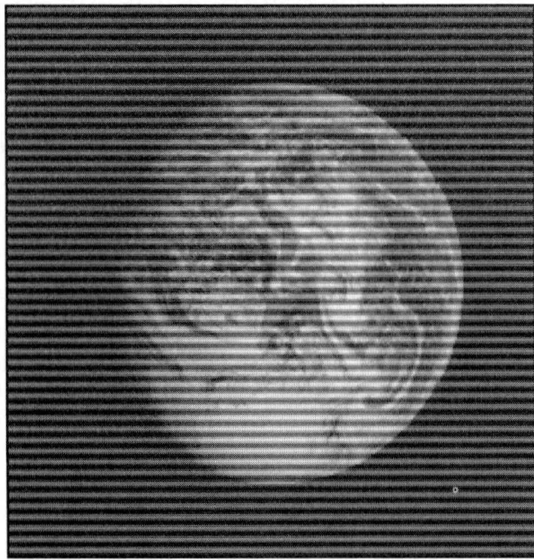

a. Image with distortion.

b. Restored image.

Figure 1.3-2 Contrast Stretching

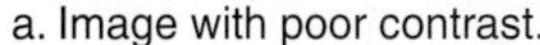

a. Image with poor contrast.

b. Image enhanced by contrast stretching.

images. Although image compression is used in computer vision systems, it is included as an image processing topic because much of the work being done in the field is in areas where we want to compress images that are to be examined by people, so we want to understand exactly what part of the image data is important for human perception. By taking advantage of the physiological and psychological aspects of the human visual system, still image data can be reduced 10 to 50 times, and motion image data (video) can be reduced by factors of 100 or even 200.

The medical community has many important applications for image processing, often involving various types of diagnostic imaging. The beauty of the diagnostic imaging modalities, including PET (Positron Emission Tomography), CT (Computerized Tomography), and MRI (Magnetic Resonance Imaging) scanning, is that they allow the medical professional to look into the human body without the need to cut it open. Image processing is also widely used in many different types of biological research, for example, to enhance microscopic images to bring out features that are otherwise undiscernible. The entertainment industry uses image processing for assembling special effects, editing, and creating artificial scenes and beings (computer animation, closely allied with the field of computer graphics). Image processing is being used to enable people to see what they would look like with a new haircut, a new pair of eyeglasses, or even a new nose. Computer-aided design, which uses tools from image processing and computer graphics, allows the user to design a new building or spacecraft and explore it from the inside out. This type of capability can be used, for example, by people wanting to explore different modifications to their homes, from a new room to new carpeting, and will let them see the end result before the work has even begun. Virtual reality is one application that exemplifies future possibilities, where applications are without bound, and image processing techniques, combined with new developments in allied areas, will continue to affect our lives in ways we can scarcely imagine.

1.4 Computer Imaging Systems

Computer imaging systems are comprised of two primary component types, hardware and software. The hardware components, as seen in Figure 1.4-1, can be divided into the image acquisition subsystem, the computer itself, and the display devices. The software allows us to manipulate the image and perform any desired processing on

Figure 1.4-1 Computer Imaging System Hardware

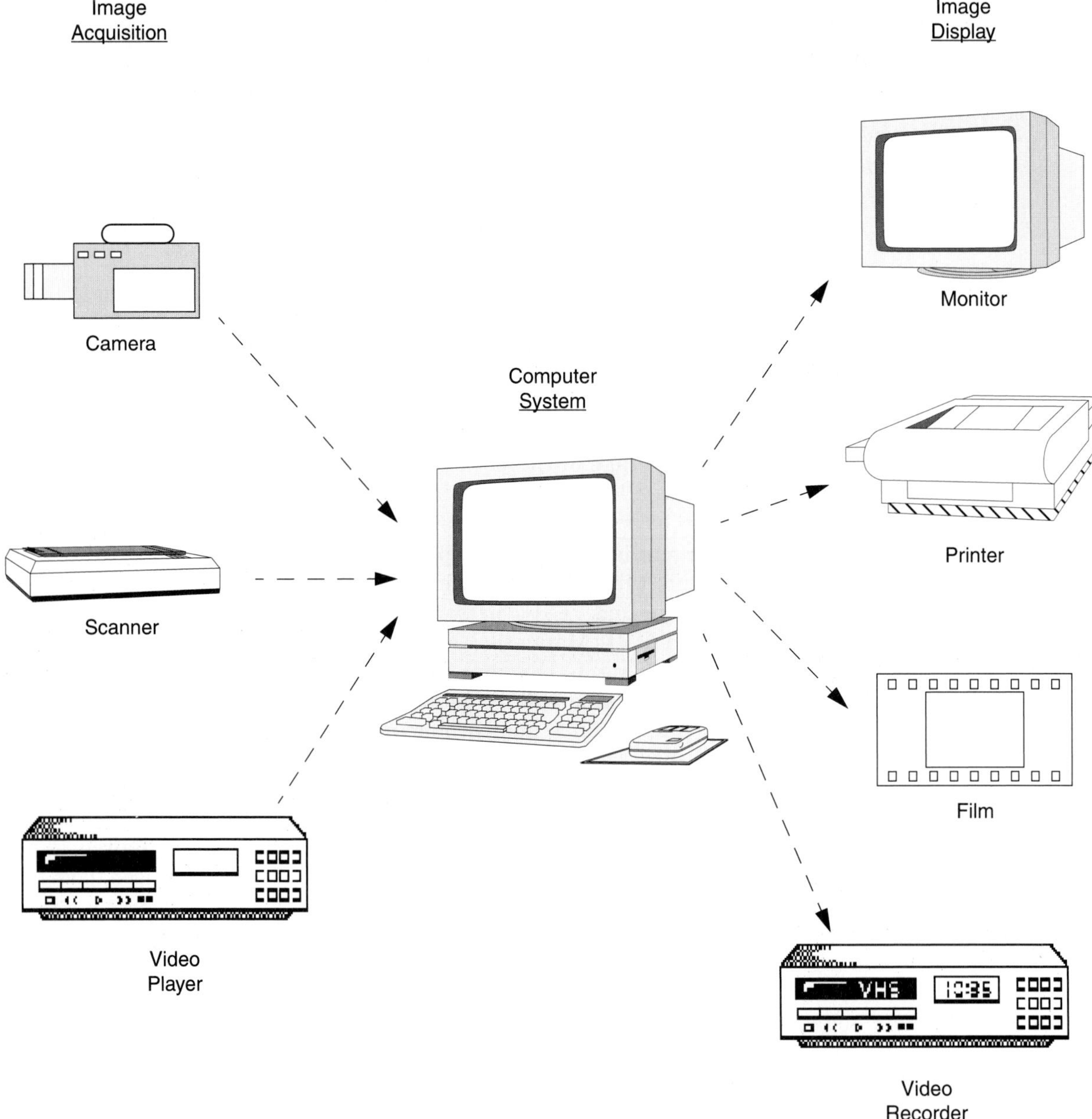

the image data. Additionally, we may also use software to control the image acquisition and storage process.

The computer system may be a general-purpose computer with a frame grabber, or image digitizer, board in it. The *frame grabber* is a special-purpose piece of hardware that accepts a standard video signal and outputs an image in the form that a computer can understand. This form is called a digital image.

The process of transforming a standard video signal into a digital image is called digitization. This transformation is necessary because the standard video signal is in analog (continuous) form, and the computer requires a digitized or sampled version of that continuous signal. A typical video signal contains frames of video information, where each *frame* corresponds to a full screen of visual information. Each frame may then be broken down into *fields*, and each field consists of lines of video information. In Figure 1.4-2a we see the typical image on a display device, where the solid lines

Figure 1.4-2 The Video Signal

a. One frame, two fields.

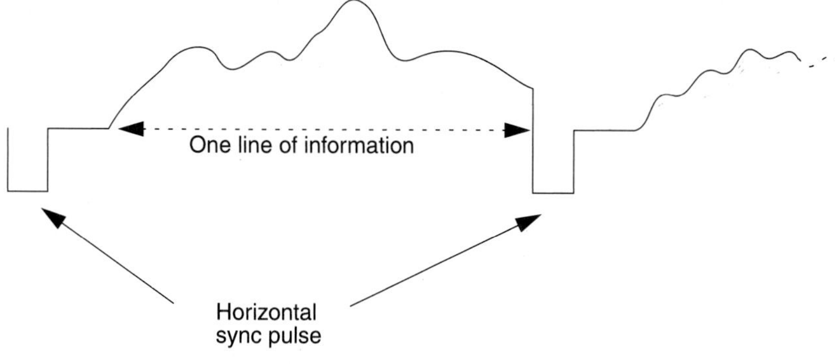

b. The video signal.

represent one field of information and the dotted lines represent the other field. These two fields make up one frame of visual information. This two-fields-per-frame model is referred to as *interlaced* video. Some types of video signals, called *noninterlaced* video, have only one field per frame.

In Figure 1.4-2b we see the electrical signal that corresponds to one line of video information. Note the *horizontal synch pulse* between each line of information; this synchronization pulse tells the display hardware to start a new line. After one frame has been displayed, a longer synchronization pulse, called the *vertical synch pulse*, tells the display hardware to start a new field or frame.

The analog video signal is turned into a digital image by sampling the continuous signal at a fixed rate. In Figure 1.4-3 we see one line of a video signal being sam-

Figure 1.4-3 Digitizing (Sampling) an Analog Video Signal

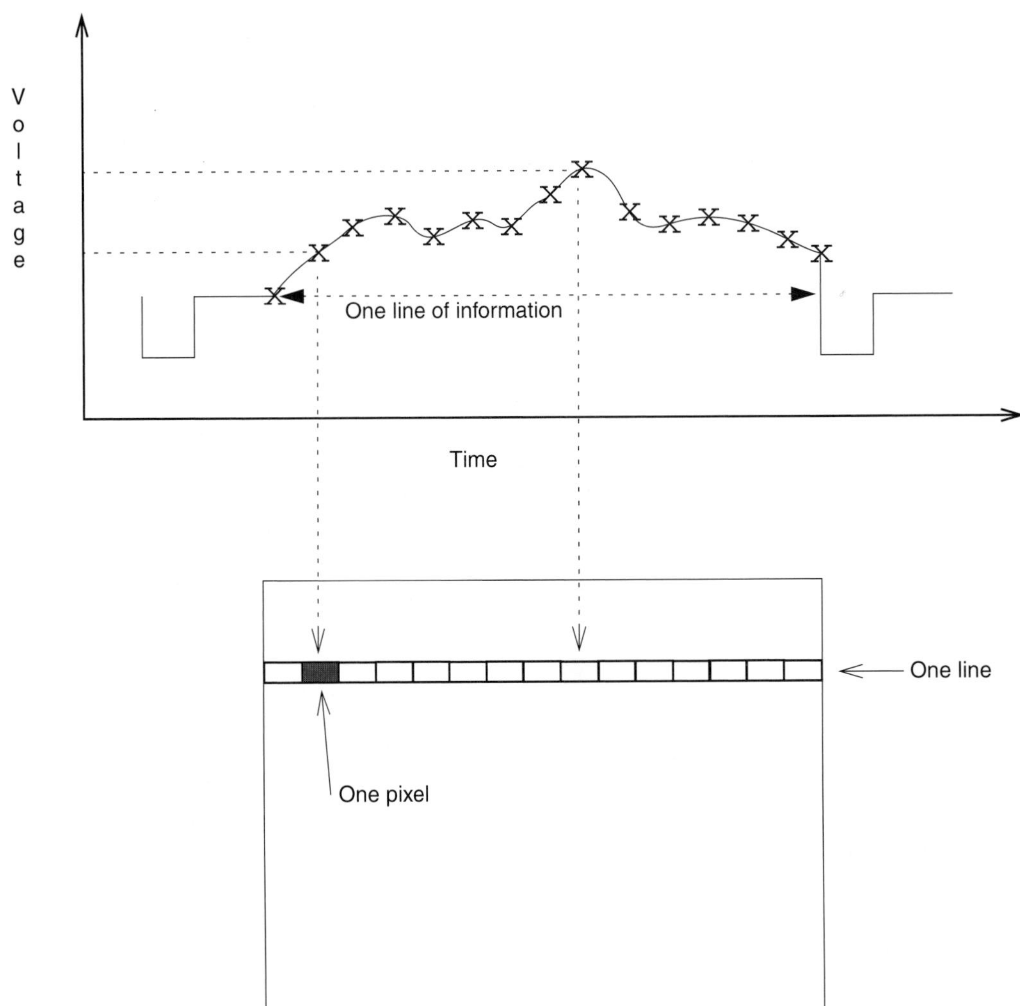

pled (digitized) by instantaneously measuring the voltage of the signal at fixed intervals in time. The value of the voltage at each instant is converted into a number that is stored, corresponding to the brightness of the image at that point. Note that the image brightness at a point depends on both the intrinsic properties of the object *and* the lighting conditions in the scene (this model is explored in Section 4.3.3). When this process has been completed for an entire frame of video information, we have "grabbed" a frame, and the computer can store it and process it as a digital image.

The image can now be accessed as a two-dimensional array of data, where each data point is referred to as a *pixel* (picture element). For digital images we will use the following notation:

$$I(r, c) = \text{the brightness of the image at the point } (r, c)$$

where r = row and c = column.

When we have the data in digital form, we can use the software to process the data. This processing can be illustrated in a hierarchical manner, as seen in Figure 1.4-4. At the very lowest level we deal with the individual pixels, where we may perform some low-level preprocessing. The next level up is the *neighborhood*, which typically consists of a single pixel and the surrounding pixels, and we may continue to perform some preprocessing operations at this level. As we continue to go up the pyramid, we get higher and higher levels of image representations and, consequently, a reduction in the amount of data. All the types of operations and image representations in Figure 1.4-4 will be explored in the following chapters.

1.5 THE CVIPTOOLS SOFTWARE

The CVIPtools (computer vision and image processing tools) software was developed at Southern Illinois University at Edwardsville and contains ANSI-compatible C functions to perform all the operations that are discussed in this book. These are divided into libraries that are explored in detail in the second part of the book. The software is on CD-ROM (see Appendix A) and can also be accessed via the World Wide Web (see Appendix B).

The philosophy underlying the development of the CVIPtools is to allow the non-programmer to have access to a wide variety of computer imaging operations (not just the "standard" ones) and to provide a platform for the exploration of these operations by allowing the user to vary all the parameters and observe the results in almost real time. This is especially facilitated by the CVIPlab program with the associated laboratory exercises and tutorials. Additionally, the function libraries allow those with programming skills to develop their own applications with a minimum of coding.

The CVIPtools software will perform computer imaging operations from simple image editing to complex segmentation or compression algorithms. One of the primary advantages of the software is that it is continually under development in a university environment, so algorithms are developed and made available for exploration long before they are available in any commercial imaging software. Another advantage is that it is being developed for educational purposes, not simply end-user results, so the focus is on *learning* about computer imaging.

Figure 1.4-4 The Hierarchical Image Pyramid

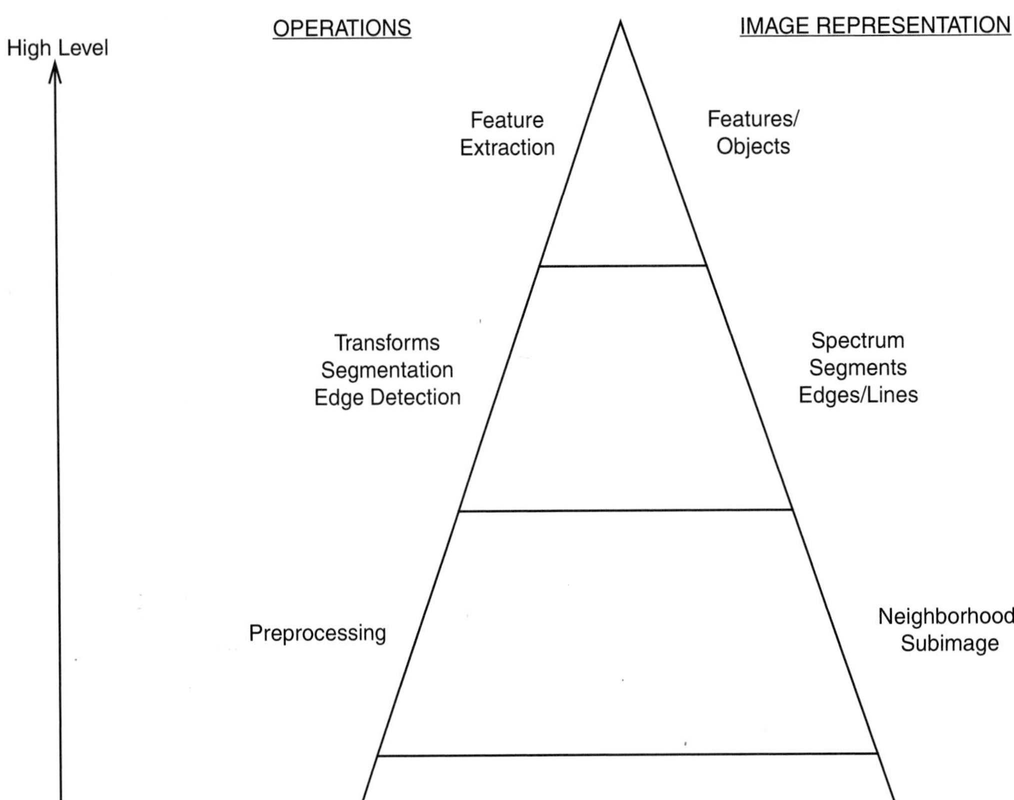

1.6 Human Visual Perception

Human visual perception is something most of us take for granted. We do not think about how the makeup of the physiological systems affects what we see and how we see it. Although human visual perception encompasses both physiological and psychological components, we will focus primarily on the physiological aspects, which are more easily quantifiable, using the current models available for understanding the systems.

The first question is Why study visual perception? We have briefly discussed the need to understand how we perceive visual information in order to design compression algorithms that compact the data as much as possible but still retain all the necessary visual information. This is desired for both transmission and storage economy. Images are often transmitted over the airwaves and will be transmitted more frequently via the World Wide Web (Internet), but people do not want to wait minutes or hours for the images. Additionally, the storage requirements can become overwhelming without compression. For example, an 8-bit monochrome image, with a resolution of 512 pixels wide by 512 pixels high, requires one quarter of a megabyte of data. If we make this a color image, it requires three quarters of a megabyte of data (one quarter for each of three color bands—red, green, and blue). With many applications requiring the capability to process thousands of images in relatively short periods of time, the need to reduce data is apparent. For the development of image enhancement algorithms, we also have the need to understand how the human visual system works. We need to know the types of operations that are likely to improve an image visually, and this can be achieved only by understanding how the information is perceived.

1.6.1 The Human Visual System

The human visual system has two primary components—the eye and the brain, which are connected by the optic nerve (see Figure 1.6-1). The structure that we know the most about is the image receiving sensor—the human eye. The brain can be thought of as being an information processing unit, analogous to the computer in our computer imaging system. These two are connected by the optic nerve, which is really a bundle of nerves that contains the pathways for the visual information to travel from the receiving sensor (the eye) to the processor (the brain). The way the human visual system works follows: 1) light energy is focused by the lens of the eye onto the sensors on the retina; 2) these sensors respond to this light energy by an electrochemical reaction that sends an electrical signal down the optic nerve to the brain; and 3) the brain uses these nerve signals to create neurological patterns that we perceive as images. The visible light energy corresponds to an electromagnetic wave that falls into the wavelength range of about 380 to 825 nanometers, although the response above 700 nanometers is minimal. How this fits in with other parts of the electromagnetic spectrum can be seen in Figure 1.6-2. Note that we cannot "see" many other parts of

Figure 1.6-1 The Human Visual System

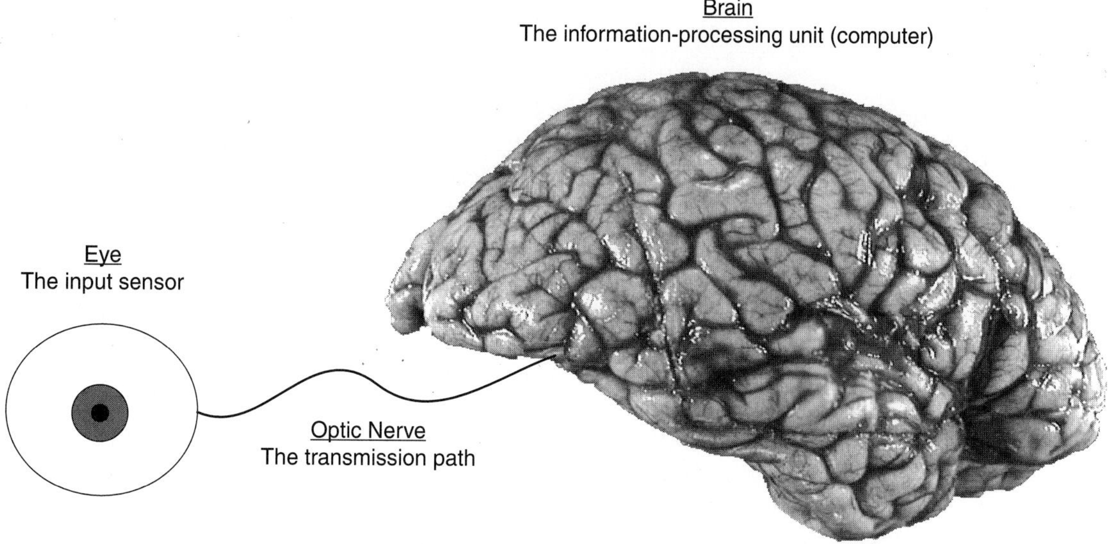

Figure 1.6-2 The Electromagnetic Spectrum

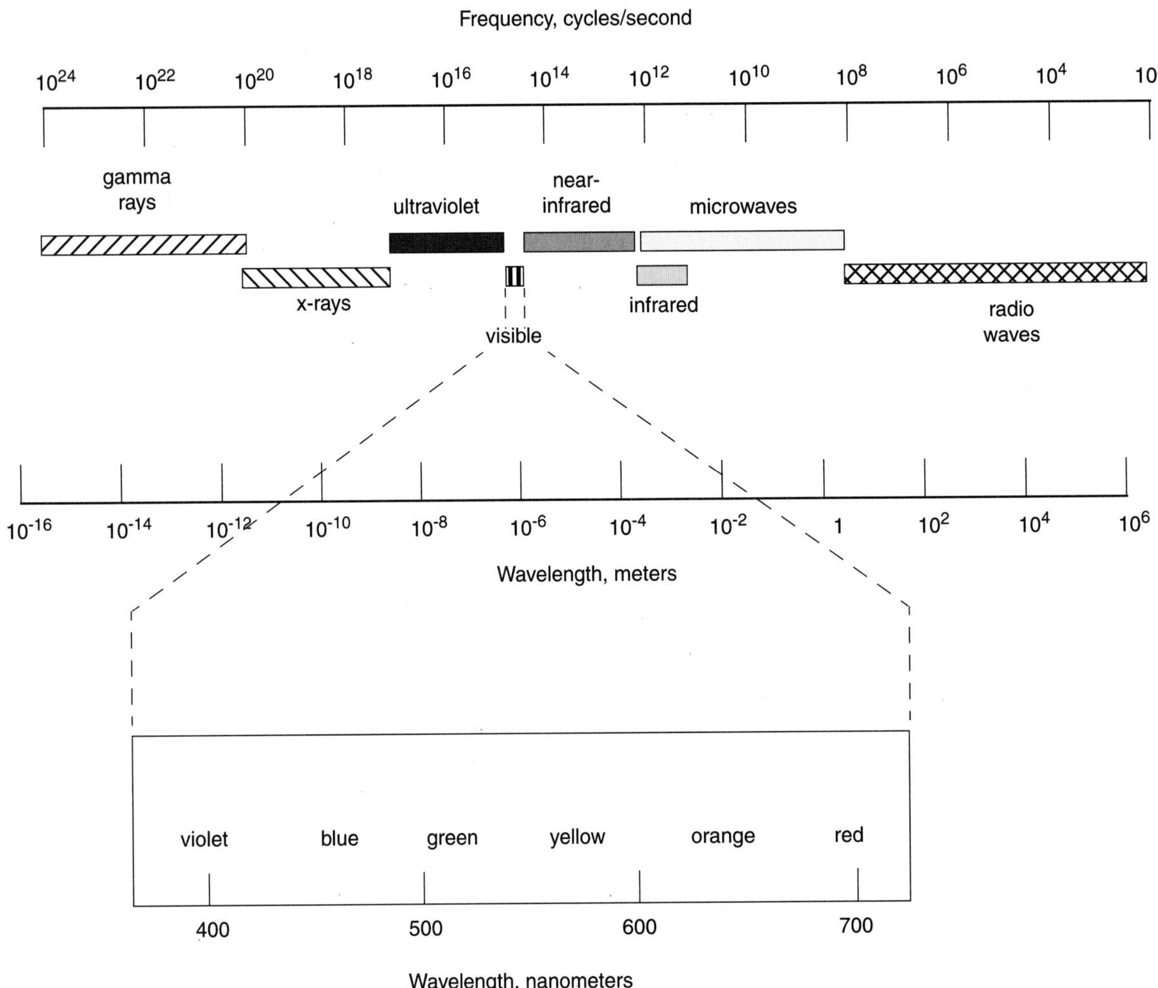

the spectrum. In imaging systems the spectrum is often divided into various *spectral bands*, where each band is defined by a range on the wavelengths (or frequency). For example, we can divide the visible spectrum into roughly three bands corresponding to "blue" (400 to 500 nm), "green" (500 to 600 nm), and "red" (600 to 700 nm).

The eye has two primary types of light energy receptors, or photoreceptors, which respond to the incoming light energy and convert it into electrical energy via a complex electrochemical process. These two types of sensors are called rods and cones. The sensors are distributed across the *retina*, the inner backside of the eye where the light energy falls after being focused by the lens (Figure 1.6-3). The *cones* are primarily used for daylight vision, are sensitive to color, are concentrated in the central region of the eye, and have a high resolution capability. The *rods* are used in night vision, see only brightness (not color), are distributed across the retina, and have medium to low level resolution. In Figure 1.6-3 we can see that there is one place on the retina where no light sensors exist; this is necessary to make a place for the optic

Figure 1.6-3 The Human Eye

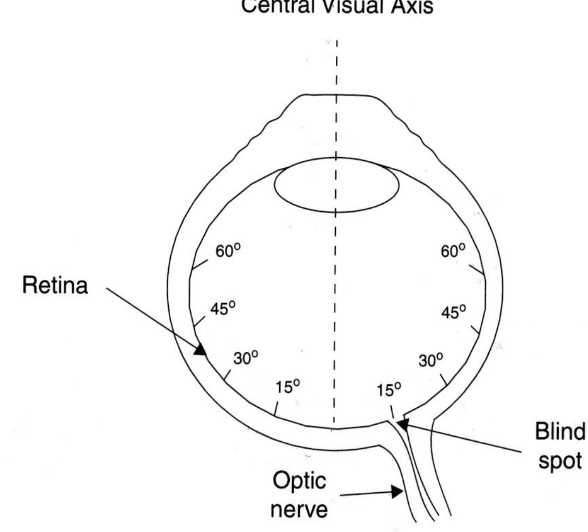

a. Basic eye structure.

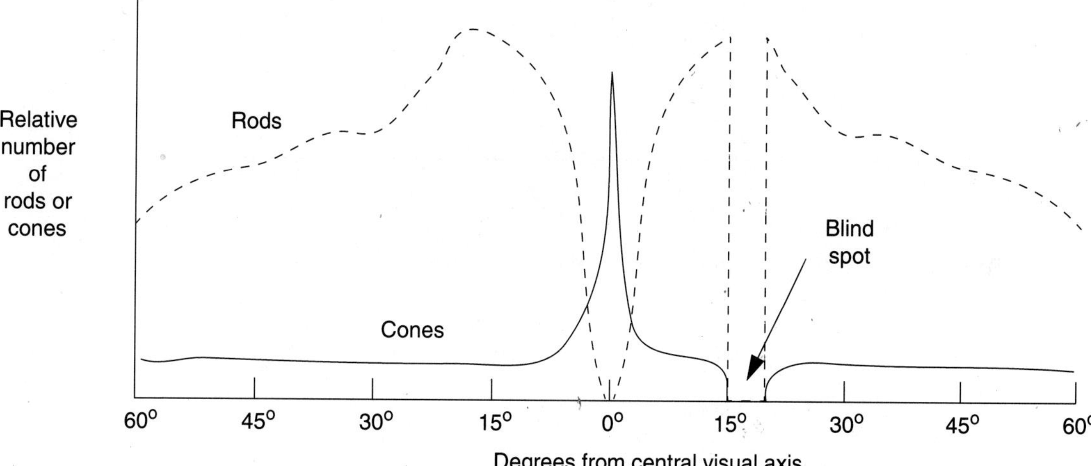

b. Concentration of rods and cones across retina.

nerve and is referred to as the *blind spot.* One of the amazing aspects of the human brain is that we do not perceive this as a blind spot; the brain fills in the missing visual information. By examining Figure 1.6-3b, we can see why an object must be in our central field of vision, which is only a few degrees wide, in order to really perceive it in fine detail. This is where the high-resolution-capability cones are concentrated. They have a higher resolution than the rods because they have individual nerves tied to each sensor, whereas the rods have multiple sensors connected to each nerve. The distribution of the rods across the retina shows us that they are more numerous than cones and that they are used for our peripheral vision—there are very few cones away from the central visual axis. The response of the rods to various wavelengths of light is shown in Figure 1.6-4a.

Figure 1.6-4 Relative Responses of Rods and Cones

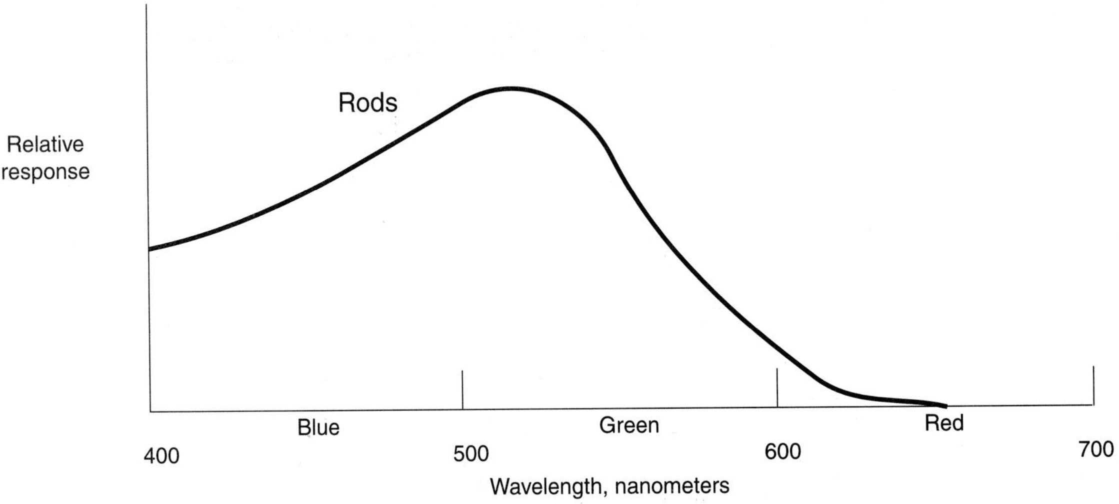

a. Rods react even in low light levels but see only a single spectral band; they cannot distinguish colors.

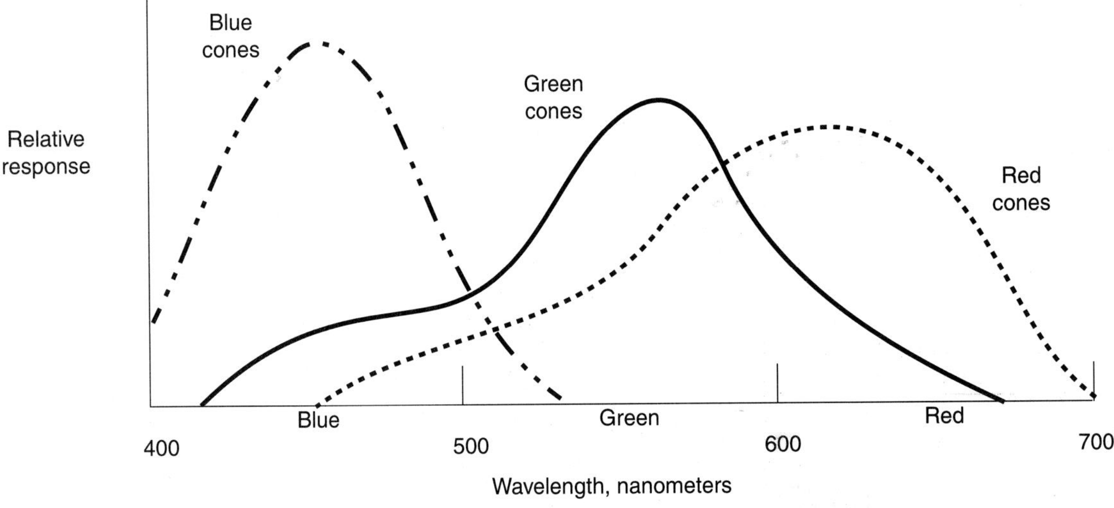

b. Cones react only to high light intensities; the three types enable us to see colors.

There are three types of cones, each responding to different wavelengths of light energy. The response to these can be seen in Figure 1.6-4b, the tristimulus (three stimuli) curves. These are called the tristimulus curves because all the colors that we perceive are the combined result of the response to these three sensors. These curves plot the wavelength versus the relative intensity of the sensor's response. About 65% of the cones respond most to wavelengths corresponding to red and yellow; 33%, to green; and 2%, to blue.

1.6.2 Spatial Frequency Resolution

In order to understand the concept of spatial frequency, we first need to define exactly what we mean by resolution. *Resolution* has to do with the ability to separate two adjacent pixels—if we can see two adjacent pixels as being separate, then we can say that we can resolve the two. If the two appear as one and cannot be seen as separate, then we cannot resolve the two. The concept of resolution is closely tied to the concept of spatial frequency, as can be seen in Figure 1.6-5. In Figure 1.6-5a we use a square wave to illustrate the *spatial frequency* concept, where *frequency* refers to how rapidly the signal is changing in *space*, and the signal has two values for the brightness—0 and Maximum. If we use this signal for one line (row) of an image and then repeat the line down the entire image, we get an image of vertical stripes, as in Figure 1.6-5b. If we increase this frequency, the stripes get closer and closer together (Figure 1.6-5c), until they finally blend together as in Figure 1.6-5d. (Remember that we are discussing the resolution of the visual system, not the display device. Here we are assuming that the display device has enough resolution to separate the lines; the resolution of the display device must also be considered in many applications.)

By looking at Figure 1.6-5d and moving it away from our eyes, we can see that the spatial frequency concept must include the distance from the viewer to the object as part of the definition. With a typical television image, we cannot resolve the individual pixels unless we get very close, so the distance from the object is important when defining spatial frequency. We can eliminate the necessity to include distance by defining spatial frequency in terms of cycles per degree, which provides us with a relative measure. A cycle is one complete change in the signal. For example, in the square wave it corresponds to one high point and one low point, thus we need at least two pixels for a cycle. When we use cycles per degree, the degree refers to field of view (the width of your thumb held at arm's length is about one degree, and television sets are typically designed for fields of view of about 5 to 15 degrees). This is illustrated in Figure 1.6-6a, where the same spatial frequency (in cycles per degree) must have larger cycles as we get farther away from the eye. In other words, as in Figure 1.6-6b, in

Figure 1.6-5 Resolution and Spatial Frequency

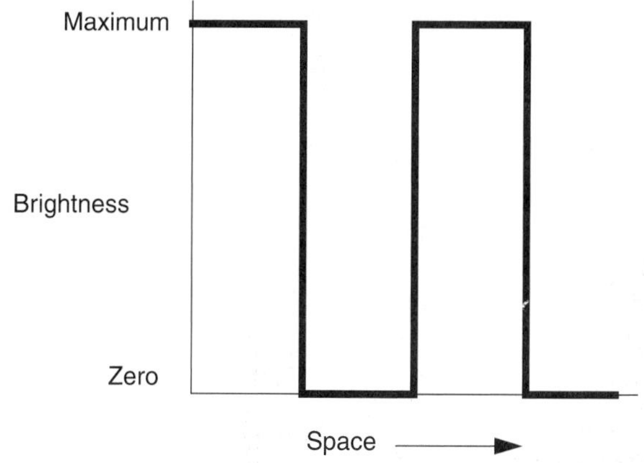

a. Square wave used to generate (b).

b. Low frequency ($f = 2$).

Figure 1.6-5 (Continued)

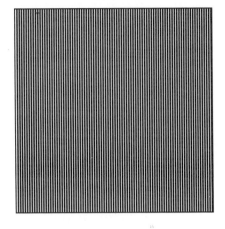

c. High frequency ($f = 10$).

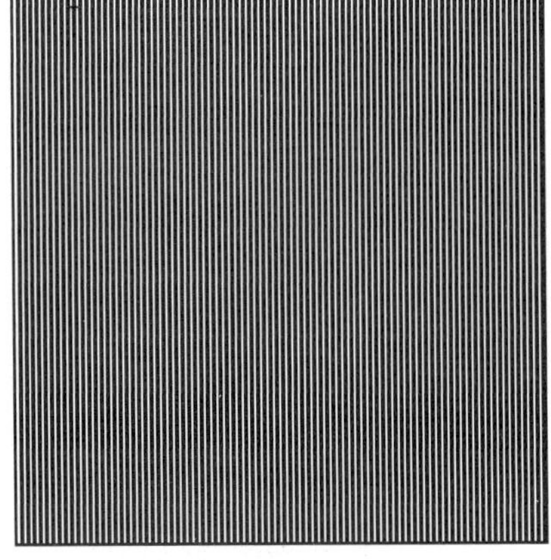

d. Very high frequencies are difficult to resolve ($f = 75$).

Figure 1.6-6 Cycles per Degree

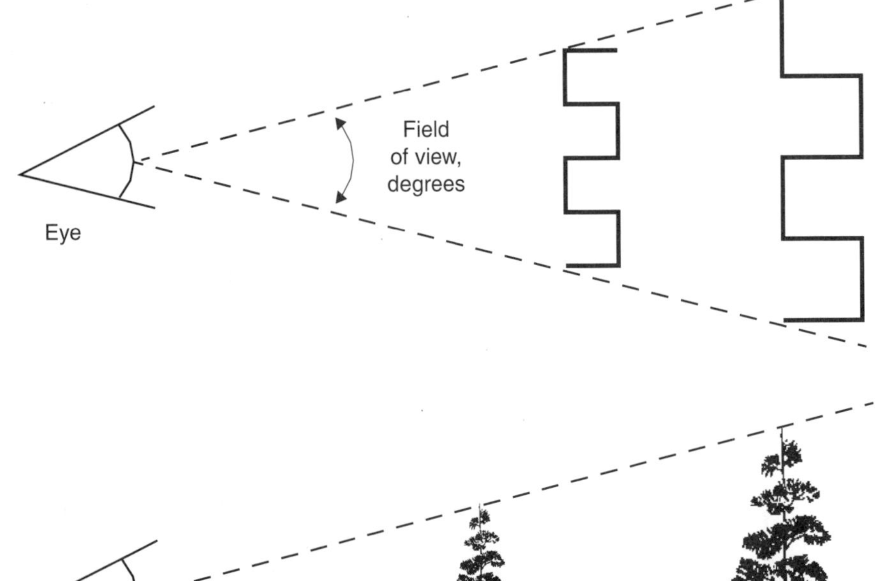

a. With a fixed field of view of a given number of cycles, the farther from the eye, the larger each cycle must be.

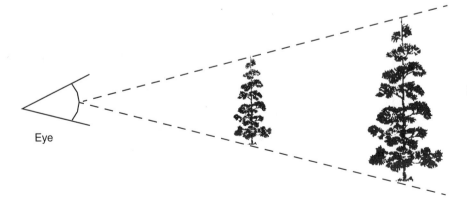

b. A larger, more distant object can appear to be the same size as a smaller, closer object.

order for a larger object to appear the same size, it must be farther away. This definition decouples the distance of the observer from consideration and provides a metric for measuring the spatial resolution of the human visual system.

The physical mechanisms that affect the spatial frequency response of the visual system are both optical and neural. We are limited in spatial resolution by the physical size of the image sensors, the rods and cones; we cannot resolve things smaller than the individual sensor. The primary optical limitations are caused by the lens itself—it is of finite size, which limits the amount of light it can gather and typically contains imperfections that cause distortion in our visual perception. Although gross imperfections can be corrected with lenses (glasses or contacts), subtle flaws cannot be corrected. Additionally, certain factors such as the lens being slightly yellow (which progresses with age) limit the eye's response to various wavelengths of light.

The spatial resolution of the visual system has been empirically determined and is plotted in Figure 1.6-7. Note that the spatial resolution is affected by the average (background) brightness of the display, the two plots correspond to different bright-

Figure 1.6-7 Spatial Resolution

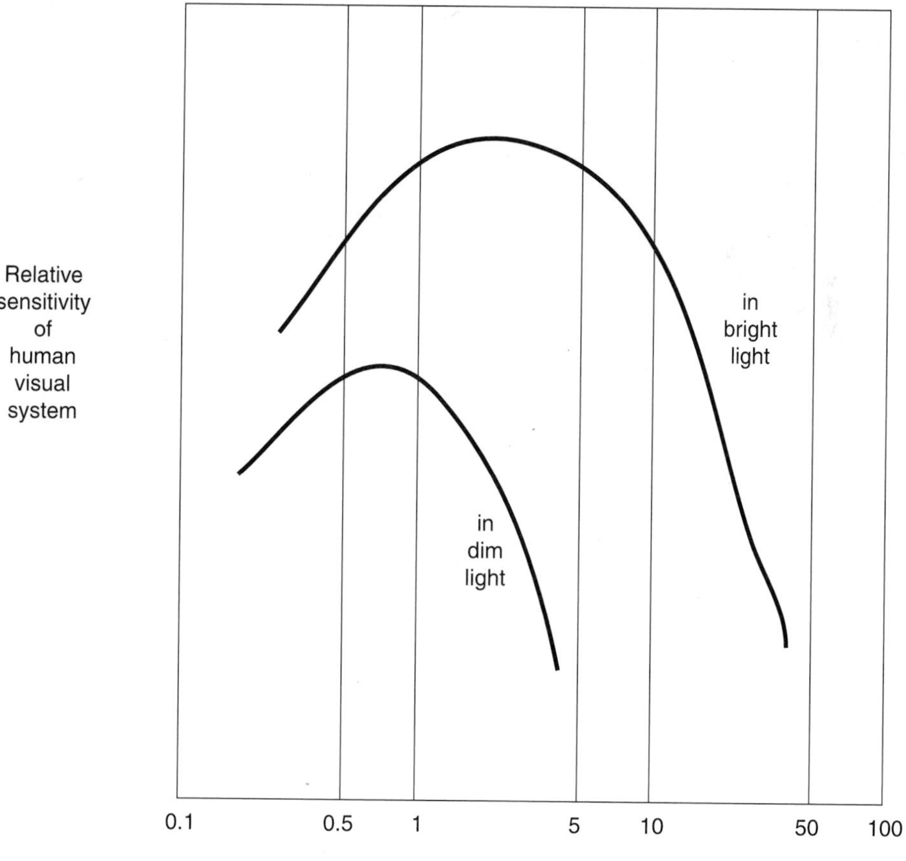

ness levels. In general, we have the ability to perceive higher spatial frequencies at brighter levels, but overall the cutoff frequency is about 50 cycles per degree, peaking at around 4 cycles per degree. These plots are for monochrome images; for color information that has been decoupled from the brightness information (as in television signals), the visual system has less spatial resolution.

Figure 1.6-8 Log of Light Intensity

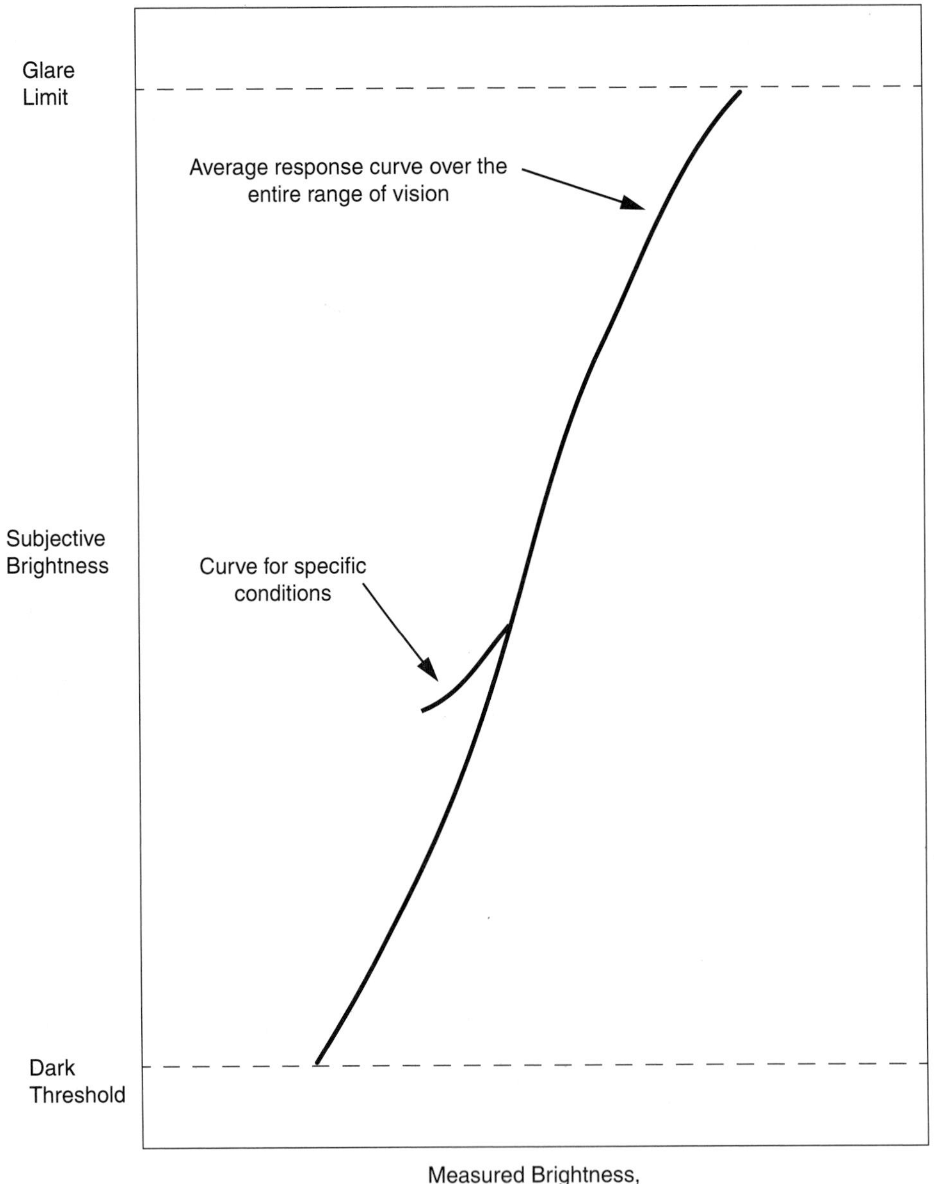

1.6.3 Brightness Adaptation

The vision system responds to a wide range of brightness levels. The response actually varies based on the average brightness observed and is limited by the dark threshold and the glare limit. Light intensities below the dark threshold or above the glare limit are either too dark to see or blinding. We cannot see across the entire range at any one time, but our system will adapt to existing light conditions. This is largely a result of the pupil in the eye, which acts as a diaphragm on the lens by controlling the amount it is open or closed and thus controls the amount of light that can enter.

Subjective brightness is a logarithmic function of the light intensity incident on the eye. Figure 1.6-8 is a plot of the range and variation of the system. The vertical axis shows the entire range of subjective brightness over which the system responds, and the horizontal corresponds to the measured brightness. The horizontal axis is actually the log of the light intensity, so this results in an approximately linear response. A typical response curve for a specific lighting condition can be seen in the smaller curve plotted; any brightness levels below this curve will be seen as black. This small curve can be extended to higher levels (above the main curve), but if the lighting conditions change, the entire small curve will simply move upward.

In images we observe many brightness levels, and the vision system can adapt to a wide range, as we have seen. However, it has been experimentally determined that we can detect only about 20 changes in brightness in a small area within a complex image. But, for an entire image, due to the brightness adaptation that our vision system exhibits, it has been determined that about 100 different gray levels are necessary to create a realistic image. For 100 gray levels in a digital image, we need at least 7 bits/pixel ($2^7 = 128$). If fewer gray levels are used, we observe false contours (bogus lines) resulting from gradually changing light intensity not being accurately represented, as in Figure 1.6-9.

Figure 1.6-9 False Contouring

a. Original image (8 bits/pixel).

b. False contours (3 bits/pixel).

An interesting phenomenon that our vision system exhibits related to brightness is called the Mach Band effect. This effect creates an optical illusion, as can be seen in Figure 1.6-10. Here we observe that when there is a sudden change in intensity, our vision system response overshoots the edge, thus creating a scalloped effect. This phenomenon accentuates edges and helps us to distinguish, and separate, objects within

Figure 1.6-10 Mach Band Effect

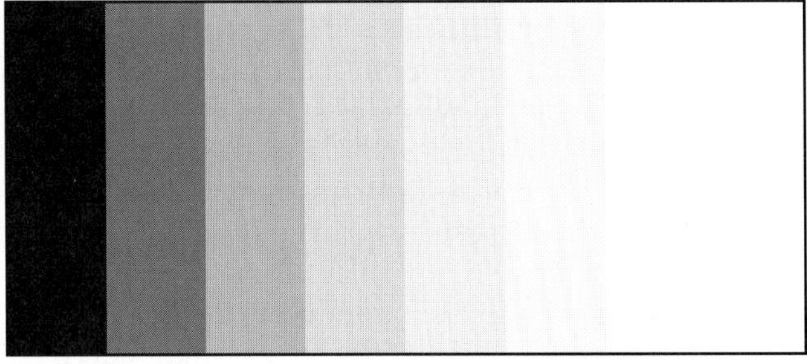

a. Image with gray levels that appear uniformly spaced.

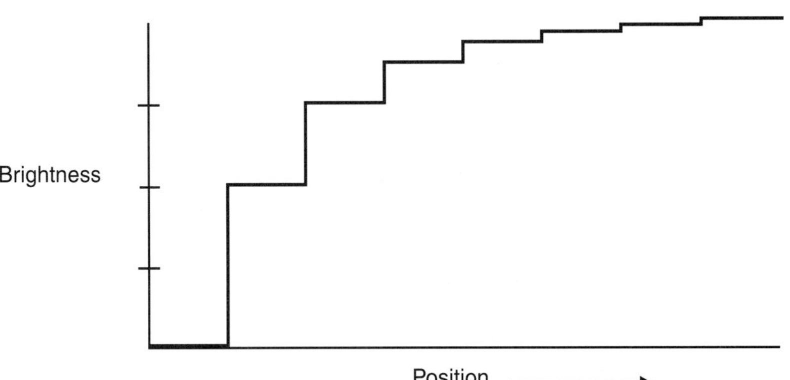

b. Actual brightness values are logarithmically spaced.

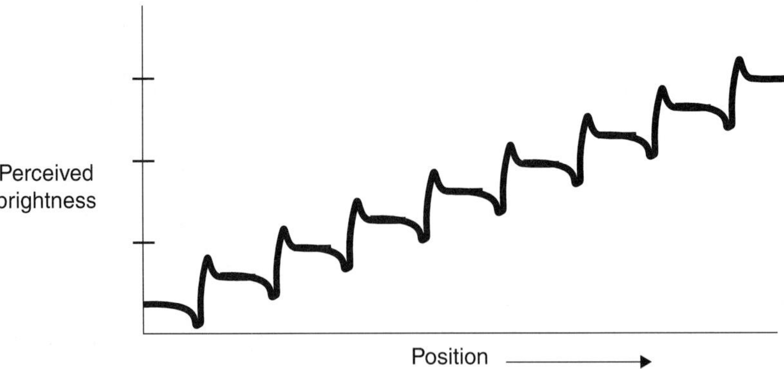

c. Because of the Mach band effect, the human visual system perceives overshoot at edges.

an image. This ability, combined with our brightness adaptation response, allows us to see outlines even in dimly lit areas.

1.6.4 Temporal Resolution

Temporal response of the human visual system deals with how we respond to visual information as a function of time. This is most useful when considering video and motion in images. Although we will deal primarily with two-dimensional still images, a basic understanding of the temporal response of the human visual system is necessary to have a complete overview of human vision.

In Figure 1.6-11 we see a plot of temporal contrast sensitivity on the vertical axis versus frequency (in time, not space) on the horizontal. This is a plot of what is known as flicker sensitivity. *Flicker sensitivity* refers to our ability to observe a flicker in a video signal displayed on a monitor. The variables here are brightness and frequency. We show two plots to illustrate the variation in response due to image brightness. Here, as with the frequency response, the brighter the display, the more sensitive we

Figure 1.6-11 Temporal Resolution

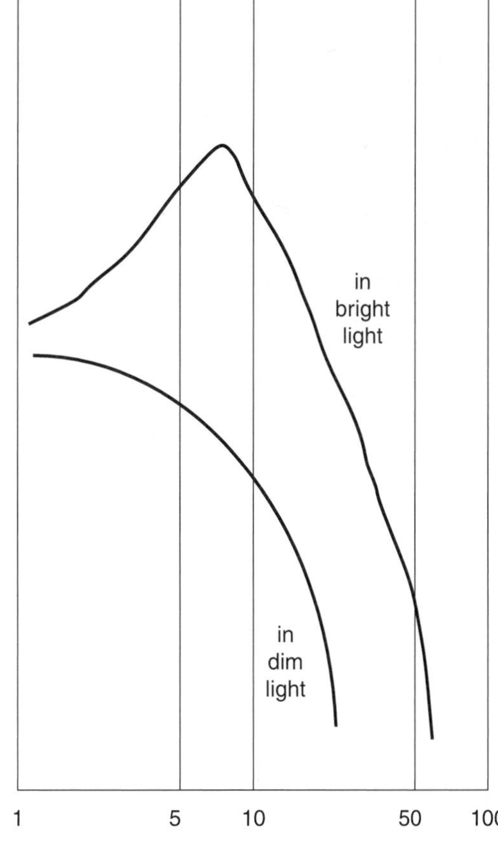

are to changes. Note that the changes are relative, that is, a percentage change based on the average brightness. The primary point here is that the human visual system has a temporal cutoff frequency of about 50 hertz (cycles per second). This is what allows us to view video signals that have a field rate of 60 hertz without any perceivable flicker.

1.7 Image Representation

We have seen that the human visual system receives an input image as a collection of spatially distributed light energy; this form is called an *optical image*. Optical images are the types we deal with everyday—cameras capture them, monitors display them, and we see them. We know that these optical images are represented as video information in the form of analog electrical signals and have seen how these are sampled to generate the digital image $I(r, c)$.

The digital image $I(r, c)$ is represented as a two-dimensional array of data, where each pixel value corresponds to the brightness of the image at the point (r, c). In linear algebra terms, a two-dimensional array like our image model $I(r, c)$ is referred to as a *matrix*, and one row (or column) is called a *vector*. This image model is for monochrome (one-color, this is what we normally refer to as black and white) image data, but we have other types of image data that require extensions or modifications to this model. Typically, these are multiband images (color, multispectral), and they can be modeled by a different $I(r, c)$ function corresponding to each separate band of brightness information. The image types we will consider are: 1) binary, 2) gray-scale, 3) color, and 4) multispectral.

1.7.1 Binary Images

Binary images are the simplest type of images and can take on two values, typically black and white, or '0' and '1.' A binary image is referred to as a 1 bit/pixel image because it takes only 1 binary digit to represent each pixel. These types of images are most frequently used in computer vision applications where the only information required for the task is general shape, or outline, information. For example, to position a robotic gripper to grasp an object, to check a manufactured object for deformations, for facsimile (FAX) images, or in optical character recognition (OCR).

Binary images are often created from gray-scale images via a threshold operation, where every pixel above the threshold value is turned white ('1'), and those below it are turned black ('0'). In Figure 1.7-1, we see examples of binary images. Figure 1.7-1a is a page of text, such as might be used in an OCR application; Figure 1.7-1b is the outline of an object; and in Figure 1.7-1c we have the results of an edge detection operation (see Section 2.3).

1.7.2 Gray-Scale Images

Gray-scale images are referred to as monochrome, or one-color, images. They contain brightness information only, no color information. The number of bits used for each pixel determines the number of different brightness levels available. The typical image contains 8 bits/pixel data, which allows us to have 256 (0–255) different bright-

Figure 1.7-1 Binary Images

```
wht2d(3)              C Library Functions              wht2d(3)

NAME
     wht2d - performs Walsh or Hadamard transform

SYNOPSIS
      #include <stdio.h>
      #include <stdlib.h>
      #include <math.h>
      #include "CVIPtools.h"
      #include "CVIPimage.h"
      #include "CVIPdef.h"

     IMAGE *wht2d(IMAGE *in_IMAGE, int ibit, int block_size)

      <in_IMAGE> - pointer to an IMAGE structure
      <ibit> - 0=inverse Walsh transform, 1=Walsh transform
               2=inverse Hadamard transform, 3=Hadamard transform
      <block_size> - block size (4,8,16,...largest_dimension/2)

PATH
     $CVIPHOME/TRANSFORMS/wht2d.c

DESCRIPTION
     This function  performs a fast Hadamard-ordered Walsh-
     Hadamard Transform on an image. The result is then reor-
     dered for display in sequency order. The routine works on
     any image with dimensions that are powers of 2. Optional
     zero-padding may be performed if input image has different
     dimensions.
```

a. Binary text.

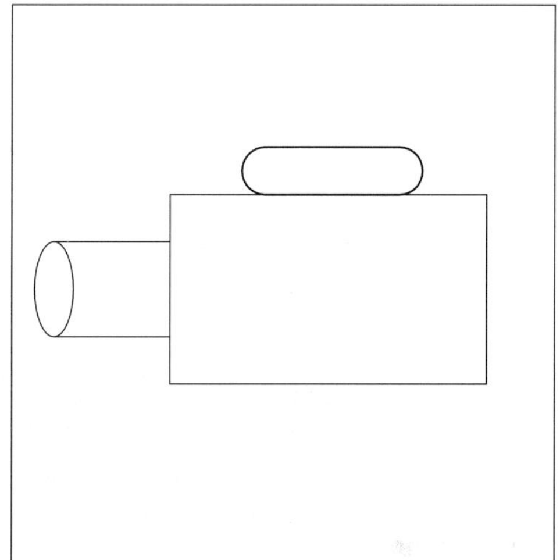

b, Object outline.

c. Edge detection and threshold operation.

ness (gray) levels. This representation provides more than adequate brightness resolution, in terms of the human visual system's requirements, and provides a "noise margin" by allowing for approximately twice as many gray levels as required. This noise margin is useful in real-world applications because of the many different types of noise (false information in the signal) inherent in real systems. Additionally, the 8-bit representation is typical due to the fact that the *byte*, which corresponds to 8-bits of data, is the standard small unit in the world of digital computers.

In certain applications, such as medical imaging or astronomy, 12 or 16 bits/pixel representations are used. These extra brightness levels become useful only when the image is "blown-up," that is, when a small section of the image is made much larger. In this case we may be able to discern details that would be missing without this additional brightness resolution. Of course, to be useful, this also requires a higher level of spatial resolution (number of pixels). If we go beyond these levels of brightness resolution, we typically divide the light energy into different bands, where each *band* refers to a specific subsection of the visible image spectrum.

1.7.3 Color Images

Color images can be modeled as three-band monochrome image data, where each band of data corresponds to a different color. The actual information stored in the digital image data is the brightness information in each spectral band. When the image is displayed, the corresponding brightness information is displayed on the screen by picture elements that emit light energy corresponding to that particular color. Typical color images are represented as red, green, and blue, or RGB images. Using the 8-bit monochrome standard as a model, the corresponding color image would have 24 bits/pixel—8-bits for each of the three color bands (red, green, and blue). In Figure 1.7-2a we see a representation of a typical RGB color image. Figure 1.7-2b illustrates that, in addition to referring to a row or column as a vector, we can refer to a single pixel's red, green, and blue values as a *color pixel vector*—(R, G, B).

For many applications, RGB color information is transformed into a mathematical space that decouples the brightness information from the color information. After this is done, the image information consists of a one-dimensional brightness, or luminance, space and a two-dimensional color space. Now the two-dimensional color space does not contain any brightness information, but it typically contains information regarding the relative amounts of the different colors. An additional benefit of modeling the color information in this manner is that it creates a more people-oriented way of describing the colors.

For example, the hue/saturation/lightness (HSL) color transform allows us to describe colors in terms that we can more readily understand (see Figure 1.7-3). The *lightness* is the brightness of the color, and the *hue* is what we normally think of as "color" (for example green, blue, or orange). The *saturation* is a measure of how much white is in the color (for example, pink is red with more white, so it is less saturated than a pure red). Most people can relate to this method of describing color. For example, "a deep, bright orange" would a have a large intensity ("bright"), a hue of "orange," and a high value of saturation ("deep"). We can picture this color in our minds, but if we defined this color in terms of its RGB components, $R = 245$, $G = 110$, and $B = 20$, most people would have no idea how this color appears. Because the HSL color space was developed based on heuristics relating to human perception, various methods are available to transform RGB pixel values into the HSL color space. Most of these are algorithmic in nature and are geometric approximations to mapping the RGB color cube into some HSL-type color space (see Figure 1.7-4).

A color transform based on a geometrical coordinate transform is the spherical transform. This transform decouples the brightness information from the color information. The spherical coordinate transform (SCT) has been successfully used in a

Figure 1.7-2 Color Image Representation

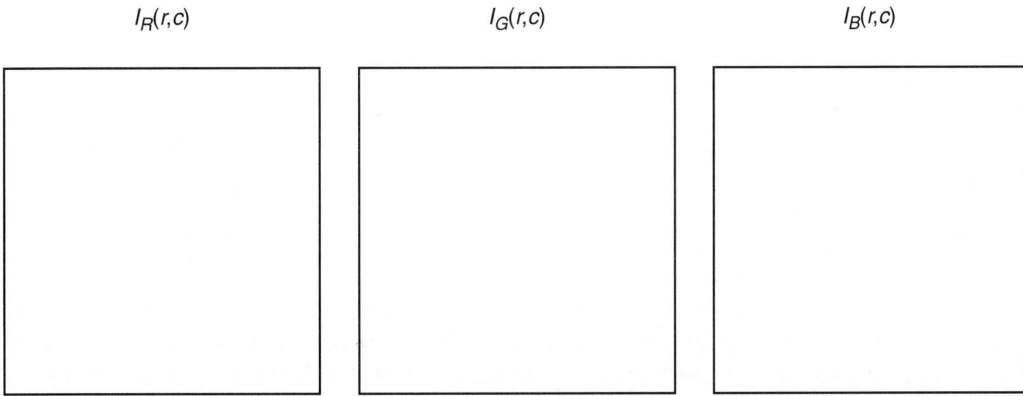

a. A typical RGB color image can be thought of as three separate images: $I_R(r,c)$, $I_G(r,c)$, and $I_B(r,c)$.

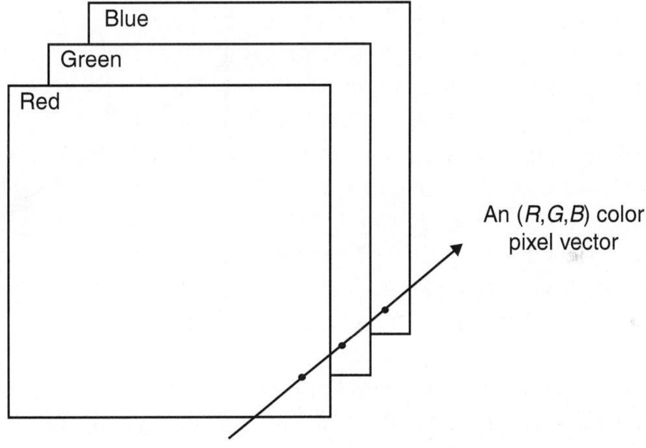

b. A color pixel vector consists of the red, green, and blue pixel values (R,G,B) at one given row/column pixel coordinate (r,c).

color segmentation algorithm described in Chapter 2. The equations relating the SCT to the RGB components follow:

$$L = \sqrt{R^2 + G^2 + B^2}$$

$$\angle A = \cos^{-1}\left[\frac{B}{L}\right]$$

$$\angle B = \cos^{-1}\left[\frac{R}{L\sin(\angle A)}\right]$$

Figure 1.7-3 HSL Color Space

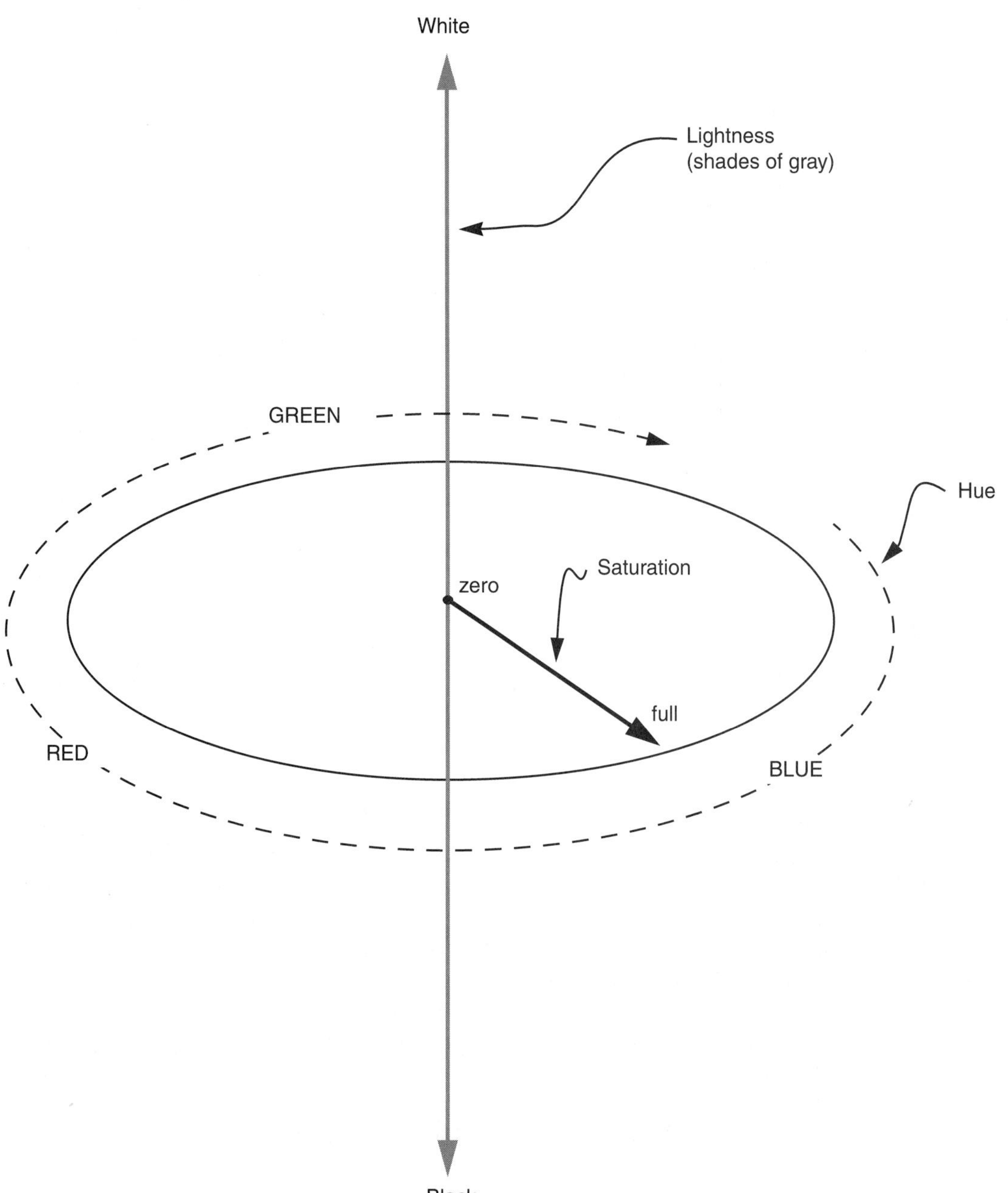

where L is the length of the RGB vector, angle A is the angle from the B-axis to the RG plane, and angle B is the angle between the R- and G-axes. Here, L contains the brightness information, and the two angles contain the color information.

One problem associated with the color spaces previously described is that they are not perceptually uniform. This means that two different colors in one part of the

Figure 1.7-4 RGB to HSL Mapping

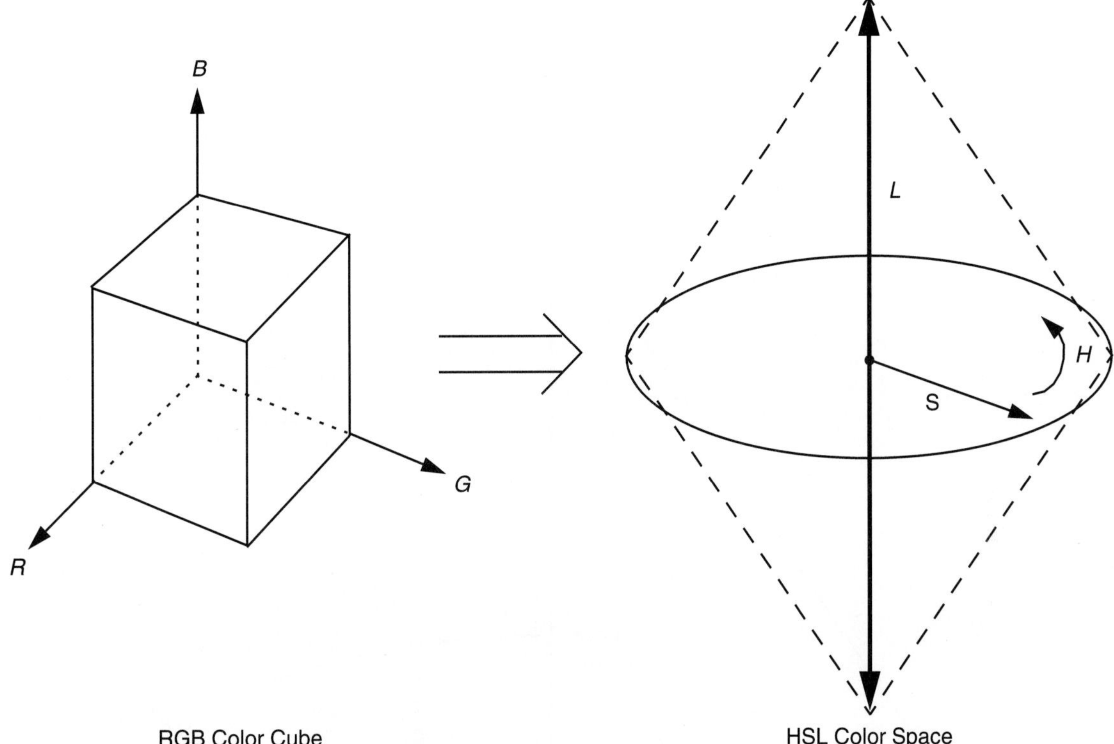

color space will not exhibit the same degree of perceptual difference as two colors in another part of the color space, even though they are the same "distance" apart (see Figure 1.7-5). Therefore, we cannot define a metric to tell us how close, or far apart, two colors are in terms of human perception. In computer imaging applications a perceptually uniform color space could be very useful. For example, if we are trying to identify objects for a computer vision system by color information, we need some method to compare the object's color to a database of the colors of the available objects.

The science of color and how the human visual system perceives color has been studied extensively by an international body, the Commission Internationale de l'Eclairage (CIE). The CIE has defined internationally recognized color standards. One of the basic concepts developed by the CIE involves chromaticity coordinates. For our RGB color space, *chromaticity coordinates* are defined as follows:

$$r = \frac{R}{R + G + B}$$

$$g = \frac{G}{R + G + B}$$

$$b = \frac{B}{R + G + B}$$

These equations basically normalize the individual color components to the sum of the three, which we have seen is one way to represent the brightness information. This decouples the brightness information from the coordinates, and the CIE uses

Figure 1.7-5 Color Perception

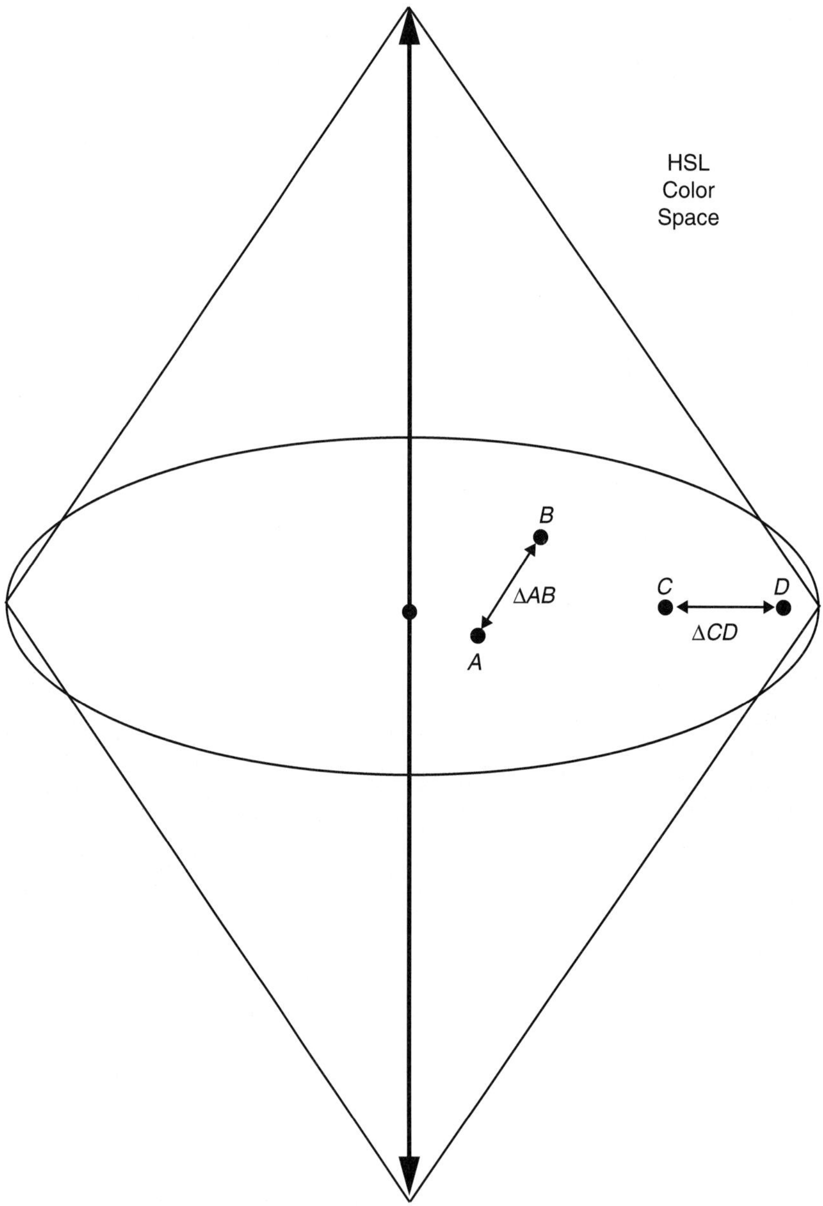

Color *A* may be orange and *B* may be green; colors *C* and *D* may be slightly different shades of the same color. However, $\Delta AB = \Delta CD$.

chromaticity coordinates as the basis of the color transforms they define. These include the standard CIE XYZ color space and the perceptually uniform L*u*v*, L*a*b* color spaces. The science of color and human perception is a fascinating topic and can be explored in greater depth with the references.

Another important international committee for developing standards of interest to those involved in computer imaging is the International Telecommunications Union-Radio (ITU-R, previously CCIR). This committee has specified the standard for digital video known as ITU-R 601. This standard is based on one luminance signal (*Y*) and two color difference signals (*Cr* and *Cb*). Given a 24 bits/pixel RGB signal, we can find the *Y*, *Cr*, and *Cb* values as follows:

$$\begin{aligned} Y &= 0.299R + 0.587G + 0.114B \\ Cb &= -0.1687R - 0.3313G + 0.5B + 128 \\ Cr &= 0.5R - 0.4187G - 0.0813B + 128 \end{aligned}$$

This transform is used in many color image compression algorithms, implemented in both hardware and software.

The final color transform we will discuss is called the principal components transform. This mathematical transform allows us to apply statistical methods to put as much of the three-dimensional color information as possible into only one band. The *principal components transform* (PCT) works by examining all the RGB vectors within an image and finding the linear transform that aligns the coordinate axes so that most of the information is along one axis, the principal axis. Often, we can get 90% or more of the information into one band. The PCT is used in image compression schemes and is discussed in more detail in Chapter 2 where it is used as part of an image segmentation algorithm.

1.7.4 Multispectral Images

Multispectral images typically contain information outside the normal human perceptual range. This may include infrared, ultraviolet, X-ray, acoustic, or radar data. These are not images in the usual sense because the information represented is not directly visible by the human system. However, the information is often represented in visual form by mapping the different spectral bands to RGB components. If more than three bands of information are in the multispectral image, the dimensionality is reduced by applying a principal components transform.

Sources for these types of images include satellite systems, underwater sonar systems, various types of airborne radar, infrared imaging systems, and medical diagnostic imaging systems. The number of bands into which the data are divided is strictly a function of the sensitivity of the imaging sensors used to capture the images. For example, even the visible spectrum can be divided into many more than three bands; three are used because this mimics our visual system. Most of the satellites currently in orbit collect image information in two to seven spectral bands; typically one to three are in the visible spectrum, one or more are in the infrared region, and some have sensors that operate in the radar range (see Figure 1.6-2). The newest satellites have sensors that collect image information in 30 or more bands. As the amount of data that needs to be transmitted, stored, and processed increases, the importance of topics such as compression becomes more and more apparent.

1.8 Digital Image File Formats

Why do we need so many different types of image file formats? The short answer is that there are many different types of images and applications with varying requirements. A more complete answer (which we will not go into here) also considers market share, proprietary information, and a lack of coordination within the imaging industry. However, some standard file formats have been developed, and the ones presented here are widely available. Many other image types can be readily converted to one of the types presented here by easily available image conversion software.

A field related to computer imaging is that of computer graphics. *Computer graphics* is a specialized field within the computer science realm that refers to the reproduction of visual data through the use of the computer. This includes the creation of computer images for display or print, and the process of generating and manipulating any images (real or artificial) for output to a monitor, printer, camera, or any other device that will provide us with an image. Computer graphics can be considered a part of computer imaging, insofar as many of the same tools the graphics artist uses may be used by the computer imaging specialist.

In computer graphics, types of image data are divided into two primary categories: bitmap and vector. *Bitmap images* (also called raster images) can be represented by our image model $I(r, c)$, where we have pixel data and the corresponding brightness values stored in some file format. *Vector images* refer to methods of representing lines, curves, and shapes by storing only the key points. These *key points* are sufficient to define the shapes, and the process of turning these into an image is called *rendering*. After the image has been rendered, it can be thought of as being in bitmap format, where each pixel has specific values associated with it.

Most of the types of file formats discussed fall into the category of bitmap images, although some are compressed, so that the $I(r, c)$ values are not directly available until the file is decompressed. In general, these types of images contain both header information and the raw pixel data. The header must contain information regarding: 1) the number of rows (height), 2) the number of columns (width), 3) the number of bands, 4) the number of bits per pixel, and 5) the file type. Additionally, with some of the more complex file formats, the header may contain information about the type of compression used and any other necessary parameters to create the image, $I(r, c)$.

The simplest file formats are the BIN and the PPM file formats. The BIN format is simply the raw image data $I(r, c)$. This file contains no header information, so the user must know the necessary parameters—size, number of bands, and bits per pixel—to use the file as an image. The PPM formats are widely used, and a set of conversion utilities are freely available (pbmplus). They basically contain raw image data with the simplest header possible. The PPM format includes PBM (binary), PGM (gray-scale), PPM (color), and PNM (handles any of the previous types). The headers for these image file formats contain a "magic number" that identifies the file type, the image width and height, the number of bands, and the maximum brightness value (which determines the required number of bits per pixel for each band).

Two image file formats commonly used on many different computer platforms, as well as on the World Wide Web, are the TIFF (Tagged Image File Format) and GIF (Graphics Interchange Format) file formats. GIF files are limited to a maximum of 8 bits/pixel and allow for a type of compression called LZW (Lempel-Ziv-Welch, see Chapter 5). The 8 bits/pixel limitation does not mean that it does not support color images, it simply means that no more than 256 colors (2^8) are allowed in an image.

This is typically implemented by means of a lookup table (LUT), where the 256 colors are stored in a table, and 1 byte (8 bits) is used as an index (address) into that table for each pixel (see Figure 1.8-1). The GIF image header is 13 bytes long and contains the basic information required.

The TIFF file format is more sophisticated than GIF and has many more options and capabilities. TIFF files allow a maximum of 24 bits/pixel and support five types of compression, including RLE (run-length encoding), LZW, and JPEG (Joint Photographic Experts Group) (see Chapter 5). The TIFF header is of variable size and is arranged in a hierarchical manner. TIFF is one of the most comprehensive formats available and is designed to allow the user to customize it for specific applications.

Figure 1.8-1 Lookup Table

8-bit Index	RED	GREEN	BLUE
0	R_0	G_0	B_0
1	R_1	G_1	B_1
2	R_2	G_2	B_2
⋮	⋮	⋮	⋮
254	R_{254}	G_{254}	B_{254}
255	R_{255}	G_{255}	B_{255}

One byte is stored for each pixel in $I(r,c)$. When displayed, this 8-bit value is used as an index into the LUT, and the corresponding RGB values are displayed for that pixel.

JPEG File Interchange Format (JFIF) is rapidly becoming a standard that allows images compressed with the JPEG algorithm to be used in many different computer platforms. The JFIF files have a Start of Image (SOI) and an application (APPO) marker that serve as a file header. JPEG image compression is being used extensively on the WWW and is expected to become the de facto standard for many applications. The JPEG algorithm is very flexible, so it can create relatively large files with excellent image quality or very small files with relatively poor image quality. The user can define the amount of compression desired dependent on the needs of the application.

Two formats that were initially computer specific, but have become commonly used throughout the industry, are the Sun Raster and the SGI (Silicon Graphics, Inc.) file formats. The Sun Raster file format is much more ubiquitous than the SGI, but SGI has become the leader in state-of-the-art graphics computers. The SGI format handles up to 16 million colors and supports RLE compression. The SGI image header is 512 bytes (with the majority of the bytes not used, presumably for future extensions) followed by the image data. The Sun Raster format is defined to allow for any number of bits per pixel and also supports RLE compression and color LUTs. It has a 32-byte header, followed by the image data.

One file format discussed here, EPS (encapsulated PostScript), is not of the bitmap variety. It is actually a language that supports more than images; it is commonly used in desktop publishing. EPS is directly supported by many printers (in the hardware itself), so it is commonly used for data interchange across hardware and software platforms. It is a commonly used standard that allows output devices, monitors, printers, and computer software to communicate regarding both graphics and text. The primary advantage of the EPS format is its wide acceptance. The disadvantage of using EPS is that the files are very big because it is a general-purpose language designed for much more than just images. In computer imaging, EPS is used primarily as a means to generate printed images. The EPS files actually contain text and can be created by any text editor but are typically generated by applications software. The language itself is very complex and continually evolving.

The final image file type discussed here is the VIP (Visualization in Image Processing) format, developed specifically for the CVIPtools software. When performing computer imaging tasks, temporary images are often created that use floating point representations that are beyond the standard 8 bits/pixel capabilities of most display devices. The process of representing this type of data as an image is referred to as *data visualization* and can be achieved by remapping the data to the 8-bit range, 0 to 255. *Remapping* is the process of taking the original data and defining an equation to translate the original data to the output data range, typically 0 to 255 for 8-bit display. The two most commonly used methods in computer imaging are linear and logarithmic mapping. In Figure 1.8-2 we see a graphical representation and example of how this process is performed. In this example the original data ranges outside the bounds of a standard image from –200 to 440. It is remapped to the 8-bit range from 0 to 255. An equation is found that will map the lowest value (–200) to 0 and the highest value (440) to 255, while all the intermediate values are remapped to values within this range (0–255). We can see that this process may result in a loss of information.

The VIP file format was required because we needed to support many nonstandard image formats. This format was defined to allow disk file support for the image data structure used within the CVIPtools software (see Chapter 8). It allows any data type, including floating point and complex numbers, any image size, and any number

Figure 1.8-2 Remapping for Display

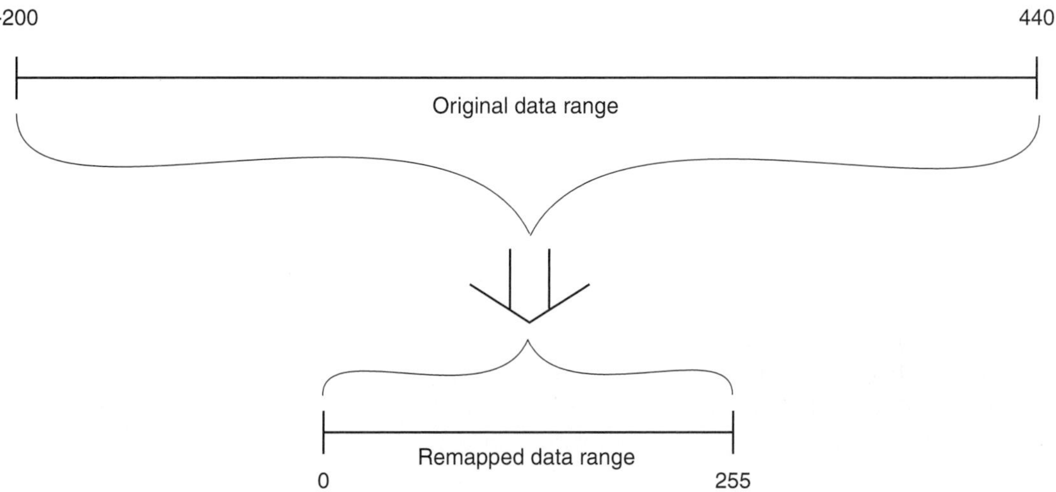

of bands and has a special history data structure built into it that allows the maintenance of a record of operations that have been performed on the image. More details on the VIP format are included in the second part of the book, which covers the CVIP-tools software.

1.9 References

A comprehensive treatment of computer vision can be found in [Ballard/Brown 82], [Haralick/Shapiro 92], [Horn 86], [Granland/Knutsson 95], and [Jain/Kasturi/Schnuck 95]. Comprehensive image processing texts include [Castleman 96], [Gonzalez/Woods 92], [Jain 89], [Pratt 91], [Bracewell 95], and [Rosenfeld/Kak 82].

Books that bring computer vision and image processing together include [Schalkoff 89], [Granland/Knutsson 95], and [Banks 90]. One book that takes a more practical, lab-oriented approach to computer vision and image processing is [Galbiati 90]. A good conceptual and practical approach to computer imaging is taken by [Baxes 94] and [Myler/Weeks 93]. [Russ 92] provides a good handbook for the computer imaging specialist.

For further study of digital video processing, see [Tekalp 95] and [Sid-Ahmed 95]. [Tekalp 95] has much information on motion estimation methods not available in other texts. For other sources of software see [Myler/Weeks 93], [Baxes 94], and [Sid-Ahmed 95]. Additionally, the Khoros (www.khoros.unm.edu) and Vista (www.cs.ubc.ca/labs/lci/vista) software available on the World Wide Web (Internet) are of interest.

Books that integrate computer imaging topics with human vision include [Levine 85], [Deutsch 93], [Marr 82], and [Arbib/Hanson 90]. For more on color in computing systems as related to human vision, see [Durrett 87]. A comprehensive treatment of color science can be found in [Wyszecki/Stiles 82], including all the details on the CIE color systems. [Sid-Ahmed 95] contains more details relating the human visual system to television signal processing. [Hill 90] provides algorithms for transforming RGB data into HSL color spaces.

The most comprehensive source for information on image and graphics file formats is [Murray/VanRyper 94]. Some of the information on multispectral and satellite images is in [Bell 95].

The applications discussed can be found in the trade magazines *Advanced Imaging* (May 1995 and June 1993), *Biophotonics International* (Jan/Feb 1995), *Design News* (9/23/91), and *Photonics Spectra* (April 1992).

Arbib, M. A., and Hanson, A. R. (Eds.), *Vision, Brain and Cooperative Computation*, Cambridge, MA: MIT Press, 1990.

Ballard, D. H., and Brown, C. M., *Computer Vision*, Englewood Cliffs, NJ: Prentice Hall, 1982.

Banks, S., *Signal Processing, Image Processing and Pattern Recognition*, Cambridge, UK: Prentice Hall International (UK) Ltd., 1990.

Baxes, G. A., *Digital Image Processing: Principles and Applications*, New York: Wiley, 1994.

Bell, T. E., "Remote Sensing," *IEEE Spectrum*, pp. 25–31, March 1995.

Bracewell, R. N., *Two-Dimensional Imaging*, Englewood Cliffs, NJ: Prentice Hall, 1995.

Castleman, K. R., *Digital Image Processing*, Englewood Cliffs, NJ: Prentice Hall, 1996.

Deutsch, S., and Deutsch, A., *Understanding the Nervous System—An Engineering Perspective*, New York: IEEE Press, 1993.

Durrett, H. J. (Ed.), *Color and the Computer*, San Diego, CA: Academic Press, 1987.

Galbiati, L. J., *Machine Vision and Digital Image Processing Fundamentals*, Englewood Cliffs, NJ: Prentice Hall, 1990.

Gonzalez, R. C., and Woods, R. E., *Digital Image Processing*, Reading, MA: Addison-Wesley, 1992.

Granlund, G., and Knutsson, H., *Signal Processing for Computer Vision*, Boston: Kluwer Academic Publishers, 1995.

Haralick, R. M., and Shapiro, L. G., *Computer and Robot Vision*, Reading, MA: Addison-Wesley, 1992.

Hill, F. S., *Computer Graphics*, New York: Macmillan, 1990.

Horn, B. K. P., *Robot Vision*, Cambridge, MA: The MIT Press, 1986.

Jain, A. K., *Fundamentals of Digital Image Processing*, Englewood Cliffs, NJ: Prentice Hall, 1989.

Jain, R., Kasturi, R., and Schnuck, B. G., *Machine Vision*, New York: McGraw Hill, 1995.

Levine, M. D., *Vision in Man and Machine*, New York: McGraw Hill, 1985.

Marr, D., *Vision*, New York: Freeman and Company, 1982.

Murray, J. D., and VanRyper, W., *Encyclopedia of Graphics File Formats*, Sebastopol, CA: O'Reilly and Associates, 1994.

Myler, H. R., and Weeks, A. R., *Computer Imaging Recipes in C*, Englewood Cliffs, NJ: Prentice Hall, 1993.

Pratt, W. K., *Digital Image Processing*, New York: Wiley, 1991.

Rosenfeld, A., and Kak, A. C., *Digital Picture Processing*, San Diego, CA: Academic Press, 1982.

Russ, J. C., *The Image Processing Handbook*, Boca Raton, FL: CRC Press, 1992.

Schalkoff, R. J., *Digital Image Processing and Computer Vision*, New York: Wiley, 1989.

Sid-Ahmed, M. A. *Image Processing: Theory, Algorithms, and Architectures*, Englewood Cliffs, NJ: Prentice Hall, 1995.

Tekalp, A. M., *Digital Video Processing*, Englewood Cliffs, NJ: Prentice Hall, 1995.

Wyszecki, G., and Stiles, W. S., *Color Science: Concepts and Methods, Quantitative Data and Formulae*, New York: Wiley, 1982.

CHAPTER 2

Image Analysis

2.1 Introduction

Image analysis involves manipulating the image data to determine exactly the information necessary to help solve a computer imaging problem. This analysis is typically part of a larger process, is iterative in nature, and allows us to answer application-specific questions: Do we need color information? Do we need to transform the image data into the frequency domain? Do we need to segment the image to find object information? What are the important features in the images?

2.1.1 Overview

Image analysis is primarily a data reduction process. As we have seen, images contain enormous amounts of data, typically on the order of hundreds of kilobytes or even megabytes. Often much of this information is not necessary to solve a specific computer imaging problem, so a primary part of the image analysis task is to determine exactly what information is necessary. Image analysis is used in both computer vision and image processing applications.

For computer vision, the end product is typically the extraction of high-level information for computer analysis or manipulation (see Computer Vision Lab Exercise #8 in Chapter 8). This high-level information may include shape parameters to control a robotic manipulator or color and texture features to help in the diagnosis of a skin tumor. Image analysis is central to the computer vision process and is often uniquely associated with computer vision; however, image analysis is an important tool for image processing applications as well.

In image processing applications, image analysis methods may be used to help determine the type of processing required and the specific parameters needed for that processing (see Image Processing Lab Exercise #8 in Chapter 8). For example, determining the degradation function for an image restoration procedure (Chapter 3), developing an enhancement algorithm (Chapter 4), and determining exactly what information is visually important for an image compression method (Chapter 5) are all image analysis tasks. The tools and concepts necessary for image analysis presented in this chapter will provide a background for the other CVIP application areas presented in this part of the book.

2.1.2 System Model

The *image analysis process*, illustrated in Figure 2.1-1, can be broken down into three primary stages: 1) Preprocessing, 2) Data Reduction, and 3) Feature Analysis. Preprocessing is used to remove noise and eliminate irrelevant, visually unnecessary information. *Noise* is unwanted information that can result from the image acquisition process. Other preprocessing steps might include gray-level or spatial quantization (reducing the number of bits per pixel or the image size) or finding regions of interest for further processing. The second stage, data reduction, involves either reducing the data in the spatial domain or transforming it into another domain called the frequency domain (Figure 2.1-2) and then extracting features for the analysis process. In the third stage, feature analysis, the features extracted by the data reduction process are examined and evaluated for their use in the application.

A more detailed diagram of this process is shown in Figure 2.1-3. After preprocessing we can perform segmentation on the image in the spatial domain or convert it into the frequency domain via a mathematical transform. After either of these processes we may choose to filter the image. This filtering process further reduces the data and allows us to extract the features that we may require for analysis. After the

Figure 2.1-1 Image Analysis

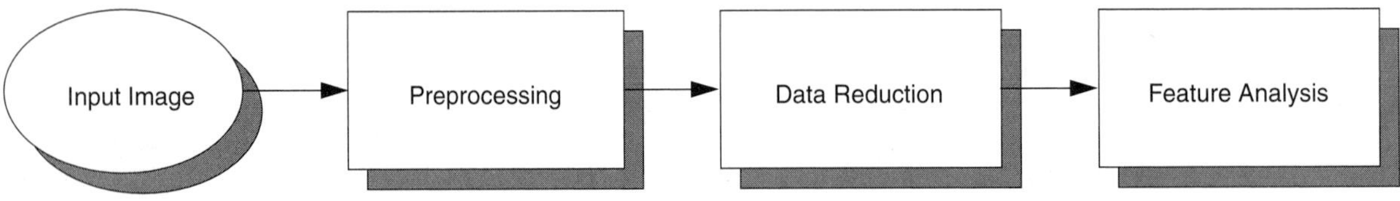

Figure 2.1-2 Image Analysis Domains

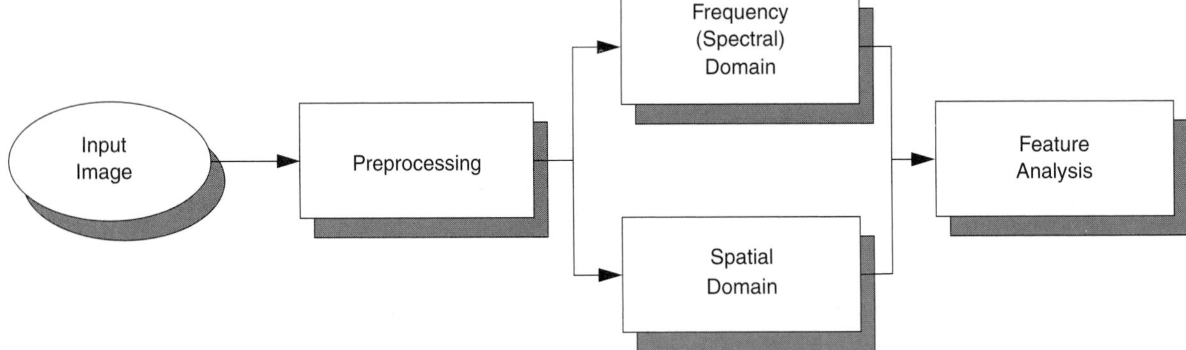

Figure 2.1-3 Image Analysis

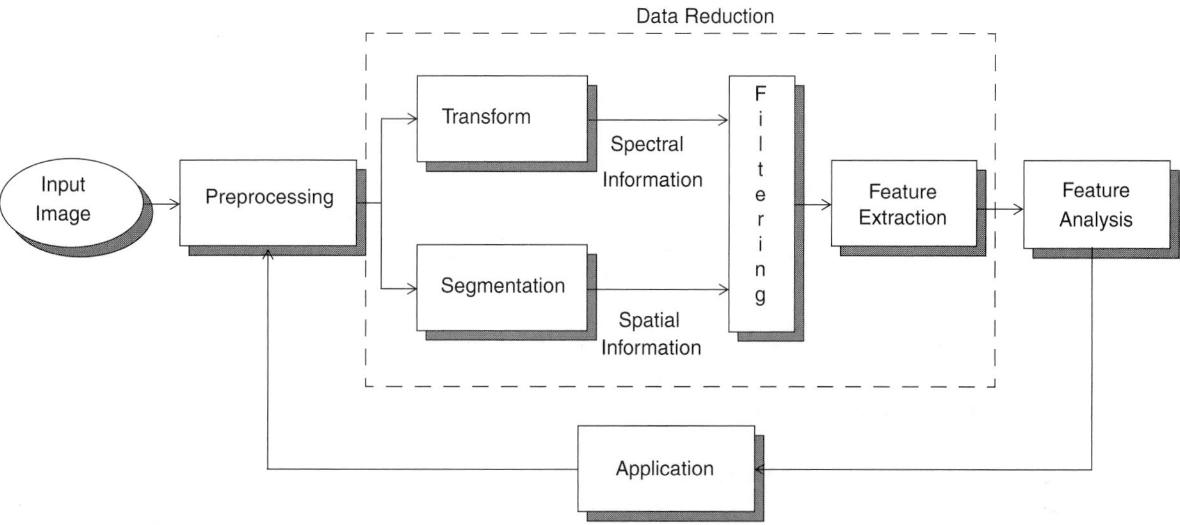

analysis we have a feedback loop that provides for an application-specific review of the analysis results. This approach often leads to an iterative process that is not complete until satisfactory results are achieved. The application feedback loop is a key aspect of the entire process.

2.2 PREPROCESSING

The preprocessing algorithms, techniques, and operators are used to perform initial processing that makes the primary data reduction and analysis task easier. They include operations related to extracting regions of interest, performing basic algebraic operations on images, enhancing specific image features, and reducing data in both resolution and brightness. Preprocessing is a stage where the requirements are typically obvious and simple, such as the removal of artifacts from images, or the elimination of image information that is not required for the application. For example, in one application we needed to eliminate borders from the images that had been digitized from film (the film frames); in another we had to mask out rulers that were present in skin tumor slides. Another example of a preprocessing step involves a robotic gripper that needs to pick and place an object; for this, we reduce a gray-level image to a binary (two-valued) image that contains all the information necessary to discern the object's outline.

2.2.1 Region-of-Interest Image Geometry

Often, for image analysis, we want to investigate more closely a specific area within the image, called a Region of Interest (ROI). To do this we need operations that modify the spatial coordinates of the image, and these are categorized as image geometry operations. The image geometry operations discussed here include crop, zoom, enlarge, shrink, translate, and rotate (see Image Processing Lab Exercise #2 in Chapter 8).

The image *crop* process is the process of selecting a small portion of the image, a subimage, and cutting it away from the rest of the image. After we have cropped a subimage from the original image, we can *zoom* in on it by enlarging it. This zoom process can be done in numerous ways, but typically a zero- or first-order hold is used. A zero-order hold is performed by repeating previous pixel values, thus creating a blocky effect. To extend the image size with a first-order hold, we do linear interpolation between adjacent pixels. A comparison of the images resulting from these two methods is shown in Figure 2.2-1.

Although the implementation of the zero-order hold is straightforward, the first-order hold is more complicated. The easiest way to do this is to find the average value between two pixels and use that as the pixel value between those two; we can do this for the rows first, as follows:

Figure 2.2-1 Zooming Methods

a. Original image. Area to be zoomed is outlined at center.

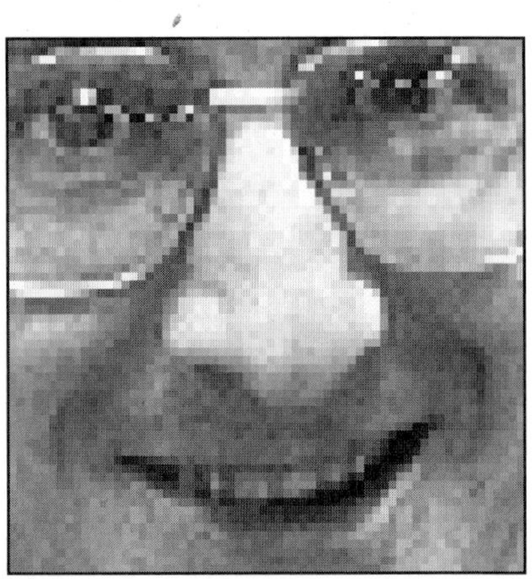

b. Image enlarged by zero-order hold. Note the blocky effect.

c. Image enlarged by first-order hold. Note the smoother effect.

ORIGINAL IMAGE ARRAY

$$\begin{bmatrix} 8 & 4 & 8 \\ 4 & 8 & 4 \\ 8 & 2 & 8 \end{bmatrix}$$

IMAGE WITH ROWS EXPANDED

$$\begin{bmatrix} 8 & 6 & 4 & 6 & 8 \\ 4 & 6 & 8 & 6 & 4 \\ 8 & 5 & 2 & 5 & 8 \end{bmatrix}$$

The first two pixels in the first row are averaged, $(8 + 4)/2 = 6$, and this number is inserted between those two pixels. This is done for every pixel pair in each row. Next, take that result and expand the columns in the same way, as follows:

IMAGE WITH ROWS AND COLUMNS EXPANDED

$$\begin{bmatrix} 8 & 6 & 4 & 6 & 8 \\ 6 & 6 & 6 & 6 & 6 \\ 4 & 6 & 8 & 6 & 4 \\ 6 & 5.5 & 5 & 5.5 & 6 \\ 8 & 5 & 2 & 5 & 8 \end{bmatrix}$$

This method allows us to enlarge an $N \times N$ sized image to a size of $(2N - 1) \times (2N - 1)$ and can be repeated as desired.

Another method that achieves the same result requires a mathematical process called convolution. With this method of image enlargement, a two-step process is required: 1) extend the image by adding rows and columns of zeros between the existing rows and columns and 2) perform the convolution. The image is extended as follows:

ORIGINAL IMAGE ARRAY

$$\begin{bmatrix} 3 & 5 & 7 \\ 2 & 7 & 6 \\ 3 & 4 & 9 \end{bmatrix}$$

IMAGE EXTENDED WITH ZEROS

$$\begin{bmatrix} 0 & 0 & 0 & 0 & 0 & 0 & 0 \\ 0 & 3 & 0 & 5 & 0 & 7 & 0 \\ 0 & 0 & 0 & 0 & 0 & 0 & 0 \\ 0 & 2 & 0 & 7 & 0 & 6 & 0 \\ 0 & 0 & 0 & 0 & 0 & 0 & 0 \\ 0 & 3 & 0 & 4 & 0 & 9 & 0 \\ 0 & 0 & 0 & 0 & 0 & 0 & 0 \end{bmatrix}$$

Next, we use a *convolution mask*, which is slid across the extended image, and perform a simple arithmetic operation at each pixel location.

CONVOLUTION MASK FOR FIRST-ORDER HOLD

$$\begin{bmatrix} \frac{1}{4} & \frac{1}{2} & \frac{1}{4} \\ \frac{1}{2} & 1 & \frac{1}{2} \\ \frac{1}{4} & \frac{1}{2} & \frac{1}{4} \end{bmatrix}$$

The *convolution process* requires us to overlay the mask on the image, multiply the coincident values, and sum all these results. This is equivalent to finding the vector inner product of the mask with the underlying subimage. The *vector inner product* is found by overlaying the mask on a subimage, multiplying coincident terms, and summing the resulting products. For example, if we put the mask over the upper-left corner of the image, we obtain (from right to left, and top to bottom):

$$\frac{1}{4}(0) + \frac{1}{2}(0) + \frac{1}{4}(0) + \frac{1}{2}(0) + 1(3) + \frac{1}{2}(0) + \frac{1}{4}(0) + \frac{1}{2}(0) + \frac{1}{4}(0) = 3$$

Note that the existing image values do not change. The next step is to slide the mask over by one pixel and repeat the process, as follows:

$$\frac{1}{4}(0) + \frac{1}{2}(0) + \frac{1}{4}(0) + \frac{1}{2}(3) + 1(0) + \frac{1}{2}(5) + \frac{1}{4}(0) + \frac{1}{2}(0) + \frac{1}{4}(0) = 4$$

Note this is the average of the two existing neighbors. This process continues until we get to the end of the row, each time placing the result of the operation in the location corresponding to center of the mask. When the end of the row is reached, the mask is moved down one row, and the process is repeated row by row until this procedure has been performed on the entire image; the process of sliding, multiplying, and summing is called convolution (see Figure 2.2-2). Note that the output image must be put in a separate image array, called a buffer, so that the existing values are not overwritten during the convolution process. If we call the convolution mask $M(r, c)$ and the image $I(r, c)$, the convolution equation is given by

$$\sum_{x=-\infty}^{\infty} \sum_{y=-\infty}^{\infty} I(r-x, c-y) M(x, y)$$

Figure 2.2-2 The Convolution Process

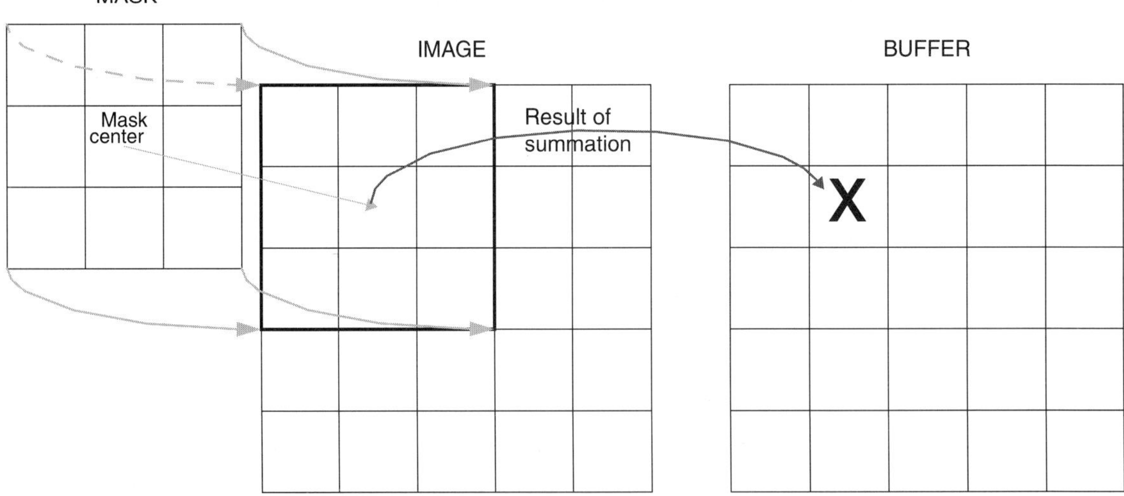

a. Overlay the convolution mask in the upper-left corner of the image. Multiply coincident terms, sum, and put the result into the image buffer at the location that corresponds to the mask's current center, which is $(r,c) = (1,1)$.

Figure 2.2-2 (Continued)

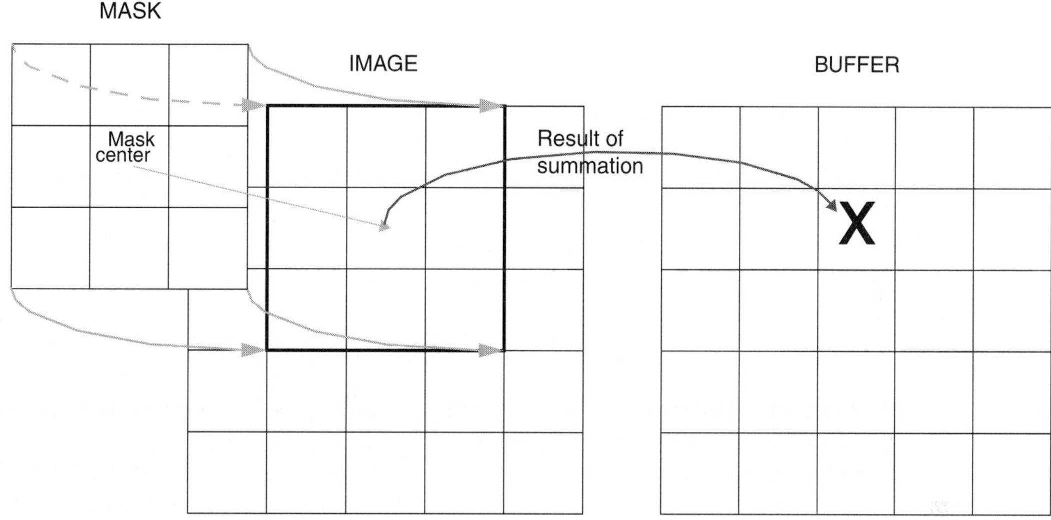

b. Move the mask one pixel to the right, multiply coincident terms, sum, and place the new result into the buffer at the location that corresponds to the new center location of the convolution mask, which is now at $(r,c) = (1,2)$. Continue to the end of the row.

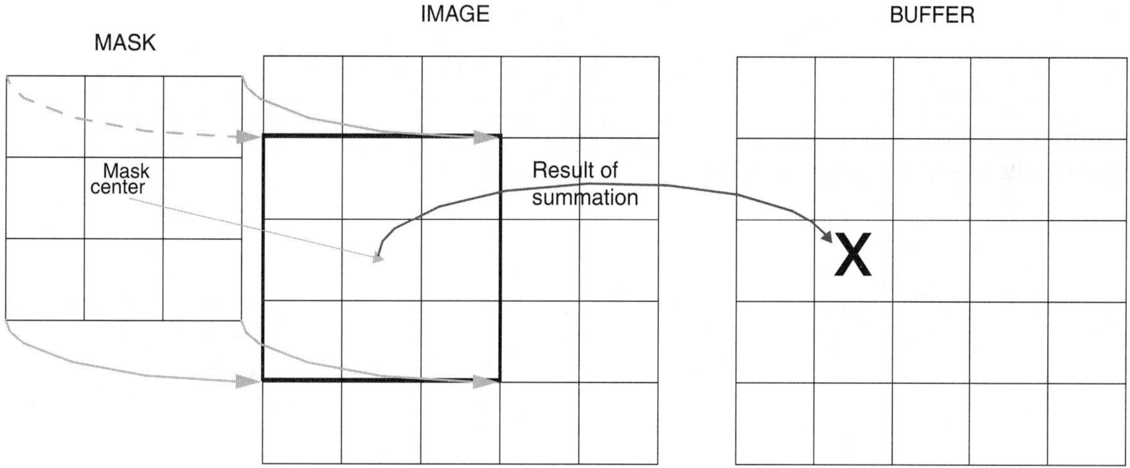

c. Move the mask down one row and repeat the process until the mask is convolved with the entire image. Note that we 'lose' the outer row(s) and column(s).

For theoretical reasons beyond the scope of this discussion, this equation assumes that the image and mask are extended with zeros infinitely in all directions and that the origin of the mask is at its center. Also, for theoretical reasons, the previous description of convolution assumes that the convolution mask is symmetric; in other words, if it is flipped about its center, it will remain the same. If it is not symmetric, it must be flipped before the procedure given can be followed. For computer imaging applications these convolution masks are typically symmetric.

At this point a good question would be Why use this convolution method when it requires so many more calculations than the basic averaging-of-neighbors method? The answer is that many computer imaging boards can perform convolution in hardware, which is generally very fast, typically much faster than applying a faster algorithm in software. Not only can first-order hold be performed via convolution, but zero-order hold can also be achieved by extending the image with zeros and using the following convolution mask:

ZERO-ORDER HOLD CONVOLUTION MASK

$$\begin{bmatrix} 1 & 1 \\ 1 & 1 \end{bmatrix}$$

Note that for this mask we will need to put the result in the pixel location corresponding to the lower-right corner because there is no center pixel.

These methods will only allow us to enlarge an image by a factor of $(2N - 1)$, but what if we want to enlarge an image by something other than a factor of $(2N - 1)$? To do this we need to apply a more general method; we take two adjacent values and linearly interpolate more than one value between them. This is done by defining an enlargement number K and then following this process: 1) subtract the two adjacent values, 2) divide the result by K, 3) add that result to the smaller value, and keep adding the result from the second step in a running total until all $(K - 1)$ intermediate pixel locations are filled.

EXAMPLE 2-1

We want to enlarge an image to three times its original size, and we have two adjacent pixel values 125 and 140.

1. Find the difference between the two values, $140 - 125 = 15$.
2. The desired enlargement is $K = 3$, so we get $15/3 = 5$.
3. Next determine how many intermediate pixel values we need: $K - 1 = 3 - 1 = 2$. The two pixel values between the 125 and 140 are $125 + 5 = 130$ and $125 + 2 * 5 = 135$.

We do this for every pair of adjacent pixels, first along the rows and then along the columns. This will allow us to enlarge the image by any factor of $K(N - 1) + 1$, where K is an integer and $N \times N$ is the image size. Typically, N is large and K is small, so this is approximately equal to KN.

The process opposite to enlarging an image is shrinking it. This is not typically done to examine a ROI more closely but to reduce the amount of data that needs to be processed. Shrinking is explored more in Section 2.2.4.

Two other operations of interest for the ROI image geometry are translation and rotation. These processes may be performed for many application-specific reasons, for example to align an image with a known template in a pattern matching process or to make certain image details easier to see. The translation process can be done with the following equations:

$$r' = r + r_0$$
$$c' = c + c_0$$

where r' and c' are the new coordinates, r and c are the original coordinates, and r_0 and c_0 are the distances to move, or translate, the image.

The rotation process requires the use of these equations:

$$\hat{r} = r(\cos\theta) + c(\sin\theta)$$
$$\hat{c} = -r(\sin\theta) + c(\cos\theta)$$

where $\hat{r}$ and $\hat{c}$ are the new coordinates, r and c are the original coordinates, and θ is the angle to rotate the image. θ is defined in a clockwise direction from the horizontal axis at the image origin in the upper-left corner.

The rotation and translation process can be combined into one set of equations:

$$\hat{r}' = (r + r_0)(\cos\theta) + (c + c_0)(\sin\theta)$$
$$\hat{c}' = -(r + r_0)(\sin\theta) + (c + c_0)(\cos\theta)$$

where $\hat{r}'$ and $\hat{c}'$ are the new coordinates and r, c, r_0, c_0, and θ are previously defined. There are some practical difficulties with the direct application of these equations. When translating, what is done with the "leftover" space? If we move everything one row down, what do we put in the top row? There are two basic options: fill the top row with a constant value, typically black (0) or white (255), or wrap-around by shifting the bottom row to the top, shown in Figure 2.2-3. Rotation also creates some practical difficulties. As Figure 2.2-4a illustrates, the image may be rotated off the "screen" (image plane). Although this can be fixed by a translation back to the center (Figure 2.2-4b, c), we have leftover space in the corners. We can fill this space with a constant or extract the central, rectangular portion of the image and enlarge it to the original image size.

Figure 2.2-3 Translation

1
2
3
4

a. BEFORE: A four-row image translating down by one row, $r_0 = 1$.

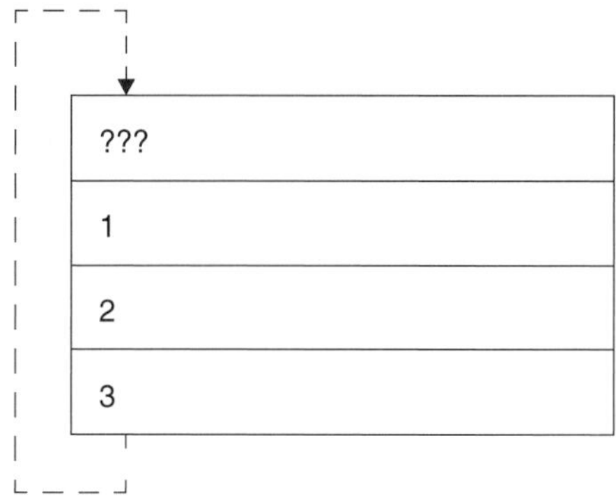

b. AFTER: If we wrap-around, row 4 goes into ???. Otherwise, the top row is filled with a constant, typically zero.

Figure 2.2-4 Rotation

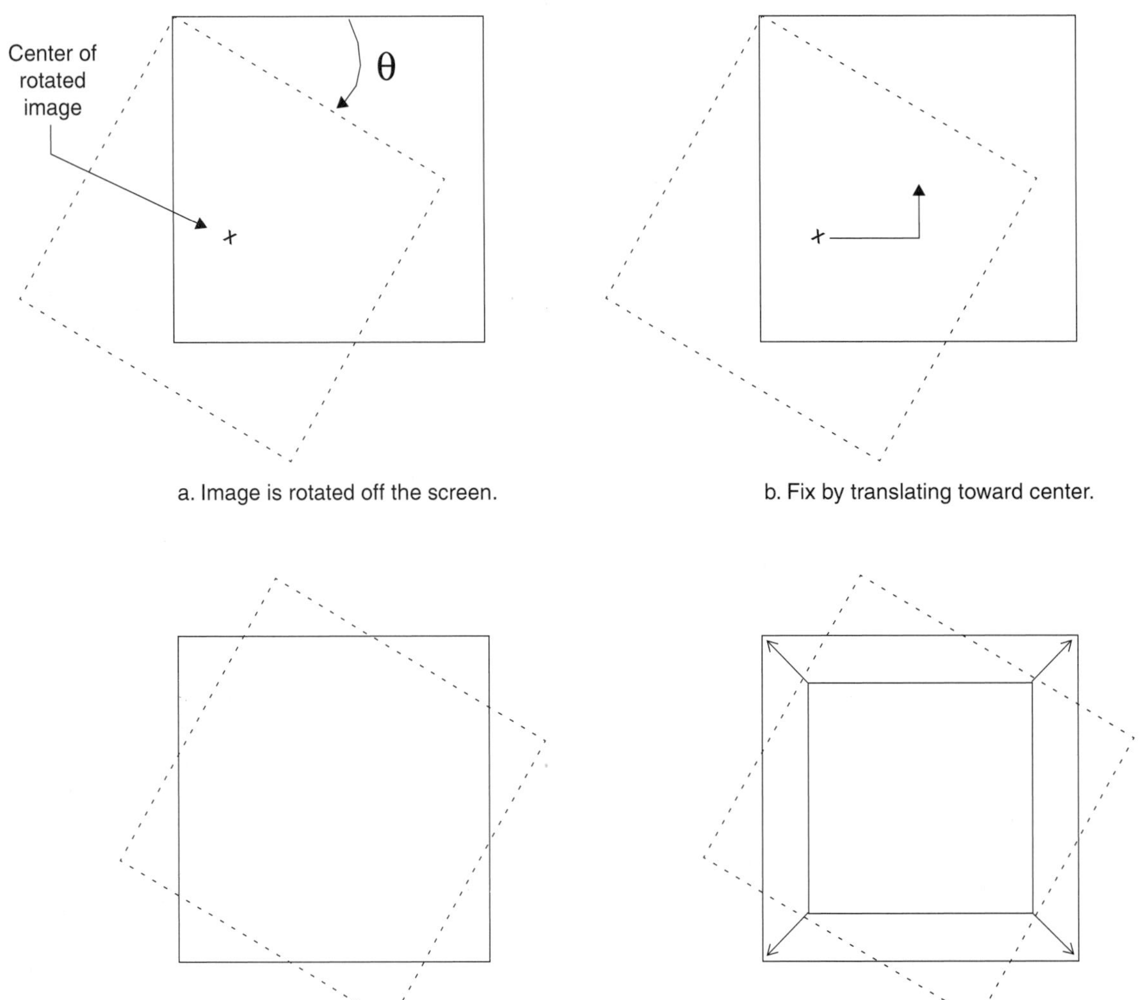

a. Image is rotated off the screen.

b. Fix by translating toward center.

c. Translation complete.

d. Crop and enlarge if desired.

2.2.2 Image Algebra

There are two primary categories of algebraic operations applied to images: arithmetic and logic (see Image Processing Exercise #1 in Chapter 8). Addition, subtraction, division, and multiplication comprise the arithmetic operations, while AND, OR, and NOT make up the logic operations. These operations are performed on two images, except for the NOT logic operation, which requires only one image, and is done on a pixel-by-pixel basis.

To apply the arithmetic operations to two images, we simply operate on corresponding pixel values. For example, to add images I_1 and I_2 to create I_3:

$$I_1(r,c) + I_2(r,c) = I_3(r,c)$$

$$I_1 = \begin{bmatrix} 3 & 4 & 7 \\ 3 & 4 & 5 \\ 2 & 4 & 6 \end{bmatrix} \quad I_2 = \begin{bmatrix} 6 & 6 & 6 \\ 4 & 2 & 6 \\ 3 & 5 & 5 \end{bmatrix} \quad I_3 = \begin{bmatrix} 3+6 & 4+6 & 7+6 \\ 3+4 & 4+2 & 5+6 \\ 2+3 & 4+5 & 6+5 \end{bmatrix} = \begin{bmatrix} 9 & 10 & 13 \\ 7 & 6 & 11 \\ 5 & 9 & 11 \end{bmatrix}$$

Addition is used to combine the information in two images. Applications include development of image restoration algorithms for modeling additive noise, and special effects, such as image morphing, in motion pictures (Figure 2.2-5). Note that true *image morphing* may also require the use of geometric transformations (see Section 3.5), to align the two images. Image morphing is also usually a time-based operation, so that a proportionally increasing amount of the second image is usually added to the first image over time.

Subtraction of two images is often used to detect motion. Consider the case where nothing has changed in a scene; the image resulting from subtraction of two sequential images is filled with zeros—a black image. If something has moved in the

Figure 2.2-5 Image Addition

Image Morphing Example

a. First original.

b. Second original.

c. Addition of images (a) and (b).

Additive Noise Example

d. Original image.

e. Gaussian noise, 400 variance, zero mean.

f. Addition of images (d) and (e).

scene, subtraction produces a nonzero result at the location of movement. Figure 2.2-6 illustrates the use of subtraction for motion detection.

Multiplication and division are used to adjust the brightness of an image. One image typically consists of a constant number greater than one. Multiplication of the pixel values by a number greater than one will brighten the image, and division by a factor greater than one will darken the image. Brightness adjustment is often used as a preprocessing step in image enhancement and is shown in Figure 2.2-7.

Figure 2.2-6 Subtraction

a. Original scene.

b. Same scene at a later time.

c. Subtraction of original scene from later time. Note that only objects that moved in the scene appear in the resultant image.

Figure 2.2-7 Multiplication and Division

a. Original image.

b. Image multiplied by 2.

c. Image divided by 2.

The logic operations AND, OR, and NOT form a complete set, meaning that any other logic operation (XOR, NOR, NAND) can be created by a combination of these basic elements. They operate in a bit-wise fashion on pixel data.

EXAMPLE 2-2

We are performing a logic AND on two images. Two corresponding pixel values are 111_{10} in one image and 88_{10} in the second image. The corresponding bit strings are $111_{10} = 01101111_2$ and $88 = 01011000_2$.

$$\begin{array}{r} 01101111_2 \\ \text{AND } 01011000_2 \\ \hline 01001000_2 \end{array}$$

The logic operations AND and OR are used to combine the information in two images. This may be done for special effects, but a more useful application for image analysis is to perform a masking operation. Use AND and OR as a simple method to extract a Region of Interest from an image, if more sophisticated graphical methods are not available. For example, a white square ANDed with an image will allow only the portion of the image coincident with the square to appear in the output image, with the background turned black; and a black square ORed with an image will allow only the part of the image corresponding to the black square to appear in the output image but will turn the rest of the image white. This process is called image masking, and Figure 2.2-8 illustrates the results of these operations. The NOT operation creates a negative of the original image, by inverting each bit within each pixel value, and is shown in Figure 2.2-9.

2.2.3 Spatial Filters

Spatial filtering is typically done for noise removal or to perform some type of image enhancement. These operators are called spatial filters to distinguish them from frequency domain filters, which are discussed in Section 2.5. The three types of filters discussed here include: 1) mean filters, 2) median filters, and 3) enhancement

Figure 2.2-8 Image Masking

a. Original image.

b. Square for AND mask.

c. Resulting image, (a) AND (b).

d. Square for OR mask.

e. Resulting image, (a) OR (d).

Figure 2.2-9 Complement Image

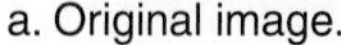

a. Original image.

b. NOT operator applied to image.

filters. The first two are used primarily to conceal or remove noise, although they may also be used for special applications. For instance, a mean filter adds a "softer" look to an image, as in Figure 2.2-10. The enhancement filters highlight edges and details within the image.

Many spatial filters are implemented with convolution masks. Because a convolution mask operation provides a result that is a weighted sum of the values of a pixel and its neighbors, it is called a *linear filter*. One interesting aspect of convolution masks is that the overall effect can be predicted based on their general pattern. For example, if the coefficients of the mask sum to one, the average brightness of the image will be retained. If the coefficients sum to zero, the average brightness will be lost and will return a dark image. Furthermore, if the coefficients are alternating positive and negative, the mask is a filter that returns edge information only; if the coefficients are all positive, it is a filter that will blur the image.

Figure 2.2-10 Mean Filter

a. Original image.

b. Mean-filtered image, 3×3 kernel. Note the softer appearance.

The *mean filters* are essentially averaging filters. They operate on local groups of pixels called neighborhoods and replace the center pixel with an average of the pixels in this neighborhood. This replacement is done with a convolution mask such as the following 3×3 mask:

$$\begin{bmatrix} 1/9 & 1/9 & 1/9 \\ 1/9 & 1/9 & 1/9 \\ 1/9 & 1/9 & 1/9 \end{bmatrix}$$

Note that the coefficients of this mask sum to one, so the image brightness will be retained, and the coefficients are all positive, so it will tend to blur the image. There are other more complex mean filters that are designed to deal with specific types of noise. These are discussed in Chapter 3.

The median filter is a nonlinear filter. A *nonlinear filter* has a result that cannot be found by a weighted sum of the neighborhood pixels, such as is done with a convolution mask. However, the median filter does operate on a local neighborhood. After the size of the local neighborhood is defined, the center pixel is replaced with the median, or center, value present among its neighbors, rather than by their average.

EXAMPLE 2-3

Given the following 3×3 neighborhood:

$$\begin{bmatrix} 5 & 5 & 6 \\ 3 & 4 & 5 \\ 3 & 4 & 7 \end{bmatrix}$$

we first sort the values in order of size—(3,3,4,4,5,5,5,6,7); then we select the middle value, in this case it is 5. This 5 is then placed in the center location.

A median filter can use a neighborhood of any size, but 3×3, 5×5, and 7×7 are typical. Note that the output image must be written to a separate image (a buffer), so that the results are not corrupted as this process is performed. Figure 2.2-11 illustrates the use of a median filter for noise removal.

The enhancement filters considered here include laplacian-type and difference filters. These types of filters will tend to bring out, or enhance, details in the image. Two 3×3 convolution masks for the laplacian-type filters are

$$\begin{bmatrix} 0 & -1 & 0 \\ -1 & 5 & -1 \\ 0 & -1 & 0 \end{bmatrix} \quad \begin{bmatrix} 1 & -2 & 1 \\ -2 & 5 & -2 \\ 1 & -2 & 1 \end{bmatrix}$$

The laplacian-type filters will enhance details in all directions equally. The difference filters will enhance details in the direction specific to the mask selected. There are four difference filter convolution masks, corresponding to lines in the vertical, horizontal, and two diagonal directions:

Figure 2.2-11 Median Filter

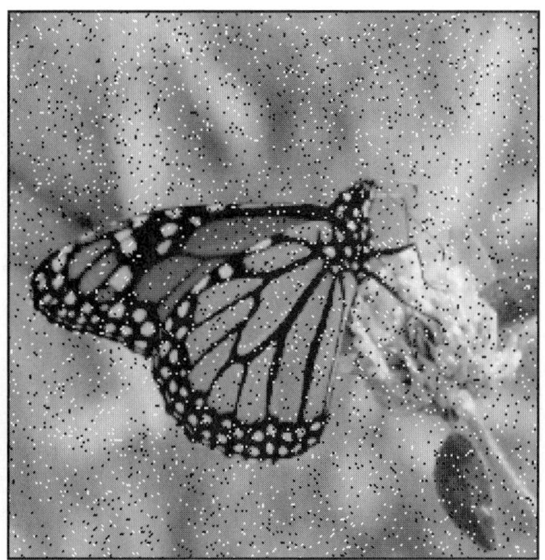

a. Original image with added salt-and-pepper noise.

b. Median-filtered image using a 3 × 3 kernel.

VERTICAL	HORIZONTAL	DIAGONAL1	DIAGONAL2
$\begin{bmatrix} 0 & 1 & 0 \\ 0 & 1 & 0 \\ 0 & -1 & 0 \end{bmatrix}$	$\begin{bmatrix} 0 & 0 & 0 \\ 1 & 1 & -1 \\ 0 & 0 & 0 \end{bmatrix}$	$\begin{bmatrix} 1 & 0 & 0 \\ 0 & 1 & 0 \\ 0 & 0 & -1 \end{bmatrix}$	$\begin{bmatrix} 0 & 0 & 1 \\ 0 & 1 & 0 \\ -1 & 0 & 0 \end{bmatrix}$

The results of applying these filters are shown in Figure 2.2-12. A more detailed discussion of enhancement filters is presented in Chapter 4.

2.2.4 Image Quantization

Image quantization is the process of reducing the image data by removing some of the detail information by mapping groups of data points to a single point. This can be done either to the pixel values themselves $I(r, c)$ or to the spatial coordinates (r, c). Operation on the pixel values is referred to as *gray-level reduction*, while operating on the spatial coordinates is called *spatial reduction*.

The simplest method of gray-level reduction is thresholding. We select a threshold gray level and set everything above that value equal to '1' (255 for 8-bit data) and everything at or below the threshold equal to '0'. This effectively turns a gray-level image into a binary (two-level) image and is often used as a preprocessing step in the extraction of object features such as shape, area, or perimeter.

A more versatile method of gray-level reduction is the process of taking the data and reducing the number of bits per pixel. This can be done very efficiently by masking the lower bits via an AND operation. With this method, the number of bits that are masked determines the number of gray levels available.

Figure 2.2-12 Enhancement Filters

a. Original image 1.

b. Image 1 after laplacian filter application.

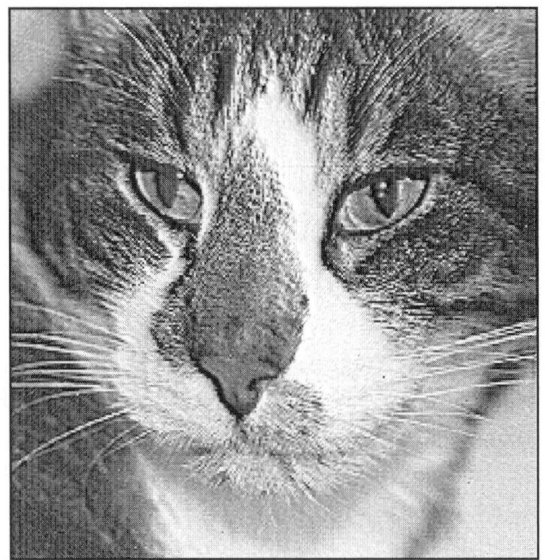

c. Contrast-enhanced version of laplacian-filtered image. Compare with (a) and note the improvement in fine detail information, which can be seen in the cat's fur.

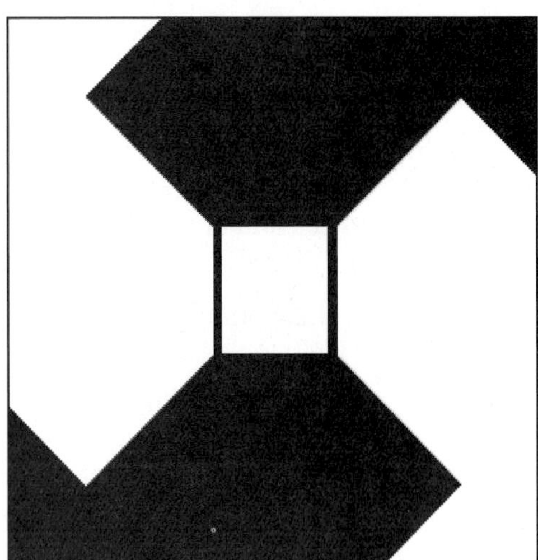

d. Original image 2.

Figure 2.2-12 (Continued)

e. Horizontal difference filter result.

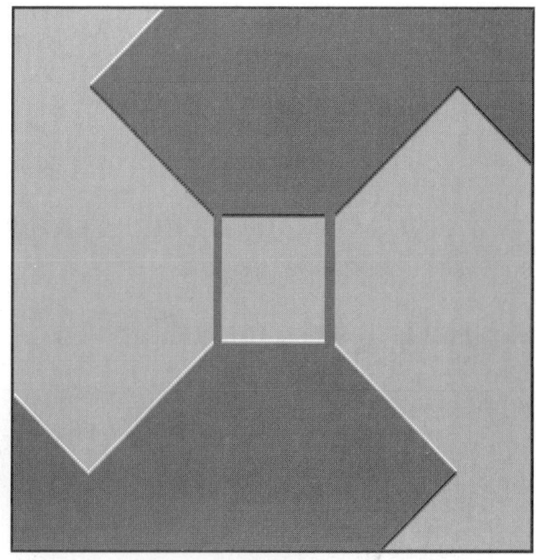

f. Vertical difference filter result.

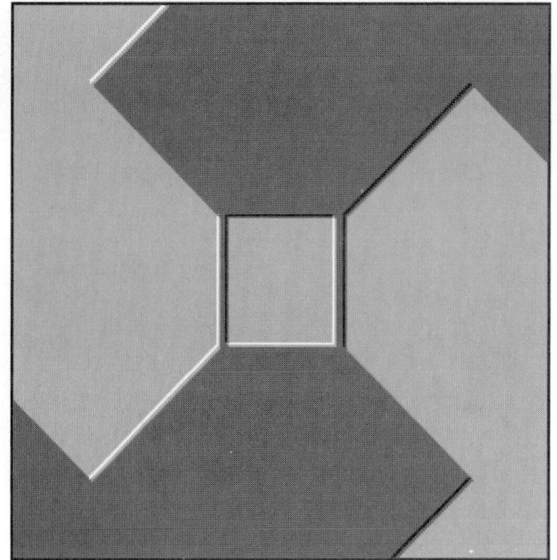

g. Diagonal1 difference filter result.

h. Diagonal2 difference filter result.

EXAMPLE 2-4

We want to reduce 8-bit information containing 256 possible gray-level values down to 32 possible values. This can be done by ANDing each 8-bit value with the bit string 11111000. This is equivalent to dividing by eight (2^3), corresponding to the lower three bits that we are masking and then shifting the result left three times. Now, gray-level values in the range of 0–7 are mapped to 0, gray levels in the range of 8–15 are mapped to 8, and so on.

Figure 2.2-13 False Contouring

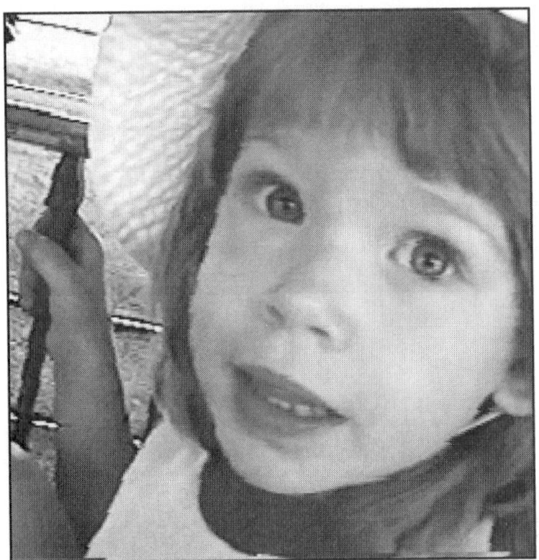

a. Original image.

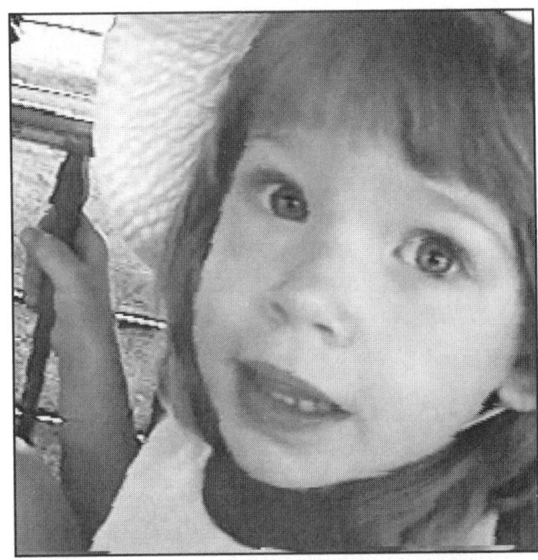

b. Quantized by masking to 128 gray levels.

c. Quantized by masking to 64 gray levels.

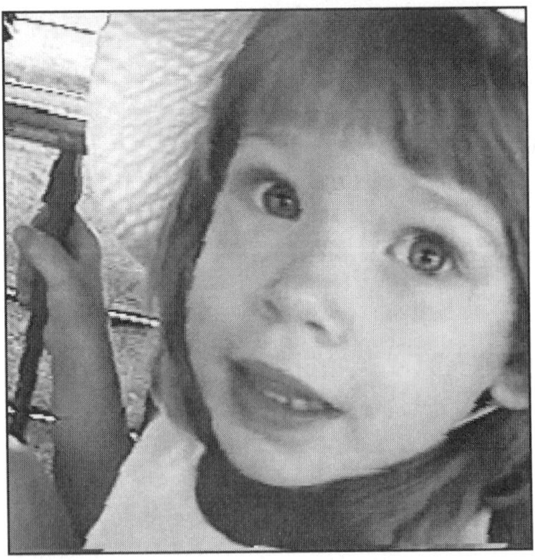

d. Quantized by masking to 32 gray levels.

Figure 2.2-13 (Continued)

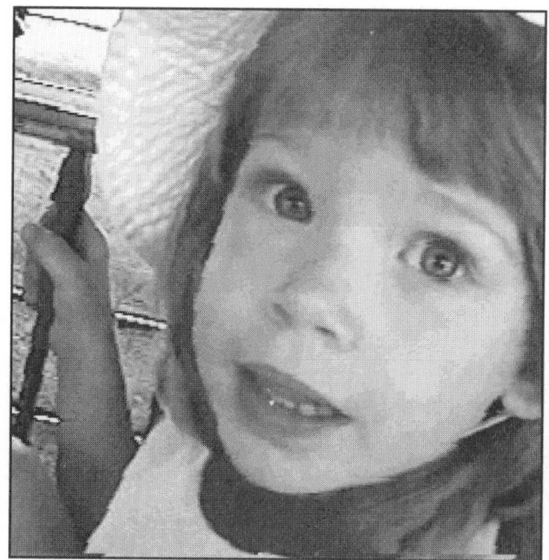

e. Quantized by masking to 16 gray levels.

f. Quantized by masking to 8 gray levels.

g. Quantized by masking to 4 gray levels.

h. Quantized by masking to 2 gray levels.

We can see that by masking the lower three bits we reduce 256 gray levels to 32 gray levels: $256 \div 8 = 32$. The general case requires us to mask k bits, where 2^k is divided into the original gray-level range to get the quantized range desired. Using this method, we can reduce the number of gray levels to any power of 2: 2, 4, 8, 16, 32, 64, or 128. This is illustrated in Figure 2.2-13; as the number of gray levels decreases, we see an increase in a phenomenon called contouring. Contouring appears in the images as false edges, or lines, as a result of the gray-level quantization. This false contouring effect can be visually improved upon by using an IGS (improved gray-scale) quantization method. The IGS method takes advantage of the human visual system's sensitivity to edges by adding a small random number to each pixel before quantization, which results in a more visually pleasing appearance (see Figure 2.2-14).

The AND-based method maps the quantized gray-level values to the low end of each range. Alternately, if we want to map the quantized gray-level values to the high end of each range we use an OR operation. The number of 1 bits in the OR mask determines how many quantized gray levels are available.

Figure 2.2-14 IGS Quantization

a. Original image.

b. Uniform quantization, eight levels.

c. IGS quantization, eight levels.

EXAMPLE 2-5

To reduce 256 gray levels down to 32 we use a mask of 00000111. Now, values in the range of 0–7 are mapped to 7, those ranging from 8 to 15 are mapped to 15, and so on.

EXAMPLE 2-6

To reduce 256 gray levels down to 16 we use a mask of 00001111. Now, values in the range of 0–15 are mapped to 15, those ranging from 16 to 31 are mapped to 31, and so on.

To determine the number of 1 bits in our OR mask, we apply a method similar to the AND mask method. We set k bits equal to 1, where 2^k is divided into the original gray-level range to get the quantized range desired.

Another potentially useful variation is to map the quantized values to the midpoint of the range. This is done by an AND after the OR operation, or an OR after the AND operation, to either shift the values up or down.

EXAMPLE 2-7

If we performed the quantization down to 16 levels by an OR with a mask of 00001111, which maps the values to the high end of the range, we could shift the values down to the middle of the range by ANDing with a mask of 11111100.

Although this AND/OR method is very efficient, it is not flexible since the size of the quantization bins is not variable (see Figure 2.2-15a).

There are other methods of gray-level quantization that allow for variable bin sizes (Figure 2.2-15b). These methods are more complicated than, and not as fast as, those used with uniform bins. One such use is in simulating the response of the human visual system by using logarithmically spaced bins. The use of variable bin size is application dependent and requires application-specific information. For example, in Figure 2.2-16 we can see the result of an application where four gray levels provided optimal results. Here we are applying varying bin sizes and mapping them to specific gray levels. In Figure 2.2-16 the gray levels in the range 0–101 were mapped to 79, 102–188 were mapped to 157, 189–234 were mapped to 197, and 235–255 were mapped to 255. These numbers were determined as the result of application-specific feedback (see Figure 2.1-3). For this application, the second brightest gray level (197) was used to identify fillings in dental X-rays.

Quantization of the spatial coordinates results in reducing the actual size of the image. This is accomplished by taking groups of pixels that are spatially adjacent and mapping them to one pixel. This can be done in one of three ways: 1) averaging, 2) median, or 3) decimation. For the averaging method, we take all the pixels in each group and find the average gray level by summing the values and dividing by the number of pixels in the group. With the second method, median, we sort all the pixel values from lowest to highest and then select the middle value. The third approach, decimation, also known as subsampling, entails simply eliminating some of the data.

Figure 2.2-15 Quantization Bins

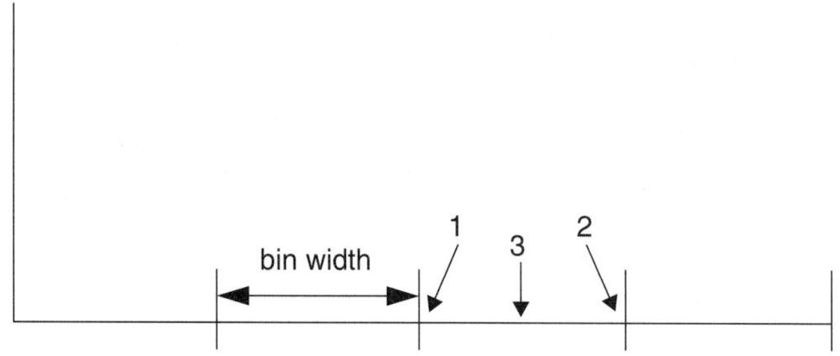

a. Uniform quantization bins: all bins are the same width. Values that fall within the same bin can be mapped to the low end (1), high end (2), or middle (3).

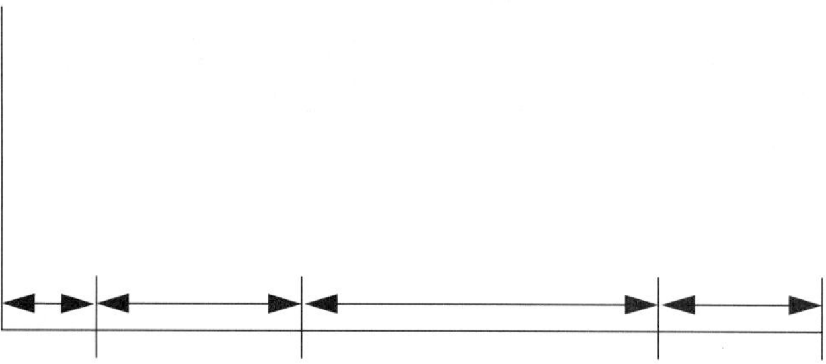

b. Variable quantization bins are of different widths.

Figure 2.2-16 Variable Bin-Width Quantization

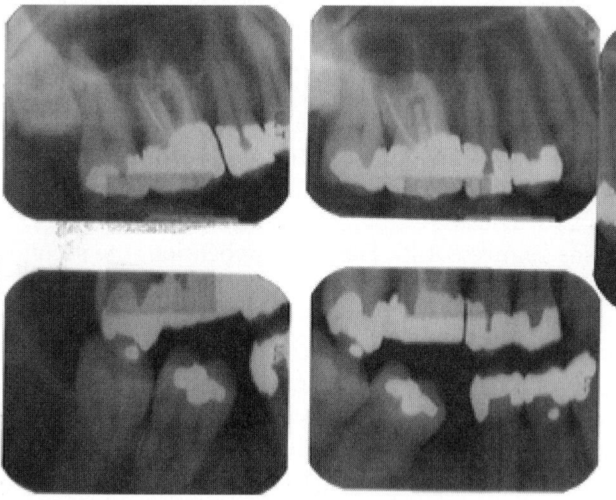

a. Original image.

Figure 2.2-16 (Continued)

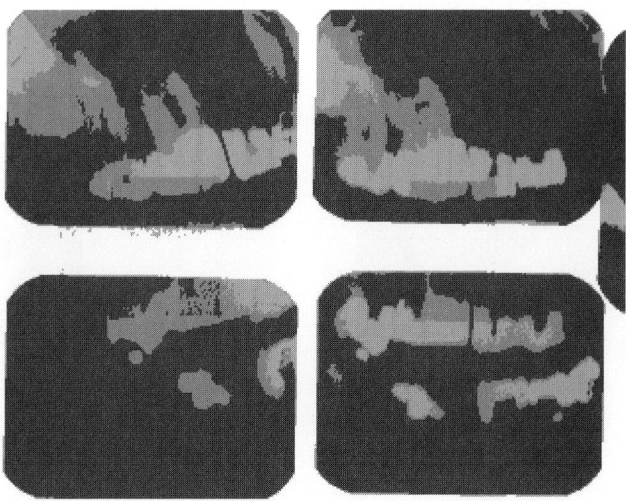

b. After variable bin-width quantization.

For example, to reduce the image by a factor of two, we simply take every other row and column and delete them. To improve the image quality when applying the decimation technique, we may want to preprocess the image with an averaging, or mean, spatial filter—this type of filtering is called *anti-aliasing filtering.*

To perform spatial quantization we specify the desired size, in pixels, of the resulting image. For example, to reduce a 512×512 image to one quarter its size, we specify that we want the output image to be 256×256 pixels. We now take every 2×2 pixel block in the original image and apply one of the three methods listed previously to create a reduced image. It should be noted that this method of spatial reduction allows for simple forms of geometric distortion, specifically, stretching or shrinking along the horizontal or vertical axis. Geometric distortion is explored more fully in Chapter 3. If we take a 256×256 image and reduce it to a size of 128×64, we will have shrunk the image as well as squeezed it horizontally. This result is shown in Figure 2.2-17.

2.3 EDGE/LINE DETECTION

The edge and line detection operators presented here represent the various types of operators in use today (see Computer Vision Lab Exercises #5 and #6 in Chapter 8). Many are implemented with convolution masks, and most are based on discrete approximations to differential operators. Differential operations measure the rate of change in a function, in this case, the image brightness function. A large change in image brightness over a short spatial distance indicates the presence of an edge. Some edge detection operators return orientation information (information about the direction of the edge), whereas others only return information about the existence of an

Figure 2.2-17 Spatial Reduction

a. Original image 256 × 256.

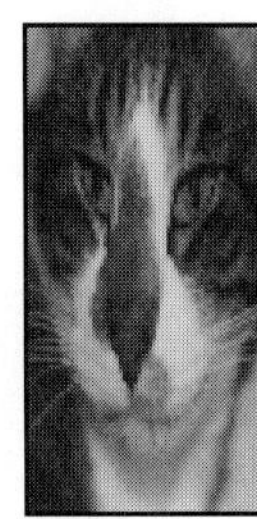

b. Spatially reduced image now 128 × 64.
Note the distortion.

edge at each point. Also included in this section is a special transform, the Hough Transform, which is specifically defined to find lines.

Edge detection methods are used as a first step in the line detection process. Edge detection is also used to find complex object boundaries by marking potential edge points corresponding to places in an image where rapid changes in brightness occur. After these edge points have been marked, they can be merged to form lines and object outlines.

With many of these operators, noise in the image can create problems. That is why it is best to preprocess the image to eliminate, or at least minimize, noise effects. To deal with noise effects, we must make tradeoffs between the sensitivity and the accuracy of an edge detector. For example, if the parameters are set so that the edge detector is very sensitive, it will tend to find many potential edge points that are attributable to noise. If we make it less sensitive, it may miss valid edges. The parameters that we can set include the size of the edge detection mask and the value of the gray-level threshold. A larger mask is less sensitive to noise; a lower gray-level threshold will tend to reduce noise effects.

Edge detection operators are based on the idea that edge information in an image is found by looking at the relationship a pixel has with its neighbors. If a pixel's gray-level value is similar to those around it, there is probably not an edge at that point. However, if a pixel has neighbors with widely varying gray levels, it may represent an edge point. In other words, an edge is defined by a discontinuity in gray-level values. Ideally, an edge separates two distinct objects. In practice, apparent edges are caused by changes in color or texture or by the specific lighting conditions present during the image acquisition process.

Figure 2.3-1 illustrates the differences between an ideal edge and a real edge. Figure 2.3-1a shows a representation of one row in an image of an ideal edge. The ver-

Figure 2.3-1 Ideal vs. Real Edge

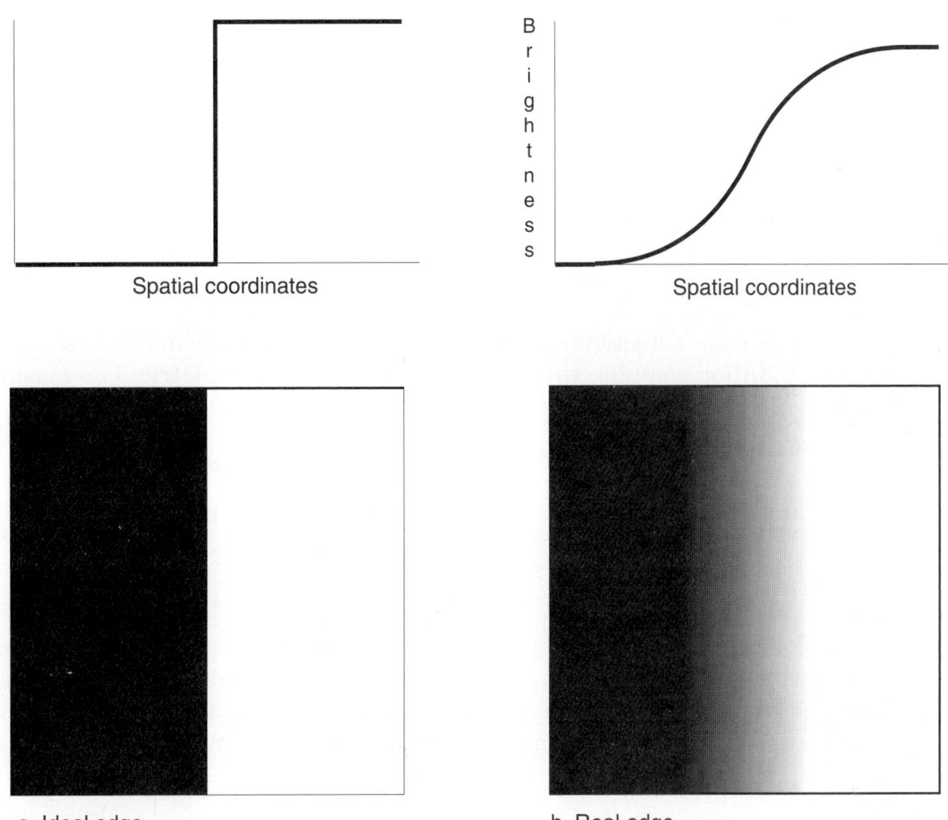

a. Ideal edge. b. Real edge.

tical axis represents brightness, and the horizontal axis shows the spatial coordinate. The abrupt change in brightness characterizes an ideal edge. In the corresponding image, the edge appears very distinct. In Figure 2.3-1b we see the representation of a real edge, which changes gradually. This gradual change is a minor form of blurring caused by the imaging device, the lenses, or the lighting, and it is typical for real-world (as opposed to computer-generated) images. In the image in Figure 2.3-1b, where the edge has been exaggerated for illustration purposes, note that from a visual perspective this image contains the same information as does the ideal image: black on one side, white on the other, with a line down the center.

2.3.1 Roberts Operator

The Roberts operator marks edge points only; it does not return any information about the edge orientation. It is the simplest of the edge detection operators and will work best with binary images (gray-level images can be made binary by a threshold operation). There are two forms of the Roberts operator. The first consists of the square root of the sum of the differences of the diagonal neighbors squared, as follows:

$$\sqrt{[I(r,c) - I(r-1,c-1)]^2 + [I(r,c-1) - I(r-1,c)]^2}$$

The second form of the Roberts operator is the sum of the magnitude of the differences of the diagonal neighbors, as follows:

$$|I(r,c) - I(r-1,c-1)| + |I(r,c-1) - I(r-1,c)|$$

The second form of the equation is often used in practice due to its computational efficiency—it is typically faster for a computer to find an absolute value than to find square roots.

2.3.2 Sobel Operator

The Sobel edge detection masks look for edges in both the horizontal and vertical directions and then combine this information into a single metric. The masks are as follows:

ROW MASK COLUMN MASK

$$\begin{bmatrix} -1 & -2 & -1 \\ 0 & 0 & 0 \\ 1 & 2 & 1 \end{bmatrix} \qquad \begin{bmatrix} -1 & 0 & 1 \\ -2 & 0 & 2 \\ -1 & 0 & 1 \end{bmatrix}$$

These masks are each convolved with the image (see Figure 2.2-2). At each pixel location we now have two numbers: s_1, corresponding to the result from the row mask, and s_2, from the column mask. We use these numbers to compute two metrics, the edge magnitude and the edge direction, which are defined as follows:

EDGE MAGNITUDE

$$\sqrt{s_1^2 + s_2^2}$$

EDGE DIRECTION

$$\tan^{-1}\left[\frac{s_1}{s_2}\right]$$

The edge direction is perpendicular to the edge itself because the direction specified is the direction of the gradient, along which the gray levels are changing.

2.3.3 Prewitt Operator

The Prewitt is similar to the Sobel, but with different mask coefficients. The masks are defined as follows:

ROW MASK COLUMN MASK

$$\begin{bmatrix} -1 & -1 & -1 \\ 0 & 0 & 0 \\ 1 & 1 & 1 \end{bmatrix} \qquad \begin{bmatrix} -1 & 0 & 1 \\ -1 & 0 & 1 \\ -1 & 0 & 1 \end{bmatrix}$$

These masks are each convolved with the image. At each pixel location we find two numbers: p_1, corresponding to the result from the row mask, and p_2, from the column mask. We use these results to determine two metrics, the edge magnitude and the edge direction, which are defined as follows:

EDGE MAGNITUDE

$$\sqrt{p_1^{\,2} + p_2^{\,2}}$$

EDGE DIRECTION

$$\tan^{-1}\left[\frac{p_1}{p_2}\right]$$

As with the Sobel edge detector, the direction lies 90° from the apparent direction of the edge.

2.3.4 Kirsch Compass Masks

The Kirsch edge detection masks are called compass masks because they are defined by taking a single mask and rotating it to the eight major compass orientations: North, Northwest, West, Southwest, South, Southeast, East, and Northeast. The masks are defined as follows:

$$k_0 = \begin{bmatrix} -3 & -3 & 5 \\ -3 & 0 & 5 \\ -3 & -3 & 5 \end{bmatrix} \quad k_1 = \begin{bmatrix} -3 & 5 & 5 \\ -3 & 0 & 5 \\ -3 & -3 & -3 \end{bmatrix} \quad k_2 = \begin{bmatrix} 5 & 5 & 5 \\ -3 & 0 & -3 \\ -3 & -3 & -3 \end{bmatrix} \quad k_3 = \begin{bmatrix} 5 & 5 & -3 \\ 5 & 0 & -3 \\ -3 & -3 & -3 \end{bmatrix}$$

$$k_4 = \begin{bmatrix} 5 & -3 & -3 \\ 5 & 0 & -3 \\ 5 & -3 & -3 \end{bmatrix} \quad k_5 = \begin{bmatrix} -3 & -3 & -3 \\ 5 & 0 & -3 \\ 5 & 5 & -3 \end{bmatrix} \quad k_6 = \begin{bmatrix} -3 & -3 & -3 \\ -3 & 0 & -3 \\ 5 & 5 & 5 \end{bmatrix} \quad k_7 = \begin{bmatrix} -3 & -3 & -3 \\ -3 & 0 & 5 \\ -3 & 5 & 5 \end{bmatrix}$$

The edge magnitude is defined as the maximum value found by the convolution of each of the masks with the image. The edge direction is defined by the mask that produces the maximum magnitude; for instance, k_0 corresponds to a vertical edge, whereas k_5 corresponds to a diagonal edge in the Northwest/Southeast direction. We also see that the last four masks are actually the same as the first four, but flipped about a central axis.

2.3.5 Robinson Compass Masks

The Robinson compass masks are used in a manner similar to the Kirsch masks but are easier to implement because they rely only on coefficients of 0, 1, and 2, and are symmetrical about their directional axis—the axis with the zeros. We only need to compute the results on four of the masks; the results from the other four can be obtained by negating the results from the first four. The masks are as follows:

$$r_0 = \begin{bmatrix} -1 & 0 & 1 \\ -2 & 0 & 2 \\ -1 & 0 & 1 \end{bmatrix} \quad r_1 = \begin{bmatrix} 0 & 1 & 2 \\ -1 & 0 & 1 \\ -2 & -1 & 0 \end{bmatrix} \quad r_2 = \begin{bmatrix} 1 & 2 & 1 \\ 0 & 0 & 0 \\ -1 & -2 & -1 \end{bmatrix} \quad r_3 = \begin{bmatrix} 2 & 1 & 0 \\ 1 & 0 & -1 \\ 0 & -1 & -2 \end{bmatrix}$$

$$r_4 = \begin{bmatrix} 1 & 0 & -1 \\ 2 & 0 & -2 \\ 1 & 0 & -1 \end{bmatrix} \quad r_5 = \begin{bmatrix} 0 & -1 & -2 \\ 1 & 0 & -1 \\ 2 & 1 & 0 \end{bmatrix} \quad r_6 = \begin{bmatrix} -1 & -2 & -1 \\ 0 & 0 & 0 \\ 1 & 2 & 1 \end{bmatrix} \quad r_7 = \begin{bmatrix} -2 & -1 & 0 \\ -1 & 0 & 1 \\ 0 & 1 & 2 \end{bmatrix}$$

The edge magnitude is defined as the maximum value found by the convolution of each of the masks with the image. The edge direction is defined by the mask that produces the maximum magnitude. It is interesting to note that masks r_0 and r_6 are the same as the Sobel masks. We can see that any of the edge detection masks can be extended by rotating them in a manner like these compass masks, which will allow us to extract explicit information about edges in any direction.

2.3.6 Laplacian Operators

The laplacian operators described here are similar to the ones used for preprocessing as described in Section 2.2.3. The three laplacian masks that follow represent different approximations of the laplacian operator. Unlike compass masks, the laplacian masks are rotationally symmetric, which means edges at all orientations contribute to the result. They are applied by selecting one mask and convolving it with the image. The sign of the result (positive or negative) from two adjacent pixel locations provides directional information, and tells us which side of the edge is brighter.

LAPLACIAN MASKS

$$\begin{bmatrix} 0 & -1 & 0 \\ -1 & 4 & -1 \\ 0 & -1 & 0 \end{bmatrix} \quad \begin{bmatrix} 1 & -2 & 1 \\ -2 & 4 & -2 \\ 1 & -2 & 1 \end{bmatrix} \quad \begin{bmatrix} -1 & -1 & -1 \\ -1 & 8 & -1 \\ -1 & -1 & -1 \end{bmatrix}$$

These masks differ from the laplacian type previously described in that the center coefficients have been decreased by one. With these masks we are trying to find edges and are not interested in the image itself. An easy way to picture the difference is to consider the effect each mask would have when applied to an area of constant value. The preceding convolution masks would return a value of zero. If we increase the center coefficients by one, each would return the original gray level. Therefore, if we are only interested in edge information, the sum of the coefficients should be zero. If we want to retain most of the information that is in the original image, the coefficients should sum to a number greater than zero. The larger this sum, the less the processed image is changed from the original image. Consider an extreme example in which the center coefficient is very large compared with the other coefficients in the mask. The resulting pixel value will depend most heavily upon the current value, with only minimal contribution from the surrounding pixel values.

2.3.7 Frei-Chen Masks

The Frei-Chen masks are unique in that they form a complete set of basis vectors (see Section 2.5). This means that we can represent any 3×3 subimage as a weighted sum of the nine Frei-Chen masks (Figure 2.3-2). These weights are found by projecting a 3×3 subimage onto each of these masks. This projection process is similar to the convolution process in that both overlay the mask on the image, multiply coincident terms, and sum the results (a vector inner product). This is best illustrated by example.

EXAMPLE 2-8

Suppose that we have the following subimage, I_s:

$$I_s = \begin{bmatrix} 1 & 0 & 1 \\ 1 & 0 & 1 \\ 1 & 0 & 1 \end{bmatrix}$$

To project this subimage onto the Frei-Chen masks, start by finding the projection onto f_1. Overlay the subimage on the mask and consider the first row. The 1 in the upper-left corner of the subimage coincides with the 1 in the upper-left corner of the mask, the 0 is over the $\sqrt{2}$, and the 1 on the upper-right corner of the subimage coincides with the 1 in the mask. Note that all these must be summed and then multiplied by the $1/(2\sqrt{2})$ factor to normalize the masks. The projection of I_s onto f_1 is equal to

$$\frac{1}{2\sqrt{2}}[1(1) + 0(\sqrt{2}) + 1(1) + 1(0) + 0(0) + 1(0) + 1(-1) + 0(-\sqrt{2}) + 1(-1)] = 0$$

If we follow this process and project the subimage I_s onto each of the Frei-Chen masks, we get the following:

$$f_1 \rightarrow 0, f_2 \rightarrow 0, f_3 \rightarrow 0, f_4 \rightarrow 0, f_5 \rightarrow -1, f_6 \rightarrow 0, f_7 \rightarrow 0, f_8 \rightarrow -1, f_9 \rightarrow 2$$

We can now see what is meant by a complete set of basis vectors allowing us to represent a subimage by a weighted sum. The basis vectors in this case are the Frei-Chen masks, and the weights are the projection values. Take the weights and multiply them by each mask; then sum the corresponding values. For this example the only nonzero terms correspond to masks f_5, f_8, and f_9, and we find the following:

$$(-1)(\frac{1}{2})\begin{bmatrix} 0 & 1 & 0 \\ -1 & 0 & -1 \\ 0 & 1 & 0 \end{bmatrix} + (-1)(\frac{1}{6})\begin{bmatrix} -2 & 1 & -2 \\ 1 & 4 & 1 \\ -2 & 1 & -2 \end{bmatrix} + (2)(\frac{1}{3})\begin{bmatrix} 1 & 1 & 1 \\ 1 & 1 & 1 \\ 1 & 1 & 1 \end{bmatrix} = \begin{bmatrix} 1 & 0 & 1 \\ 1 & 0 & 1 \\ 1 & 0 & 1 \end{bmatrix} = I_s$$

Figure 2.3-2 Frei-Chen Masks

$$f_1 = \frac{1}{2\sqrt{2}}\begin{bmatrix} 1 & \sqrt{2} & 1 \\ 0 & 0 & 0 \\ -1 & -\sqrt{2} & -1 \end{bmatrix} \quad f_2 = \frac{1}{2\sqrt{2}}\begin{bmatrix} 1 & 0 & -1 \\ \sqrt{2} & 0 & -\sqrt{2} \\ 1 & 0 & -1 \end{bmatrix} \quad f_3 = \frac{1}{2\sqrt{2}}\begin{bmatrix} 0 & -1 & \sqrt{2} \\ 1 & 0 & -1 \\ -\sqrt{2} & 1 & 0 \end{bmatrix}$$

$$f_4 = \frac{1}{2\sqrt{2}}\begin{bmatrix} \sqrt{2} & -1 & 0 \\ -1 & 0 & 1 \\ 0 & 1 & -\sqrt{2} \end{bmatrix} \quad f_5 = \frac{1}{2}\begin{bmatrix} 0 & 1 & 0 \\ -1 & 0 & -1 \\ 0 & 1 & 0 \end{bmatrix} \quad f_6 = \frac{1}{2}\begin{bmatrix} -1 & 0 & 1 \\ 0 & 0 & 0 \\ 1 & 0 & -1 \end{bmatrix}$$

$$f_7 = \frac{1}{6}\begin{bmatrix} 1 & -2 & 1 \\ -2 & 4 & -2 \\ 1 & -2 & 1 \end{bmatrix} \quad f_8 = \frac{1}{6}\begin{bmatrix} -2 & 1 & -2 \\ 1 & 4 & 1 \\ -2 & 1 & -2 \end{bmatrix} \quad f_9 = \frac{1}{3}\begin{bmatrix} 1 & 1 & 1 \\ 1 & 1 & 1 \\ 1 & 1 & 1 \end{bmatrix}$$

The first four masks comprise the edge subspace.
The next four masks comprise the line subspace.
The final mask is the average subspace.

We have seen how the Frei-Chen masks can be used to represent a subimage as a weighted sum, but how are they used for edge detection? The Frei-Chen masks can be grouped into a set of four masks for an edge subspace, four masks for a line subspace, and one mask for an average subspace. To use them for edge detection, select a particular subspace of interest and find the relative projection of the image onto the particular subspace. This is given by the following equation:

$$\cos \Theta = \sqrt{\frac{M}{S}}$$

where

$$M = \sum_{k \in \{e\}} (I_s, f_k)^2$$

$$S = \sum_{k=1}^{9} (I_s, f_k)^2$$

The set $\{e\}$ consists of the masks of interest. The (I_s, f_k) notation refers to the process of overlaying the mask on the subimage, multiplying coincident terms, and summing the results (a vector inner product). An illustration of this is shown in Figure 2.3-3. The advantage of this method is that we can select particular edge or line masks of interest and consider the projection of those masks only.

2.3.8 Edge Operator Performance

In evaluating the performance of many processes, we can consider both objective and subjective evaluation. The objective metric allows us to compare different techniques with fixed analytical methods, whereas the subjective methods often have unpredictable results. However, for many computer imaging applications, the subjective measures tend to be the most useful. We will examine the types of errors encoun-

Figure 2.3-3 Frei-Chen Projection

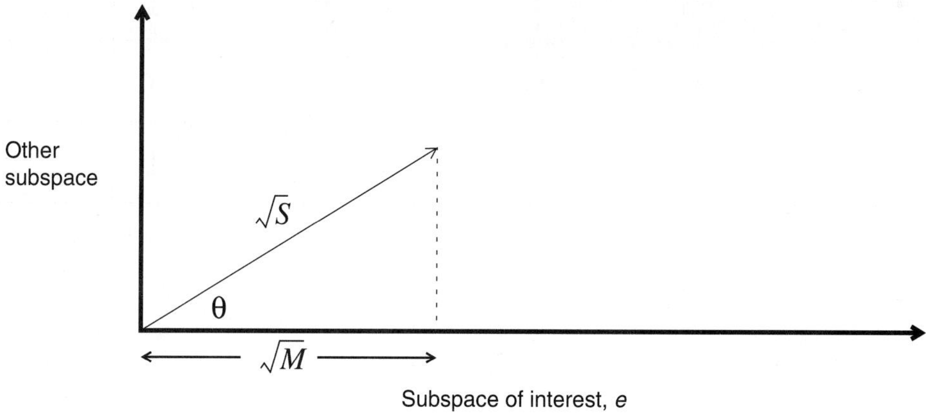

$$\cos \theta = \sqrt{\frac{M}{S}}$$

A two-dimensional representation of the Frei-Chen projection concept. The actual Frei-Chen space is nine-dimensional, where each dimension is given by one of the masks f_k.

tered with edge detection, look at an objective measure based on these criteria, and review results of the various edge detectors for our own subjective evaluation.

To develop a performance metric for edge detection operators, we need to define what constitutes success and what types of errors can occur. The types of errors are: 1) missing valid edge points, 2) classifying noise pulses as valid edge points, and 3) smearing edges (see Figure 2.3-4). If these errors do not occur, we can say that we have achieved success.

One metric, called the Pratt Figure of Merit Rating Factor, is defined as follows:

$$R = \frac{1}{I_N} \sum_{i=1}^{I_F} \frac{1}{1 + \alpha d^2}$$

$I_N =$ the maximum of I_I and I_F

$I_I =$ the number of ideal edge points in the image

Figure 2.3-4 Errors in Edge Detection

a. Original image.

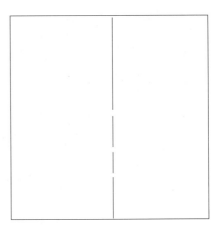

b. Missed edge points.

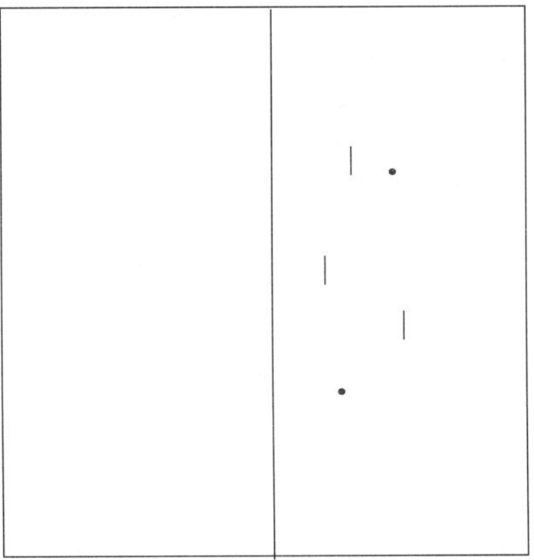

c. Noise misclassified as edge points.

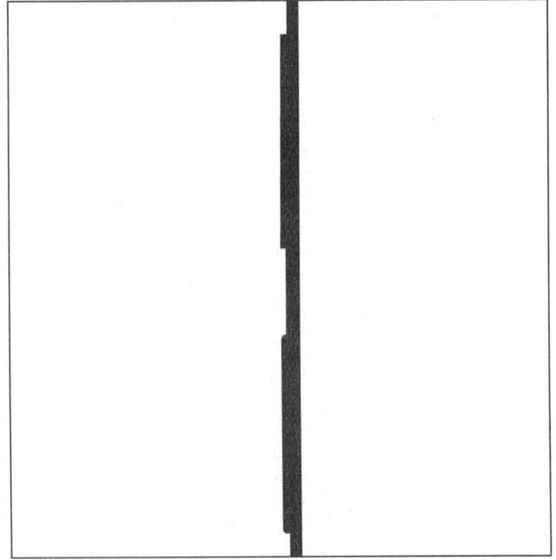

d. Smeared edge.

I_F = the number of edge points found by the edge detector

α = a scaling constant that can be adjusted to adjust the penalty for offset edges

d = the distance of a found edge point to an ideal edge point

For this metric, R will be 1 for a perfect edge. Normalizing to the maximum of the ideal and found edge points guarantees a penalty for smeared edges or missing edge points. In general, this metric assigns a better rating to smeared edges than to offset or missing edges. This is done because techniques exist to thin smeared edges, but it is difficult to determine when an edge is missed.

Figure 2.3-5 Edge Detection Examples

a. Original image.

b. Sobel operator .

c. Prewitt operator.

d. Frei-Chen operator, edge subspace.

Figure 2.3-5 (Continued)

e. Laplacian operator.

f. Kirsch operator.

g. Roberts operator.

h. Robinson operator.

As previously mentioned, the objective metrics are often of limited use in practical applications, so we will take a subjective look at the results of the edge detectors. The human visual system is still superior, by far, to any computer vision system that has yet been devised and is often used as the final judge in application development. Figure 2.3-5 shows the results of the edge detection operators. Here we see similar results from all the operators but the laplacian. This occurs because the laplacian returns positive *and* negative numbers that get linearly remapped to 0 to 255 (for 8-bit display), which means that the background value of 0 is mapped to some intermediate gray level. For the other edge detection operators, only the magnitude is used for displaying the results. Similar results can be achieved with the laplacian if we apply a threshold operation to the results (thresholding for boundary detection is explored in Section 2.4.4).

If we add noise to the image, the edge detection results are not as good. As mentioned before, we can preprocess the image with various spatial filters to remove, or hide, some of the effects from noise (this is explored more in Chapters 3 and 4), or we can expand the edge detection operators themselves to mitigate noise effects. One way to do this is to extend the size of the edge detection masks. An example of this method is to extend the Prewitt edge mask as follows:

EXTENDED PREWITT EDGE DETECTION MASK

$$\begin{bmatrix} 1 & 1 & 1 & 0 & -1 & -1 & -1 \\ 1 & 1 & 1 & 0 & -1 & -1 & -1 \\ 1 & 1 & 1 & 0 & -1 & -1 & -1 \\ 1 & 1 & 1 & 0 & -1 & -1 & -1 \\ 1 & 1 & 1 & 0 & -1 & -1 & -1 \\ 1 & 1 & 1 & 0 & -1 & -1 & -1 \\ 1 & 1 & 1 & 0 & -1 & -1 & -1 \end{bmatrix}$$

We then can rotate this by 90° and have both row and column masks that can be used like the Prewitt operators to return the edge magnitude and gradient. These types of operators are called boxcar operators and can be extended to any size, although 7×7, 9×9, and 11×11 are typical. The Sobel operator can be extended in a similar manner:

EXTENDED SOBEL EDGE DETECTION MASK

$$\begin{bmatrix} -1 & -1 & -1 & -2 & -1 & -1 & -1 \\ -1 & -1 & -1 & -2 & -1 & -1 & -1 \\ -1 & -1 & -1 & -2 & -1 & -1 & -1 \\ 0 & 0 & 0 & 0 & 0 & 0 & 0 \\ 1 & 1 & 1 & 2 & 1 & 1 & 1 \\ 1 & 1 & 1 & 2 & 1 & 1 & 1 \\ 1 & 1 & 1 & 2 & 1 & 1 & 1 \end{bmatrix}$$

We can also define a truncated pyramid operator as follows:

$$\begin{bmatrix} 1 & 1 & 1 & 0 & -1 & -1 & -1 \\ 1 & 2 & 2 & 0 & -2 & -2 & -1 \\ 1 & 2 & 3 & 0 & -3 & -2 & -1 \\ 1 & 2 & 3 & 0 & -3 & -2 & -1 \\ 1 & 2 & 3 & 0 & -3 & -2 & -1 \\ 1 & 2 & 2 & 0 & -2 & -2 & -1 \\ 1 & 1 & 1 & 0 & -1 & -1 & -1 \end{bmatrix}$$

This operator provides weights that decrease as we get away from the center pixel. These operators are used in the same method as the Prewitt and Sobel—we define a row and column mask and then find a magnitude and direction at each point. A comparison of applying the extended operators and the standard operators to a noisy image is shown in Figure 2.3-6. With noisy images, the extended operators exhibit better performance than the 3×3 masks but require more computations and tend to blur the edges slightly. The improved performance is most noticeable in the girl's facial features, which are almost completely obscured by the noise when using the 3×3 masks.

Figure 2.3-6 Edge Detection Examples—Noise

a. Original image.

b. Image with added gaussian and salt-and-pepper noise.

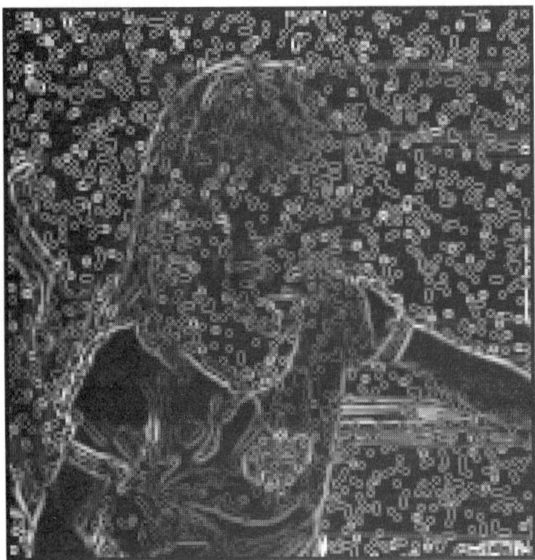

c. Sobel with 3×3 mask.

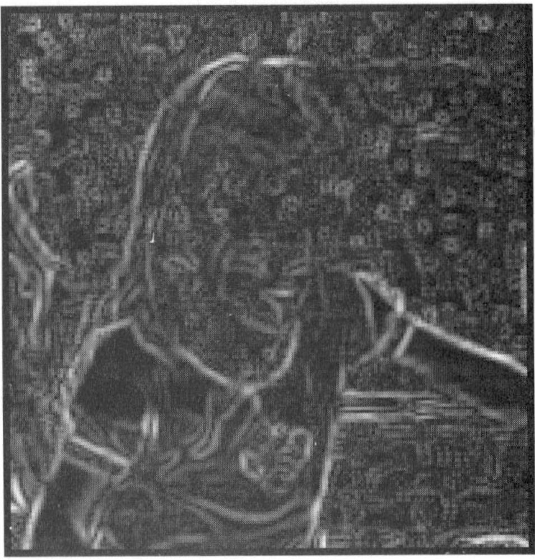

d. Sobel with 7×7 mask.

Figure 2.3-6 (Continued)

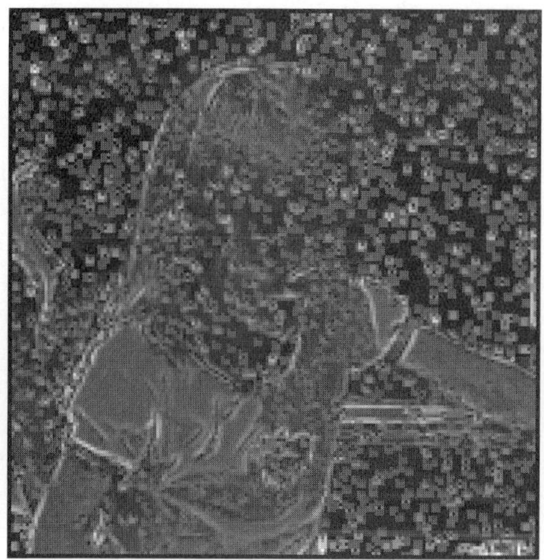

e. Prewitt with 3×3 mask.

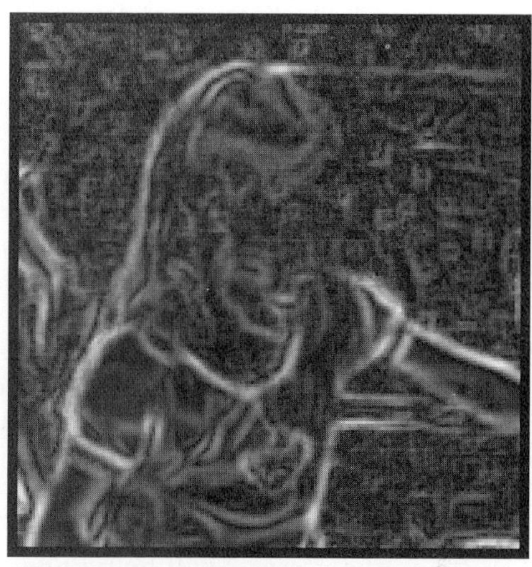

f. Prewitt with 7×7 mask.

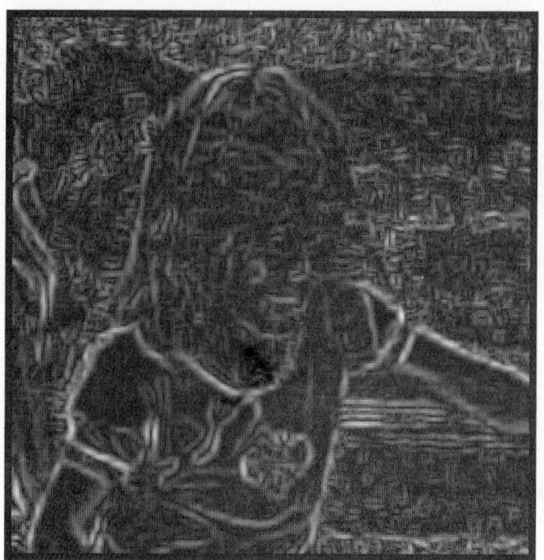

g. Truncated pyramid with 7×7 mask.

2.3.9 Hough Transform

The Hough transform is designed specifically to find lines. Until this point we have not really differentiated between lines and edges, but we now define a line as a collection of edge points that are adjacent and have the same direction. The Hough transform is an algorithm that will take a collection of edge points, as found by an edge detector, and find all the lines on which these edge points lie. Although a brute force search method can be used that will check every point with every possible line, the primary advantage of the Hough transform is that it reduces the search time for finding lines and the corresponding set of points.

In order to understand the Hough transform, we will first consider the normal (perpendicular) representation of a line:

$$\rho = r \cos\theta + c \sin\theta$$

If we have a line in our row and column (rc)-based image space, we can define that line by ρ, the distance from the origin to the line along a perpendicular to the line, and θ, the angle between the r-axis and the ρ-line (see Figure 2.3-7). Now, for each pair of values of ρ and θ we have defined a particular line. The range on θ is $\pm 90°$ and ρ ranges from 0 to $\sqrt{2}N$, where N is the image size. Next, we can take this $\rho\theta$ parameter space and quantize it, to reduce our search time. We quantize the $\rho\theta$ parameter space, as shown in Figure 2.3-8, by dividing the space into a specific number of blocks. Each block corresponds to a line, or group of possible lines, with ρ and θ varying across the increment as defined by the size of the block. The size of these blocks corresponds to the coarseness of the quantization; bigger blocks provide less line resolution.

The algorithm used for the Hough transform (see Figure 2.3-9 for a flowchart of the process) will help understand what this means. The algorithm consists of three primary steps:

1. Define the desired increments on ρ and θ, Δ_ρ and Δ_θ, and quantize the space accordingly.
2. For every point of interest (typically points found by edge detectors that exceed some threshold value), plug the values for r and c into the line equation:

$$\rho = r \cos\theta + c \sin\theta$$

Figure 2.3-7 Hough Transform

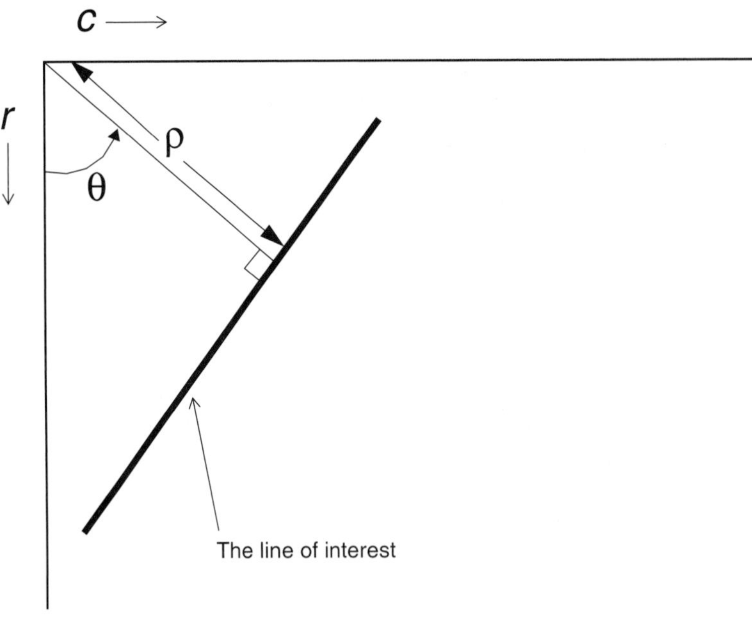

Figure 2.3-8 Hough Space

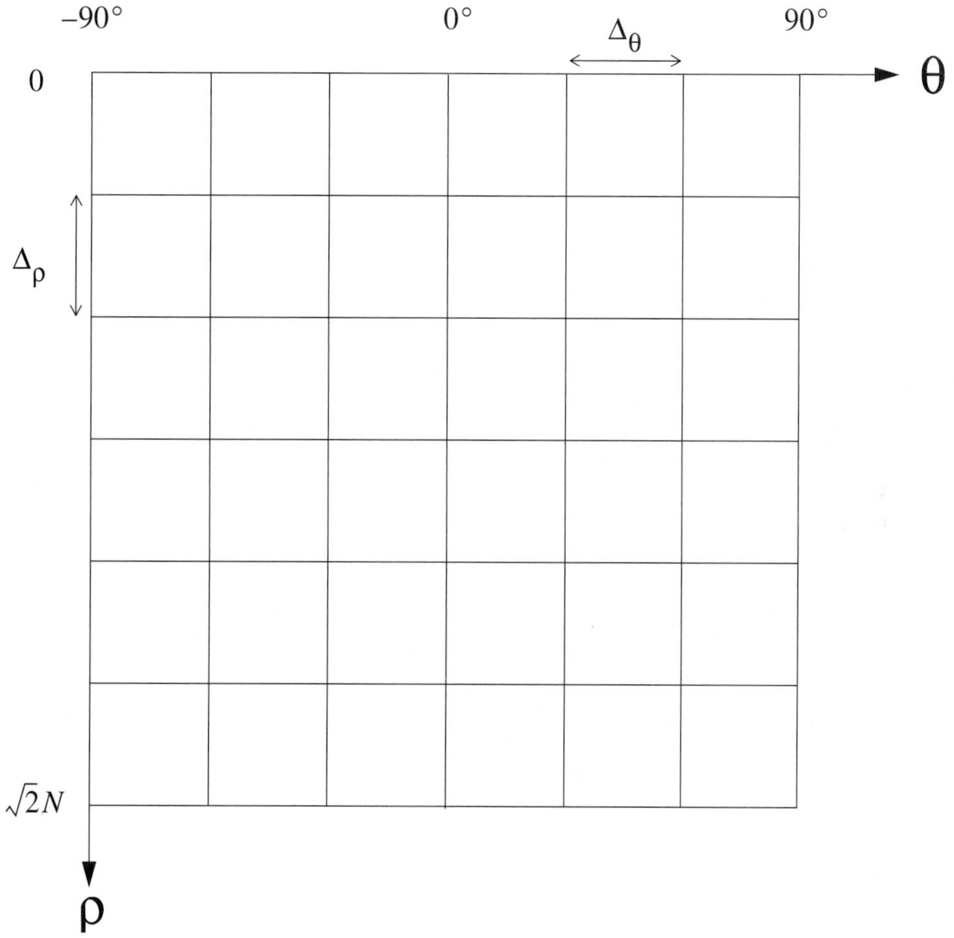

Then, for each value of θ in the quantized space, solve for ρ.

3. For each $\rho\theta$ pair from step 2, record the *rc* pair in the corresponding block in the quantized space. This constitutes a hit for that particular block.

When this process is completed, the number of hits in each block corresponds to the number of pixels on the line as defined by the values of ρ and θ in that block. The advantage of large quantization blocks is that the search time is reduced, but the price paid is less line resolution in the image space. Examining Figure 2.3-10, we can see that this means that the line of interest in the image space can vary more. One block in the Hough Space corresponds to all the solid lines in this figure—this is what we mean by reduced line resolution.

Next, select a threshold and examine the quantization blocks that contain more points than the threshold. Here, we look for continuity by searching for gaps in the line by looking at the distance between points on the line (remember that the points on a line correspond to points recorded in the block). When this process is completed,

Figure 2.3-9 Hough Transform Flowchart

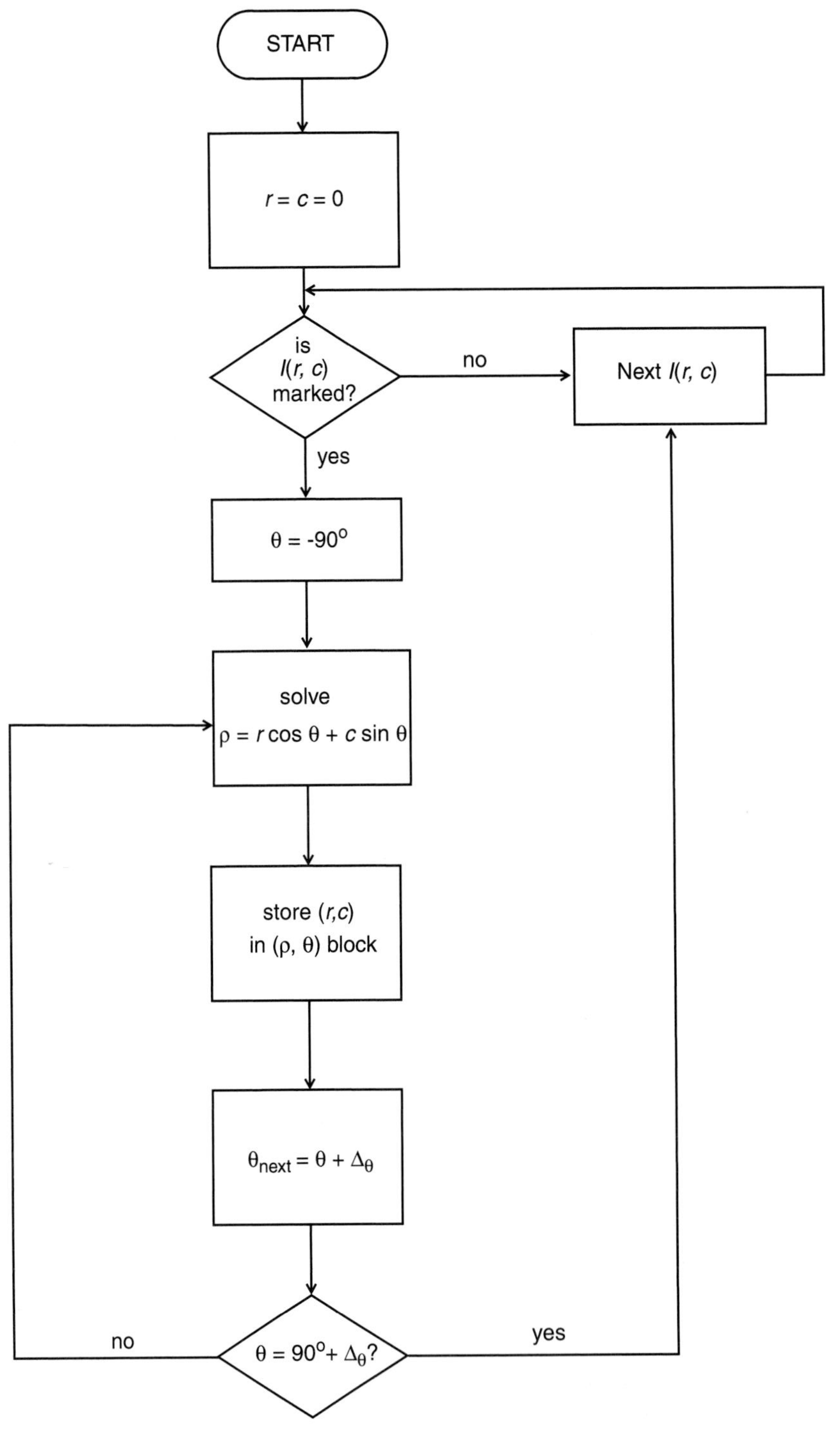

The flowchart is followed until all $I(r, c)$ have been examined.

Figure 2.3-10 Effects of Quantization Block Size for Hough Transform

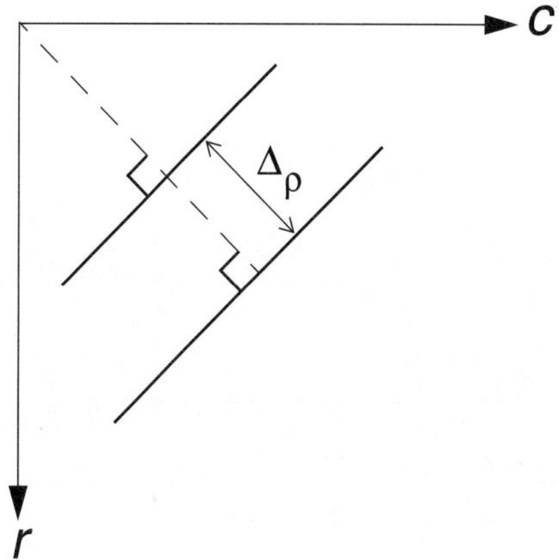

a. Range of lines included by choice of Δ_ρ.

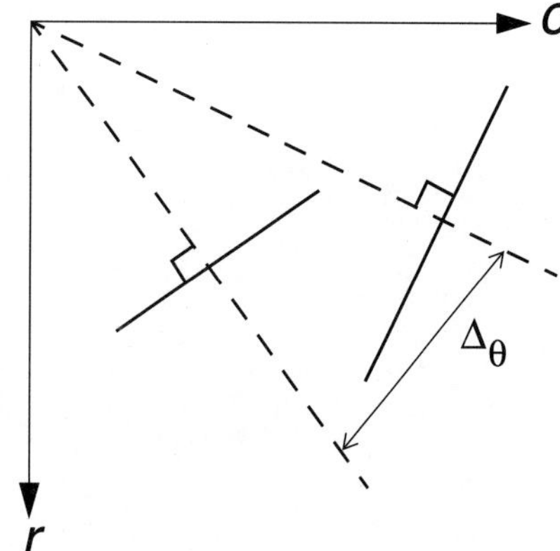

b. Range of lines included by choice of Δ_θ.

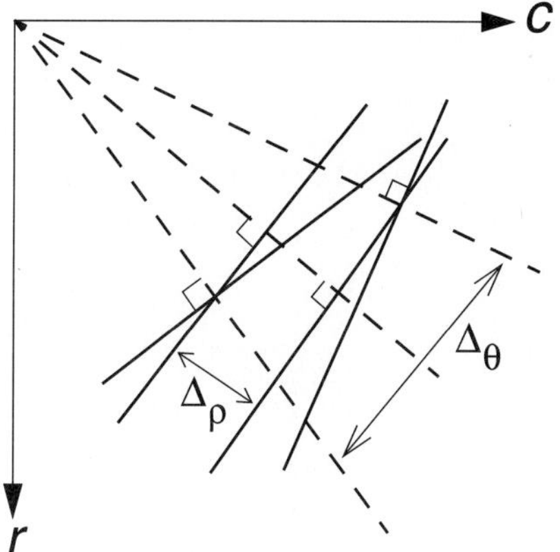

c. Range of lines included by choice of block size.

the lines are marked in the output image. Note that the Hough transform will allow us to look for lines of specific orientation, if desired. The result of applying the Hough transform to an aircraft image is shown in Figure 2.3-11. The Sobel edge detection operator was used on the original image to provide input to the Hough transform. The Sobel edge detection results were thresholded at a gray level of 25; then the Hough transform number-of-points threshold was set at a minimum of 20 pixels per line.

2.4 SEGMENTATION

Image segmentation is important in many computer vision and image processing applications. The goal of *image segmentation* is to find regions that represent objects or meaningful parts of objects. Division of the image into regions corresponding to objects

Figure 2.3-11 Hough Transform

a. Original image.

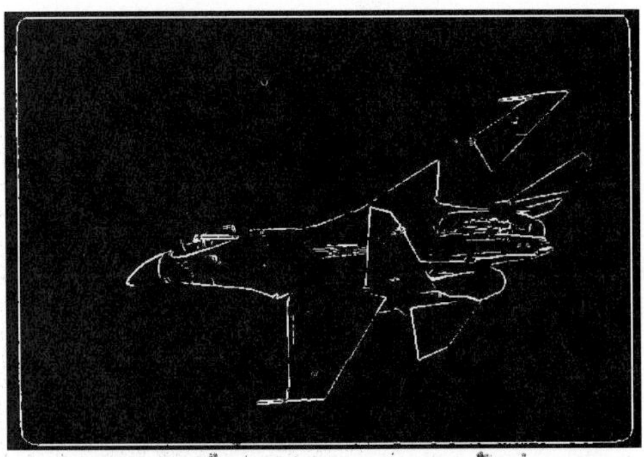

b. Sobel edge detector followed by thresholding.

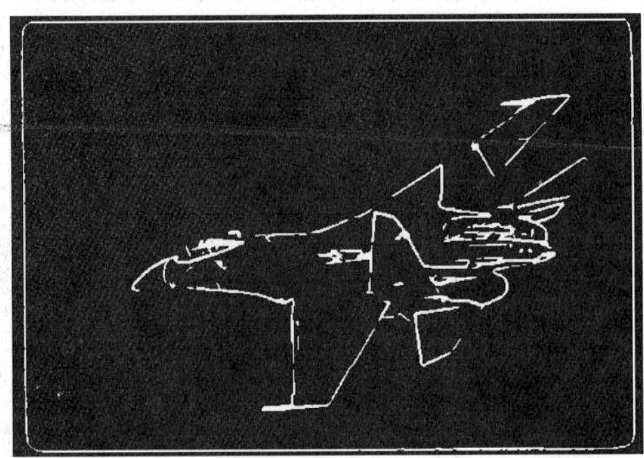

c. Hough output—$\Delta_\theta = 1°$, $\Delta_\rho = 1$, threshold = 20.

of interest is necessary before any processing can be done at a level higher than that of the pixel. Identifying real objects, pseudo-objects, and shadows or actually finding anything of interest within the image requires some form of segmentation.

2.4.1 Overview

Image segmentation methods will look for objects that either have some measure of homogeneity within themselves or have some measure of contrast with the objects on their border. Most image segmentation algorithms are modifications, extensions, or combinations of these two basic concepts. The homogeneity and contrast measures can include features such as gray level, color, and texture. After we have performed some preliminary segmentation, we may incorporate higher-level object properties, such as perimeter and shape, into the segmentation process.

Before we look at the different segmentation methods, we need to consider some of the problems associated with image segmentation. The major problems are a result of noise in the image and digitization of a continuous image. Noise is typically caused by the camera, the lenses, the lighting, or the signal path and can be reduced by the use of the preprocessing methods previously discussed. Spatial digitization can cause problems regarding connectivity of objects. These problems can be resolved with careful connectivity definitions and heuristics applicable to the specific domain.

Connectivity refers to the way in which we define an object. After we have segmented an image, which segments should be connected to form an object? Or, at a lower level, when searching the image for homogeneous regions, how do we define which pixels are connected? We must define which of the surrounding pixels are considered to be neighboring pixels. A pixel has eight possible neighbors: two horizontal neighbors, two vertical neighbors, and four diagonal neighbors. We can define connectivity in three different ways: 1) four-connectivity, 2) eight-connectivity, and 3) six-connectivity. Figure 2.4-1 illustrates these three definitions. With four-connectivity the

Figure 2.4-1 Connectivity

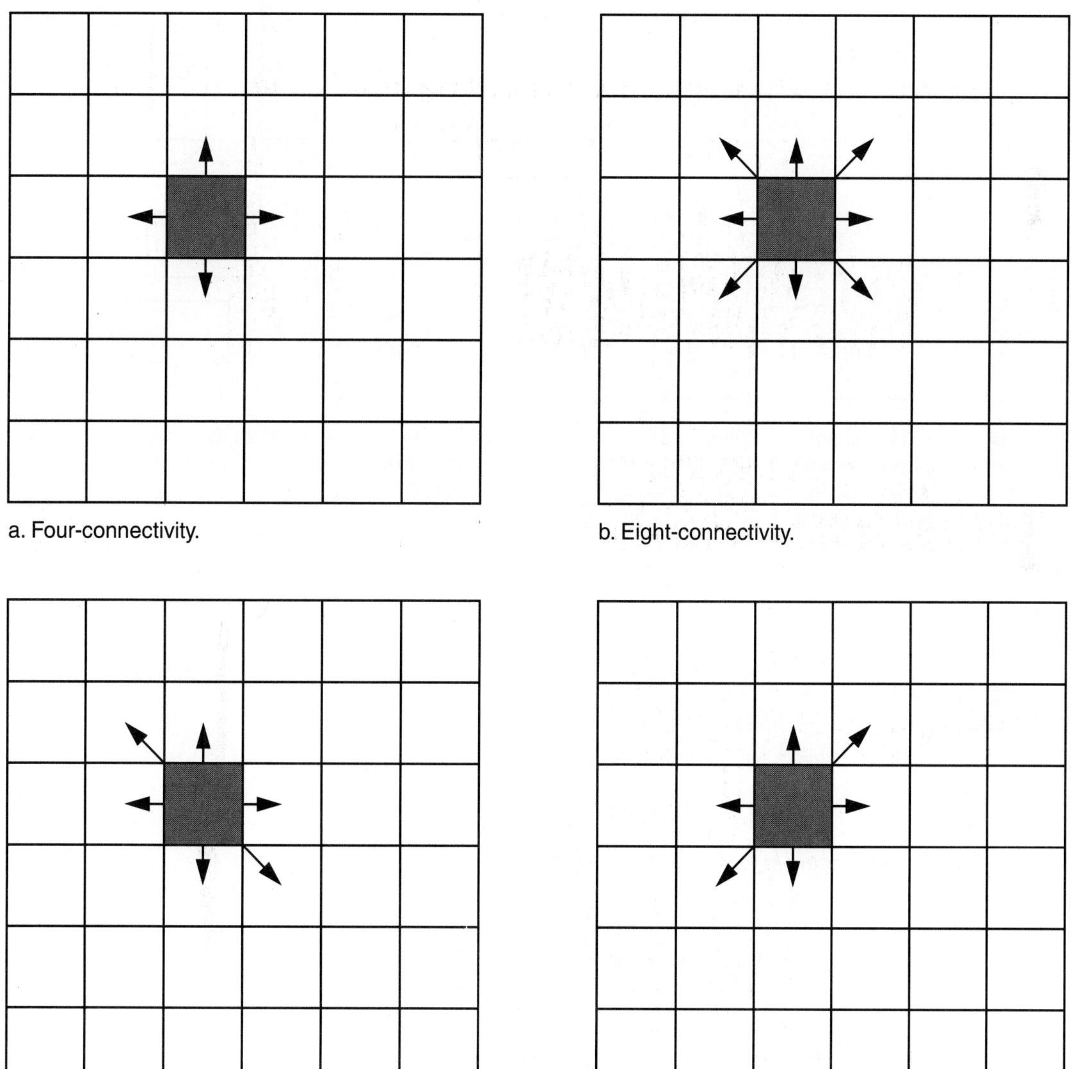

a. Four-connectivity.

b. Eight-connectivity.

c. Six-connectivity NW/SE.

d. Six-connectivity NE/SW.

only neighbors considered connected are the horizontal and vertical neighbors; with eight-connectivity all the eight possible neighboring pixels are considered connected; and with six-connectivity the horizontal, vertical, and two of the diagonal neighbors are connected. Which definition is chosen depends on the application, but the key to avoiding problems is to be consistent.

We can divide image segmentation techniques into three main categories (see Figure 2.4-2): 1) region growing and shrinking, 2) clustering methods, and 3) boundary detection. The region growing and shrinking methods use the row and column (*rc*)-based image space, whereas the clustering techniques can be applied to any domain

Figure 2.4-2 Image Segmentation

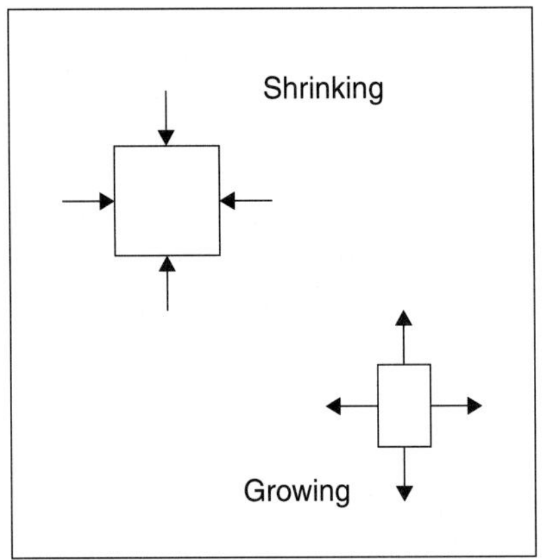

a. Region growing/shrinking is performed by finding homogeneous regions and changing them until they no longer meet the homogeneity criteria.

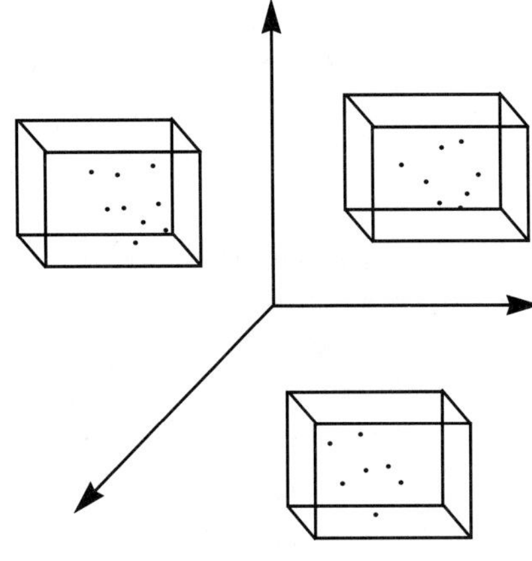

b. Clustering looks for data that can be grouped in domains other than the spatial domain.

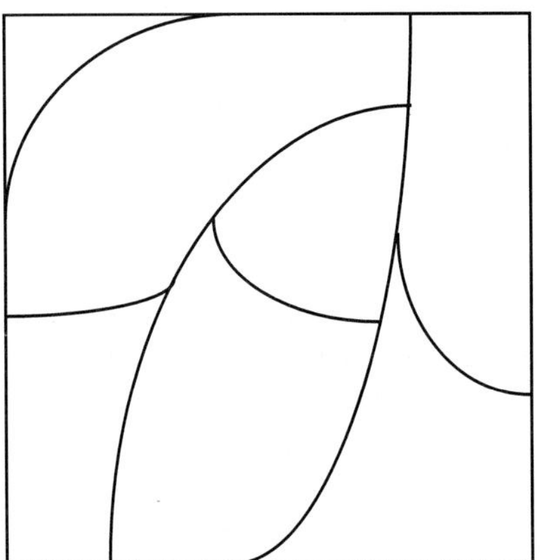

c. Boundary detection is often achieved using a differentiation operator to find lines or edges, followed by postprocessing to connect the points into borders.

(spatial domain, color space, feature space, etc.). The boundary detection methods are extensions of the edge detection techniques discussed in the previous section, and the necessary postprocessing for image segmentation will be discussed in Section 2.4.4.

2.4.2 Region Growing and Shrinking

Region growing and shrinking methods segment the image into regions by operating principally in the *rc*-based image space. Some of the techniques used are local, in which small areas of the image are processed at a time; others are global, with the entire image considered during processing. Methods that can combine local and global techniques, such as split and merge, are referred to as state space techniques and use graph structures to represent the regions and their boundaries.

Various split and merge algorithms have been described, but they all are most effective when heuristics applicable to the domain under consideration can be applied. This gives a starting point for the initial split. In general, the split and merge technique proceeds as follows:

1. Define a homogeneity test. This involves defining a homogeneity measure, which may incorporate brightness, color, texture, or other application-specific information, and determining a criterion the region must meet to pass the homogeneity test.
2. Split the image into equally sized regions.
3. Calculate the homogeneity measure for each region.
4. If the homogeneity test is passed for a region, then a merge is attempted with its neighbor(s). If the criterion is not met, the region is split.
5. Continue this process until all regions pass the homogeneity test.

There are many variations of this algorithm. For example, we can start out at the global level, where we consider the entire image as our initial region, and then follow an algorithm similar to the preceding algorithm, but without any region merging. Algorithms based on splitting only are called *multiresolution* algorithms. Alternately, we can start at the smallest level and only merge, with no region splitting. This merge-only approach is one example of region growing methods. Often the results from all these approaches will be quite similar, with the differences apparent only in computation time. Parameter choice, such as the minimum block size allowed for splitting, will heavily influence the computational burden as well as the resolution available in the results.

The user-defined homogeneity test is largely application dependent, but the general idea is to look for features that will be similar within an object and different from the surrounding objects. In the simplest case we might use gray level as our feature of interest. Here we could use the gray-level variance as our homogeneity measure and define a homogeneity test that required the gray-level variance within a region to be less than some threshold. We can define *gray-level variance* as

$$\frac{1}{N-1}\sum_{(r,c)\in \text{REGION}}[I(r,c)-\bar{I}]^2$$

$$\bar{I}=\frac{1}{N}\sum_{(r,c)\in \text{REGION}}I(r,c)$$

The variance is basically a measure of how widely the gray levels within a region vary. Higher-order statistics can be used for features such as texture.

A similar approach involves searching the image for a homogeneous region and growing it until it no longer meets the homogeneity criteria. At this point, a new region that exhibits homogeneity is found and is grown. This process continues until the entire image is divided into regions.

2.4.3 Clustering Techniques

Clustering techniques are image segmentation methods by which individual elements are placed into groups; these groups are based on some measure of similarity within the group. The major difference between these techniques and the region growing techniques is that domains other than the *rc*-based image space (the spatial domain) may be considered as the primary domain for clustering. Some of these other domains include color spaces, histogram spaces, or complex feature spaces. (Note that the terms *domain* and *space* are used interchangeably here. These terms both refer to some abstract *N*-dimensional mathematical space, not to be confused with the spatial domain, which refers to the row-column [*rc*] image space.)

What is done is to look for clusters in the space of interest. The simplest method is to divide the space of interest into regions by selecting the center or median along each dimension and splitting it there; this can be done iteratively until the space is divided into the specific number of regions needed. This method is used in the SCT/Center and PCT/Median segmentation algorithms. This method will be effective only if the space we are using and the entire algorithm is designed intelligently because the center or median split alone may not find good clusters.

The next level of complexity uses an adaptive and intelligent method to decide where to divide the space. These methods include histogram thresholding and other, more complex feature-space-based statistical methods. Representative algorithms will be discussed conceptually here, and a detailed look will be taken at two application-specific algorithms.

Recursive region splitting is a clustering method that has become a standard technique. This method uses a thresholding of histograms technique to segment the image. A set of histograms is calculated for a specific set of features, and then each of these histograms is searched for distinct peaks (see Figure 2.4-3). The best peak is selected and the image is split into regions based on this thresholding of the histogram. One of the first algorithms based on these concepts proceeds as follows:

1. Consider the entire image as one region and compute histograms for each component of interest (for example, red, green, and blue for a color image).
2. Apply a peak finding test to each histogram. Select the best peak and put thresholds on either side of the peak. Segment the image into two regions based on this peak.
3. Smooth the binary thresholded image so that only a single connected subregion is left.
4. Repeat steps 1–3 for each region until no new subregions can be created, that is, no histograms have significant peaks.

Figure 2.4-3 Histogram Peak Finding

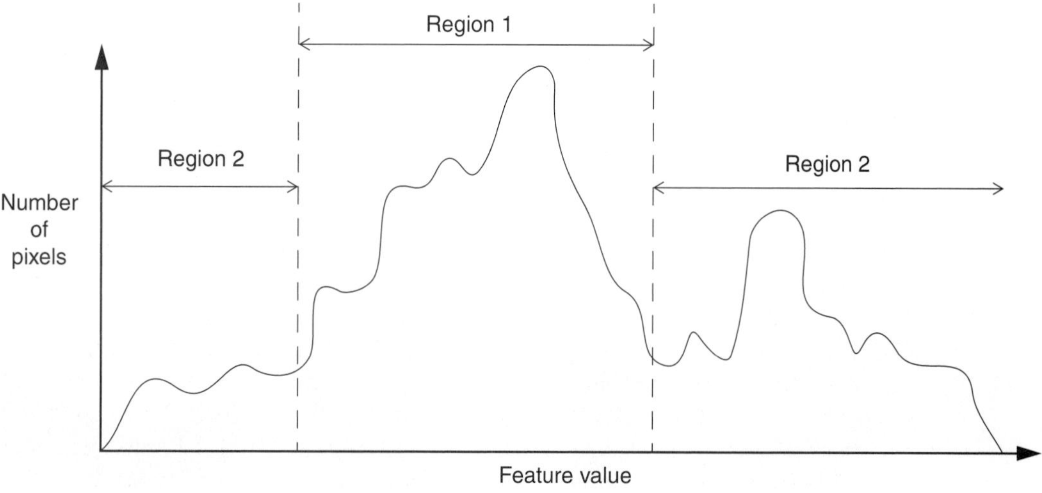

Two thresholds are selected, one on each side of the best peak. The image is then split into two regions. Region 1 corresponds to those pixels with feature values between the selected thresholds, known as those in the peak. Region 2 consists of those pixels with feature values outside the threshold.

Many of the parameters of this algorithm are application specific. For example, what peak-finding test do we use and what is a "significant" peak? An example of histogram-thresholding-based image segmentation is shown in Figure 2.4-4.

The SCT/Center algorithm was initially developed for the identification of variegated coloring in skin tumor images. Variegated coloring is a feature believed to be highly predictive in the diagnosis of melanoma, the deadliest form of skin cancer. The spherical coordinate transform (SCT) was chosen for this segmentation method because it decouples the color information from the brightness information. The brightness levels may vary with changing lighting conditions, so by using the two-dimensional color subspace defined by two angles (Figure 2.4-5a) we have a more robust algorithm.

If we slice a plane through the RGB color space, we can model a color triangle (Figure 2.4-5b). The vertices of the color triangle were chosen to bear some correlation to the human visual system. The placement of blue at the top of the triangle, and the way in which the spherical transform was defined, relates to the physiological fact that the cones in the human visual system that see blue are more discriminatory than the red- or green-sensitive cones.

We can segment the image by taking the color triangle and dividing it into blocks based on limits on the two angles. Figure 2.4-5c shows the shape of the resulting blocks. We can see that for a region defined by a range of minima and maxima on the two angles, the side of the region that is closest to the blue vertex is shorter than the side that is closest to the line that joins the red and green vertices.

Also, the distortion caused by the transform facilitates the perception-based aspect of the image segmentation; the closer to the perimeter of the triangle, the larger the region that is defined by a fixed angle range. This is analogous to the observation

Figure 2.4-4 Histogram Thresholding Segmentation

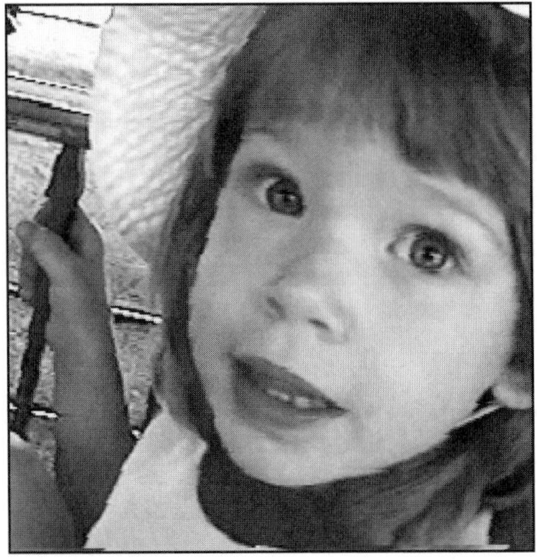

a. Original image.

b. Image after histogram thresholding segmentation using four gray levels.

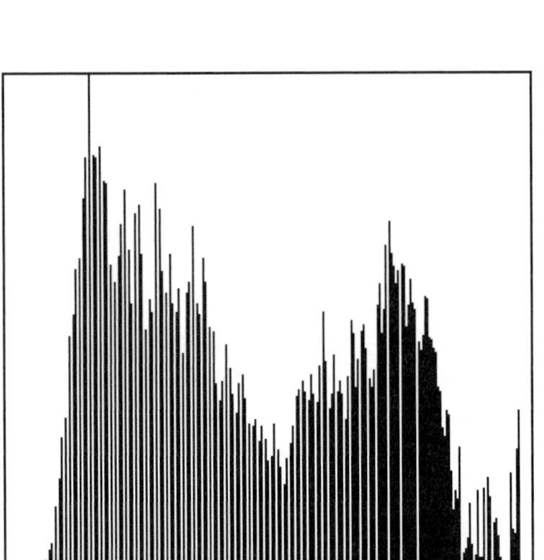

c. Histogram of image (a).

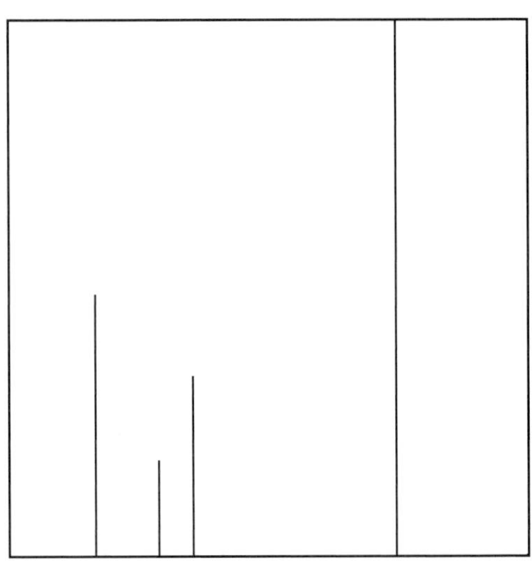

d. Histogram of image (b).

that as the white point is approached in the color space, a greater number of hues will be observable in a fixed area by the human visual system than on the perimeter of the color triangle. This observation is application specific because it only applies to colors from white (in the center of the triangle) to the green and red vertices. Skin tumor colors typically range from white out to the red vertex. The SCT/Center segmentation algorithm is outlined as follows:

1. Convert the (R,G,B) triple into spherical coordinates (L, angle A, angle B).

Figure 2.4-5 SCT/Center and Color Triangles

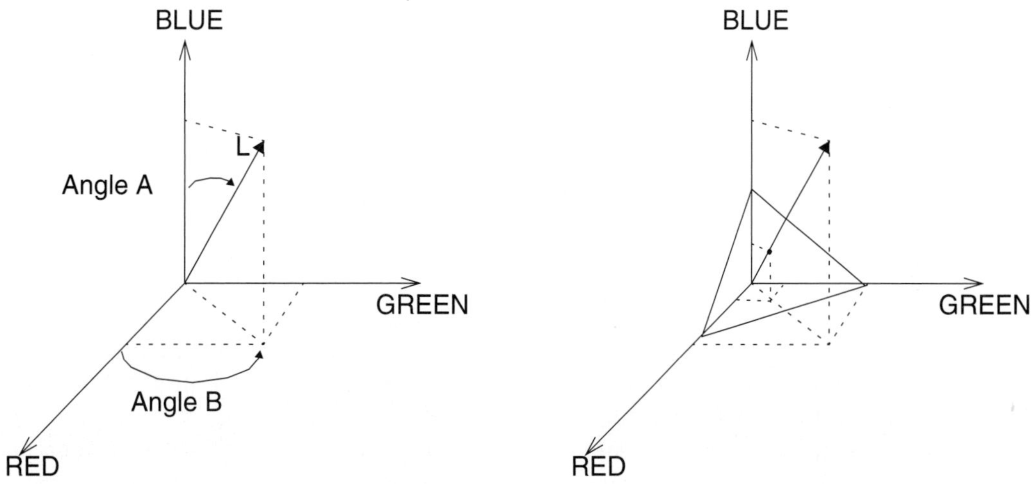

a. The spherical coordinate transform separates the red, green, and blue information into a two-dimensional color space defined by angles *A* and *B*, and a one-dimensional brightness space defined by *L*.

b. The color triangle.

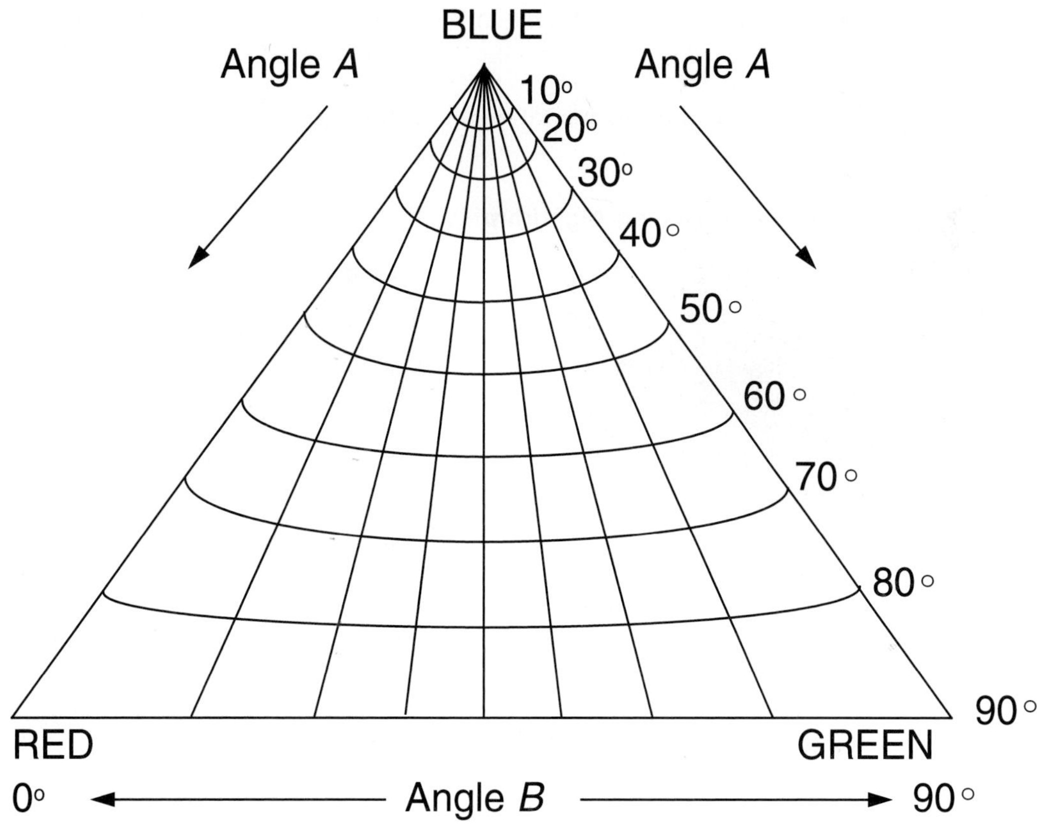

c. The color triangle showing regions defined by 10° increments on angle *A* and angle *B*.

2. Find the minima and maxima of angles A and B.
3. Divide the subspace, defined by the maxima and minima, into equally sized blocks.
4. Calculate the RGB means for the pixel values in each block.
5. Replace the original pixel values with the corresponding RGB means.

For the identification of variegated coloring in the skin tumor application, it was determined that segmenting the image into four colors was optimal. An example of this segmentation method is shown in Figure 2.4-6.

The PCT/Median color segmentation algorithm was developed because, for certain features other than variegated coloring, the results provided by the previously described algorithm were not totally satisfactory. This algorithm is based around the principal components transform (PCT). The median split part of the algorithm is based on an algorithm developed for color compression to map 24 bits/pixel color images into images requiring an average of 2 bits/pixel.

The principal components transform is based on statistical properties of the image and can be applied to any K-dimensional mathematical space. In this case, the PCT is applied to the three-dimensional color space. It was believed that the PCT used in conjunction with the median split algorithm would provide a satisfactory color image segmentation because the PCT aligns the main axis along the maximum variance path in the data set (see Figure 2.4-7). In pattern recognition theory a feature with large variance is said to have large discriminatory power. After we have transformed the color data so that most of the information (variance) lies along a principal axis, we proceed to divide the image into different colors by using a median split on the transformed data.

The PCT/Median segmentation algorithm proceeds as follows:

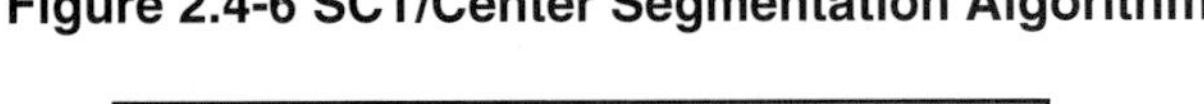

Figure 2.4-6 SCT/Center Segmentation Algorithm

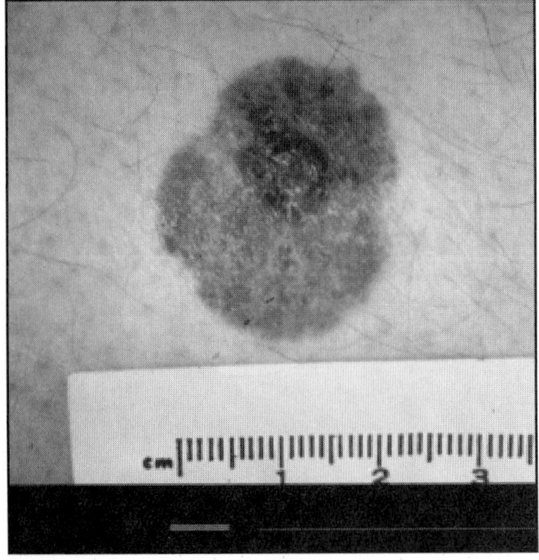

a. Original image.

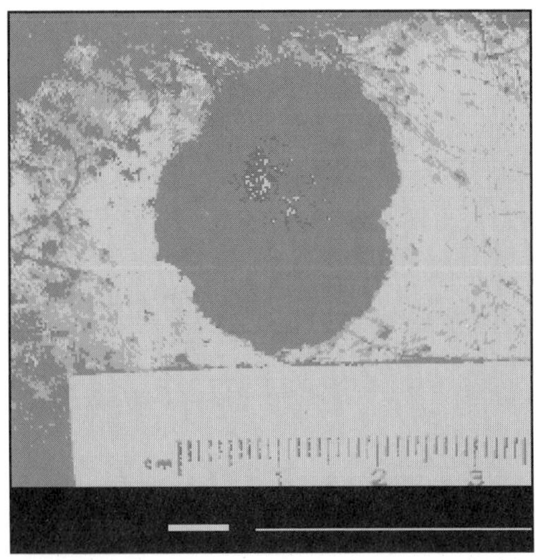

b. SCT/Center segmentation of skin tumor using four colors.

1. Find the PCT for the RGB image. Transform the RGB data using the PCT.
2. Perform the median split algorithm: find the axis that has the maximal range (initially it will be the PCT axis). Divide the data along this axis so that there are equal numbers of points on either side of the split—the median point. Continue until the desired number of colors is reached.
3. Calculate averages for all the pixels falling within a single parallelepiped (box).
4. Map each pixel to the closest average color values, based on a Euclidean distance measure.

Figure 2.4-7 Principal Components Transform

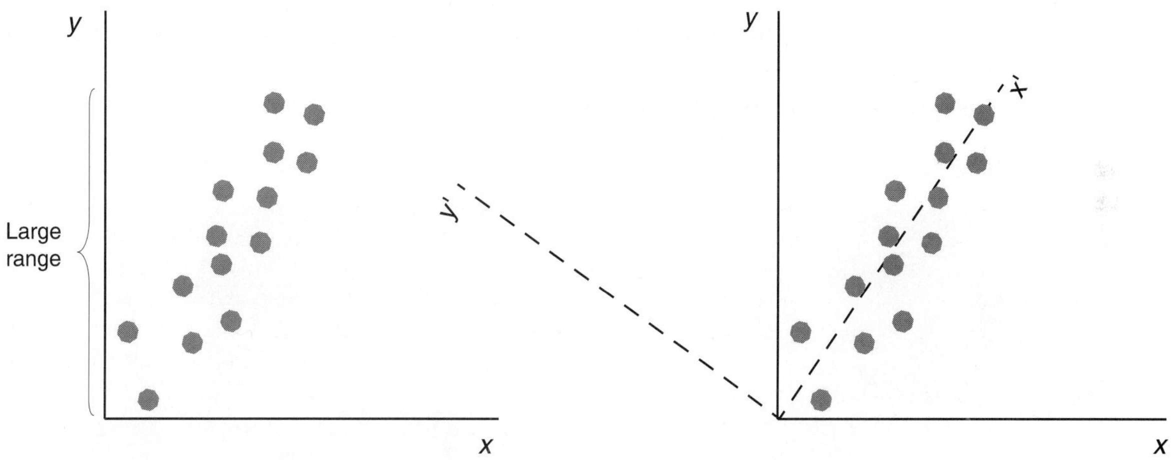

a. Original data exhibit a large range.

b. PCT aligns the main axis (x′) along the maximum variance path.

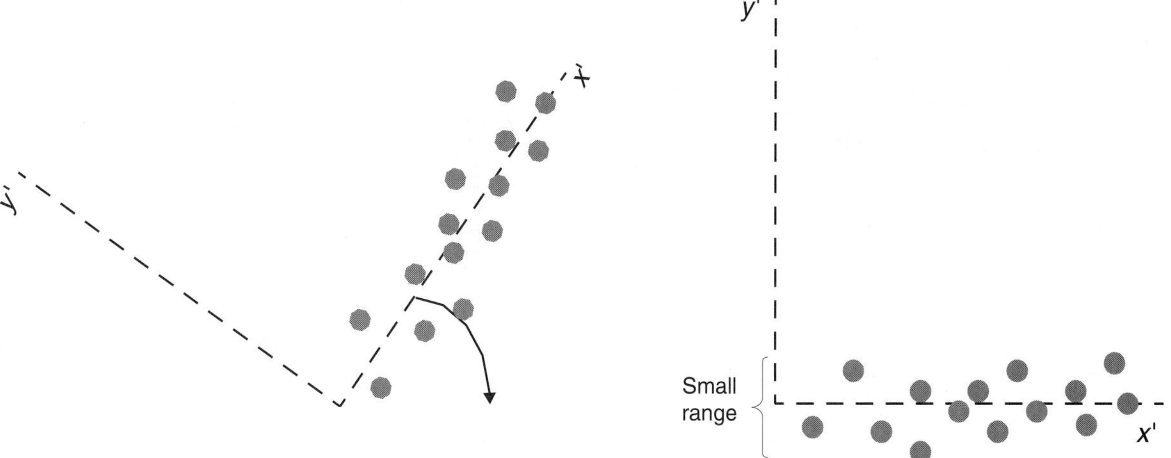

c. The new axes are rotated.

d. Transformed data now have a small range. Most of the variance, or information, is along the x′-axis, in one dimension rather than two, as in (a).

For the skin tumor application it was determined that the optimum number of colors depended upon the feature of interest. Results of this segmentation algorithm are shown in Figure 2.4-8. Here we observe that if the image is segmented with more colors, then more of the details in the image are retained (as expected), whereas a smaller number of colors will segment the image on a coarser scale, leaving only relatively large features. Selection of the number of colors for segmentation has a significant impact on the difficulty of the feature identification task—if the proper number of colors is selected for a specific feature it can make the feature identification process relatively easy.

It is interesting to note that the PCT is also used in image compression (coding), since this transform is optimal in the least-square-error sense. This means that most of the information, assumed to be directly correlated with variance, is in a reduced dimensionality. In the case of the skin tumor images, it was experimentally determined that the dimension with the largest variance after the PCT was performed con-

Figure 2.4-8 PCT/Median Segmentation Algorithm

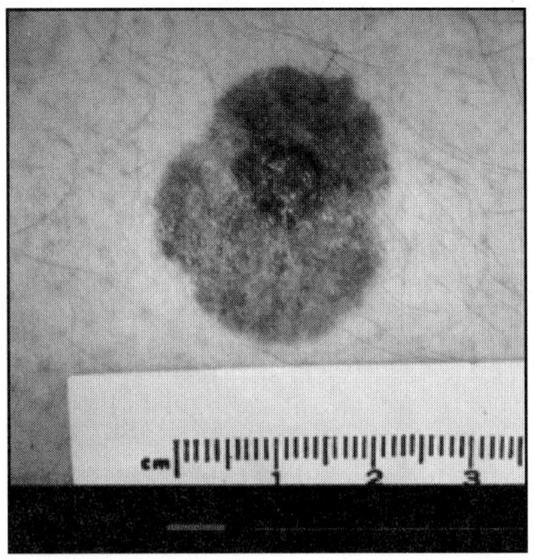

a. Original image.

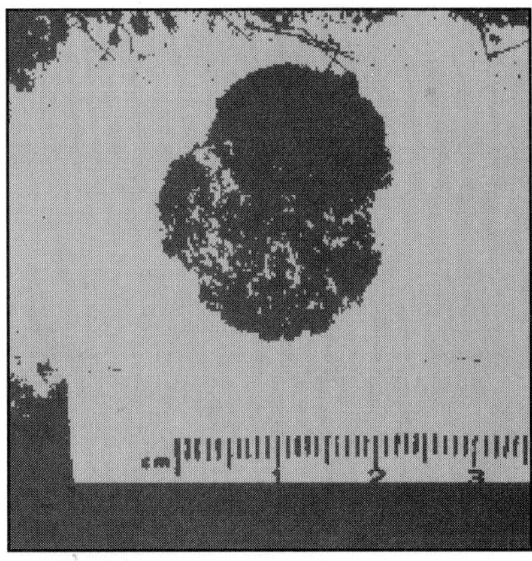

b. PCT/Median segmented image with two colors.

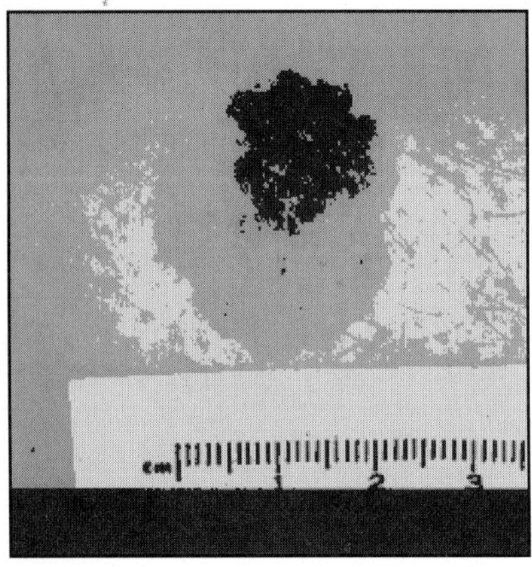

c. PCT/Median segmented image with four colors.

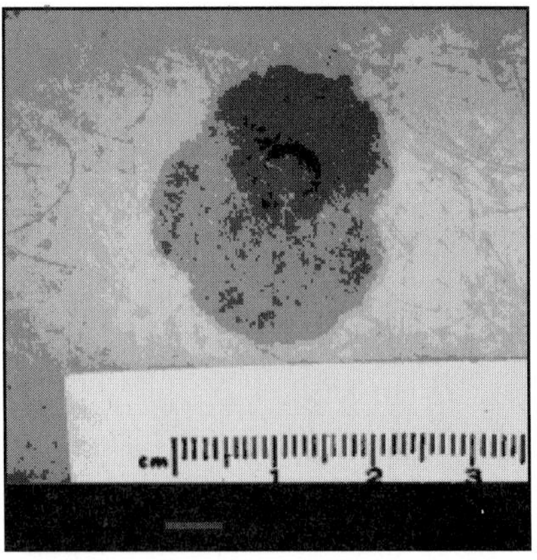

d. PCT/Median segmented image with eight colors.

tained approximately 91% of the variance. This would allow at least a 3:1 compression and still retain 91% of the information! Image segmentation is really a data reduction process, so it is not surprising that many of the procedures and methods used are also used in image compression.

2.4.4 Boundary Detection

Boundary detection, as a method of image segmentation, is performed by finding the boundaries between objects, thus indirectly defining the objects. This method is usually begun by marking points that may be a part of an edge. These points are then merged into line segments, and the line segments are then merged into object boundaries. The edge detectors previously described are used to mark points of rapid change, thus indicating the possibility of an edge. These edge points represent local discontinuities in specific features, such as brightness, color, or texture.

After the edge detection operation has been performed, the next step is to threshold the results. One method to do this is to consider the histogram of the edge detection results, looking for the best valley (Figure 2.4-9). Often, the histogram of an image that has been operated on by an edge detector is unimodal (one peak), so it may be difficult to find a good valley. This method works best with a bimodal histogram. Another method that provides reasonable results is to use the average value for the threshold, as in Figure 2.4-10. With very noisy images, a good rule of thumb is to use 10–20% of the peak value as a threshold.

After we have determined a threshold for the edge detection, we need to merge the existing edge segments into boundaries. This is done by edge linking. The simplest approach to edge linking involves looking at each point that has passed the threshold test and connecting it to all other such points that are within a maximum distance. This method tends to connect many points and is not useful for images where too many points have been marked; it is most applicable to simple images.

Instead of thresholding and then edge linking, we can perform edge linking on the edge-detected image before we threshold it. If this approach is used, we look at

Figure 2.4-9 Edge Detection Threshold

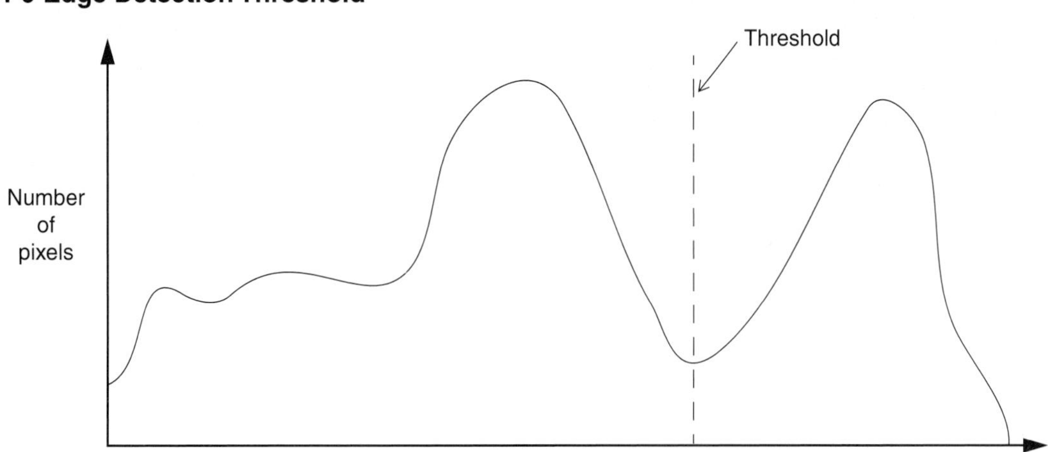

Figure 2.4-10 Average Value Thresholding

a. Image after Sobel edge detection.

b. Thresholding with average value.

small neighborhoods (3×3 or 5×5) and link similar points. Similar points are defined as having close values for both magnitude and direction. The entire image undergoes this process, while keeping a list of the linked points. When the process is complete, the boundaries are determined by the linked points.

2.4.5 Combined Approaches

Image segmentation methods may actually be a combination of region growing methods, clustering methods, and boundary detection. We could consider the region growing methods to be a subset of the clustering methods, by allowing the space of interest to include the row and column parameters. Quite often, in boundary detection, heuristics applicable to the specific domain must be employed in order to find the true object boundaries. What is considered noise in one application may be the feature of interest in another application. Finding boundaries of different features, such as texture, brightness, or color, and applying artificial intelligence techniques at a higher level to correlate the feature boundaries found to the specific domain may give the best results. Optimal image segmentation is likely to be achieved by focusing on the application and on how the different methods can be used, singly or in combination, to achieve the desired results.

2.4.6 Morphological Filtering

Morphology relates to the structure or form of objects. Morphological filtering simplifies a segmented image to facilitate the search for objects of interest. This is done by smoothing out object outlines, filling small holes, eliminating small projections, and using other similar techniques. Even though this section will focus on applications to binary images, the extension of the concepts to gray-level images will also be discussed. We will look at the different types of operations available and at some examples of their use.

The two principal morphological operations are dilation and erosion. *Dilation* allows objects to expand, thus potentially filling in small holes and connecting disjoint objects. *Erosion* shrinks objects by etching away (eroding) their boundaries. These operations can be customized for an application by the proper selection of the *structuring element*, which determines exactly how the objects will be dilated or eroded.

The dilation process is performed by laying the structuring element on the image and sliding it across the image in a manner similar to convolution. The difference is in the operation performed. It is best described in a sequence of steps:

1. If the origin of the structuring element coincides with a '0' in the image, there is no change; move to the next pixel.
2. If the origin of the structuring element coincides with a '1' in the image, perform the OR logic operation on all pixels within the structuring element.

An example is shown in Figure 2.4-11. Note that with a dilation operation, all the '1' pixels in the original image will be retained, any boundaries will be expanded, and small holes will be filled

The erosion process is similar to dilation, but we turn pixels to '0', not '1'. As before, slide the structuring element across the image and then follow these steps:

1. If the origin of the structuring element coincides with a '0' in the image, there is no change; move to the next pixel.
2. If the origin of the structuring element coincides with a '1' in the image, and any of the '1' pixels in the structuring element extend beyond the object ('1' pixels) in the image, then change the '1' pixel in the image to a '0'.

In Figure 2.4-12, the only remaining pixels are those that coincide to the origin of the structuring element where the entire structuring element was contained in the

Figure 2.4-11 Dilation

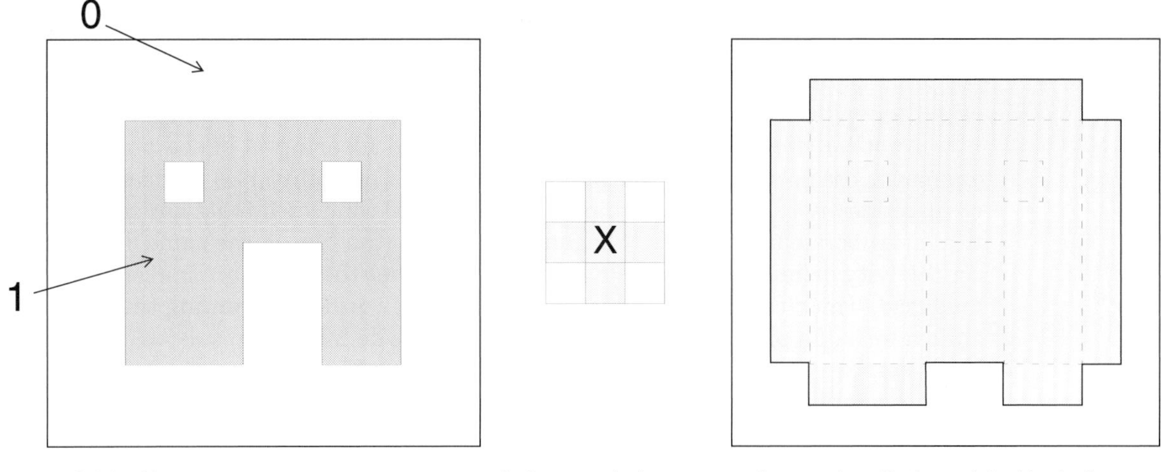

a. Original image.

b. Structural element; x = origin.

c. Image after dilation; original in dashes.

Figure 2.4-12 Erosion

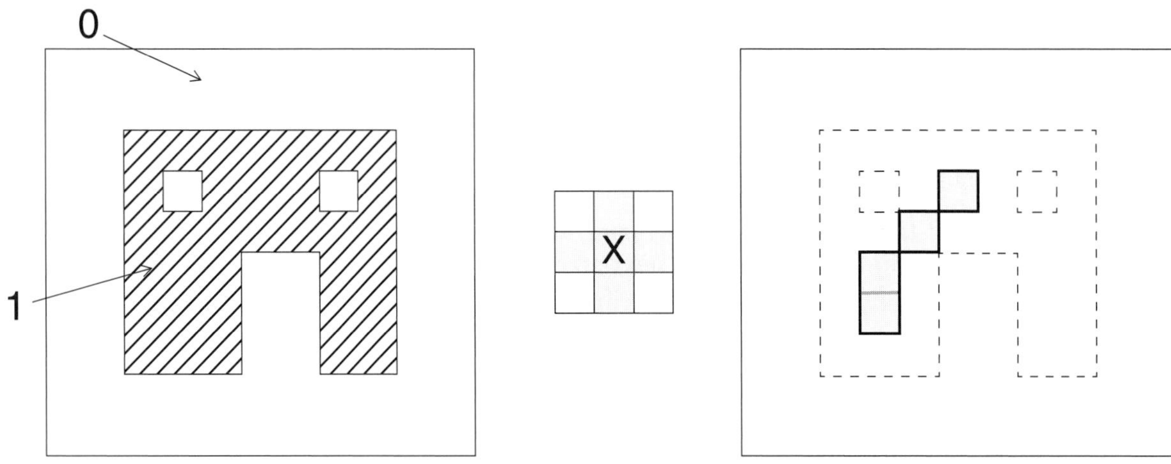

a. Original image. b. Structural element; x = origin c. Image after erosion; original in dashes.

existing object. Because the structuring element is 3 pixels wide, the 2-pixel-wide right leg of the image object was eroded away, but the 3-pixel-wide left leg retained some of its center pixels.

These two basic operations, dilation and erosion, can be combined into more complex sequences. The most useful of these for morphological filtering are called opening and closing. *Opening* consists of an erosion followed by a dilation and can be used to eliminate all pixels in regions that are too small to contain the structuring element. In this case the structuring element is often called a probe, because it is probing the image looking for small objects to filter out of the image. See Figure 2.4.13 for an example of opening.

Closing consists of a dilation followed by erosion and can be used to fill in holes and small gaps. In Figure 2.4-14 we see that the closing operation has the effect of filling in holes and closing gaps. Comparing Figure 2.4-14 to Figure 2.4-13, we see that the order of operation is important. Closing and opening will have different results even though both consist of an erosion and a dilation.

Another approach to binary morphological filtering is based on an iterative approach. The usefulness of this approach lies in its flexibility. It is based on a definition of six-connectivity, in which each pixel is considered connected to its horizontal and vertical neighbors but to only two diagonal neighbors (the two on the same diagonal). This connectivity definition is equivalent to assuming that the pixels are laid out on a hexagonal grid, which can be simulated on a rectangular grid by assuming that each row is shifted by half a pixel (see Figure 2.4-15). With this definition a pixel can be surrounded by 14 possible combinations of 1's and 0's, as seen in Figure 2.4-16; we call these different combinations *surrounds*. For this approach to morphological filtering, we define:

1. The set of surrounds S, where $a = 1$.
2. A logic function, $L(a, b)$, where b is the current pixel value, specifies the output of the morphological function.

Figure 2.4-13 Opening

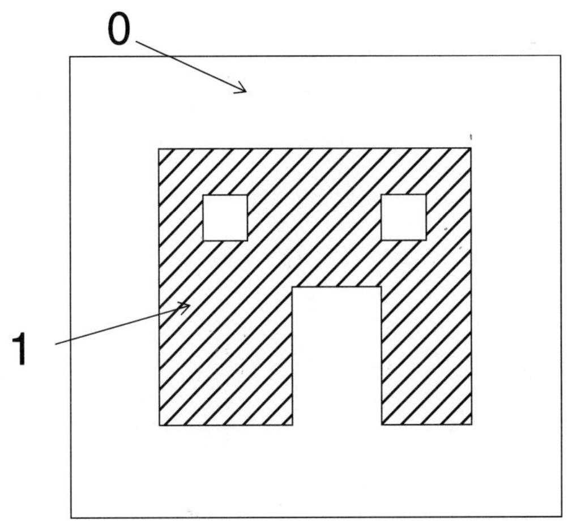

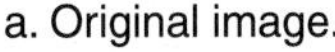

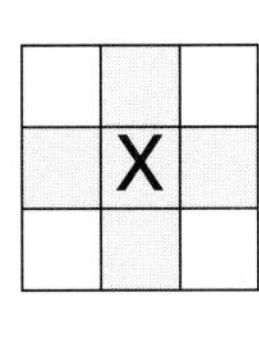

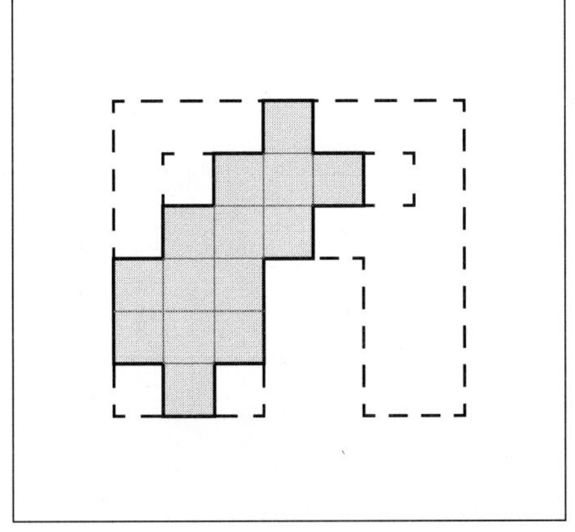

a. Original image.

b. Structural element; x = origin.

c. Image after opening; erosion followed by dilation.

Figure 2.4-14 Closing

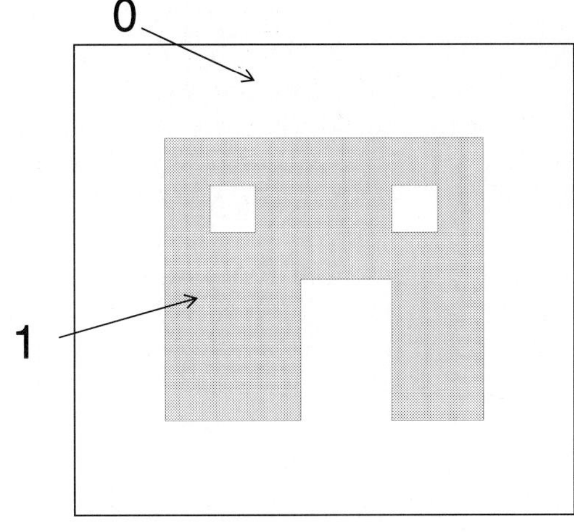

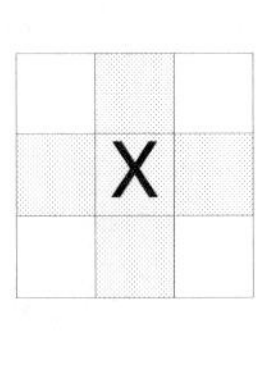

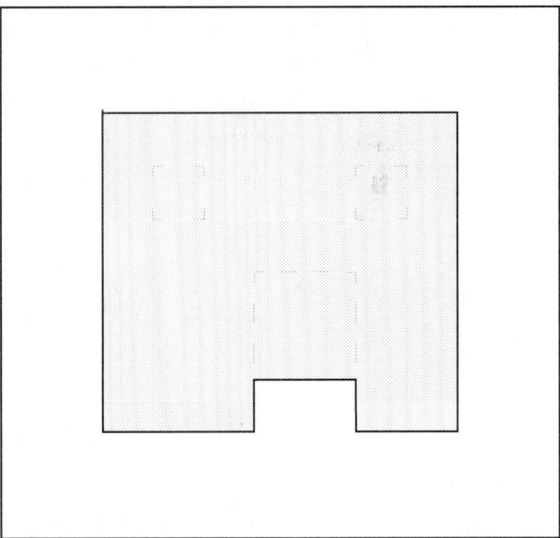

a. Original image.

b. Structural element; x = origin.

c. Image after closing; dilation followed by erosion; original in dashes.

3. The number of iterations n.

The function $L(\ )$ and the values of a and b are all functions of the row and column, (r, c), but for concise notation this is implied. Set S can contain any or all of the 14 surrounds defined in Figure 2.4-16. $L(a, b)$ can be any logic function, but it turns out that the most useful are the AND and OR functions. The AND function tends to etch away at object boundaries (erosion), and the OR function tends to grow objects (dilation).

Figure 2.4-15 Hexagonal Grid

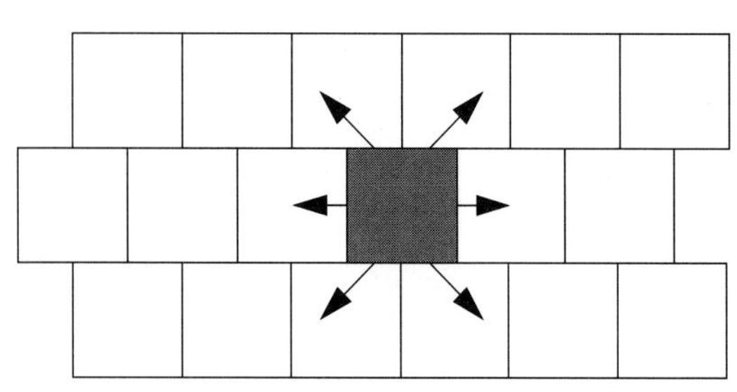

a. Rectangular image grid with every other row shifted by one-half pixel.

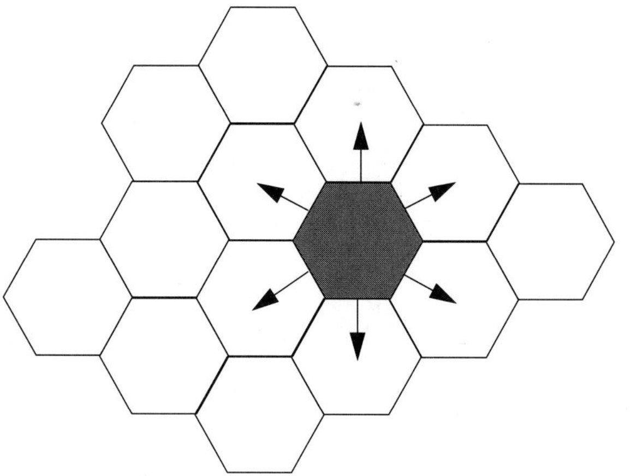

b. Hexagonal grid.

Figure 2.4-16 Surrounds for Iterative Morphological Filtering

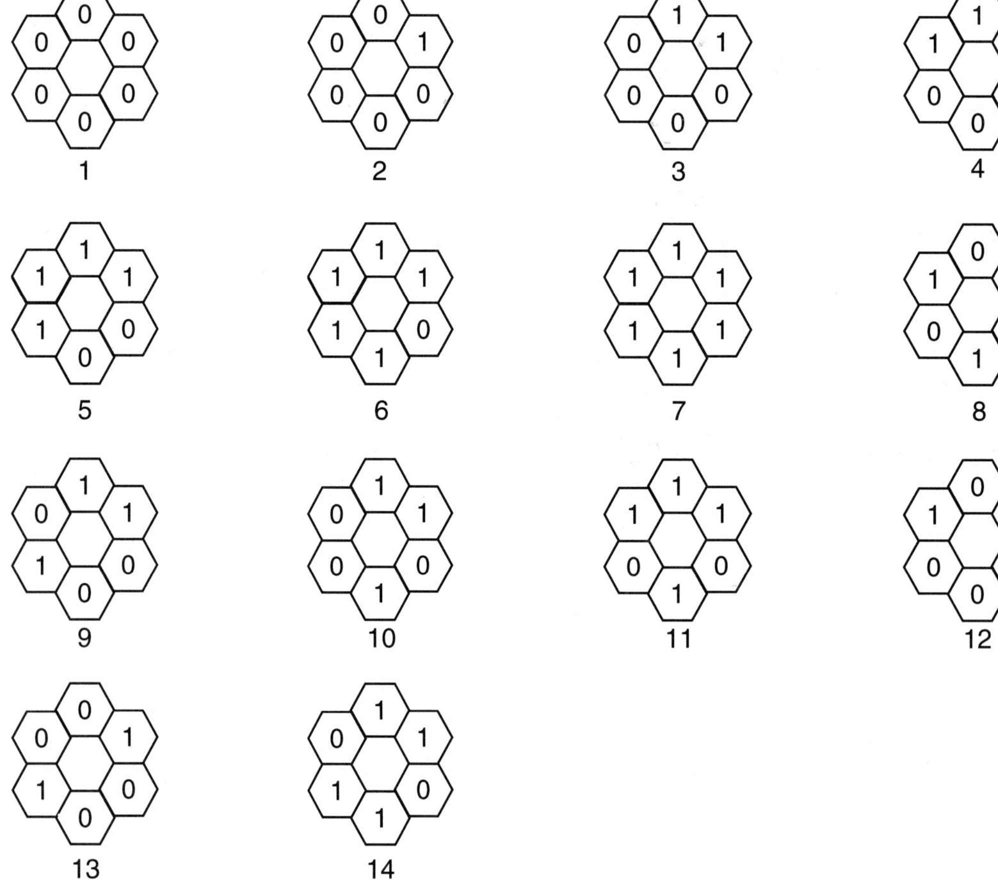

EXAMPLE 2-9

Let $S = \{2, 3, 4, 5, 6\}$ and $L = a + b$ (+ = OR). Because $L(a, b)$ is an OR operation, all pixels that are 1 in the original will remain 1. The only pixels that will change are those that are 0 in the original image and have a surround that is S (this means that $a = 1$). If we examine the set S, we see that this set contains all pixels that are surrounded by a connected set of 1's. This operation will expand the object, but because the surrounds of disconnected 1 pixels are not included in S, disjoint objects will not connect.

We can see from this example that this method is more flexible than the methods described earlier. We can use this technique to define methods for dilation, erosion, opening, and closing, as well as others. For this technique the selection of the set S is comparable to defining the structuring element in the previously described approaches, and the operation $L(a, b)$ defines the type of filtering that occurs (see Computer Vision Lab Exercise #4 in Chapter 8 for examples of iterative morphological filtering).

The morphological operations described (dilation, erosion, opening, and closing) can be extended to gray-level images in different ways. The easiest method is to simply threshold the gray-level image to create a binary image and then apply the existing operators. For many applications this is not desired because too much information is lost during the thresholding process. Another method that allows us to retain more information is to treat the image as a sequence of binary images by operating on each gray level as if it were the 1 value and assuming everything else to be 0. The resulting images can then be combined by laying them on top of each other and "promoting" each pixel to the highest gray-level value coincident with that location.

An example of results from gray-level morphological filtering is shown in Figure 2.4-17. For this application an opening operation followed by a closing operation was performed. A circular structuring element was used, as the object of interest was the tumor border. The opening procedure served to smooth the contours of the object, break narrow isthmuses, and eliminate thin protrusions and small objects. Next, the closing was performed to fill in gaps and eliminate small holes. To understand gray-level morphology fully, we must remember that with two adjacent gray levels, the brighter one is considered to be the object (the equivalent of '1' in a binary image), and the darker one is the background (the '0' equivalent in binary morphology). In this figure we see the tremendous data reduction achieved, thus simplifying the process of identifying the tumor features of interest.

2.5 DISCRETE TRANSFORMS

The transforms considered here provide information regarding the spatial frequency content of an image. In general, a transform maps image data into a different mathematical space via a transformation equation. In Section 2.4, we transformed our image data into alternate color spaces to achieve image segmentation. However, the color transforms mapped data from one color space to another color space with a one-to-one correspondence between a pixel in the input and the output. Here, we are mapping the image data from the spatial domain to the frequency domain (also called the

Figure 2.4-17 Gray-Level Morphological Filtering

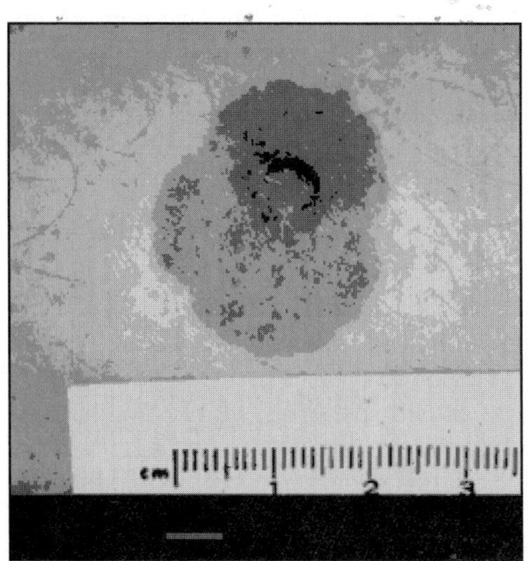

a. Original segmented tumor image.

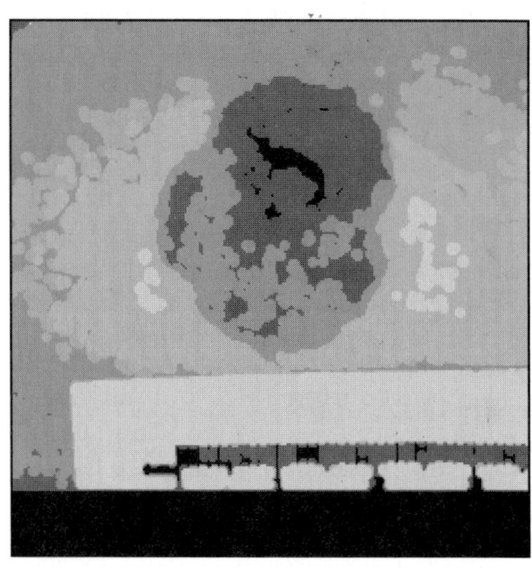

b. Image (a) after morphological opening using a 5×5 circular structuring element.

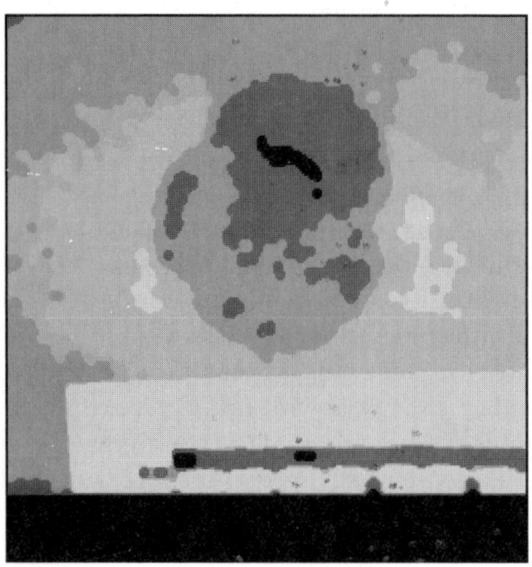

c. Image (b) after morphological closing using a 5×5 circular structuring element.

spectral domain), where *all* the pixels in the input (spatial domain) contribute to *each* value in the output (frequency domain). This is illustrated in Figure 2.5-1.

These transforms are used as tools in many areas of engineering and science, including computer imaging. Originally defined in their continuous forms, they are commonly used today in their discrete (sampled) forms. The large number of arithmetic operations required for the discrete transforms, combined with the massive amounts of data in an image, requires a great deal of computer power. The ever-

Figure 2.5-1 Discrete Transforms

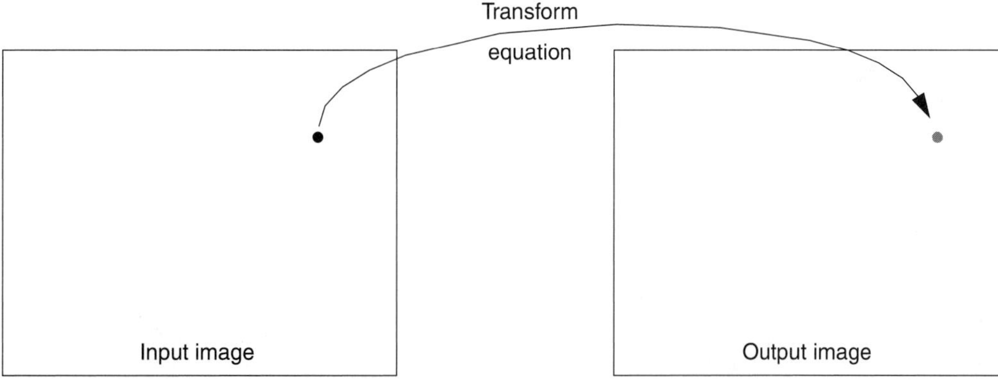

a. Color transforms use a single-pixel to single-pixel mapping.

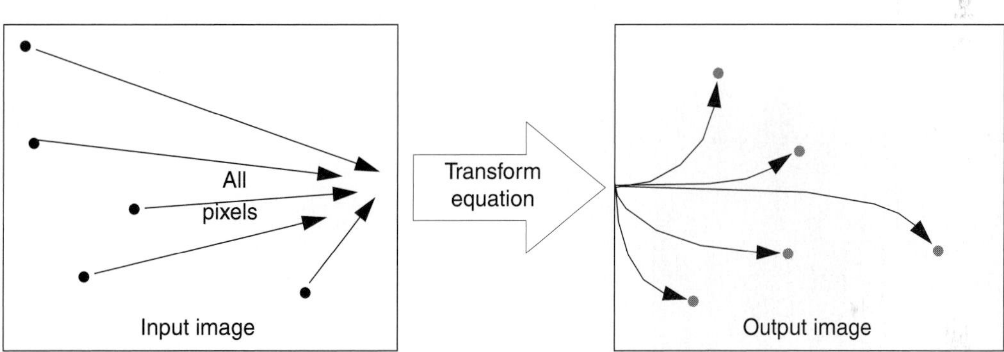

b. All pixels in the input image contribute to each value in the output image for frequency transforms.

increasing compute power, memory capacity, and disk storage available today make the use of these transforms much more feasible than in recent years.

The discrete form of these transforms is created by sampling the continuous form of the functions on which these transforms are based, that is, the *basis functions*. The functions used for these transforms are typically sinusoidal or rectangular, and the sampling process, for the one-dimensional (1-D) case, provides us with *basis vectors*. When we extend these into two-dimensions, as we do for images, they are *basis matrices* or *basis images* (see Figure 2.5-2). The process of transforming the image data into another domain, or mathematical space, amounts to projecting the image onto the basis images. The mathematical term for this projection process is an *inner product* and is identical to what was done with Frei-Chen edge and line masks in Section 2.3. The Frei-Chen projections are performed to uncover edge and line information in the image and use 3×3 image blocks. The frequency transforms considered here use the entire image, or blocks that are typically at least 8×8, and are used to discover spatial frequency information.

Figure 2.5-2 Basis Vectors and Images

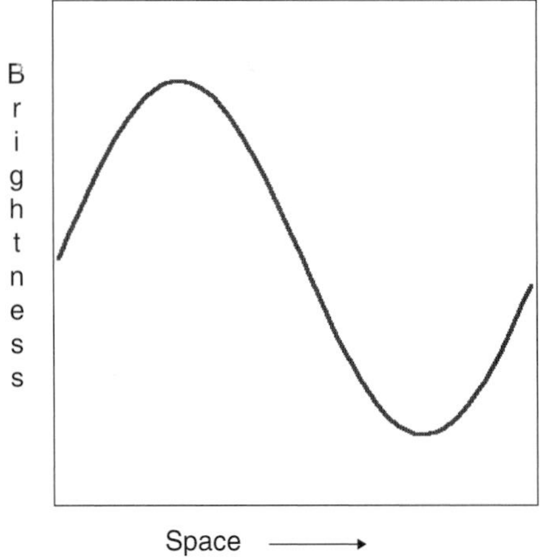

a. A basis function: a 1-D sinusoid.

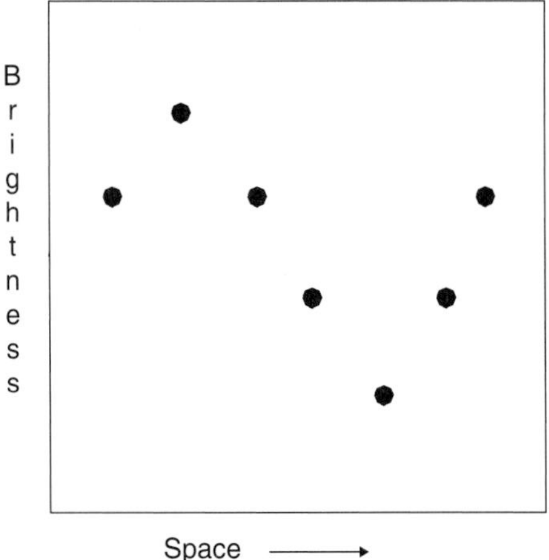

b. A basis vector: a sampled 1-D sinusoid.

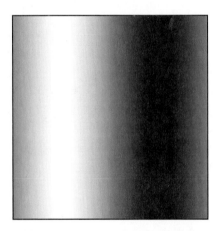

c. A basis image: a sampled sinusoid shown in 2-D as an image. The pixel brightness in each row corresponds to the sampled values of the 1-D sinusoids, which are repeated along each column.

The ways in which the image brightness levels change in space define the spatial frequency. For example, rapidly changing brightness corresponds to high spatial frequency, whereas slowly changing brightness levels relate to low-frequency information. The lowest spatial frequency, called the zero frequency term, corresponds to an image with a constant value. These concepts are illustrated in Figure 2.5-3, using square waves and sinusoids as basis vectors.

The general form of the transformation equation, assuming an $N \times N$ image, is given by:

Figure 2.5-3 Spatial Frequency

a. Frequency = 0, gray level = 54.

c. Frequency = 1, horizontal sine wave.

e. Frequency = 20, horizontal sine wave.

b. Frequency = 0, gray level = 202.

d. Frequency = 1, horizontal square wave.

f. Frequency = 20, horizontal square wave.

$$T(u, v) = \sum_{r=0}^{N-1} \sum_{c=0}^{N-1} I(r, c) B(r, c; u, v)$$

Here u and v are the frequency domain variables, $T(u, v)$ are the transform coefficients, and $B(r, c; u, v)$ correspond to the basis images. The notation $B(r, c; u, v)$ defines a set of basis images, corresponding to each different value for u and v, and the size of each is r by c (Figure 2.5-4). The transform coefficients $T(u, v)$ are the projections of $I(r, c)$ onto each $B(u, v)$. This is illustrated in Figure 2.5-5. These coefficients tell us how similar the image is to the basis image; the more alike they are, the bigger the coefficient. This transformation process amounts to decomposing the image into a weighted sum of the basis images, where the coefficients $T(u, v)$ are the weights. To obtain the image from the transform coefficients, we apply the inverse transform equation:

$$I(r, c) = T^{-1}[T(u, v)] = \sum_{u=0}^{N-1} \sum_{v=0}^{N-1} T(u, v) B^{-1}(r, c; u, v)$$

Here the $T^{-1}[T(u, v)]$ represents the inverse transform, and the $B^{-1}(r, c; u, v)$ represents the inverse basis images. In many cases the inverse basis images are the same as the forward ones, but possibly weighted by a constant.

Figure 2.5-4 A Set of Basis Vectors $B(r, c; u, v)$

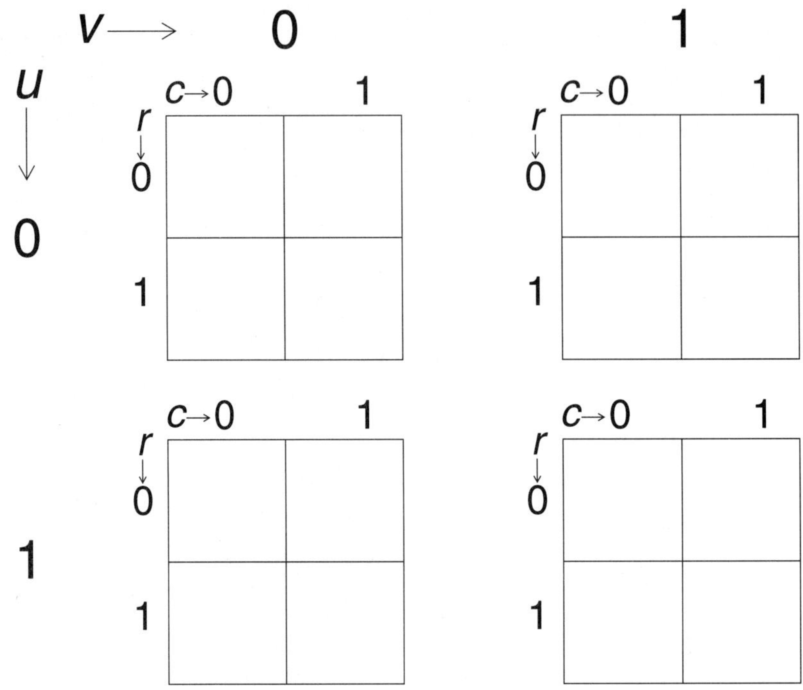

Size of generic basis vectors for a 2 × 2 transform.

Figure 2.5-5 Transform Coefficients

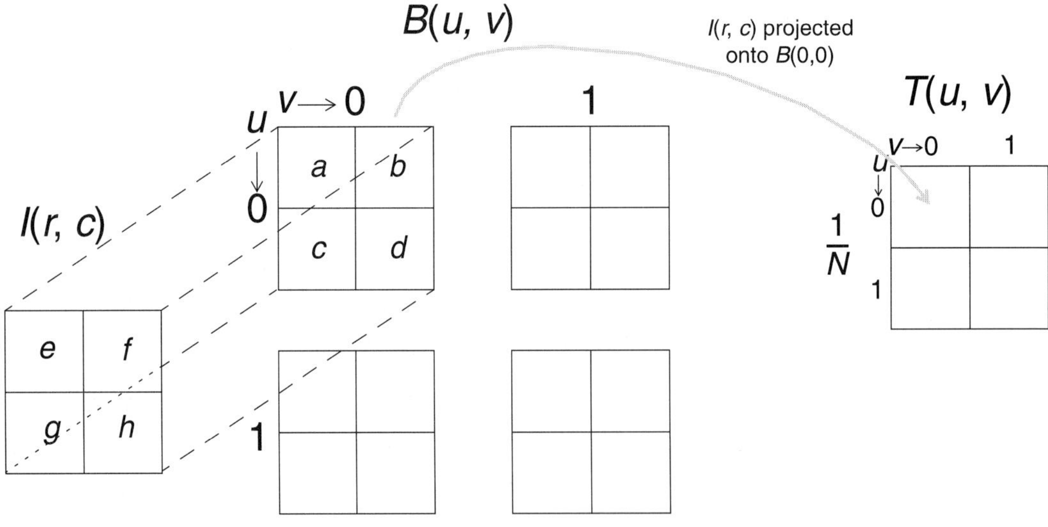

To find $T(u, v)$, we project $I(r, c)$ onto the basis vectors of $B(u, v)$.
For example, $T(0, 0)$ is the projection of $I(r, c)$ onto $B(0, 0)$, which equals $(ea + fb + gc + hd)$.

2.5.1 Fourier Transform

The Fourier transform is the most well known, and the most widely used, transform (see Image Processing exercises #3 and #5 in Chapter 8). It was developed by Baptiste Joseph Fourier (1768–1830) to explain the distribution of temperature and heat conduction. Since that time the Fourier transform has found numerous uses, including vibration analysis in mechanical engineering, circuit analysis in electrical engineering, and here in computer imaging. This transform allows for the decomposition of an image into a weighted sum of 2-D sinusoidal terms. Assuming an $N \times N$ image, the equation for the 2-D discrete Fourier transform is

$$F(u, v) = \frac{1}{N}\sum_{r=0}^{N-1}\sum_{c=0}^{N-1} I(r, c)e^{-j2\pi\frac{(ur+vc)}{N}}$$

The base of the natural logarithmic function e is about 2.71828; j, the imaginary coordinate for a complex number, equals $\sqrt{-1}$. The basis functions are sinusoidal in nature, as can be seen by Euler's identity:

$$e^{jx} = \cos x + j \sin x$$

So we can also write the Fourier transform equation as

$$F(u, v) = \frac{1}{N}\sum_{r=0}^{N-1}\sum_{c=0}^{N-1} I(r, c)[\cos(\frac{2\pi}{N}(ur + vc)) + j \sin(\frac{2\pi}{N}(ur + vc))]$$

In this case, $F(u, v)$ is also complex, with the real part corresponding to the cosine terms and the imaginary part corresponding to the sine terms. If we represent a complex spectral component by $F(u, v) = R(u, v) + jI(u, v)$, where $R(u, v)$ is the real part and $I(u, v)$ is the imaginary part, then we can define the magnitude and phase of a complex spectral component as

$$\text{MAGNITUDE} = |F(u, v)| = \sqrt{[R(u, v)]^2 + [I(u, v)]^2}$$

and

$$\text{PHASE} = \phi(u, v) = \tan^{-1}\left[\frac{I(u, v)}{R(u, v)}\right]$$

The magnitude of a sinusoid is simply its peak value, and the phase determines where the origin is or where the sinusoid starts (see Figure 2.5-6a). Figure 2.5.6b–d shows an image recovered with the phase information only, which illustrates its importance in images. Although we lose the relative magnitudes, which results in a loss of contrast (Figure 2.5-6c), we retain the relative placement of objects—in other words, the phase data contain information about *where objects are* in an image.

After we perform the transform, if we want to get our original image back, we need to apply the *inverse transform*. The inverse Fourier transform is given by

$$F^{-1}[F(u, v)] = I(r, c) = \frac{1}{N}\sum_{u=0}^{N-1}\sum_{v=0}^{N-1} F(u, v)e^{j2\pi\frac{(ur+vc)}{N}}$$

The $F^{-1}[]$ notation represents the inverse transform. This equation illustrates that the function $I(r, c)$ is represented by a weighted sum of the basis functions and that the transform coefficients $F(u, v)$ are the weights. With the inverse Fourier transform, the sign on the basis functions' exponent is changed from –1 to +1. However, this corresponds only to the phase and not the frequency and magnitude of the basis functions (see Figure 2.5-6 and the magnitude and phase equations previously).

Figure 2.5-6 Magnitude and Phase of the Fourier Transform

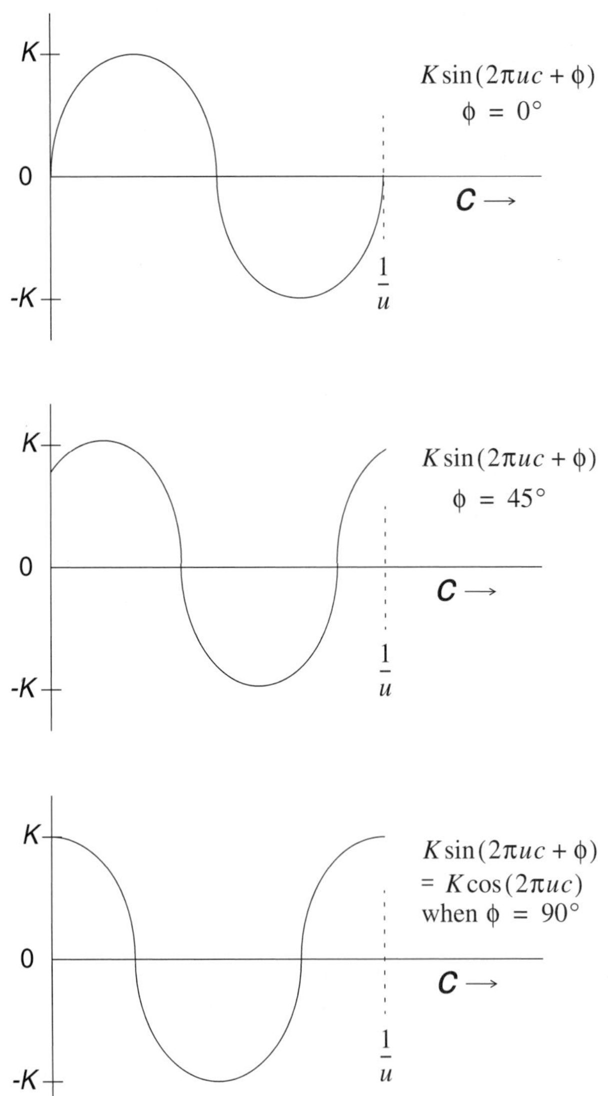

a. Magnitude and phase of sinusoidal waves.

Figure 2.5-6 (Continued)

b. Original image.

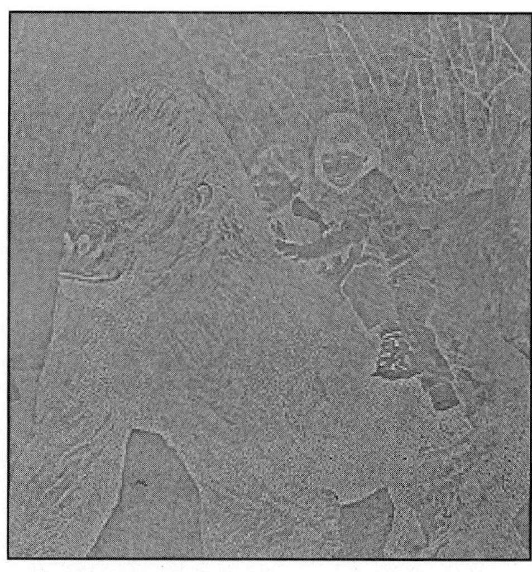

c. Phase-only image.

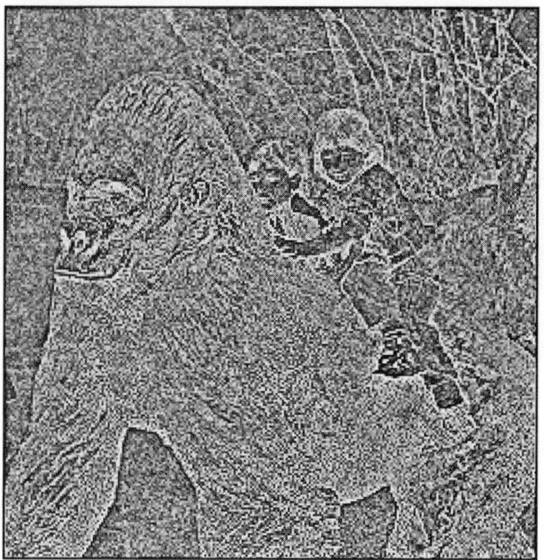

d. Histogram equalization of image (c) to show detail.

One important property of the Fourier transform is called separability. If a two-dimensional transform is *separable*, then the result can found by successive application of two one-dimensional transforms. This is illustrated by first separating the basis image term (also called the transform kernel) into a product, as follows:

$$e^{\left[-j2\pi\frac{ur+vc}{N}\right]} = e^{\left[-j2\pi\frac{ur}{N}\right]} e^{\left[-j2\pi\frac{vc}{N}\right]}$$

Next, we write the Fourier transform equation in the following form:

$$F(u, v) = \frac{1}{N}\sum_{r=0}^{N-1}\left(e^{-j2\pi\frac{ur}{N}}\right)\sum_{c=0}^{N-1} I(r, c)e^{-j2\pi\frac{vc}{N}}$$

The advantage of the separability property is that $F(u, v)$ or $I(r, c)$ can be obtained in two steps by successive applications of the one-dimensional Fourier transform or its inverse.

Expressing the equation as

$$F(u, v) = \frac{1}{N}\sum_{r=0}^{N-1} F(r, v)e^{\left[-j2\pi\frac{ur}{N}\right]}$$

where

$$F(r, v) = (N)\left(\frac{1}{N}\right)\sum_{c=0}^{N-1} I(r, c)e^{\left[-j2\pi\frac{vc}{N}\right]}$$

For each value of r, the expression inside the brackets is a one-dimensional transform with frequency values $v = 0, 1, 2, 3, \ldots, N-1$. Hence the two-dimensional function $F(r, v)$ is obtained by taking a transform along each row of $I(r, c)$ and multiplying the result by N. The desired result $F(u, v)$ is obtained by taking a transform along each column of $F(r, v)$.

2.5.2 Cosine Transform

The cosine transform, like the Fourier transform, uses sinusoidal basis functions. The difference is that the cosine transform basis functions are not complex; they use only cosine functions and not sine functions. Assuming an $N \times N$ image, the discrete cosine transform equation is given by

$$C(u, v) = \alpha(u)\alpha(v)\sum_{r=0}^{N-1}\sum_{c=0}^{N-1} I(r, c)\cos\left[\frac{(2r+1)u\pi}{2N}\right]\cos\left[\frac{(2c+1)v\pi}{2N}\right]$$

where

$$\alpha(u), \alpha(v) = \begin{cases} \sqrt{\frac{1}{N}} & \text{for } u, v = 0 \\ \sqrt{\frac{2}{N}} & \text{for } u, v = 1, 2, \ldots, N-1 \end{cases}$$

Because this transform uses only the cosine function, it can be calculated using only real arithmetic (not complex). This also affects the implied symmetry of the transform, which is explored in the filtering section. The cosine transform is often used in image compression, in particular the Joint Photographic Experts Group (JPEG) image compression method, which has been established as an international standard. In computer imaging we often represent the basis matrices as images, called basis images, where we use various gray values to represent the different values in the basis matrix. The 2-D basis images for the cosine transform are given in Figure 2.5-7 for a 4×4 image (at the bottom of the figure are the actual basis image values and the

Figure 2.5-7 Discrete Cosine Transform Basis Images

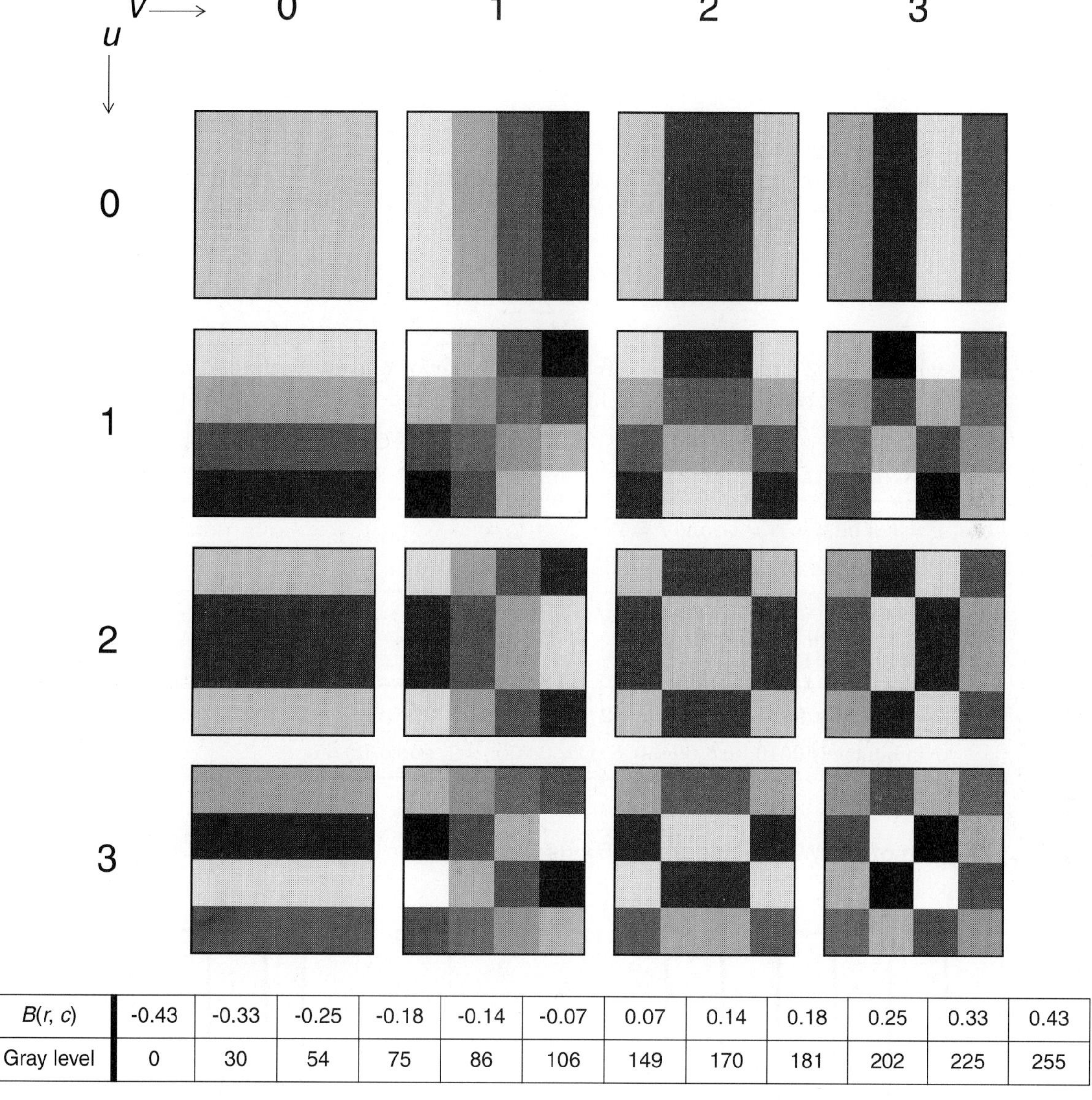

$B(r, c)$	-0.43	-0.33	-0.25	-0.18	-0.14	-0.07	0.07	0.14	0.18	0.25	0.33	0.43
Gray level	0	30	54	75	86	106	149	170	181	202	225	255

gray-level values to which they were mapped). Remember that the transform actually projects the image onto each of these basis images (see Figure 2.5-5), so the transform coefficients $C(u, v)$ tell us the amount of that particular basis image that the original image $I(r, c)$ contains.

The inverse cosine transform is given by

$$C^{-1}[C(u, v)] = I(r, c) = \sum_{u=0}^{N-1}\sum_{v=0}^{N-1} \alpha(u)\alpha(v)C(u,v)\cos\left[\frac{(2r+1)u\pi}{2N}\right]\cos\left[\frac{(2c+1)v\pi}{2N}\right]$$

2.5.3 Walsh-Hadamard Transform

The Walsh-Hadamard transform differs from the Fourier and cosine transforms in that the basis functions are not sinusoids. The basis functions are based on square or rectangular waves with peaks of ±1 (see Figure 2.5-8). Here the term *rectangular wave* refers to any function of this form, where the width of the pulse may vary. One primary advantage of a transform with these types of basis functions is that the computations are very simple. When we project the image onto the basis functions, all we need to do is multiply each pixel by ±1, as is seen is the Walsh-Hadamard transform equation (assuming an $N \times N$ image):

$$WH(u, v) = \frac{1}{N}\sum_{r=0}^{N-1}\sum_{c=0}^{N-1} I(r, c)(-1)^{\sum_{i=0}^{n-1}[b_i(r)p_i(u) + b_i(c)p_i(v)]}$$

where $N = 2^n$, the exponent on the (–1) is performed in modulo 2 arithmetic, and $b_i(r)$ is found by considering r as a binary number and finding the ith bit.

EXAMPLE 2-10

$n = 3$ (3 bits, so $N = 8$), and $r = 4$

r in binary is 100, so $b_2(r) = 1$, $b_1(r) = 0$, and $b_0(r) = 0$.

EXAMPLE 2-11

$n = 4$, (4 bits, so $N = 16$), and $r = 2$

r in binary is 0010, so $b_3(r) = 0$, $b_2(r) = 0$, $b_1(r) = 1$, and $b_0(r) = 0$.

Figure 2.5-8 Form of the Walsh-Hadamard Basis Functions

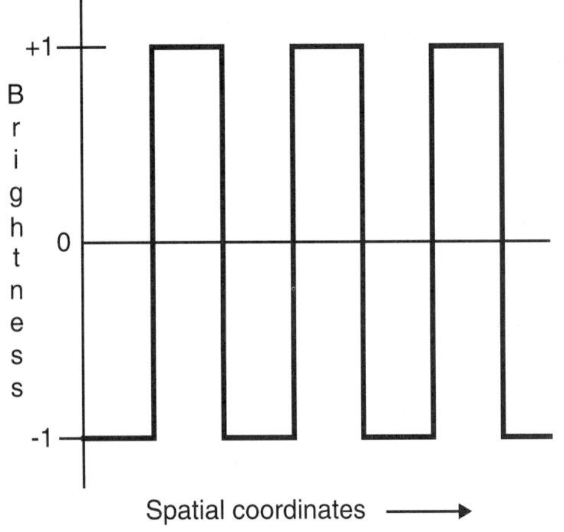

a. A square wave.

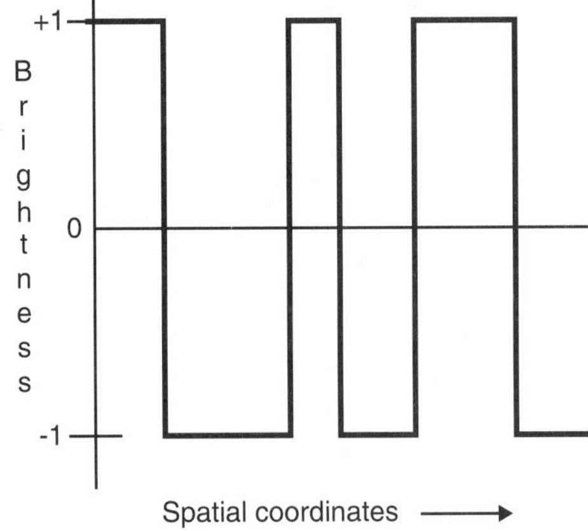

b. Representation of a rectangular wave. The width of each pulse may vary.

$p_i(u)$ is found as follows:

$$p_0(u) = b_{n-1}(u)$$
$$p_1(u) = b_{n-1}(u) + b_{n-2}(u)$$
$$p_2(u) = b_{n-2}(u) + b_{n-3}(u)$$
$$\vdots$$
$$p_{n-1}(u) = b_1(u) + b_0(u)$$

The sums are performed in modulo 2 arithmetic, and the values for $b_i(c)$ and $p_i(v)$ are found in a similar manner. Strictly speaking we cannot call the Walsh-Hadamard transform a frequency transform because the basis functions do not exhibit the frequency concept in the manner of sinusoidal functions. However, we define an analogous term for use with these types of functions. If we consider the number of zero crossings (or sign changes), we have a measure that is comparable to frequency, and we call this *sequency*. In Figure 2.5-9 we see the 1-D Walsh-Hadamard basis functions

Figure 2.5-9 1-D Walsh-Hadamard Basis Functions

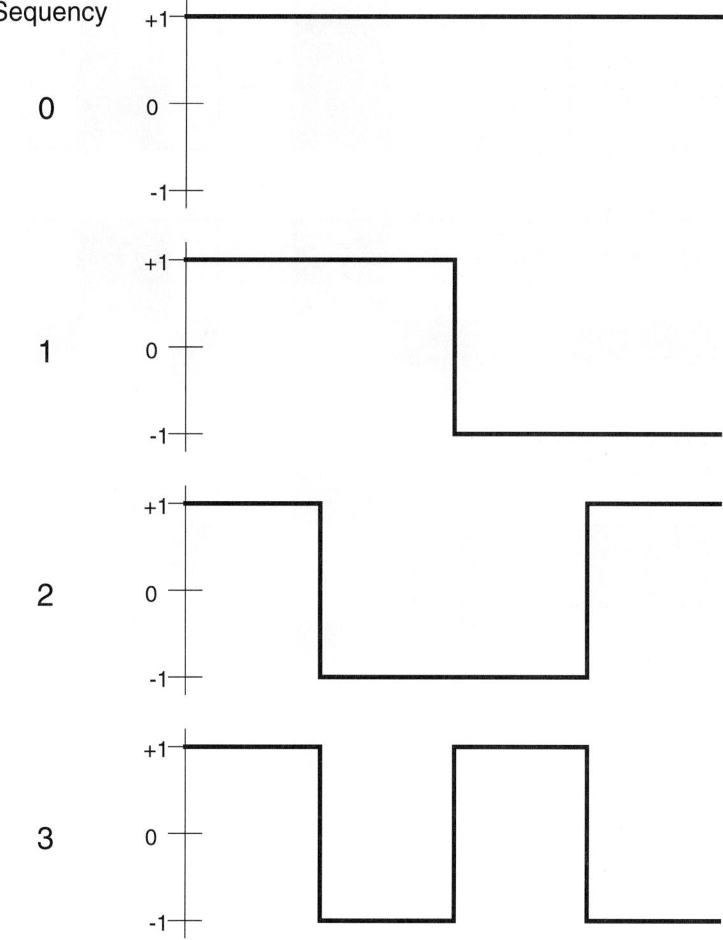

for $N = 4$ and the corresponding sequency. We can see that the basis functions are in the order of increasing sequency, much like the sinusoidal functions are in order of increasing frequency. In Figure 2.5-10, we have the basis images for the Walsh-Hadamard transform for a 4×4 image; we use white for the +1 and black for the –1.

It may be difficult to see how the 2-D basis images are generated from the 1-D basis vectors. For the terms that are along the u- or v-axis, we simply repeat the 1-D function along all the rows or columns. For the basis images that are not along the u- or v-axis, we perform a *vector outer product* on the corresponding 1-D vectors. We have seen that a vector inner product is also called a projection and is performed by over-

Figure 2.5-10 Walsh-Hadamard Basis Images

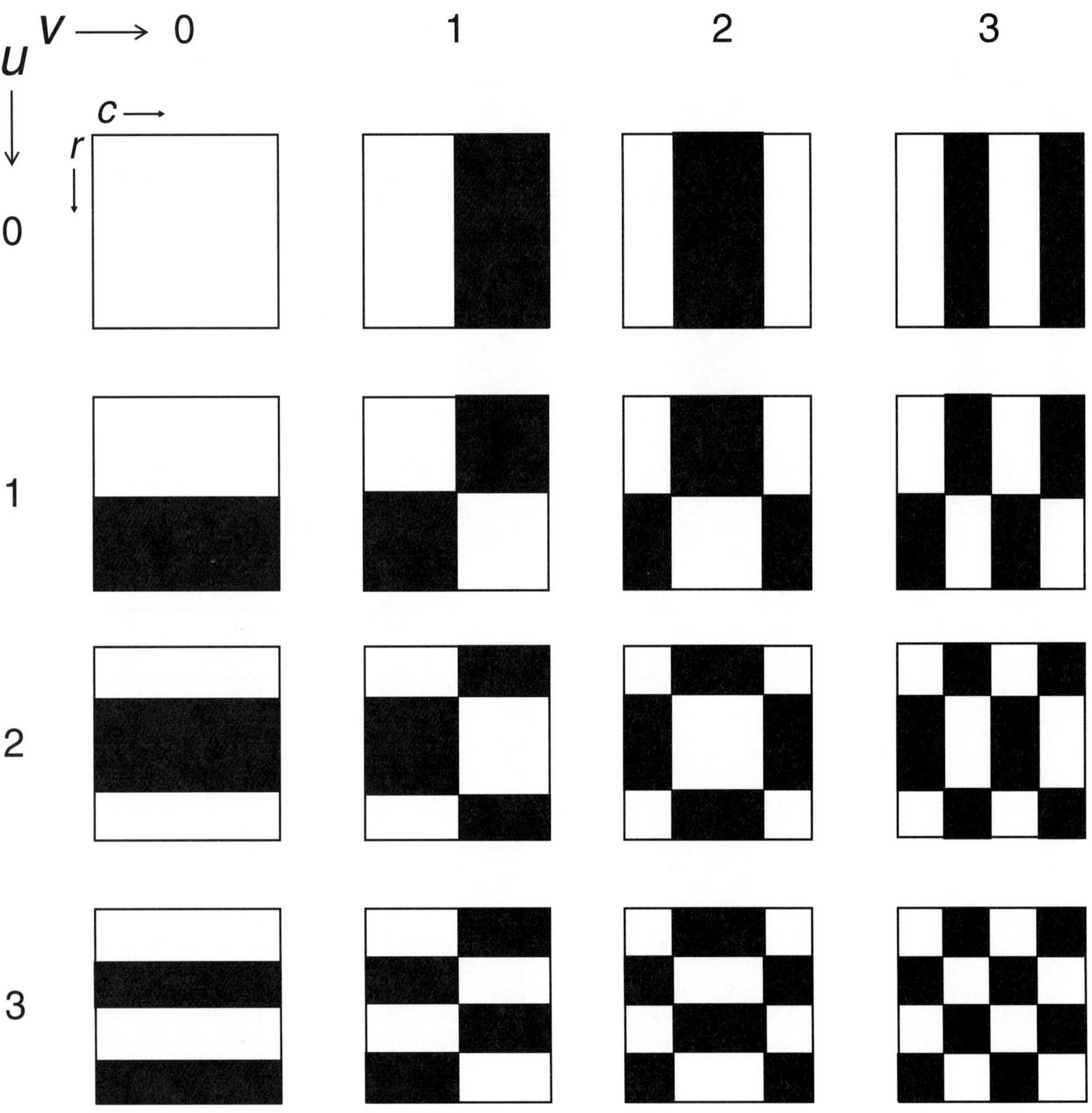

laying, multiplying coincident terms, and summing the results—this gives us a scalar, or a single number, for a result. The vector outer product gives us a matrix, which is obtained as follows:

EXAMPLE 2-12

For $(u, v) = (3, 2)$, see Figure 2.5-11. If we look along one row of the $v = 2$ $(u = 0)$ basis image in Figure 2.5-10, we find the following numbers: +1 –1 –1 +1. Then if we look along one column in the u direction for $u = 3$ $(v = 0)$, we see +1 –1 +1 –1. These are the corresponding 1-D basis vectors. We then put the row vector across the top and the column vector down the left side and fill in the matrix by multiplying the column by the corresponding row element, as in Figure 2.5-11. The resulting matrix is the vector outer product. Compare this to the corresponding basis image in Figure 2.5-10.

This process can be used to generate the 2-D basis images for any function that has a separable basis. Remember that *separable* means that the basis function can be expressed as a product of terms that depend only on one of the variable pairs, *ru* or *cv*. Note that all the functions we have considered in this section have separable basis functions. This separability also allows us to perform the 2-D transformation by two 1-D transforms, first doing a 1-D transform on the rows and then performing the 1-D transform on the resulting columns.

Figure 2.5–11 Vector Outer Product

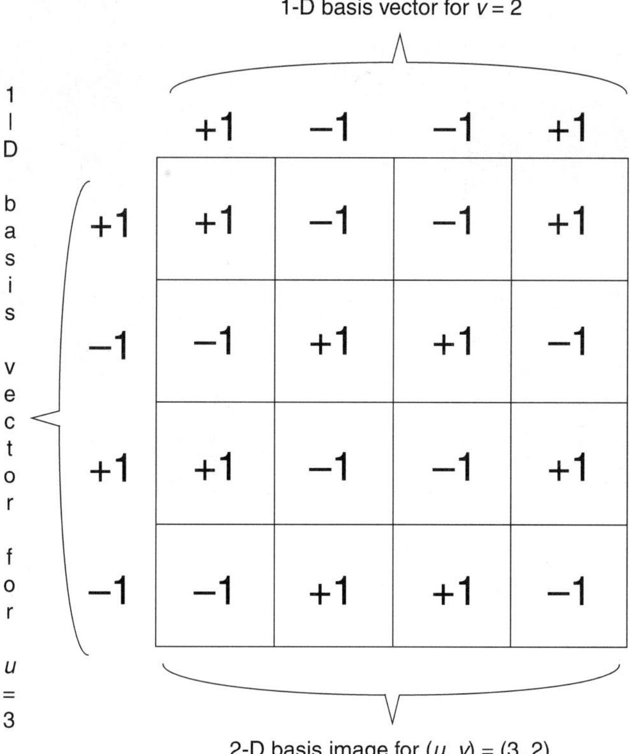

It is interesting to note that with the Walsh-Hadamard transform there is another visual method to find the off axis basis images, by assuming that the black in Figure 2.5-10 corresponds to 0 and the white corresponds to 1. The basis images not along the u- or v-axis can be obtained by taking the corresponding basis images on these axes, overlaying them, and performing an XOR followed by a NOT. For example, to find the Walsh-Hadamard basis image corresponding to $(u, v) = (3, 2)$, we take the basis image along the u-axis for $u = 3$, and the basis image along the v-axis for $v = 2$, overlay them, XOR the images, and then perform a NOT. This is illustrated in Figure 2.5-12.

The inverse Walsh-Hadamard transform equation is

$$WH^{-1}[WH(u, v)] = I(r, c) = \frac{1}{N}\sum_{u=0}^{N-1}\sum_{v=0}^{N-1} WH(u, v)(-1)^{\sum_{i=0}^{n-1}[b_i(r)p_i(u) + b_i(c)p_i(v)]}$$

Figure 2.5-12 Finding an Off-Axis Walsh-Hadamard Basis Image

a. Basis $u = 3$ $(v = 0)$.

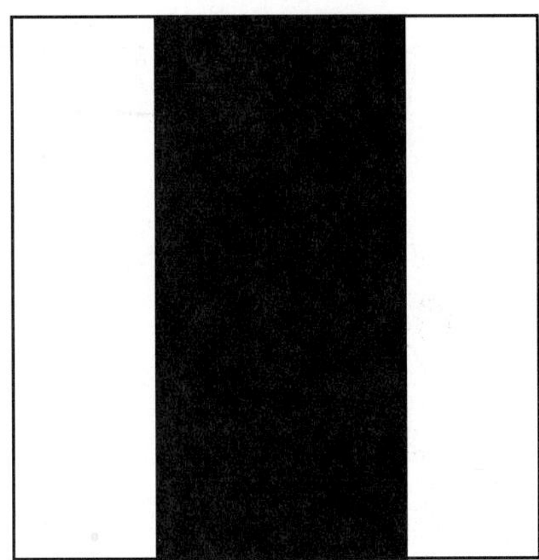

b. Basis $v = 2$ $(u = 0)$.

c. XOR of images (a) and (b).

d. NOT of image (c). This is Walsh-Hadamard basis image for $(u, v) = (3, 2)$.

2.5.4 Filtering

After the image has been transformed into the frequency or sequency domain, we may want to modify the resulting spectrum (see Image Processing exercise #5 in Chapter 8). High-frequency information can be removed with a lowpass filter, which will have the effect of blurring an image, or low-frequency information can be removed with a highpass filter, which will tend to sharpen the image. We may want to extract the frequency information in specific parts of the spectrum by bandpass filtering. Alternately, bandreject filtering can be employed to eliminate specific parts of the spectrum, for example to remove unwanted noise. All these types of filters will be explored here.

Before looking at the different methods of filtering, we want to examine the spectrum that results from the transforms under consideration—the Fourier, cosine, and Walsh-Hadamard transforms. To that end we need to consider the implied symmetry in the spectrum that results from certain theoretical considerations, which have not been rigorously developed here. We will, however, examine the practical implications of these theoretical aspects.

Basically, two types of spectral symmetries result from taking the continuous functions and sampling them to develop the discrete transforms. With the Fourier transform the implied symmetry of the transform is to repeat the discrete $N \times N$ spectrum in all directions to infinity, as illustrated in Figure 2.5-13a. The spectral components are only increasing in frequency up to the $N/2$ term and then start to decrease due to the *periodicity and conjugate symmetry* properties of the Fourier transform (see Figure 2.5-13b). With this type of symmetry, we often shift the origin of the spectrum to the center for display and filtering purposes, as in Figures 2.5-13c, d. This shifting is done by using a property of the Fourier transform called the *frequency translation property*, which will shift the spectral origin to the center by multiplying the original image $I(r, c)$ by $(-1)^{r+c}$ before we perform the transform. This is illustrated in Figures 2.5-14a, b. As a result of the way the human visual system responds to brightness, we can greatly enhance the visual information available by displaying the following log transform of the spectrum, as shown in Figures 2.5-14c, d:

$$\log(u, v) = k \log[1 + |F(u, v)|]$$

The actual dynamic range of the Fourier spectrum is much greater than the 256 gray levels (8-bits) available with most image display devices. Thus, when we remap it to 256 levels, we can see only the largest values, which are typically the low-frequency terms around the origin. The log function compresses the data, and the scaling factor k remaps the data to the 0–255 range.

The second type of implied spectral symmetry, which is evident in the cosine transform, is shown in Figure 2.5-15a. The spectrum is folded about the origin, creating a $2N \times 2N$ pattern. We can see in Figure 2.5-15b that the frequency increases from the origin outward (also see Figure 2.5-7), thus we do not need to shift it to the center. If we did shift the spectrum to the center to extract the central portion, as we did with the Fourier spectrum, we would lose half the frequency information (Figure 2.5-16). The Walsh-Hadamard transform also exhibits increasing sequency from the origin out, as was seen in Figure 2.5-10.

We are now ready to consider the various filtering methods. Lowpass filters tend to blur images. They pass low frequencies and attenuate or eliminate the high-frequency information. They are used for image compression or for hiding effects caused by noise. Visually they blur the image, although this blur is sometimes considered an enhancement because it imparts a softer effect to the image (see Figure 2.5-17). Lowpass filtering is performed by multiplying the spectrum by a filter and then applying the inverse transform to obtain the filtered image. The ideal filter function is shown in

Figure 2.5-13 Implied Symmetry for the Fourier Transform

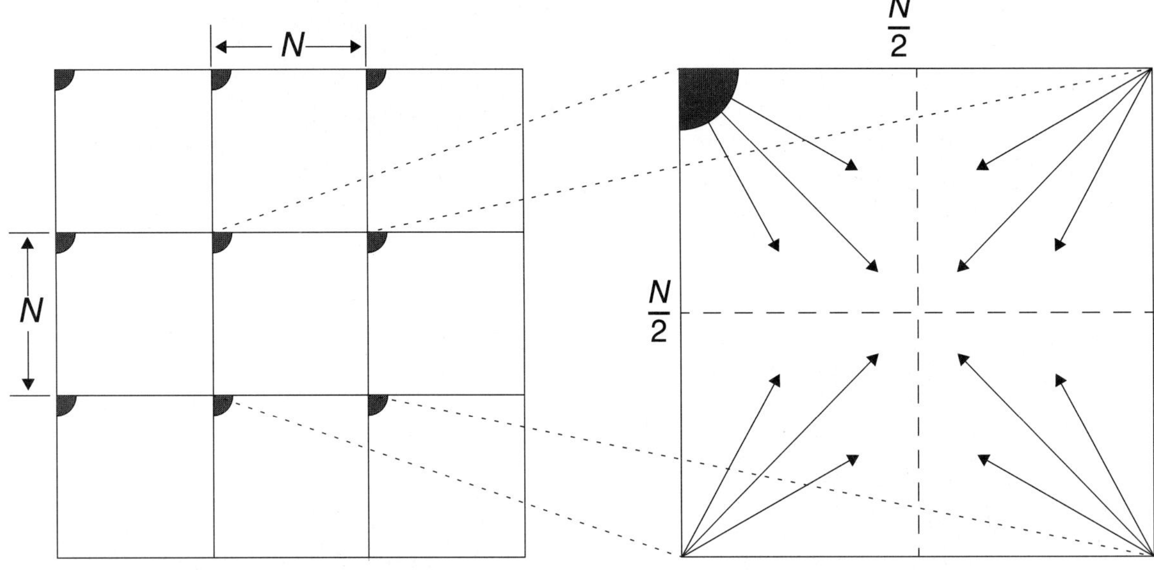

a. Implied symmetry with origin in upper-left corner. Each N × N block represents all the transform coefficients and is repeated infinitely in all directions.

b. Increasing frequency in direction of arrows.

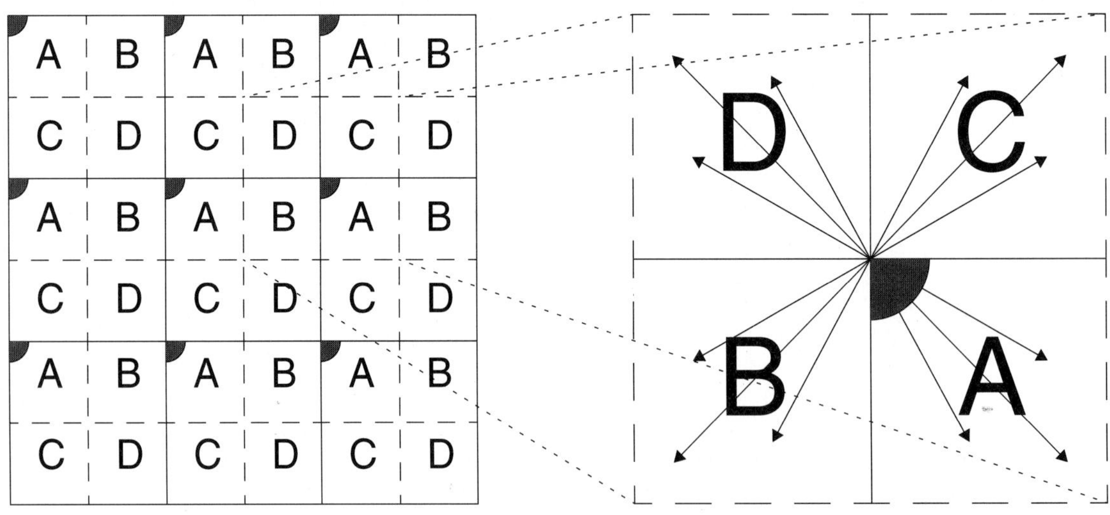

c. Periodic spectrum, with quadrants labeled A,B,C,D.

d. Spectrum shifted to center. Frequency increases in all directions as we move away from the origin.

Figure 2.5-14 Display of Fourier Spectrum

a. Original spectrum of plane.ras; note dot at upper-left corner.

b. Spectrum shifted to center.

c. Log-remapped original spectrum.

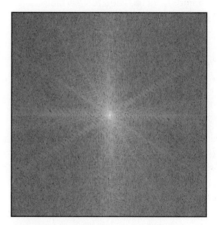

d. Log-remapped spectrum shifted to center.

Figure 2.5-18; note the two types of symmetry in the filter to match the type of symmetry in the spectrum. The frequency at which we start to eliminate information is called the *cutoff frequency*, f_0. The frequencies in the spectrum that are not filtered out are in the *passband*, while the spectral components that do get filtered out are in the *stopband*. We can represent the filtering process by the following equation:

$$I_{\text{fil}}(r, c) = T^{-1}[T(u, v)H(u, v)]$$

where $I_{\text{fil}}(r, c)$ is our filtered image, $H(u, v)$ is the filter function, $T(u, v)$ is the transform, and $T^{-1}[]$ represents the inverse transform. The multiplication, $T(u, v)H(u, v)$, is performed with a point-by-point method. That is, $T(0, 0)$ is multiplied by $H(0, 0)$, $T(0, 1)$ is multiplied by $H(0, 1)$, and so on. The resulting products are placed into an array at the same (r, c) location.

Figure 2.5-15 Cosine Symmetry

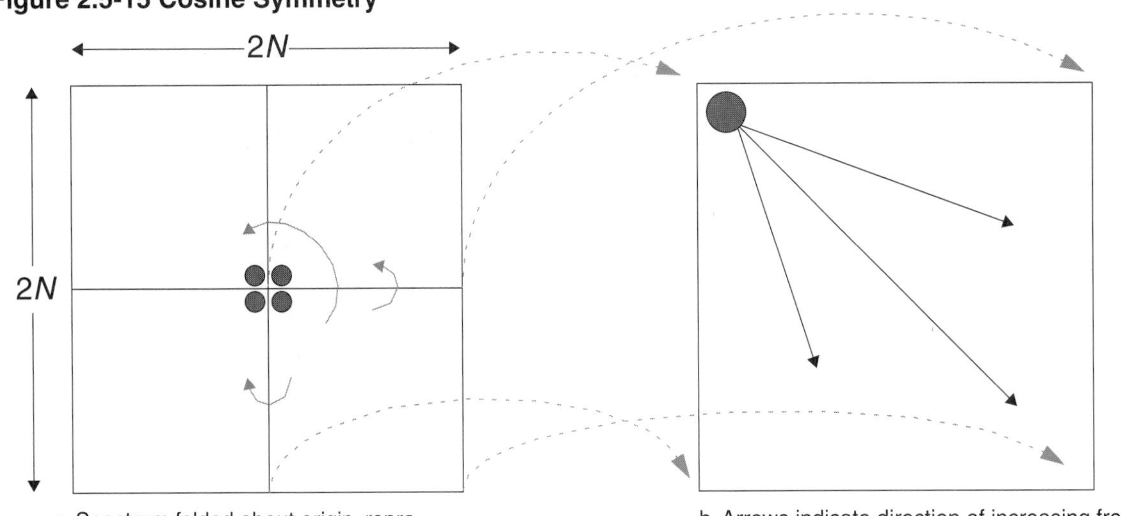

a. Spectrum folded about origin, represented by the ●. The $2N \times 2N$ block is repeated infinitely in all directions.

b. Arrows indicate direction of increasing frequency for cosine spectrum.

Figure 2.5-16 Cosine Spectrum Should Not Be Shifted to Center

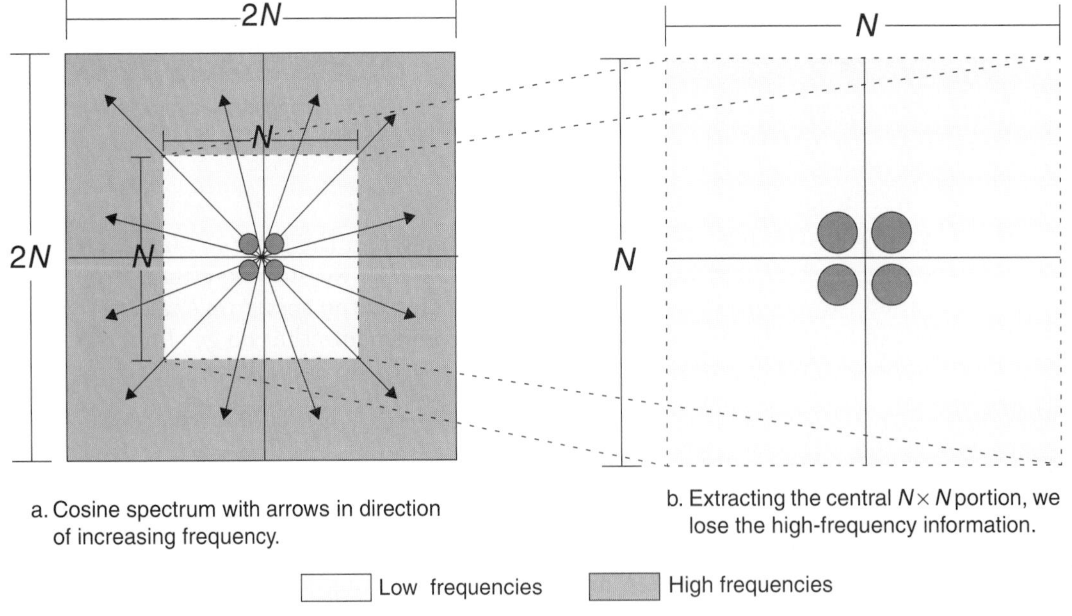

a. Cosine spectrum with arrows in direction of increasing frequency.

b. Extracting the central $N \times N$ portion, we lose the high-frequency information.

Figure 2.5-17 Lowpass Filtering

a. Original image.

b. Filtered image,using a nonideal low-pass filter. Note the blurring that softens the image.

c. Ideal lowpass-filtered image shows the ripple artifacts at boundaries. Frequency cutoff = 32.

EXAMPLE 2-13

Let $H(u, v)$ and $T(u, v)$ be the following 2×2 images.

$$H(u,v) = \begin{bmatrix} 2 & -3 \\ 4 & 1 \end{bmatrix} \qquad T(u,v) = \begin{bmatrix} 4 & 6 \\ -5 & 8 \end{bmatrix}$$

Then $T(u, v)H(u, v)$ is equal to

$$\begin{bmatrix} 8 & -18 \\ -20 & 8 \end{bmatrix}$$

Figure 2.5-18 Ideal Lowpass Filters

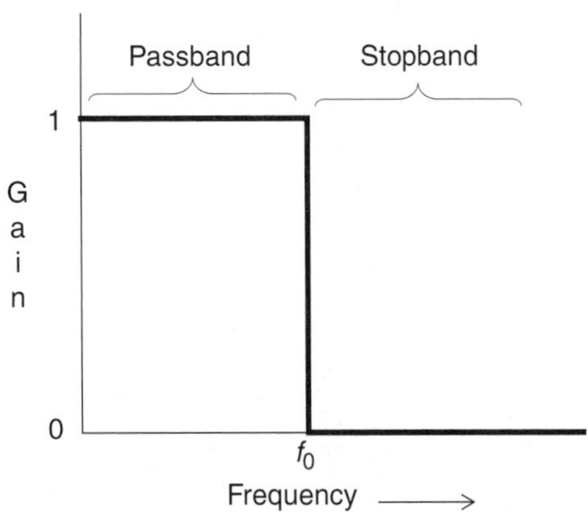

a. 1-D lowpass ideal filter.

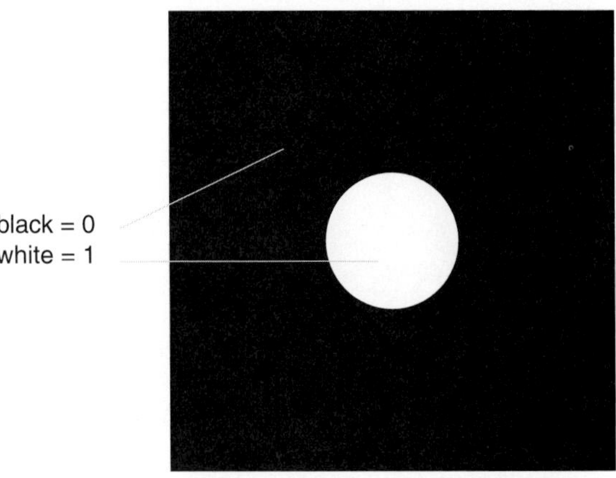

b. 2-D lowpass ideal filter shown as an image.

c. 2-D lowpass ideal filter for Walsh-Hadamard and cosine functions.

Note that for ideal filters in Figure 2.5-18 the $H(u, v)$ matrix will contain only 1's and 0's, but, as in the preceding example, the matrix can contain any numbers.

The ideal filter is called ideal because the transition from the passband to the stopband in the filter is perfect—it goes from 0 to 1 instantly. Although this type of filter is not realizable in physical systems, such as with electronic filters, it is a reality for digital image processing applications because all we are doing is multiplying numbers in software. However, the ideal filter leaves undesirable artifacts in images. This artifact appears in the lowpass filtered image in Figure 2.5-17c as ripples, or waves, wherever there is a boundary in the image. This problem can be avoided by using a "nonideal" filter that does not have perfect transition, as is shown in Figure 2.5-19. The image created in Figure 2.5-17b was generated using a nonideal filter of a type called a Butterworth filter.

Figure 2.5-19 Nonideal Lowpass Filters

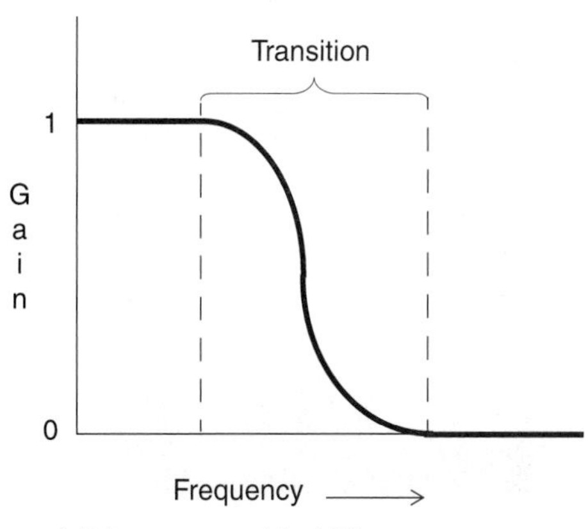

a. 1-D lowpass nonideal filter.

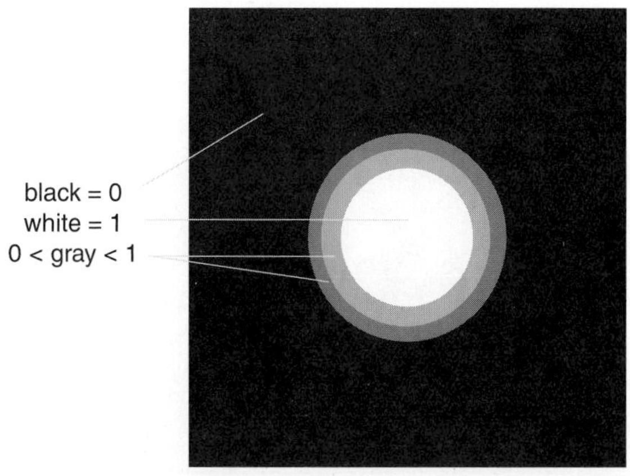

b. 2-D lowpass nonideal filter shown as an image.

c. 2-D nonideal lowpass filter for Walsh-Hadamard and cosine functions.

With the Butterworth filter we can specify the *order* of the filter, which determines how steep the slope is in the transition of the filter function. A higher order to the filter creates a steeper slope, and the closer we get to an ideal filter. In Figure 2.5-20 we compare the results of different orders of Butterworth filters. We see that as we get closer to an ideal filter, the blurring effect becomes more prominent due to the elimination of even partial high-frequency information. Another effect that is most noticeable in the eighth-order filter is the appearance of waves, or ripples, wherever boundaries occur in the image.

Next we consider highpass filters. The highpass filter function is shown in Figure 2.5-21, where we see both ideal and Butterworth filter functions. A highpass filter can be used for edge enhancement because it passes only high-frequency information,

Figure 2.5-20 Lowpass Butterworth Filters

a. Filter order = 1.

b. Filter order = 3.

c. Filter order = 6.

d. Filter order = 8.

Figure 2.5-21 Highpass Filter Functions

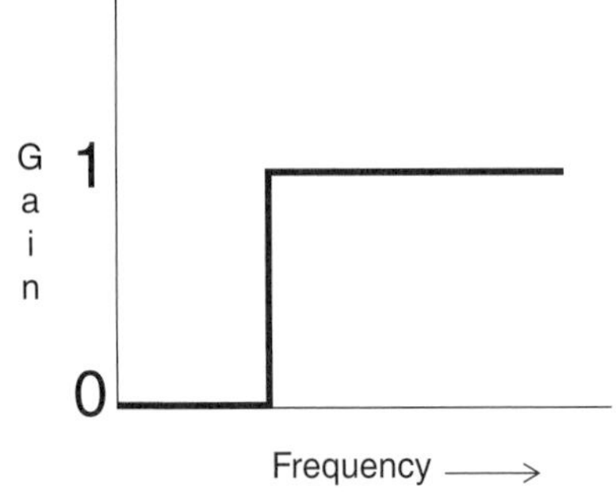

a. 1-D ideal highpass filter.

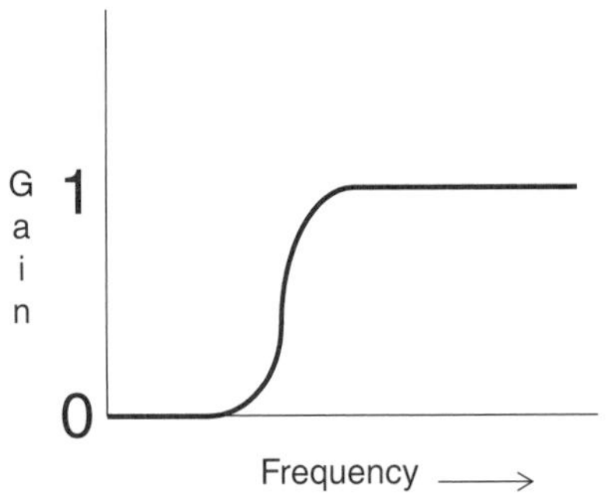

b. 1-D nonideal highpass filter.

Figure 2.5-21 (Continued)

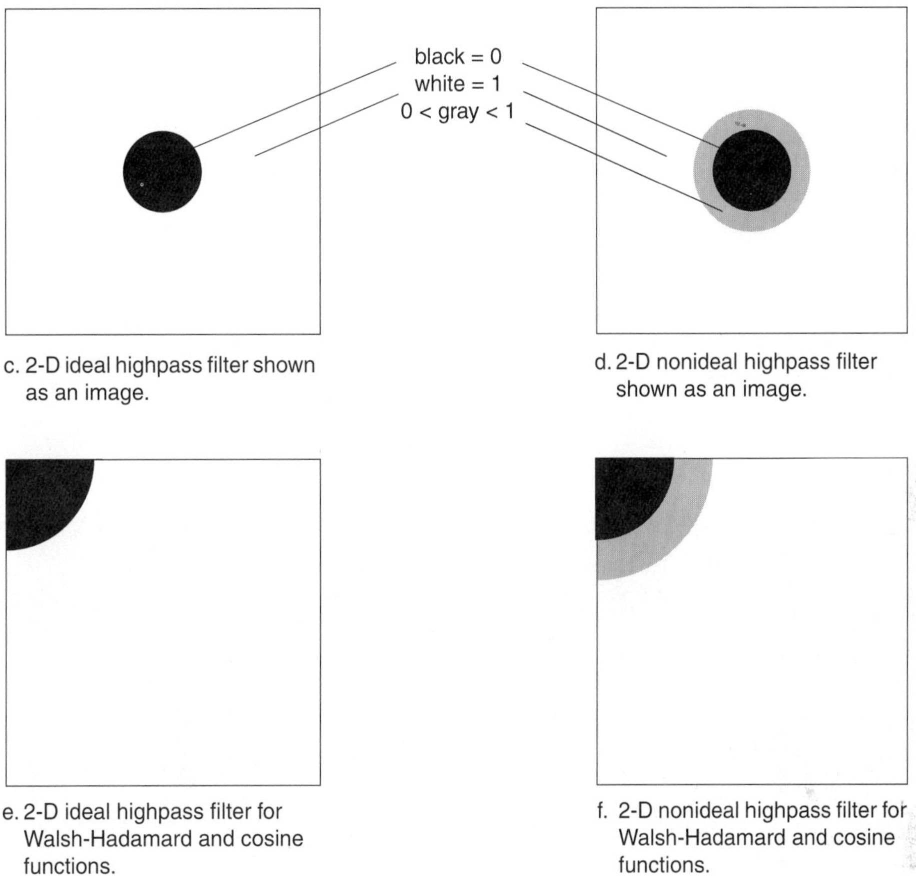

c. 2-D ideal highpass filter shown as an image.

d. 2-D nonideal highpass filter shown as an image.

e. 2-D ideal highpass filter for Walsh-Hadamard and cosine functions.

f. 2-D nonideal highpass filter for Walsh-Hadamard and cosine functions.

corresponding to places where gray levels are changing rapidly (edges in images are characterized by rapidly changing gray levels). The function for a special type of highpass filter, called a high-frequency emphasis filter, is shown in Figure 2.5-22. This filter function retains some of the low-frequency information by adding an offset value to the function, so we do not lose the overall image information. The results from applying these types of filters are shown in Figure 2.5-23. The original is shown in Figure 2.5-23a, and Figures 2.5-23b, c show the results from a Butterworth and an ideal filter function. Here we can see the edges enhanced and the ripples that occur from using an ideal filter (Figure 2.5-23c), but note a loss in the overall contrast of the image. In Figure 2.5-23d and e, we see the contrast added back to the image by using the high-frequency emphasis filter function.

The bandpass and bandreject filters are specified by two cutoff frequencies, a low cutoff and a high cutoff, shown in Figure 2.5-24. These filters can be modified into nonideal filters by making the transitions gradual at the cutoff frequencies, as was done for the lowpass filter in Figure 2.5-19 and the highpass in Figure 2.5-21. A special form of these filters is called a notch filter because it only notches out, or passes, specific frequencies (see Figures 2.5-24g, h). These three types of filters are typically used in image restoration, enhancement, and compression, and examples can be seen in Chapters 3, 4, and 5.

Figure 2.5-22 High-Frequency Emphasis Filter

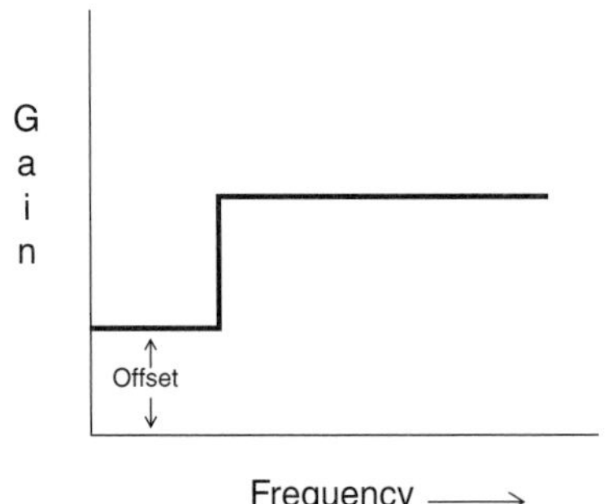

a. 1-D ideal high-frequency emphasis filter.

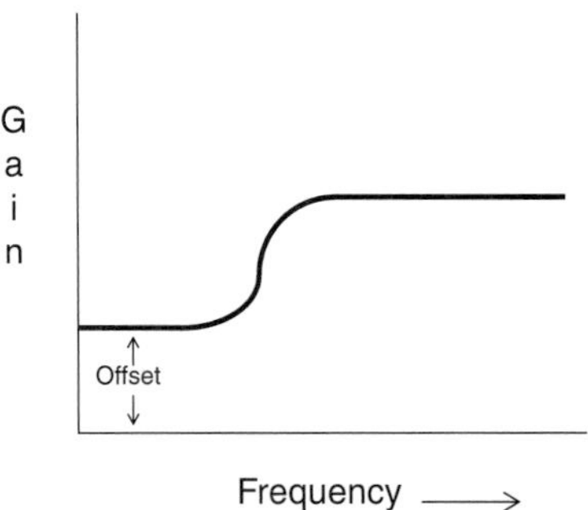

b. 1-D nonideal high-frequency emphasis filter.

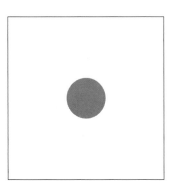

c. 2-D ideal high-frequency emphasis filter shown as an image.

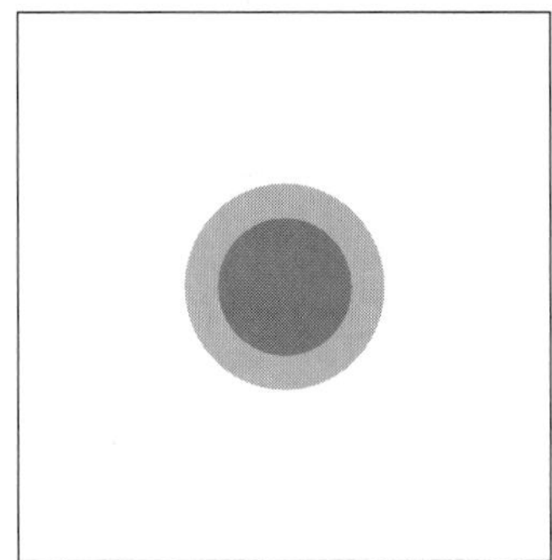

d. 2-D nonideal high-frequency emphasis filter shown as an image.

e. 2-D ideal high-frequency emphasis filter for Walsh-Hadamard and cosine functions.

f. 2-D nonideal high-frequency emphasis filter for Walsh-Hadamard and cosine functions.

Figure 2.5-23 Highpass Filtering

a. Original image.

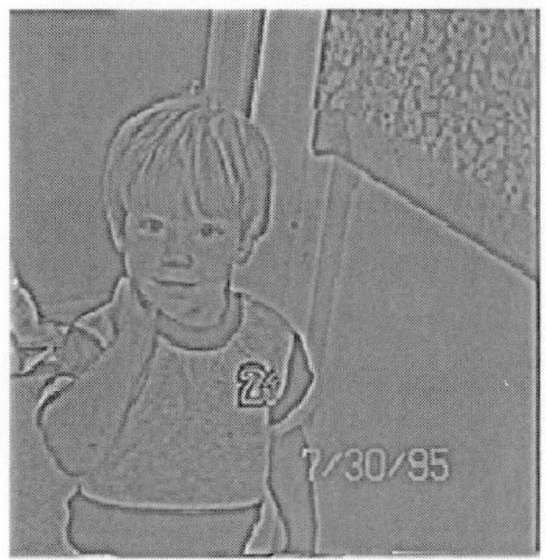

b. Butterworth filter—Order = 2; Cutoff = 32.

c. Ideal filter—Cutoff = 32.

d. High-frequency emphasis filter—Offset = 0.5; Order = 2; Cutoff = 32.

e. High-frequency emphasis filter—Offset = 1.5; Order = 2; Cutoff = 32.

Figure 2.5-24 Bandpass, Bandreject, and Notch Filters

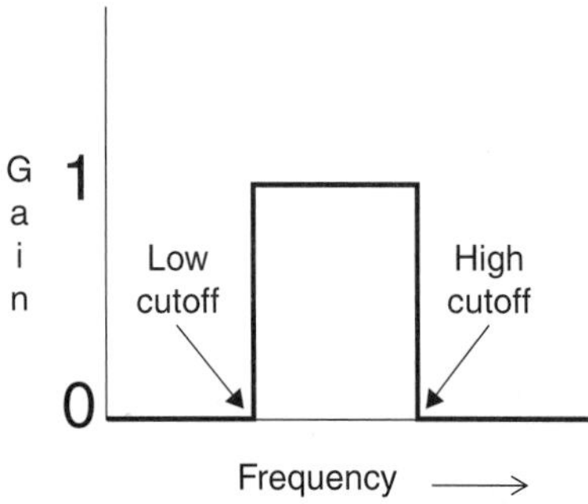

a. 1-D ideal bandpass filter.

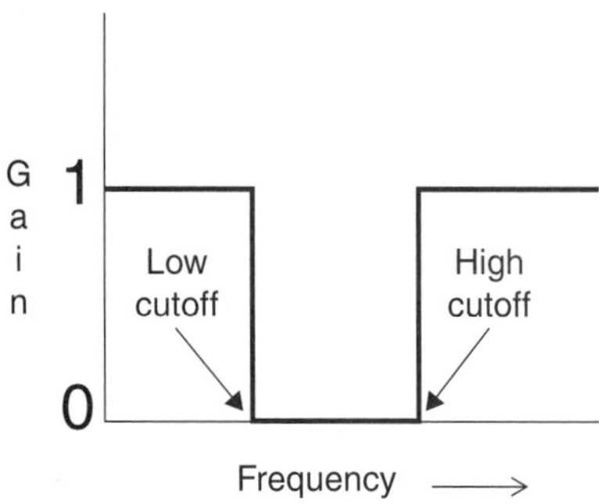

b. 1-D ideal bandreject filter.

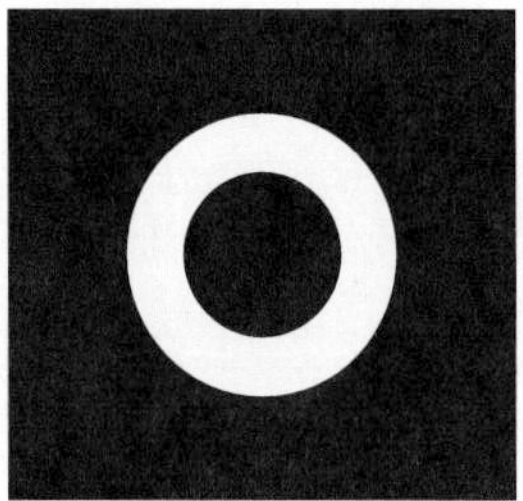

c. 2-D ideal bandpass filter shown as an image.

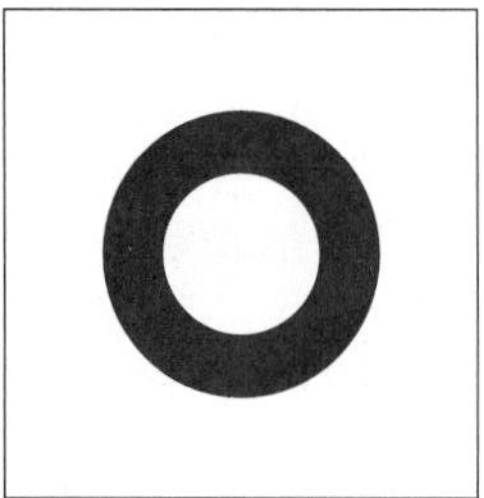

d. 2-D ideal bandreject filter shown as an image.

e. 2-D ideal bandpass filter for Walsh-Hadamard and cosine functions.

f. 2-D ideal bandreject filter for Walsh-Hadamard and cosine functions.

g. 2-D ideal notch filter for rejecting specific frequencies.

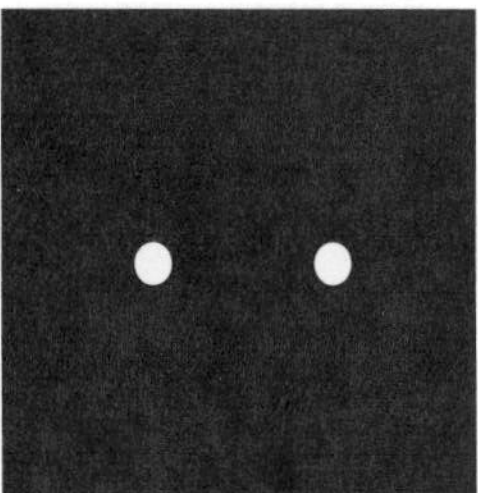

h. 2-D ideal notch filter for passing specific frequencies.

2.5.5 Wavelet Transform

The wavelet transform is really a family of transforms that satisfy specific conditions. From our perspective, the *wavelet transform* can be described as a transform that has basis functions that are shifted and expanded versions of themselves. Because of this, the wavelet transform contains not just frequency information but spatial information as well. One of the most common models for a wavelet transform uses the Fourier transform and highpass and lowpass filters. To satisfy the conditions for a wavelet transform, the filters must be *perfect reconstruction filters*, which means that any distortion introduced by the forward transform will be canceled in the inverse transform (an example of these types of filters are *quadrature mirror filters*).

The wavelet transform breaks an image down into four subsampled, or decimated, images. They are subsampled by keeping every other pixel. The results consist of one image that has been highpass filtered in both the horizontal and vertical directions, one that has been highpass filtered in the vertical and lowpass filtered in the horizontal, one that has been lowpassed in the vertical and highpassed in the horizontal, and one that has been lowpass filtered in both directions.

This transform is typically implemented in the spatial domain by using 1-D convolution filters. In the section on edge detection (Section 2.3) we looked at 2-D convolution masks that mark places in the image where the gray levels are changing rapidly. These rapid changes correspond to high-frequency information, so edge detectors are basically highpass filters. This illustrates an important Fourier transform property called the convolution theorem. The *convolution theorem* states that convolution in the spatial domain is the equivalent of multiplication in the frequency domain. We have seen that multiplication in the frequency domain is used to perform filtering; the convolution theorem tells us that we can also perform filtering in the spatial domain via convolution, such as we have already seen with spatial convolution masks. Therefore, if we can define convolution masks that satisfy the wavelet transform conditions, we can implement the wavelet transform in the spatial domain. We have also seen that if the transform basis functions are separable, we can perform the 2-D transform by using two 1-D transforms. An additional benefit of convolution versus frequency domain filtering is that, if the convolution mask is short, it is much faster.

In order to perform the wavelet transform with convolution filters, a special type of convolution called circular convolution must be used. *Circular convolution* is performed by taking the underlying image array and extending it in a periodic manner to match the symmetry implied by the discrete Fourier transform (see Figures 2.5-25a, b). The convolution process starts with the origin of the image and the convolution mask aligned, so that the first value contains contributions from the "previous" copy of the periodic image (see Figure 2.5-25c). In Figure 2.25-c we also see that the last value(s) contain contributions from the "next" copy of the extended, periodic image. Performing circular convolution allows us to retain the outer rows and columns, unlike the previously used method where the outer rows and columns were ignored. This is important since we may want to perform the wavelet transform on small blocks, and eliminating the outer row(s) and column(s) is not practical.

Numerous filters can be used to implement the wavelet transform, and two of the commonly used ones, the Daubechies and the Haar, will be explored here. These are separable, so they can be used to implement a wavelet transform by first convolving them with the rows and then the columns. The Haar basis vectors are simple:

$$LOWPASS: \quad \frac{1}{\sqrt{2}}[1 \quad 1]$$

$$HIGHPASS: \quad \frac{1}{\sqrt{2}}[1 \quad -1]$$

An example of Daubechies basis vectors (there are many others) follows:

$$LOWPASS: \quad \frac{1}{4\sqrt{2}}[1+\sqrt{3}, \quad 3+\sqrt{3}, \quad 3-\sqrt{3}, \quad 1-\sqrt{3}]$$

$$HIGHPASS: \quad \frac{1}{4\sqrt{2}}[1-\sqrt{3}, \quad \sqrt{3}-3, \quad 3+\sqrt{3}, \quad -1-\sqrt{3}]$$

To use the basis vectors to implement the wavelet transform, they must be zero-padded to be the same size as the image (or subimage). Also note that the origin of the

Figure 2.5-25 Circular Convolution

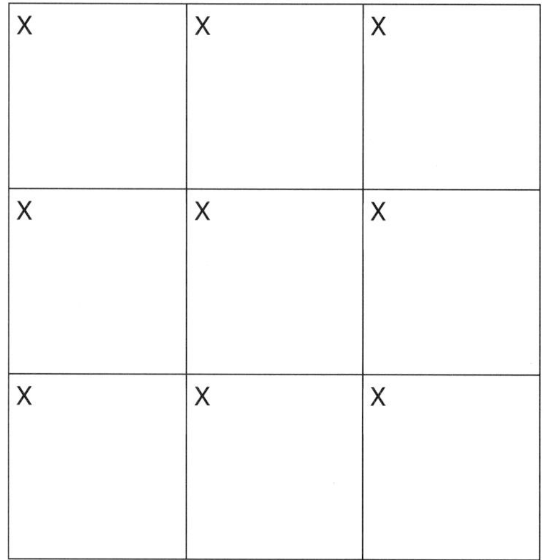

a.Extended, periodic image (X = origin).

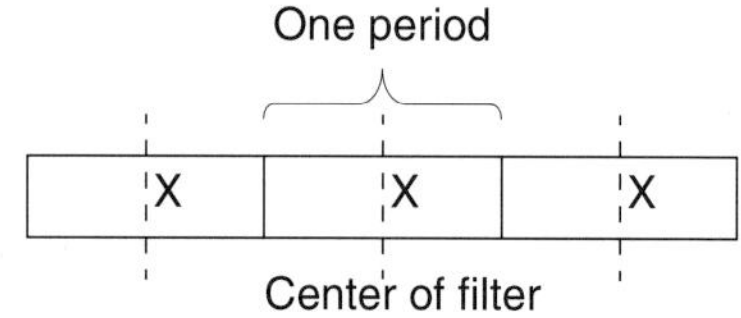

b. Extended, periodic 1-D convolution filter (X = origin).

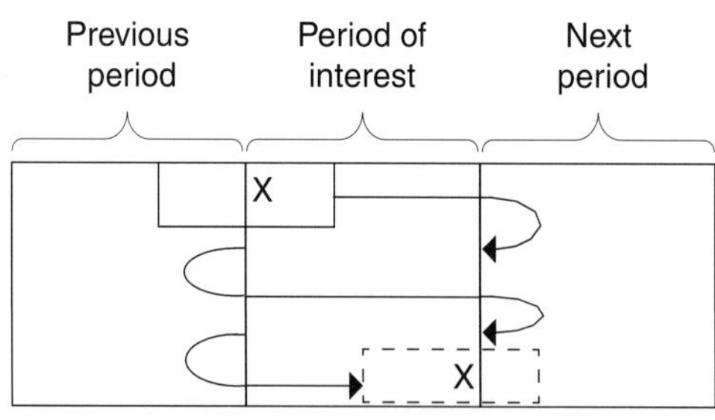

c. With circular convolution, the outer rows and columns include products of both the previous and next periods.

basis vectors is in the center, corresponding to the value to the right of the middle of the vector.

EXAMPLE 2-14

We want to use the Haar basis vectors to perform a wavelet transform on an image by dividing it into 4×4 blocks. The basis vectors need to be zero-padded so that they have a length of 4, as follows:

$$\textit{LOWPASS:} \quad \frac{1}{\sqrt{2}}[1 \quad 1 \quad 0 \quad 0]$$
$$\textit{HIGHPASS:} \quad \frac{1}{\sqrt{2}}[1 \quad -1 \quad 0 \quad 0]$$

↑ *origin* (under the second element)

These are aligned with the image so that the origins coincide, and the result from the first vector inner product is placed into the location corresponding to the origin. Note that when the vector is zero-padded on the right, the origin is no longer to the right of the center of the resulting vector. The origin is determined by selecting the coefficient corresponding to the right of center *before* zero-padding.

EXAMPLE 2-15

To use the Daubechies basis vectors to do a wavelet transform on an image by dividing it into 8 $\times$ 8 blocks, we need to zero-pad them to a length of 8, as follows:

$$\textit{LOWPASS:} \quad \frac{1}{4\sqrt{2}}[1+\sqrt{3}, \quad 3+\sqrt{3}, \quad 3-\sqrt{3}, \quad 1-\sqrt{3}, \quad 0, \quad 0, \quad 0, \quad 0]$$
$$\textit{HIGHPASS:} \quad \frac{1}{4\sqrt{2}}[1-\sqrt{3}, \quad \sqrt{3}-3, \quad 3+\sqrt{3}, \quad -1-\sqrt{3}, \quad 0, \quad 0, \quad 0, \quad 0]$$

↑ *origin* (under the third element)

Note that the origin is the value to the right of the center of the original vector before zero-padding. Because these are assumed periodic for circular convolution, we could zero-pad equally on both ends; then the origin is to the right of the center of the zero-padded vector, as follows:

$$\textit{LOWPASS:} \quad \frac{1}{4\sqrt{2}}[0, \quad 0, \quad 1+\sqrt{3}, \quad 3+\sqrt{3}, \quad 3-\sqrt{3}, \quad 1-\sqrt{3}, \quad 0, \quad 0]$$
$$\textit{HIGHPASS:} \quad \frac{1}{4\sqrt{2}}[0, \quad 0, \quad 1-\sqrt{3}, \quad \sqrt{3}-3, \quad 3+\sqrt{3}, \quad -1-\sqrt{3}, \quad 0, \quad 0]$$

↑ *origin* (under the fifth element)

After the basis vectors have been zero-padded (if necessary), the wavelet transform is performed by doing the following:

1. Convolve the lowpass filter with the rows (remember that this is done by sliding, multiplying coincident terms, and summing the results) and save the results. (Note: For the basis vectors as given, they do *not* need to be reversed for convolution.)
2. Convolve the lowpass filter with the columns (of the results from step 1) and subsample this result by taking every other value; this gives us the lowpass-lowpass version of the image.
3. Convolve the result from step 1, the lowpass filtered rows, with the highpass filter on the columns. Subsample by taking every other value to produce the lowpass-highpass image.
4. Convolve the original image with the highpass filter on the rows and save the result.
5. Convolve the result from step 4 with the lowpass filter on the columns; subsample to yield the highpass-lowpass version of the image.
6. To obtain the highpass-highpass version, convolve the columns of the result from step 4 with the highpass filter.

In practice the convolution sum of every other pixel is not performed since the resulting values are not used. This is typically done by shifting the basis vector by 2, instead of by 1 at each convolution step. Note that with circular convolution the basis vector will overlap the extended periodic copies of the image when both the first and last convolution sums are calculated.

The convention for displaying the wavelet transform results, as an image, is shown in Figure 2.5-26. In Figure 2.5-27, we see the results of applying the wavelet transform to an image. In Figure 2.5-27b we can see the lowpass-lowpass image in the upper-left corner, the lowpass-highpass images on the diagonals, and the highpass-highpass in the lower-right corner. We can continue to run the same wavelet transform on the lowpass-lowpass version of the image to get seven subimages, as in Figure 2.5-27c, or perform it another time to get ten subimages, as in Figure 2.5-27d. This process is called *multiresolution decomposition* and can continue to achieve 13, 16, or

Figure 2.5-26 Wavelet Transform Display

LOW/ LOW	HIGH/ LOW
LOW/ HIGH	HIGH/ HIGH

Location of frequency bands in a four-band wavelet-transformed image. Designation is row/column.

Figure 2.5-27 Wavelet Transform

a. Original image.

b. Wavelet transform using Daubechies basis vectors, four bands.

c. Wavelet transform using Daubechies basis vectors, seven bands.

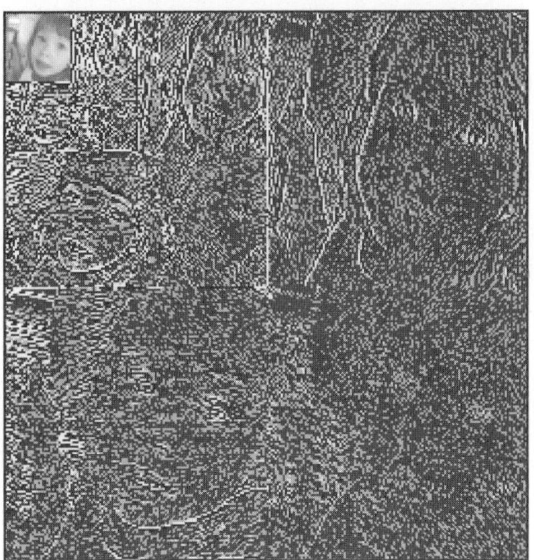

d. Wavelet transform using Daubechies basis vectors, ten bands.

as many subimages as are practical. We can see in the resulting images that the transform contains spatial information because the image itself is still visible in the transform domain. Compare this to the spectrums of the previous transforms, where there is no visible correlation to the image itself.

The inverse wavelet transform is performed by enlarging the wavelet transform data to its original size. Insert zeros between each value, convolve the corresponding (lowpass and highpass) inverse filters to each of the four subimages, and sum the results to obtain the original image. For the Haar filter, the inverse wavelet filters are identical to the forward filters; for the Daubechies example given, the inverse wavelet filters are

$$LOWPASS_{inv}: \quad \frac{1}{4\sqrt{2}}[3-\sqrt{3}, \quad 3+\sqrt{3}, \quad 1+\sqrt{3}, \quad 1-\sqrt{3}]$$

$$HIGHPASS_{inv}: \quad \frac{1}{4\sqrt{2}}[1-\sqrt{3}, \quad -1-\sqrt{3}, \quad 3+\sqrt{3}, \quad -3+\sqrt{3}]$$

The use of the wavelet transform is increasingly popular for image compression, a very active research area today. The computer revolution, along with the advent of the information superhighway, multimedia applications, and high-definition television, all contribute to the high level of interest in image compression. The multiresolution decomposition property of the wavelet transform, which separates low-resolution information from more detailed information, is useful in applications where it is desirable to have coarse information available fast such as perusing an image database or progressively transmitting images on the World Wide Web. The wavelet transform is one of the relatively new transforms being explored for image compression applications; one application described in Chapter 7 applies the wavelet transform to image compression for the transmission of medical images.

2.6 Feature Extraction and Analysis

The goal in image analysis is to extract information useful for solving application-based problems. This is done by intelligently reducing the amount of image data with the tools we have explored, including image segmentation and transforms. After we have performed these operations, we have modified the image from the lowest level of pixel data into higher-level representations. Now, we can consider extraction of features that can be useful for solving computer imaging problems. Image segmentation allows us to look at object features, and the image transforms provide us with features based on spatial frequency information—spectral features. The object features of interest include the geometric properties of binary objects, histogram features, and color features. After we have extracted the features of interest, we can analyze the image (see Section 6.3.1 for an example of feature extraction with CVIPtools).

Exactly what we do with the features will be application dependent. If we are working on a computer vision problem, the end goal may be the generation of a classification rule in order to identify objects. If we are working to develop a new image compression algorithm, we may want to determine what image data are important; the insignificant information can be compressed or eliminated completely. For image restoration we may want to determine the type of noise that exists in the image or how the image has been degraded. Image enhancement may help us to solve an image analysis problem by allowing us to analyze images that are visually pleasing.

As indicated in Figure 2.1-3, feature extraction is part of the data reduction process and is followed by feature analysis. One of the important aspects of feature analysis is to determine exactly which features are important, so the analysis is not complete until we incorporate application-specific feedback into the system. In this section we will discuss feature extraction and analysis as well as provide a brief introduction to pattern classification. Although pattern classification is used primarily in computer vision applications, it can be helpful in solving any type of computer imaging problem.

2.6.1 Feature Vectors and Feature Spaces

A feature vector is one method to represent an image, or part of an image (an object), by finding measurements on a set of features. The *feature vector* is an n-dimensional vector that contains these measurements. The measurements may be one of three types: 1) ordinal (discrete/ranked), 2) nominal (categorical), or 3) continuous numerical. This vector can be used to classify an object, or provide us with condensed higher-level image information. Associated with the feature vector is a mathematical abstraction called a *feature space*, which is also n-dimensional.

EXAMPLE 2-16

We are working on a computer vision problem for robotic control. We need to control a robotic gripper that picks parts from an assembly line and puts them into boxes. In order to do this, we need to determine: 1) where the object is in the two-dimensional plane in which the objects lie, and 2) what type of object it is—one type goes into Box A; another type goes into Box B. First, we define the feature vector that will solve this problem. We determine that knowing the area and center of area of the object, defined by an (r, c) pair, will locate it in space. We determine that if we also know the perimeter we can identify the object. So our feature vector contains four feature measures, and the feature space is four-dimensional. We can define the feature vector as [area, r, c, perimeter].

When we understand the concept of a feature vector and a feature space, we can consider methods used to compare two feature vectors. The primary methods are to either measure the difference between the two or to measure the similarity. The difference can be measured by a *distance measure* in the n-dimensional space; the bigger the distance between two vectors, the greater the difference. *Euclidean distance* is the most common metric for measuring the distance between two vectors and is given by the square root of the sum of the squares of the differences between vector components. Given two vectors **A** and **B**, where $A = [a_1\, a_2\, a_3 \ldots a_n]$ and $B = [b_1\, b_2\, b_3 \ldots b_n]$, then the Euclidean distance is given by

$$\sqrt{\sum_{i=1}^{n}(a_i - b_i)^2} = \sqrt{(a_1 - b_1)^2 + (a_2 - b_2)^2 + (a_3 - b_3)^2 + \ldots + (a_n - b_n)^2}$$

This measure may be biased as a result of the varying range on different components of the vector. For example, one component may only range from 1 to 5 and another may range from 1 to 5000, so a difference of 5 for the first component will be maximum, but a difference of 5 for the second feature may be insignificant. We can alleviate this problem by using the *range-normalized Euclidean distance*, with **A** and **B** defined as before, given by

$$\sqrt{\sum_{i=1}^{n} \frac{(a_i - b_i)^2}{R_i^2}}$$

where R_i = the range of the ith component, which is the maximum value for that component minus the minimum value for that component.

Another distance measure, called the *city block* or *absolute value metric*, is defined as follows (using A and B as before):

$$\sum_{i=1}^{n} |a_i - b_i|$$

This metric is computationally faster than the Euclidean distance but gives similar results. The city block distance can also be range-normalized to give a *range-normalized city block* distance metric, with R_i defined as before:

$$\sum_{i=1}^{n} \left| \frac{a_i - b_i}{R_i} \right|$$

The final distance metric considered here is the *maximum value* metric defined by

$$\max\{|a_1 - b_1|, |a_2 - b_2|, \ldots, |a_n - b_n|\}$$

We can see that this will measure the vector component with the maximum distance, and it can be range-normalized as with the other distance measures as follows:

$$\max\left\{\frac{|a_1 - b_1|}{R_1}, \frac{|a_2 - b_2|}{R_2}, \ldots, \frac{|a_n - b_n|}{R_n}\right\}$$

The second type of metric used for comparing two feature vectors is the *similarity measure*. The most common form of the similarity measure is one that we have already seen, the *vector inner product*. Using our definitions for the two vectors **A** and **B**, we can define the vector inner product by the following equation:

$$\sum_{i=1}^{n} a_i b_i = (a_1 b_1 + a_2 b_2 + \ldots + a_n b_n)$$

This similarity measure can also be range normalized:

$$\sum_{i=1}^{n} \frac{a_i b_i}{R_i^2} = \left(\frac{a_1 b_1}{R_1^2} + \frac{a_2 b_2}{R_2^2} + \ldots + \frac{a_n b_n}{R_n^2} \right)$$

Alternately, we can normalize this measure by dividing each vector component by the magnitude of the vector:

$$\sum_{i=1}^{n} \frac{a_i b_i}{\sqrt{\sum_{j=1}^{n} a_j^2} \sqrt{\sum_{j=1}^{n} b_j^2}} = \frac{(a_1 b_1 + a_2 b_2 + \ldots + a_n b_n)}{\sqrt{\sum_{j=1}^{n} a_j^2} \sqrt{\sum_{j=1}^{n} b_j^2}}$$

A commonly used statistically based method to normalize these measures is to take each vector component and subtract the mean and divide by the standard deviation. This method can be applied to any of the measures, both distance and similarity, but requires knowledge of the probability distribution of the feature measurements. In practice the probability distributions are often estimated by using the existing data.

When selecting features for use in a computer imaging application, an important factor is the robustness of a feature. A feature is *robust* if it will provide consistent results across the entire application domain. For example, if we are developing a system to work under any lighting conditions, we do not want to use features that are lighting dependent—they will not provide consistent results in the application domain. Another type of robustness, especially applicable to object features, is called RST-invariance, where the RST means rotation, size, and translation. A very robust feature will be RST-invariant, meaning that if the image is rotated, shrunk or enlarged, or translated (shifted left/right or up/down), the value for the feature will not change. As we explore the binary object features, consider the invariance of each feature to these simple geometric operations.

2.6.2 Binary Object Features

To extract object features, we need an image that has undergone image segmentation and any necessary morphological filtering. This will provide us with clearly defined objects, which can then be labeled and processed independently. After all the objects in the image are labeled, we can treat each object as a binary image by assuming that the labeled object has a value of '1' and that everything else is '0'. The labeling process requires us to first define the desired connectivity and then to scan the image and label connected objects with the same symbol.

If we chose six-connectivity, with the NW and SE diagonal neighbors, we can apply the algorithm given in Figure 2.6-1 to label the objects in the image (also see Computer Vision Lab Exercise #2 in Chapter 8). For this algorithm we assume that any areas not of interest in the image have been masked out by setting the pixels equal to zero. The UPDATE block in the flowchart refers to a function that will keep track of objects that have been given multiple labels. This can occur with a sequential scanning of the image if the connecting pixels are not encountered until after different parts of the object have already been labeled (see Figure 2.6-2).

After we have labeled the objects, we have an image filled with object numbers. This image is used to extract the features of interest. The binary object features we will define include area, center of area, axis of least second moment, perimeter, Euler number, projections, thinness ratio, and aspect ratio. The first four tell us something about where the object is, and the latter four tell us something about the shape of the object.

In order to provide general equations for area, center of area, and axis of least second moment, we define a function, $I_i(r, c)$:

$$I_i(r, c) = \begin{cases} 1 & \text{if } I(r, c) = i\text{th } \textit{object number} \\ 0 & \textit{otherwise} \end{cases}$$

Now we can define the *area* of the ith object as

Figure 2.6-1 Labeling Algorithm

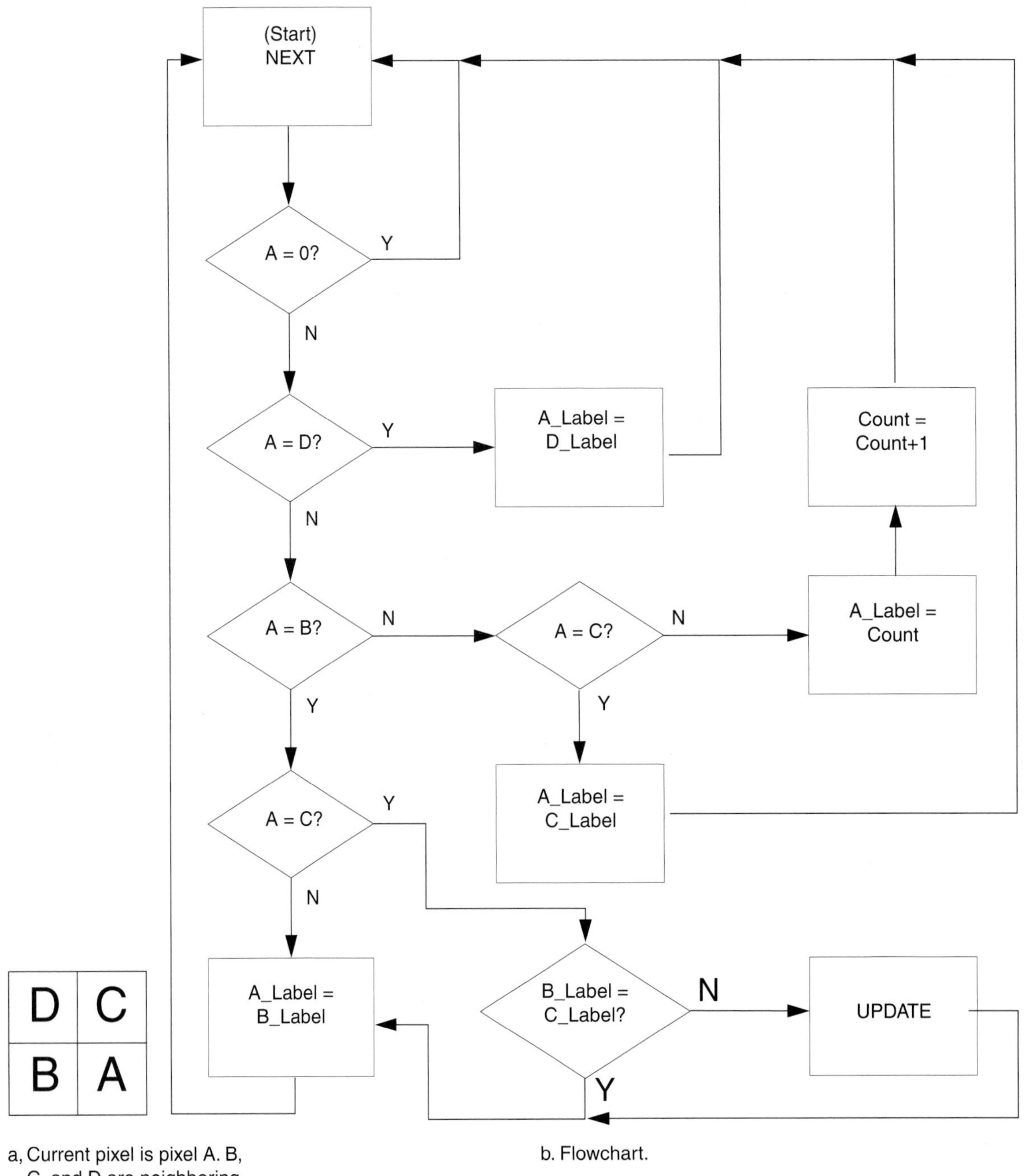

a, Current pixel is pixel A. B, C, and D are neighboring pixels, as shown.

b. Flowchart.

Figure 2.6-2 Multiple Labels

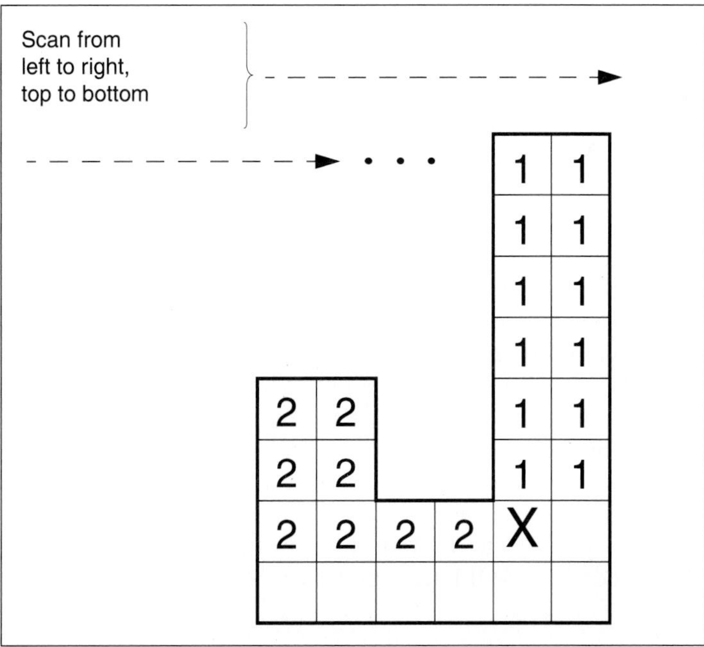

The labeling algorithm requires an UPDATE function to keep track of objects with more than one label. Multiple labeling can occur during sequential scanning, as shown above on the *J* shaped object. We label two different objects until we reach the pixel marked X, where we discover that objects 1 and 2 are connected.

$$A_i = \sum_{r=0}^{N-1} \sum_{c=0}^{N-1} I_i(r, c)$$

The area A_i is measured in pixels and indicates the relative size of the object (see Computer Vision Exercise #1 in Chapter 8). We can then define the *center of area* for an object by the pair $(\bar{r}_i, \bar{c}_i)$:

$$\bar{r}_i = \frac{1}{A_i} \sum_{r=0}^{N-1} \sum_{c=0}^{N-1} rI_i(r, c)$$

$$\bar{c}_i = \frac{1}{A_i} \sum_{r=0}^{N-1} \sum_{c=0}^{N-1} cI_i(r, c)$$

These correspond to the row coordinate of the center of area for the ith object $\bar{r}_i$ and the column coordinate of the center of area for the ith object $\bar{c}_i$. This feature will help to locate an object in the two-dimensional image plane. The next feature we will consider, the *axis of least second moment*, provides information about the object's orientation. This axis corresponds to the line about which it takes the least amount of energy to spin an object of like shape or the axis of least inertia. If we move our origin to the center of area (r, c), the axis of least second moment is defined as follows:

$$\tan(2\theta_i) = 2\frac{\sum_{r=0}^{N-1}\sum_{c=0}^{N-1} rcI_i(r, c)}{\sum_{r=0}^{N-1}\sum_{c=0}^{N-1} r^2 I_i(r, c) - \sum_{r=0}^{N-1}\sum_{c=0}^{N-1} c^2 I_i(r, c)}$$

This is shown in Figure 2.6-3. The origin is moved to the center of area for the object, and the angle is measured from the r-axis counterclockwise.

Knowing the *perimeter* of the object can help us to locate it in space and provide us with information about the shape of the object. The perimeter can be found in the original binary image by counting the number of '1' pixels that have '0' pixels as neighbors. Perimeter can also be found by applying an edge detector to the object, followed by counting the '1' pixels. Note that counting the '1' pixels is the same as finding the area, but in this case we are finding the "area" of the border. Because the digital images are typically mapped onto a square grid, curved outlines tend to be jagged, so these methods only give an estimate to the actual perimeter. An improved estimate to the perimeter can be found by multiplying the results from either of the preceding methods by $\pi/4$. If better accuracy is required, more complex methods that use chain

Figure 2.6-3 Axis of Least Second Moment

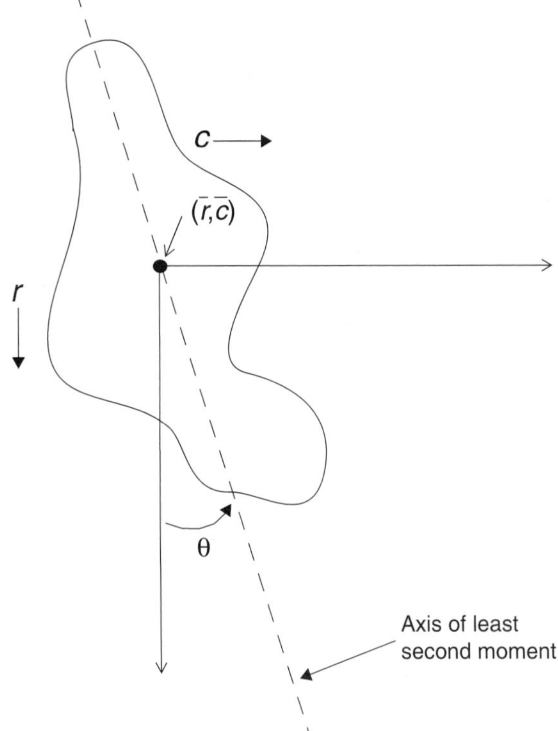

codes for finding perimeter can be used (see references). An illustration of perimeter is shown in Figure 2.6-4.

After we have found perimeter P and area A, we can calculate the *thinness ratio* T:

$$T = 4\pi\left(\frac{A}{P^2}\right)$$

This measure has a maximum value of 1, which corresponds to a circle, so this also is used as a measure of roundness. The closer to 1, the more like a circle the object is. As the perimeter becomes larger relative to the area, this ratio decreases, and the object is getting thinner. This metric is also used to determine the regularity of an object: regular objects have higher thinness ratios than similar but irregular objects. The inverse of this metric $1/T$ is sometimes called the *irregularity* or *compactness* ratio. A related feature is the *aspect ratio* (also called *elongation* or *eccentricity*), defined by the ratio of the bounding box of an object. This can be found by scanning the image and finding the minimum and maximum values on the row and columns where the object lies. This ratio is then defined by

$$\frac{c_{max} - c_{min} + 1}{r_{max} - r_{min} + 1}$$

The *Euler number* of an image is defined as the number of objects minus the number of holes. For a single object, it tells us how many closed curves the object contains. It is often useful in tasks such as optical character recognition (OCR), as shown by the example in Figure 2.6-5. Using the connectivity definition we defined when we labeled the image, we can find the Euler number by finding *convexities* and *concavities*. The Euler number will be equal to the number of convexities minus the number of concavities, which are found by scanning the image for the patterns that follow.

Figure 2.6-4 Perimeter

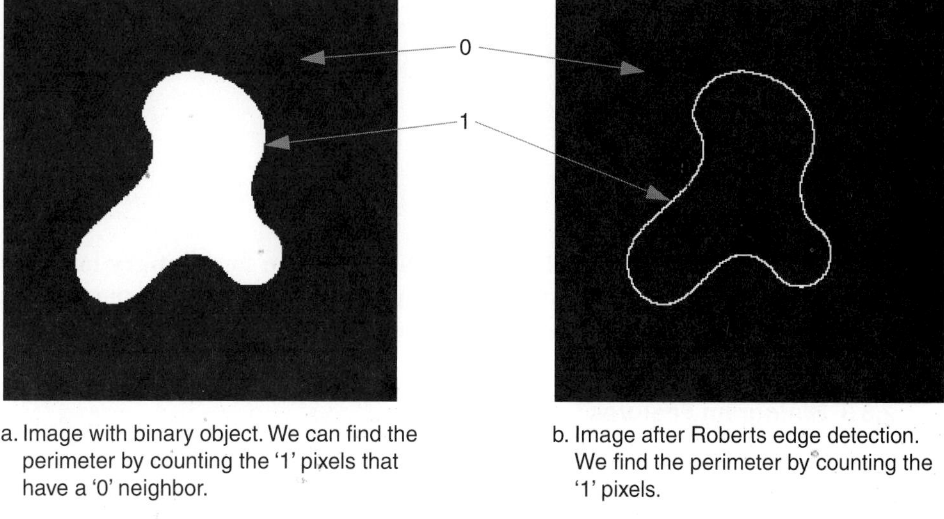

a. Image with binary object. We can find the perimeter by counting the '1' pixels that have a '0' neighbor.

b. Image after Roberts edge detection. We find the perimeter by counting the '1' pixels.

Figure 2.6-5 Euler Number

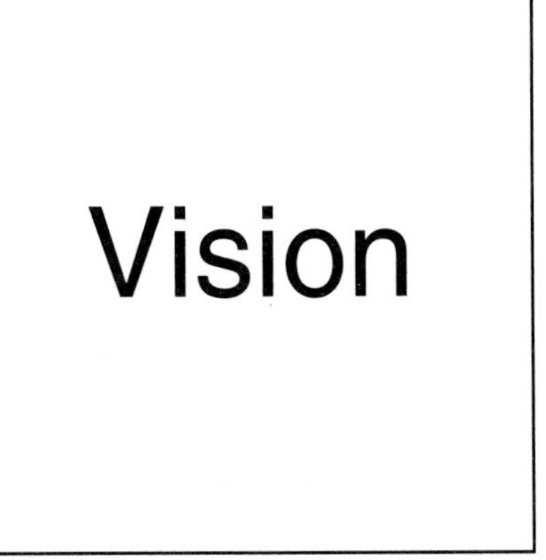

a. This image has eight objects and one hole, so its Euler number is 8 – 1 = 7. The letter *V* has Euler number of 1, *i* = 2, *s* = 1, *o* = 0, and *n* = 1.

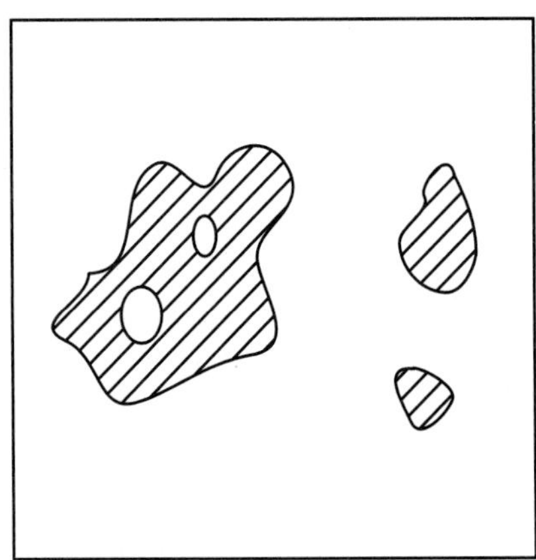

b. This image has three objects and two holes, so the Euler number is 3 – 2 = 1.

CONVEXITIES CONCAVITIES

$$\begin{bmatrix} 0 & 0 \\ 0 & 1 \end{bmatrix} \qquad \begin{bmatrix} 0 & 1 \\ 1 & 1 \end{bmatrix}$$

The number of convexities and concavities can also be useful features for binary objects (see Computer Vision Lab Exercise #3 in Chapter 8).

The *projections* of a binary object, which also provide shape information, are found by summing all the pixels along rows or columns. If we sum the rows we have the *horizontal projection*; if we sum the columns we have the *vertical projection*. We can define the horizontal projection $h_i(r)$ as

$$h_i(r) = \sum_{c=0}^{N-1} I_i(r, c)$$

and the vertical projection $v_i(c)$

$$v_i(c) = \sum_{r=0}^{N-1} I_i(r, c)$$

An example of the horizontal and vertical projection for a binary image is shown in Figure 2.6-6. Projections are useful in applications like character recognition, where the objects of interest can be normalized with regard to size.

2.6.3 Histogram Features

The *histogram* of an image is a plot of the gray-level values versus the number of pixels at that value. The shape of the histogram provides us with information about the nature of the image, or subimage if we are considering an object within the image.

Figure 2.6-6 Projection

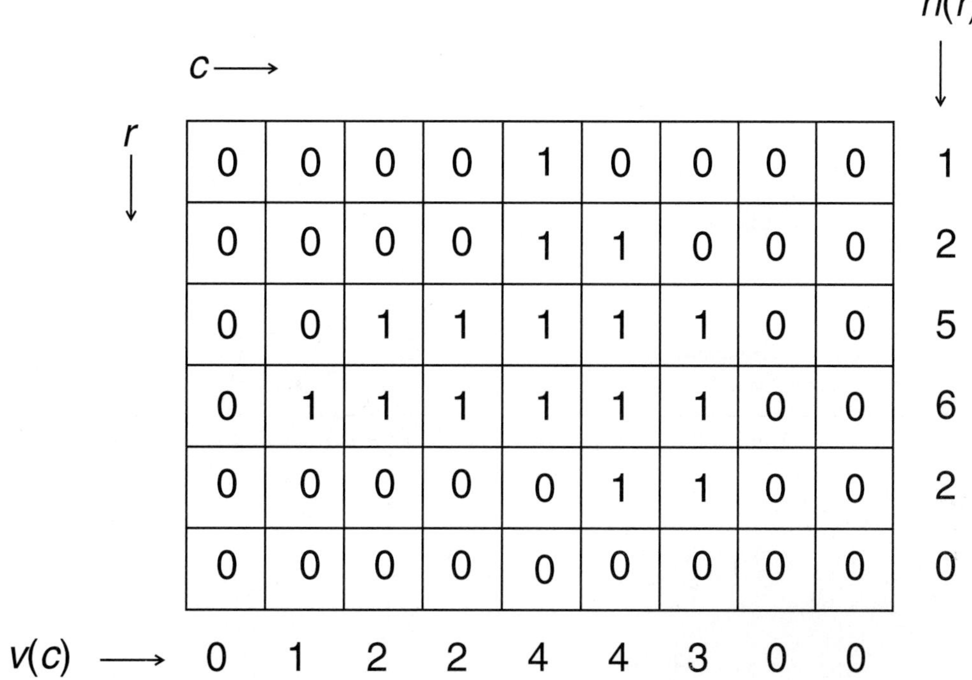

h(r)
c
r
0 0 0 0 1 0 0 0 0 1
0 0 0 0 1 1 0 0 0 2
0 0 1 1 1 1 1 0 0 5
0 1 1 1 1 1 1 0 0 6
0 0 0 0 0 1 1 0 0 2
0 0 0 0 0 0 0 0 0 0
v(c)
0 1 2 2 4 4 3 0 0

To find the projections, we sum the number of 1s in the rows and columns.

For example, a very narrow histogram implies a low-contrast image, a histogram skewed toward the high end implies a bright image, and a histogram with two major peaks, called bimodal, implies an object that is in contrast with the background. Examples of the different types of histograms are shown in Figure 2.6-7.

The histogram features that we will consider are statistically based features, where the histogram is used as a model of the probability distribution of the gray levels. These statistical features provide us with information about the characteristics of the gray-level distribution for the image or subimage. We define the first-order histogram probability $P(g)$ as

$$P(g) = \frac{N(g)}{M}$$

M is the number of pixels in the image or subimage (if the entire image is under consideration, then $M = N^2$ for an $N \times N$ image), and $N(g)$ is the number of pixels at gray level g. As with any probability distribution, all the values for $P(g)$ are less than or equal to 1, and the sum of all the $P(g)$ values is equal to 1. The features based on the first-order histogram probability are the mean, standard deviation, skew, energy, and entropy.

The *mean* is the average value, so it tells us something about the general brightness of the image. A bright image will have a high mean, and a dark image will have a low mean. We will use L as the total number of gray levels available, so the gray levels range from 0 to $L - 1$. For example, for typical 8-bit image data, L is 256 and ranges from 0 to 255. We can define the mean as follows:

Figure 2.6-7 Histograms

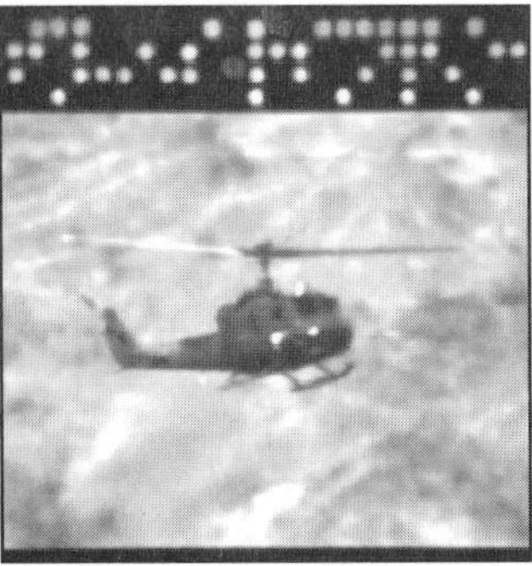

a. Object in contrast with background.

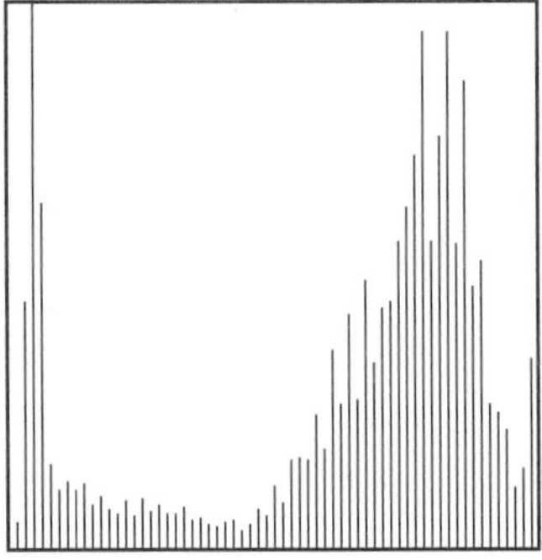

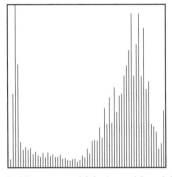

b. Histogram of (a) shows bimodal shape.

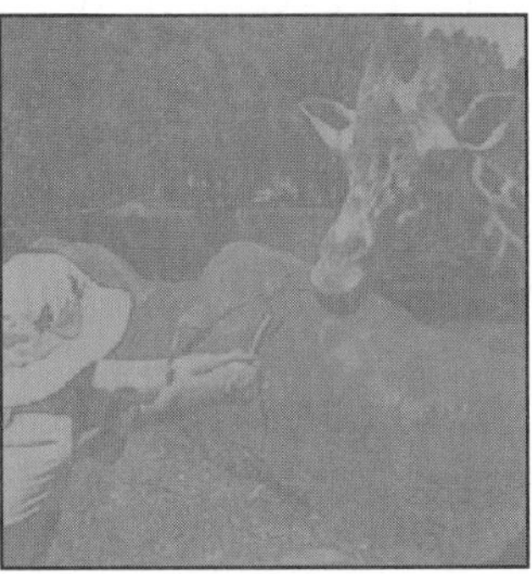

c. Low-contrast image.

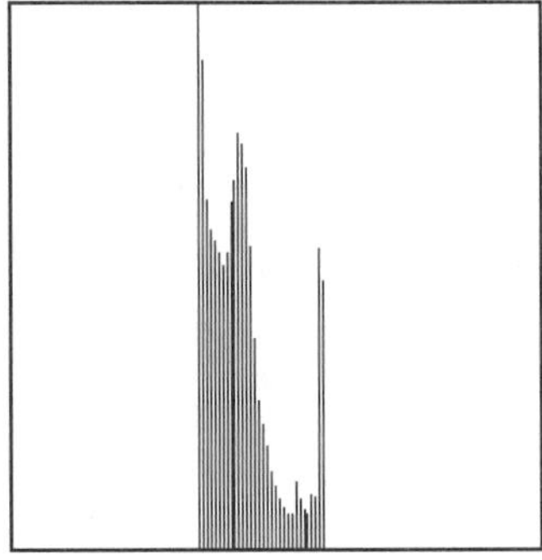

d. Histogram of (c) appears clustered.

e. High-contrast image.

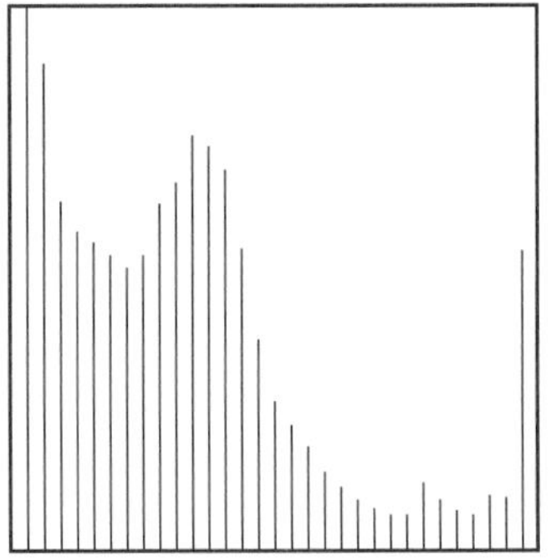

f. Histogram of (e) appears spread out.

Figure 2.6-7 (Continued)

g. Bright image.

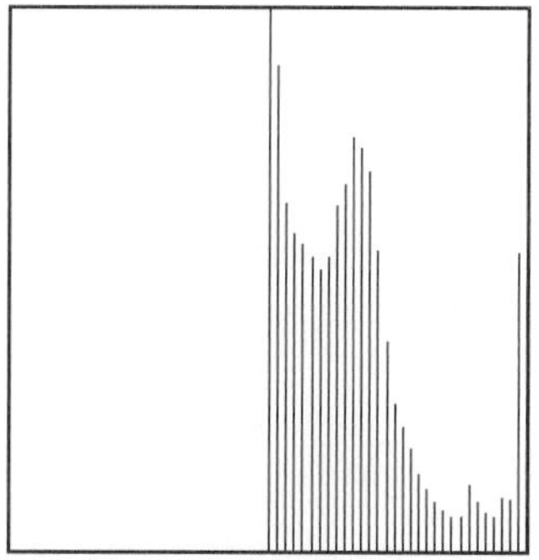

h. Histogram of (g) appears shifted to the right.

i. Dark image.

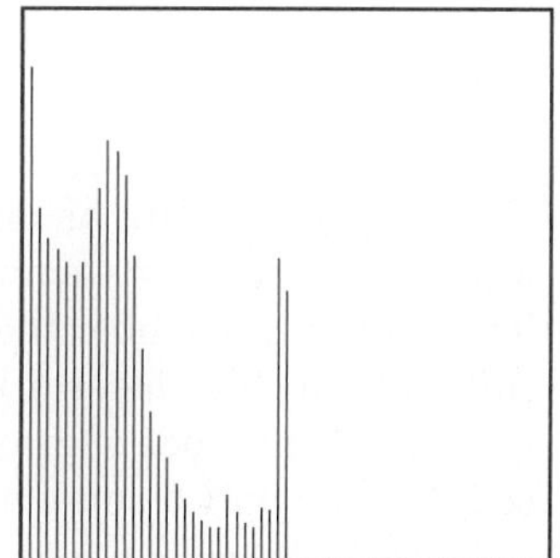

j. Histogram of (i) appears shifted to the left.

$$\bar{g} = \sum_{g=0}^{L-1} gP(g) = \sum_{r}\sum_{c} \frac{I(r, c)}{M}$$

If we use the second form of the equation, we sum over the rows and columns corresponding to the pixels in the image or subimage under consideration.

The *standard deviation*, which is also known as the square root of the variance, tells us something about the contrast. It describes the spread in the data, so a high-contrast image will have a high variance, and a low-contrast image will have a low variance. It is defined as follows:

$$\sigma_g = \sqrt{\sum_{g=0}^{L-1} (g - \bar{g})^2 P(g)}$$

The *skew* measures the asymmetry about the mean in the gray-level distribution. It is defined as

$$\text{SKEW} = \frac{1}{\sigma_g^3} \sum_{g=0}^{L-1} (g - \bar{g})^3 P(g)$$

Another method to measure the skew uses the mean, mode, and standard deviation, where the *mode* is defined as the peak, or highest, value:

$$\text{SKEW}' = \frac{\bar{g} - \text{mode}}{\sigma_g}$$

This method of measuring skew is more computationally efficient, especially considering that, typically, the mean and standard deviation have already been calculated.

The *energy* measure tells us something about how the gray levels are distributed:

$$\text{ENERGY} = \sum_{g=0}^{L-1} [P(g)]^2$$

The energy measure has a maximum value of 1 for an image with a constant value and gets increasingly smaller as the pixel values are distributed across more gray-level values (remember that all the $P(g)$ values are less than or equal to 1). The larger this value is, the easier it is to compress the image data. If the energy is high, it tells us that the number of gray levels in the image is few, that is, the distribution is concentrated in only a small number of different gray levels.

The *entropy* is a measure that tells us how many bits we need to code the image data and is given by

$$\text{ENTROPY} = -\sum_{g=0}^{L-1} P(g) \log_2[P(g)]$$

As the pixel values in the image are distributed among more gray levels, the entropy increases. This measure tends to vary inversely with the energy.

Second-order histogram features based on a joint probability distribution model can also be derived. The *second-order histogram* provides statistics based on pairs of pixels and their corresponding gray levels. The second-order histogram methods are also referred to as *co-occurrence* or *dependency matrix* methods. The features derived via these methods, such as correlation, energy, entropy, and inertia (contrast), are used in identifying texture within an image. More details on these types of features can be found in the references.

2.6.4 Color Features

Color is useful in many applications. Typical color images consisting of three color planes—red, green, and blue—can be treated as three separate gray-scale images. This approach allows us to use any of the object or histogram features previously defined, but with three times as many features, one for each color band.

Alternately, we may want to incorporate information into our feature vector pertaining to the relationship *between* the color bands. These relationships are found by considering normalized color, or color differences. This is done by using the color transforms defined in Chapter 1 and then applying to this new representation the features previously defined. For example, the chromaticity transform provides a normalized color representation, which will decouple the image brightness from the color itself. Many color transforms, including HSL, Spherical, Cylindrical, Lu*v*, and La*b*, will provide us with two color components and a brightness component. The YIQ and YCrCb provide us with color difference components that signify the relative color.

The color features chosen will be primarily application specific, but caution must be taken in selecting color features. Typically, some form of relative color is best because most absolute color measures are not very robust. In many applications the environment is not carefully controlled, so a system developed under specific color conditions using absolute color may not function properly in a different environment. Remember all the factors that contribute to the color—the lighting, the sensors, any optical filtering, any print or photographic process in the system model. If any of these factors change, then any absolute color measures such as red, green, or blue will change.

An example of the problem caused by using absolute color arose during development of a system to diagnose skin tumors automatically. An algorithm was found that seemed to always correctly identify melanoma (a deadly form of skin cancer). At one point in the research, the algorithm ceased to work. What had happened? A big mistake had been made in developing the algorithm—it had relied on some absolute color measures. The initial set of melanoma images had been digitized from Ektachrome slides, and the nonmelanoma tumor images had been digitized from Kodachrome slides. Because of the types of film involved, all the melanomas had a blue tint (Ektachrome), whereas all the other tumor images had a red tint (Kodachrome). Thus, with the first set of tumor images, the use of average color alone provided an easy way to differentiate between the melanoma and nonmelanoma tumors. As more tumor images became available, both melanoma and nonmelanoma tumors were digitized from Kodachrome (red tint), so the identification algorithm ceased to work. A senior member of the research team had a similar experience while developing a tank recognition algorithm based on Ektachrome images of Soviet tanks and Kodachrome images of U.S. tanks. Avoid absolute color measures for features, except under very carefully controlled conditions.

2.6.5 Spectral Features

With regard to spectral features, or frequency/sequency-domain-based features, the primary metric is *power*. How much power do we find in various parts of the spectrum? Texture is often measured by looking for peaks in the power spectrum, especially if the texture is periodic or directional. The power spectrum is defined by the magnitude of the spectral components squared:

$$\text{POWER} = |T(u, v)|^2$$

Figure 2.6-8 Spectral Regions

Fourier Transform Symmetry
x = origin

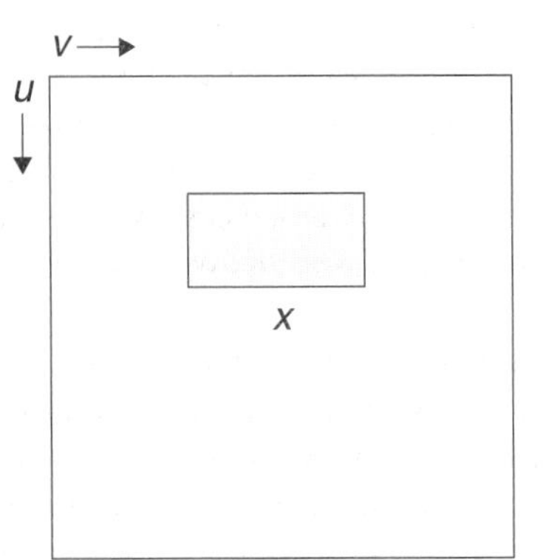

a. Box is defined by limits on u and v.

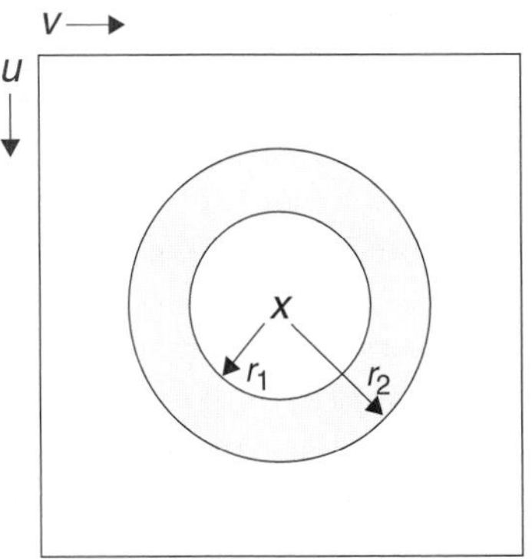

b. Ring is defined by limits on the radii from origin x.

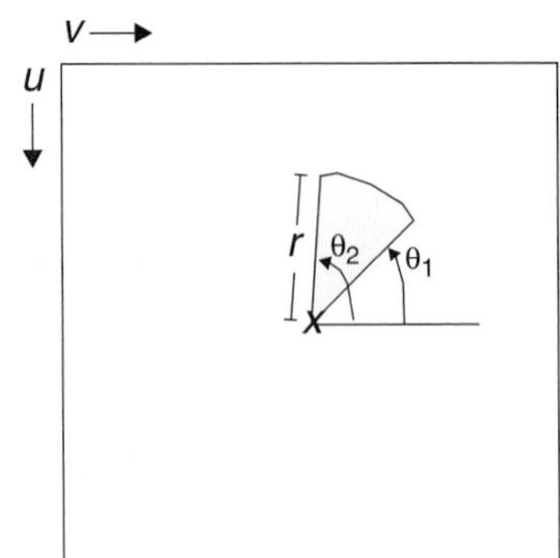

c. Sector is defined by radius r and angles θ_1 and θ_2.

Cosine and Walsh-Hadamard Transform Symmetry
x = origin

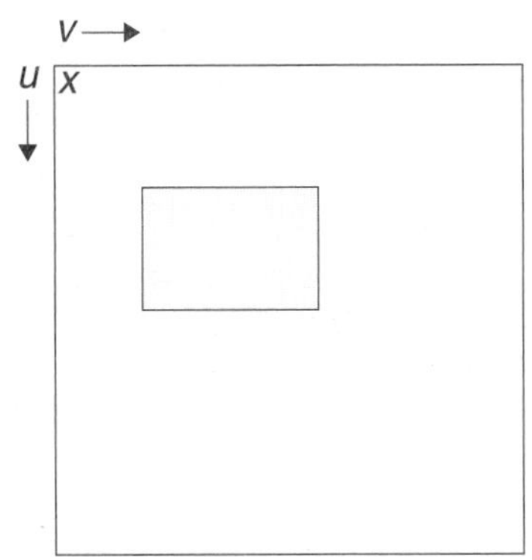

d. Box symmetry.

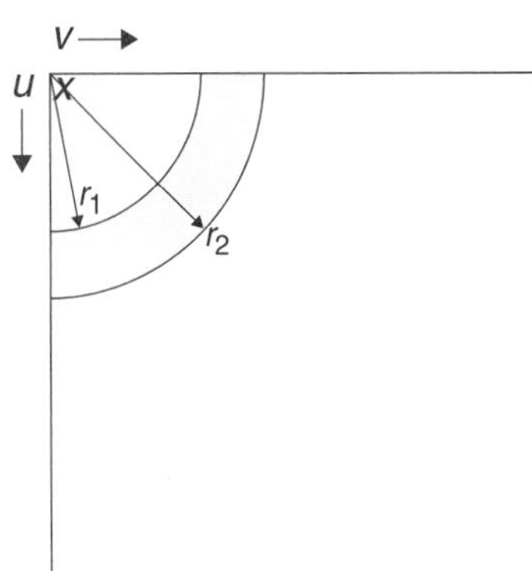

e. Ring symmetry.

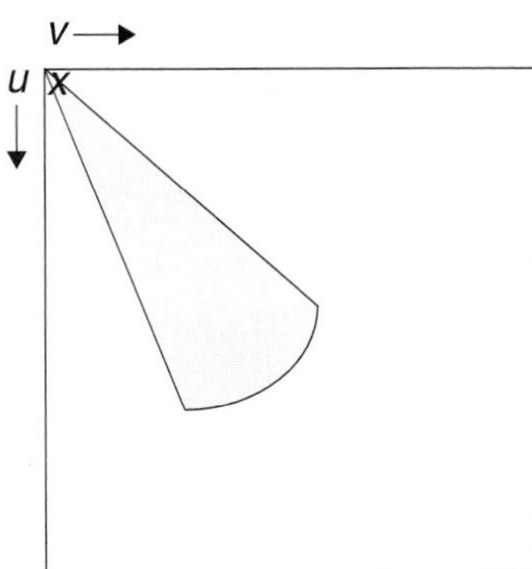

f. Sector symmetry.

It is typical to look at power in various spectral regions, and these regions can be defined as rings, sectors, or boxes. In Figure 2.6-8 we see examples of these types of spectral regions, for both types of symmetry that we have considered. We then measure the power in a region of interest by summing the power over the range of frequencies of interest:

$$\text{SPECTRAL REGION POWER} = \sum_{u \in \text{REGION}} \sum_{v \in \text{REGION}} |T(u, v)|^2$$

The *box* is the easiest to define, by setting limits on u and v.

EXAMPLE 2-17

We may be interested in all vertical spatial frequencies at a specific horizontal frequency, $v = 20$. So we define a spectral region as

$$\text{Region of Interest} = \begin{cases} -\dfrac{N}{2} < u < \dfrac{N}{2} \\ 19 < v < 21 \end{cases}$$

Then we calculate the power in this region by summing over this range of u and v. Note that u should vary from 0 to $N-1$ for non-Fourier symmetry.

The *ring* is defined by two radii, r_1 and r_2. These are measured from the origin, and the summation limits on u and v, for Fourier symmetry, are

$$-r_2 \le u < r_2$$

$$\pm\sqrt{r_1^2 - u^2} \le v < \pm\sqrt{r_2^2 - u^2}$$

(Note: For non-Fourier symmetry u will range from 0 to r_2, and v ranges over the positive square roots only.) The *sector* is defined by a radius r and two angles θ_1 and θ_2. The limits on the summation are defined by

$$\theta_1 \le \tan^{-1}\left(\frac{v}{u}\right) < \theta_2$$

$$u^2 + v^2 \le r^2$$

The sector measurement will find spatial frequency power of a specific orientation whatever the frequency (limited only by the radius), whereas the ring measure will find spatial frequency power at specific frequencies regardless of orientation. For example, the radial measure can be used to find texture. High power in small radii corresponds to smooth textures; high power in large radii corresponds to coarse texture. The sector power measure can be used to find lines or edges in a given direction, but the results are size invariant. Because of the redundancy in the Fourier spectral symmetry, we often measure sector power over half the spectrum, and the ring power over the other half of the spectrum (see Figure 2.6-9). In practice we may want to normalize these numbers, because they get very large, by dividing by the DC (average) value (note that to get the true DC power, we need to divide by $N*N$, due to the definition of the DFT).

2.6.6 Pattern Classification

Pattern classification, as related to image analysis, involves taking the features extracted from the image and using them to classify image objects automatically. This is done by developing classification algorithms that use the feature information. We use the distance or similarity measures defined in Section 2.6.1 for comparing different objects and their feature vectors.

Figure 2.6-9 Fourier Spectrum Power

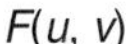

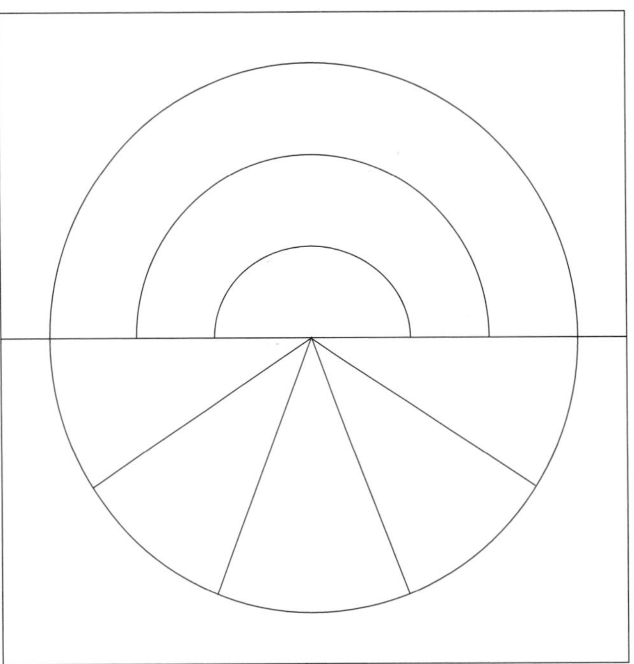

With Fourier spectrum symmetry, which contains redundant information, we often measure ring power over half the spectrum and sector power over the other half.

The primary uses of pattern classification in image analysis are for computer vision and image compression applications development. It can be considered a part of feature analysis, or as a postprocessing step to feature extraction and analysis. Pattern classification is typically the final step in the development of a computer vision algorithm because in these types of applications the goal is to identify objects (or parts of objects) in order for the computer to perform some vision-related task. These tasks range from computer diagnosis of medical images to object classification for robot control. In the case of image compression, we want to remove redundant information from the image and compress important information as much as possible. One way to compress information is to find a higher-level representation of it, which is exactly what feature analysis and pattern classification is all about—finding a single class that will represent many pixel values.

To develop a classification algorithm, we need to divide our data into a training set and a test set. This is done so that we use one set of images to develop the classification scheme and a separate set to test the classification algorithm. In order for this to work properly, both the training and test sets should represent the images that will be seen in the application domain. The use of two distinct sets of images provides us with test results that are unbiased that will be a good predictor of the success we can expect to achieve with the classification algorithm in the actual application.

The selection of the two sets should be done before development starts, to avoid biasing the test results. The size of the sets depends on many factors, but in practice we typically split the available images into two equally sized groups. Theoretically, we want to maximize the size of the training set to develop the best algorithm, but the larger the test set is, the more confidence we have that the results are indicative of application success. If time allows, it is often instructive to use increasingly larger training sets (randomly selected) and analyze the results. What we expect to achieve is a monotonically increasing success rate as the training set size increases. If this does not happen, we need to verify that our training set(s) actually represents the domain of interest.

Another important concept in the development of a pattern classification scheme is the idea of a *cost function*. Often we may not want to rely on correct classification as the sole criterion in evaluating success of a classification system because some types of misclassification may be more "costly" than others. For example, if we are developing a medical system to diagnose cancer, the cost of mistakenly identifying a malignant tumor as benign is much higher than the cost of identifying a benign tumor as malignant. In the first case, the patient dies, whereas in the second case the patient is subjected to some temporary stress but survives. Or consider a system to identify land mines. What are the relevant cost functions?

After the data have been divided into the training and test sets, work can begin on the development of the classification algorithm. Many methods are available for this; we will consider some basic representative methods. The general approach is to use the information in the training set to classify the "unknown" example in the test set. It is assumed that all the examples available have a known classification. The success rate is measured by correct classification, but how various misclassifications contribute to the failure rate must be weighted by a cost function.

The simplest method for identifying an example from the test set is called the *nearest neighbor* method. The object of interest is compared to every example in the training set, using either a distance measure, a similarity measure, or a combination of measures. The "unknown" object is then identified as belonging to the same class as the closest example in the training set. This is indicated by the smallest number if using a distance measure or the largest number if using a similarity measure. This process is computationally intensive and not very robust.

We can make the nearest neighbor method more robust by selecting not just the vector it is closest to but a group of close feature vectors. This is called the *K-nearest neighbor* method, where, for example, $K = 5$. Then we assign the unknown feature vector to the class that occurs most often in the set of K-Neighbors. This is still very computationally intensive because we must compare each unknown example to every example in the training set, and we want the training set as large as possible to maximize success.

We can reduce this computational burden by using a method called *nearest centroid*. Here, we find the centroids for each class from the examples in the training set, and then we compare the unknown samples to the representative centroids only. The centroids are calculated by finding the average value for each vector component in the training set.

EXAMPLE 2-18

Suppose that we have a training set of four feature vectors, and we have two classes.

Class 1: example 1, [3,4,7]; example 2, [1,7,6]

Class 2: example 1, {4,2,9]; example 2, [2,3,3]

The representative vector, centroid, for class 1 is

[(3 + 1)/2, (4 + 7)/2, (7 + 6)/2] = [2, 5.5, 6.5]

The representative vector, centroid, for class 2 is

[(4 + 2)/2, (2 + 3)/2, (9 + 3)/2] = [3,2 2.5, 6]

To identify an unknown example, we need only compare it to these two representative vectors, not the entire training set.

Template matching is a pattern classification method that uses the raw image data as a feature vector (see Computer Vision Lab Exercise #7 in Chapter 8). A template is devised, possibly via a training set, which is then compared to subimages by using a distance or similarity measure. Typically, a threshold is set on this measure to determine when we have found a match, that is, a successful classification.

Many more sophisticated methods for pattern classification are available, including methods that use probability density models of the classes, artificial intelligence approaches, fuzzy logic approaches, and neural networks. More information can be found on these methods in the references.

2.7 References

The method of zooming via convolution masks is described in [Sid-Ahmed 95]. For spatial filtering, [Galbiati 90] and [Myler/Weeks 93] contain additional information. More detailed information regarding edge detection can be found in [Pratt 91], [Ballard/Brown 82], [Haralick/Shapiro 92], and [Horn 86]. The Hough Transform as described here can be found in [Gonzalez/Woods 92]. Details on improved gray-scale (IGS) quantization can be found in [Gonzalez/Woods 92].

The definitions for connectivity are described in [Horn 86], and further information can be found in [Haralick/Shapiro 92]. The PCT/Median and SCT/Center image segmentation methods presented are described in [Umbaugh 90] and applied in [Umbaugh/Moss/Stoecker 89], [Umbaugh/Moss/Stoecker 93], and [Umbaugh/Moss/Stoecker 92]. More information about image segmentation methods can be found in [Haralick/Shapiro 92], [Schalkoff 89], [Castleman 96], and [Jain/Kasturi/Schnuck 95]. A good description of image morphology is found in [Jain/Kasturi/Schnuck 95], [Baxes 94], and [Myler/Weeks 93], and more extensive information is found in [Haralick/Shapiro 92], [Giardina/Dougherty 88], and [Schalkoff 89]. The iterative method to morphological filtering is described in [Horn 86].

For the section on discrete transforms, many excellent texts are available, including [Bracewell 95], [Gonzalez/Woods 92], [Jain 89], [Rosenfeld/Kak 82], and [Pratt 91]. The wavelet transform is described in [Kjoelen 95]. More information on wavelet transforms is found in [Masters 94] and [Castleman 96].

For more in-depth coverage of feature extraction and analysis, and pattern recognition, see [Duda/Hart 73], [Nadler/Smith 93], [Granlund/Knutsson 95], [Schalkoff 89], [Schalkoff 92], [Tou/Gonzalez 74], [Castleman 96], [Gose/Johnsonbaugh/Jost 96], and [Banks 90]. Information regarding chain codes can be found in [Gonzalez/Woods 92], [Jain/Kasturi/Schnuck 95], [Baxes 94], and [Ballard/Brown 82]. Specific information for texture-based features can be found in [Nadler/Smith 93], [Haralick/Shapiro 92], [Castleman 96], [Rosenfeld/Kak 82], [Granlund/Knutsson 95], and [Pratt 91]. More information on object features can be found in [Levine 85], [Castleman 96], and [Schalkoff 89]. In particular, [Schalkoff 89] has some interesting RST-invariant features. For a practical approach to the use of neural networks in pattern recognition see [Masters 94]. For an excellent handbook for image processing see [Russ 92].

Ballard, D. H., and Brown, C. M., *Computer Vision*, Englewood Cliffs, NJ: Prentice Hall, 1982.

Banks, S., *Signal Processing, Image Processing and Pattern Recognition*, Englewood Cliffs, NJ: Prentice Hall, 1990.

Baxes, G. A., *Digital Image Processing: Principles and Applications*, New York: Wiley, 1994.

Bracewell, R. N., *Two-Dimensional Imaging*, Englewood Cliffs, NJ: Prentice Hall, 1995.

Castleman, K. R., *Digital Image Processing*, Englewood Cliffs, NJ: Prentice Hall, 1996.

Duda, R. O., and Hart, P. E., *Pattern Classification and Scene Analysis*, New York: Wiley, 1973.

Galbiati, L. J., *Machine Vision and Digital Image Processing Fundamentals*, Englewood Cliffs, NJ: Prentice Hall, 1990.

Giardina, C. R., and Dougherty, E. R., *Morphological Methods in Image and Signal Processing*, Englewood Cliffs, NJ: Prentice Hall, 1988.

Gonzalez, R. C., and Woods, R. E., *Digital Image Processing*, Reading, MA: Addison-Wesley, 1992.

Gose, E., Johnsonbaugh, R., and Jost, S., *Pattern Recognition and Image Analysis*, Upper Saddle River, NJ: Prentice Hall PTR, 1996.

Granlund, G., and Knutsson, H., *Signal Processing for Computer Vision*, Boston: Kluwer Academic Publishers, 1995.

Haralick, R. M., and Shapiro, L. G., *Computer and Robot Vision*, Reading, MA: Addison-Wesley, 1992.

Horn, B. K. P., *Robot Vision*, Cambridge, MA: The MIT Press, 1986.

Jain, A. K., *Fundamentals of Digital Image Processing*, Englewood Cliffs, NJ: Prentice Hall, 1989.

Jain, R., Kasturi, R., and Schnuck, B. G., *Machine Vision*, New York: McGraw Hill, 1995.

Kjoelen, A., *Wavelet Based Compression of Skin Tumor Images*, Master's Thesis in Electrical Engineering, Southern Illinois University at Edwardsville, 1995.

Levine, M. D., *Vision in Man and Machine*, New York: McGraw Hill, 1985.

Masters, T., *Signal and Image Processing with Neural Networks*, New York: Wiley, 1994.

Myler, H. R., and Weeks, A. R., *Computer Imaging Recipes in C*, Englewood Cliffs, NJ: Prentice Hall, 1993.

Nadler, M., and Smith, E. P., *Pattern Recognition Engineering*, New York: Wiley, 1993.

Pratt, W. K., *Digital Image Processing*, New York: Wiley, 1991.

Rosenfeld, A., and Kak, A. C., *Digital Picture Processing*, San Diego, CA: Academic Press, 1982.

Russ, J. C., *The Image Processing Handbook*, Boca Raton, FL: CRC Press, 1992.

Schalkoff, R. J., *Digital Image Processing and Computer Vision*, New York: Wiley, 1989.

Schalkoff, R. J., *Pattern Recognition: Statistical, Structural and Neural Approaches*, New York: Wiley, 1992.

Sid-Ahmed, M. A., *Image Processing: Theory, Algorithms, and Architectures*, New York: McGraw Hill, 1995.

Tou, J. T., and Gonzalez, R. C., *Pattern Recognition Principles*, Reading, MA: Addison-Wesley, 1974.

Umbaugh, S. E, *Computer Vision in Medicine: Color Metrics and Image Segmentation Methods for Skin Cancer Diagnosis*, PhD dissertation, UMI Dissertation Service, 1990.

Umbaugh, S. E, Moss, R. H., and Stoecker, W. V., "Automatic Color Segmentation of Images with Application to Detection of Variegated Coloring in Skin Tumors," *IEEE Engineering in Medicine and Biology*, Vol. 8, No. 4, pp. 43–52, December 1989.

Umbaugh, S. E, Moss, R. H., and Stoecker, W. V., "An Automatic Color Segmentation Algorithm with Application to Identification of Skin Tumor Borders," *Computerized Medical Imaging and Graphics*, Vol. 16, No. 3, pp. 227–235, 1992.

Umbaugh, S. E, Moss, R. H., and Stoecker, W. V., "Automatic Color Segmentation Algorithms with Application to Skin Tumors Feature Identification," *IEEE Engineering in Medicine and Biology*, Vol. 12, No. 3, pp. 75–82, September 1993.

CHAPTER 3

Image Restoration

3.1 Introduction

Image restoration methods are used to improve the appearance of an image by application of a restoration process that uses a mathematical model for image degradation. Examples of the types of degradation include blurring caused by motion or atmospheric disturbance, geometric distortion caused by imperfect lenses, superimposed interference patterns caused by mechanical systems, and noise from electronic sources. It is assumed that the degradation model is known or can be estimated. The idea is to model the degradation process and then apply the inverse process to restore the original image. In general, image restoration is more of an art than a science; the restoration process relies on the experience of the individual to model the degradation process successfully. In this chapter we will consider the various types of degradation that can be modeled and discuss the various techniques available to restore the image. The types of degradation models include both spatial and frequency domain considerations. (See Section 6.3.3 for an example of restoration using CVIPtools.)

In practice the degradation process model is often not known and must be experimentally determined or estimated. (A number of techniques are available to estimate this degradation analytically. They are beyond the scope of the discussion and can be explored with the references.) Any available information regarding the images and the systems used to acquire and process them is helpful. This information, combined with the developer's experience, can be applied to solve the specific application. A general block diagram for the image restoration process is provided in Figure 3.1-1. Here we see that degraded image(s) and knowledge of the image creation process are provided as input to the development of the degradation model. Knowledge of the image creation process is application specific; for example, it is helpful to know how a specific lens distorts an image or how mechanical vibration from a satellite affects an image. This information may be provided by knowledge about the image acquisition process

Figure 3.1-1 Image Restoration Process

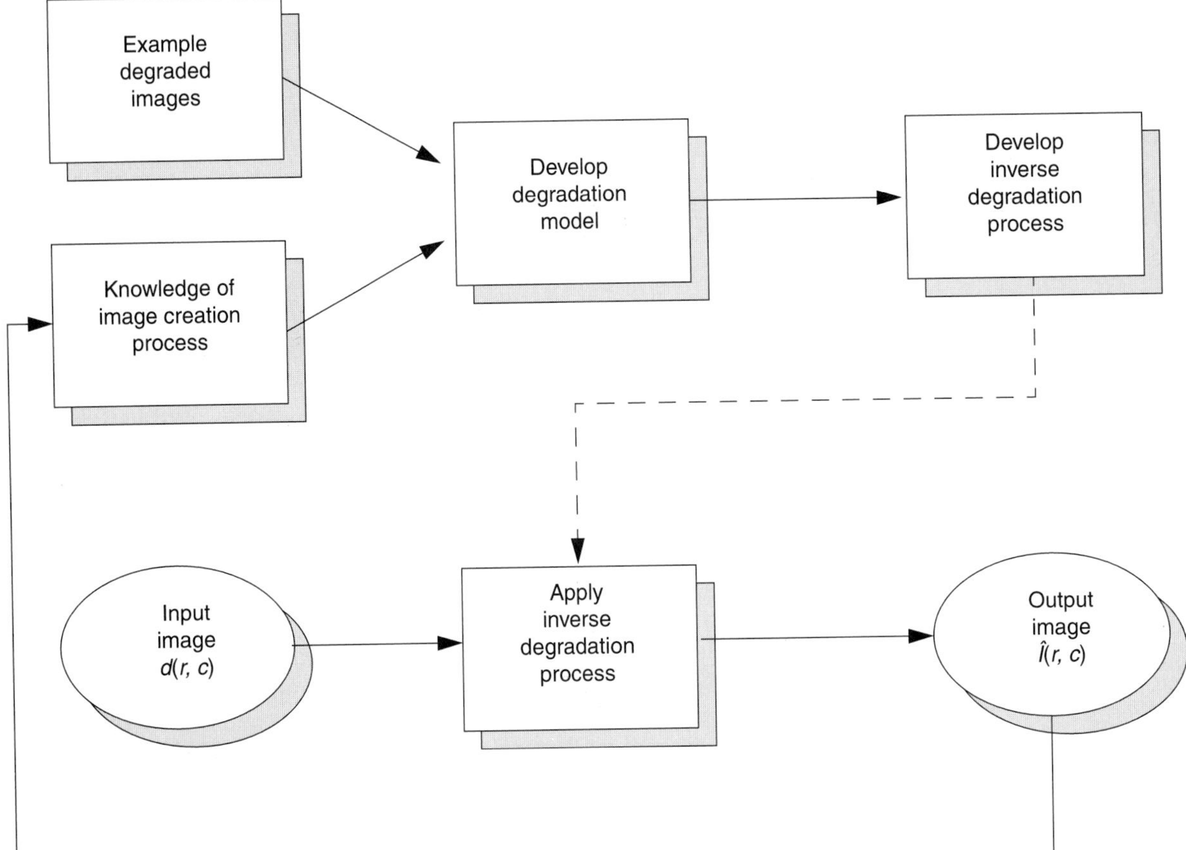

itself, or it may extracted from the degraded images by applying image analysis techniques. After the degradation process has been developed, the formulation of the inverse process follows. This inverse degradation process is then applied to the degraded image, $d(r, c)$, which results in the output image $\hat{I}(r, c)$. This output image $\hat{I}(r, c)$ is the restored image that represents an estimate of the original image $I(r, c)$. After the estimated image has been created, any knowledge gained by observation and analysis of this image is used as additional input for the development of the degradation model. This process continues until satisfactory results are achieved. With this perspective, we can define *image restoration* as the process of finding an approximation to the degradation process and finding the appropriate inverse process to estimate the original image.

3.1.1 System Model

The degradation process model consists of two parts, the degradation function and the noise function. The general model in the spatial domain follows:

$$d(r, c) = h(r, c) * I(r, c) + n(r, c)$$

where the $*$ denotes the convolution process

$d(r, c)$ = degraded image

$h(r, c)$ = degradation function

$I(r, c)$ = original image

$n(r, c)$ = additive noise function

Because convolution in the spatial domain is equivalent to multiplication in the frequency domain, the frequency domain model is

$$D(u, v) = H(u, v)I(u, v) + N(u, v)$$

where $D(u, v)$ = Fourier transform of the degraded image

$H(u, v)$ = Fourier transform of the degradation function

$I(u, v)$ = Fourier transform of the original image

$N(u, v)$ = Fourier transform of the additive noise function

Based on our definition of the image restoration process, and the preceding model, we can see that what needs to be done is to find the degradation function $h(r, c)$ (or its frequency domain representation $H(u, v)$) and the noise model $n(r, c)$ (or $N(u, v)$). Note that other models can be defined, specifically a multiplicative noise model where the noise function is not added to the image but is multiplied by the image. To handle this case, we typically take the logarithm of the degraded image, thus decoupling the noise and image functions into an additive process (see Chapter 4 on homomorphic filtering).

3.2 NOISE

What is noise? *Noise* is any undesired information that contaminates an image. Noise appears in images from a variety of sources. The digital image acquisition process, which converts an optical image into a continuous electrical signal that is then sampled, is the primary process by which noise appears in digital images. At every step in the process there are fluctuations caused by natural phenomena that add a random value to the exact brightness value for a given pixel. In typical images the noise can be modeled with either a gaussian ("normal"), uniform, or salt-and-pepper ("impulse") distribution. The shape of the distribution of these noise types as a function of gray level can be modeled as a histogram and can be seen in Figure 3.2-1. In Figure 3.2-1a we see the bell-shaped curve of the gaussian noise distribution, which can be analytically described by

$$\text{HISTOGRAM}_{\text{Gaussian}} = \frac{1}{\sqrt{2\pi\sigma^2}} e^{-(g-m)^2/2\sigma^2}$$

where g = gray level

m = mean (average)

σ = standard deviation (σ^2 = variance)

About 70% of all the values fall within the range from one standard deviation (σ) below the mean (m) to one above, and about 95% fall within two standard deviations.

Theoretically, this equation defines values from $-\infty$ to $+\infty$, but because the actual gray levels are only defined over a finite range, the number of pixels at the lower and upper values will be higher than this equation predicts. This is a result of the fact that all the noise values below the minimum will be clipped to the minimum, and those above the maximum will be clipped at the maximum value. This is a factor that must be considered with all theoretical noise models, when applied to a fixed, discrete range such as with digital images (e.g., 0 to 255). In Figure 3.2-1b is the uniform distribution:

$$\text{HISTOGRAM}_{\text{Uniform}} = \begin{cases} \dfrac{1}{b-a} & \text{for } a \le g \le b \\ 0 & \text{elsewhere} \end{cases}$$

$$\text{mean} = \frac{a+b}{2}$$

$$\text{variance} = \frac{(b-a)^2}{12}$$

With the uniform distribution, the gray-level values of the noise are evenly distributed across a specific range, which may be the entire range (0 to 255 for 8-bits), or a smaller portion of the entire range. In Figure 3.2-1c is the salt-and-pepper distribution:

$$\text{HISTOGRAM}_{\text{Salt \& Pepper}} = \begin{cases} A & \text{for } g = a \text{ (``pepper'')} \\ B & \text{for } g = b \text{ (``salt'')} \end{cases}$$

In the salt-and-pepper noise model there are only two possible values, a and b, and the probability of each is typically less than 0.1—with numbers greater than this, the noise will dominate the image. For an 8-bit image, the typical value for pepper noise is 0 and for salt-noise, 255.

The gaussian model is most often used to model natural noise processes, such as those occurring from electronic noise in the image acquisition system. The salt-and-pepper type noise is typically caused by malfunctioning pixel elements in the camera sensors, faulty memory locations, or timing errors in the digitization process. Uniform noise is useful because it can be used to generate any other type of noise distribution and is often used to degrade images for the evaluation of image restoration algorithms because it provides the most unbiased or neutral noise model. In Figure 3.2-2, we see examples of these three types of noise, and how they appear in images. Visually, the gaussian and uniform noisy images appear similar, but the image with added salt-and-pepper noise is very distinctive.

In addition to the gaussian, other noise models based on exponential distributions are useful for modeling noise in certain types of digital images. Radar range and velocity images typically contain noise that can be modeled by the Rayleigh distribution, defined by

$$\text{HISTOGRAM}_{\text{Rayleigh}} = \frac{2g}{\alpha} e^{-g^2/\alpha}$$

$$\text{where} \quad \text{mean} = \sqrt{\frac{\pi\alpha}{4}}$$

$$\text{variance} = \frac{\alpha(4-\pi)}{4}$$

Figure 3.2-1 Noise Distribution

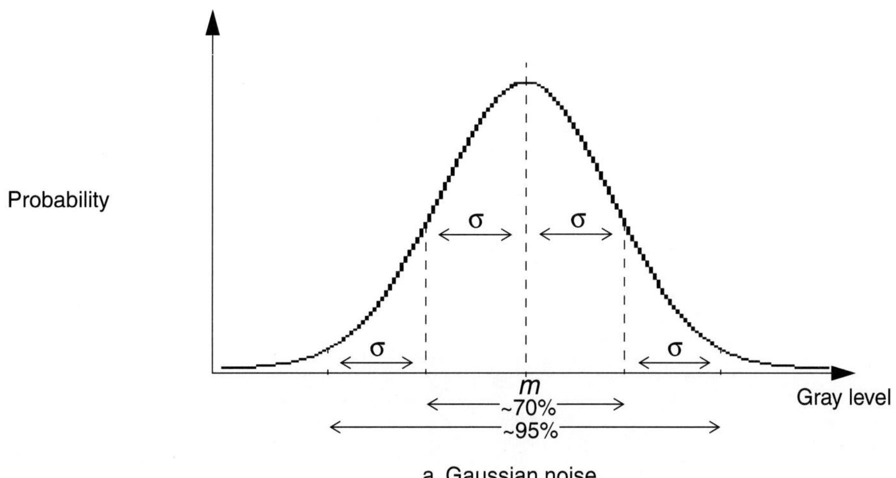

a. Gaussian noise.

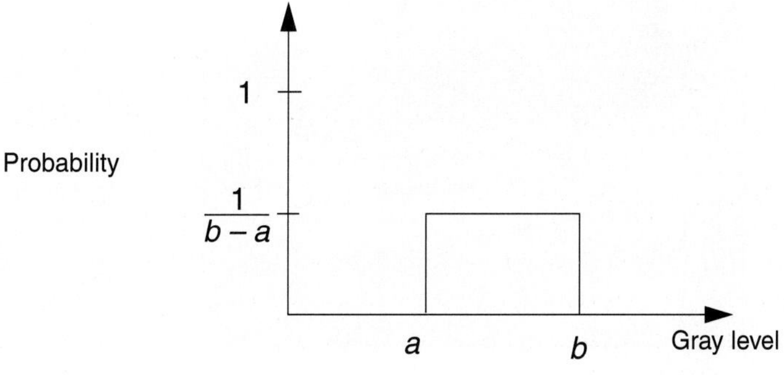

b. Uniform noise.

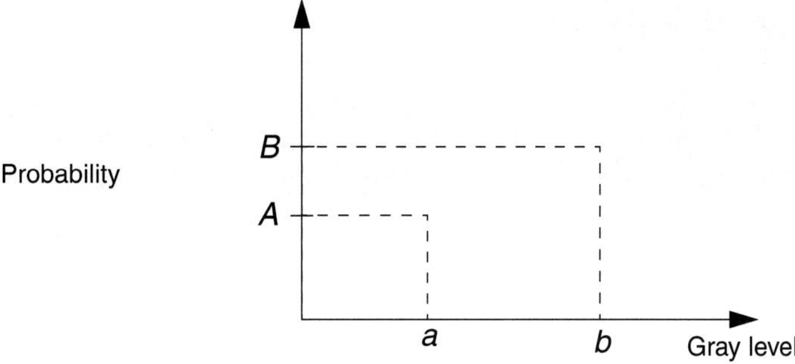

c. Salt-and-pepper noise.

The peak value for the Rayleigh distribution is at $\sqrt{\alpha/2}$. Negative exponential noise occurs in laser-based images, and if this type of image is lowpass filtered, the noise can be modeled as gamma noise. The equation for negative exponential noise:

$$\text{HISTOGRAM}_{\text{Negative Exponential}} = \frac{e^{-\frac{g}{\alpha}}}{\alpha}$$

$$\text{where} \quad \text{variance} = \alpha^2$$

Figure 3.2-2 Noise

a. Original image.

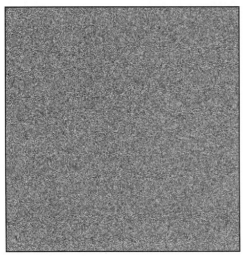

b. Gaussian noise—mean = 0; $\sigma \approx 27$.

c. Original image with added gaussian noise.

Figure 3.2-2 (Continued)

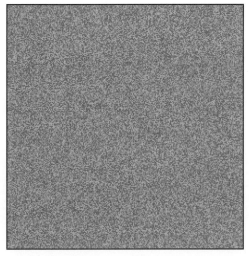

d. Uniform noise—$a = -47$; $b = +47$.

e. Original image with added uniform noise.

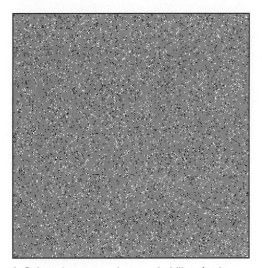

f. Salt-and-pepper noise—probability of salt = .05; probability of pepper = .05.

g. Original image with added salt-and-pepper noise.

The equation for gamma noise:

$$\text{HISTOGRAM}_{\text{Gamma}} = \frac{g^{\alpha-1}}{(\alpha-1)!a^{\alpha}} e^{-\frac{g}{a}}$$

$$\text{where} \qquad \text{variance} = a^2\alpha$$

The histograms (distributions) for these can be seen in Figure 3.2-3. Note that negative exponential noise is actually gamma noise with the peak moved to the origin ($\alpha = 1$).

Figure 3.2-3 Noise Histograms

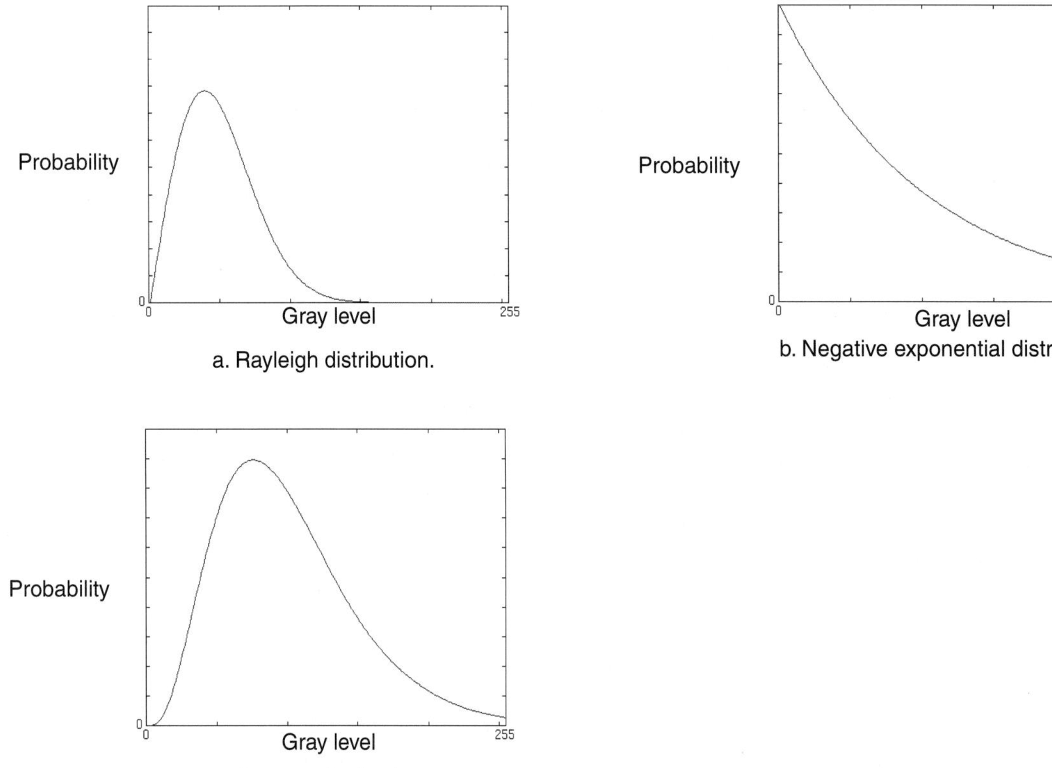

a. Rayleigh distribution.

b. Negative exponential distribution.

c. Gamma distribution.

Many of the types of noise that occur in natural phenomena can be modeled as some form of exponential noise such as those described here.

There are various approaches to determining the type of noise that has corrupted an image. Ideally, we want to find an image (or subimage) that contains only noise, and then we can use its histogram for the noise model. For example, if we have access to the system that generated the images, noise images can be acquired by aiming the imaging device (e.g., camera) at a blank wall—the resulting image will contain only an average (DC) value as a result of the lighting conditions, and any fluctuations will be from noise.

If we cannot find "noise-only" images, a portion of the image is selected that has a known histogram, and that knowledge is used to determine the noise characteristics. This may be a subimage of constant value or a well-defined line—any portion of the image where we know what to expect in the histogram. We can then subtract the known values from the histogram, and what is left is our noise model. We can then compare this noise model to the ones available and select the best match. In order to develop a valid model with any of these approaches, many such images (or subimages) need to be evaluated. For more information on the theoretical approach to noise modeling, see the references on digital signal processing, statistical or stochastic processes, and communications theory. In practice, the best model is often determined empirically.

3.3 NOISE REMOVAL USING SPATIAL FILTERS

Spatial filters can be effectively used to remove various types of noise in digital images. These spatial filters typically operate on small neighborhoods, 3×3 to 11×11, and some can be implemented as convolution masks. For this section, we will use the degradation model defined in Section 3.1.1, with the assumption that $h(r, c)$ causes no degradation, so the only corruption to the image is caused by additive noise, as follows:

$$\begin{aligned} d(r,c) &= I(r,c) + n(r,c) \\ \text{where } d(r,c) &= \text{degraded image} \\ I(r,c) &= \text{original image} \\ n(r,c) &= \text{additive noise function} \end{aligned}$$

The two primary categories of spatial filters for noise removal are order filters and mean filters. The *order filters* are implemented by arranging the neighborhood pixels in order from smallest to largest gray-level value and using this ordering to select the "correct" value, while the *mean filters* determine, in one sense or another, an average value. The mean filters work best with gaussian or uniform noise, and the order filters (specifically the median, see Figure 2.2-11) work best with salt-and-pepper, negative exponential, or Rayleigh noise.

The mean filters have the disadvantage of blurring the image edges, or details; they are essentially lowpass filters. The order filters are nonlinear, so their results are sometimes unpredictable. In general, there is a tradeoff between preservation of image detail and noise elimination. To understand this concept consider an extreme case where the entire image is replaced with the average value of the image. In one sense, we have eliminated any noise present in the image, but we have also lost all the information in the image. Practical mean and order filters also lose information in their quest for noise elimination, and the trick is to minimize this information loss while maximizing noise removal. Ideally, a filter that adapts to the underlying pixel values is desired. A filter that changes its behavior based on the gray-level characteristics (statistics) of a neighborhood is called an *adaptive filter*, and these filters are effective for use in many practical applications.

3.3.1 Order Filters

Order filters are based on a specific type of image statistics called order statistics. Typically, these filters operate on small subimages, *windows*, and replace the center pixel value (similar to the convolution process). *Order statistics* is a technique that arranges all the pixels in sequential order, based on gray-level value. The placement of the value within this ordered set is referred as the *rank*. Given an $N \times N$ window W, the pixel values can be ordered from smallest to largest, as follows:

$$I_1 \leq I_2 \leq I_3 \leq \ldots \leq I_{N^2}$$
$$\text{where } \{I_1, I_2, I_3, \ldots, I_{N^2}\}$$

are the Intensity (gray-level) values of the subset of pixels in the image, that are in the $N \times N$ window, W (that is, $(r, c) \in W$)

EXAMPLE 3-1

Given the following 3 × 3 subimage:

$$\begin{bmatrix} 110 & 110 & 114 \\ 100 & 104 & 104 \\ 95 & 88 & 85 \end{bmatrix}$$

the result from applying order statistics to arrange them is

{85, 88, 95, 100, 104, 104, 110, 110, 114}

Figure 3.3-1 Median Filtering with Sliding Window

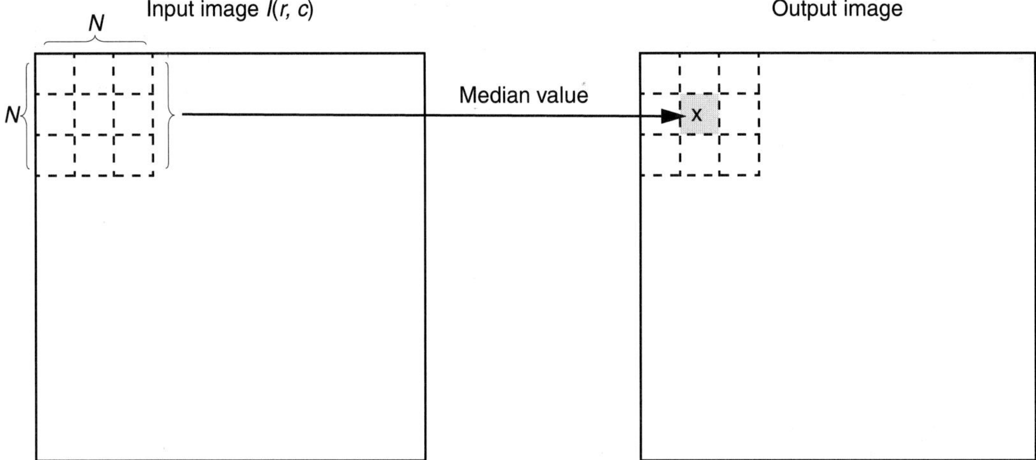

a. The input image is overlaid with an $N \times N$ window, and the median value of the pixels covered by the window is placed in the output image at location x.

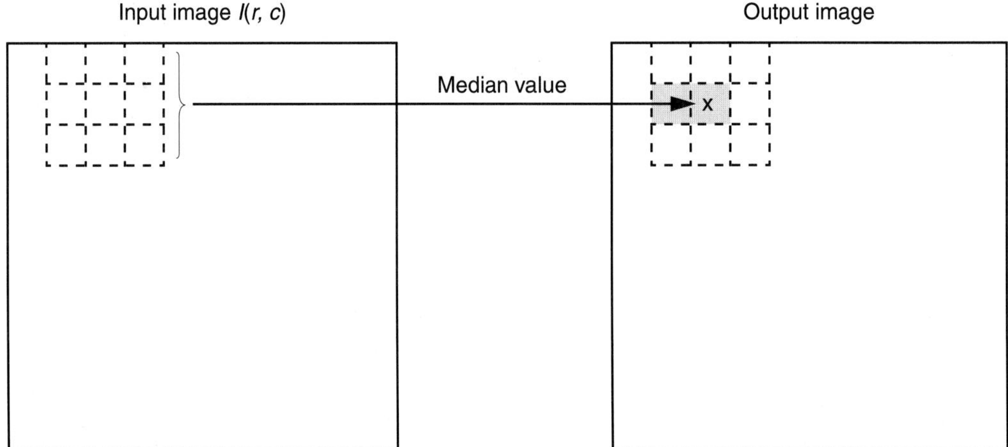

b. The window is moved one pixel to the right, and the median value of the pixels now covered by the window is placed in the output image at location x.

Figure 3.3-1 (Continued)

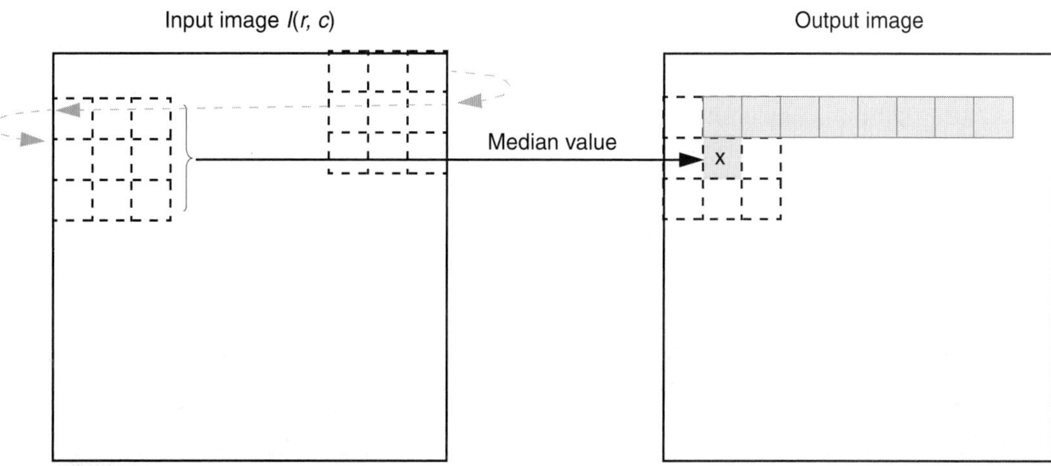

c. When the end of a row is reached, the window is moved back to the left edge of the image and down one row.

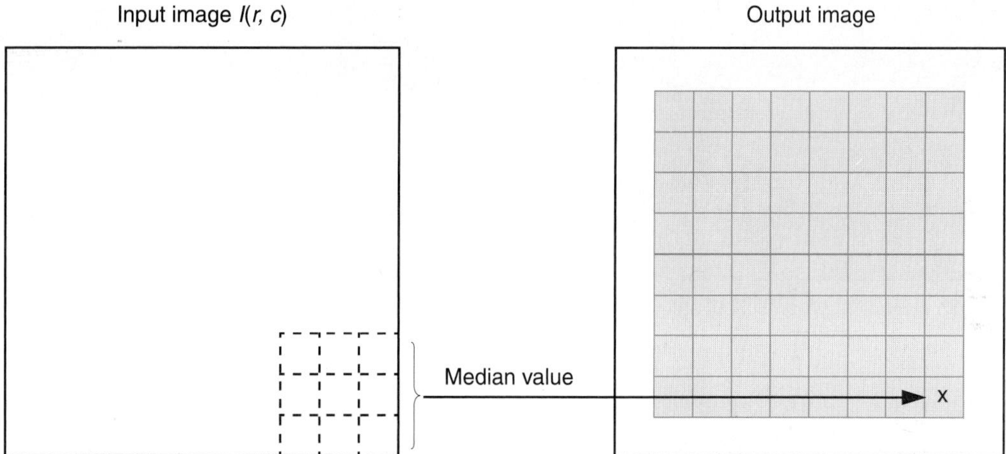

d. The entire image has been processed. Note the unprocessed outer rows and columns.

The most useful of the order filters is the median filter. The *median filter* selects the middle pixel value from the ordered set. In the preceding example, the median filter selects the value 104 because there are 4 values above it and 4 values below it. The median filtering operation is performed on an image by applying the sliding window concept, similar to what is done with convolution. In Figure 3.3-1a, the window is overlaid on the upper-left corner of the image, and the median is determined. This value is put into the output image (buffer) corresponding to the center location of the window. The window is then slid one pixel over, and the process is repeated (Figure 3.3-1b). When the end of the row is reached, the window is slid back to the left side of the image and down one row, and the process is repeated (Figure 3.3-1c). This process continues until the entire image has been processed (Figure 3.3-1d). Note that the outer rows and columns are not replaced. In practice this is usually not a problem due to the fact that the images are much larger than the masks, and these "wasted" rows and columns are often filled with zeros (or cropped off the image). For example, with a 3×3 mask, we lose one outer row and column, a 5×5 loses two rows and columns—this is not usually significant for a typical 256×256 or 512×512 image.

The maximum and minimum filters are two order filters that can be used for elimination of salt-and-pepper (impulse) noise. The *maximum filter* selects the largest value within an ordered window of pixel values, whereas the *minimum filter* selects the smallest value. The minimum filter works when the noise is primarily of the salt-type (high values), and the maximum filters works best for pepper-type noise (low values). In Figures 3.3-2a, b, the application of a minimum filter to an image contaminated

Figure 3.3-2 Minimum and Maximum Filters

a. Image with salt noise; probability of salt = .04.

b. Result of minimum filtering image (a); mask size = 3×3.

c. Image with pepper noise; probability of pepper = .04.

d. Result of maximum filtering image (c); mask size = 3×3.

with salt-type noise is shown, and in Figures 3.3-2c, d a maximum filter applied to an image corrupted with pepper-noise is shown. Here we see that these filters are excellent for this type of noise, with minimal information loss. As the size of the window gets bigger, the more information loss occurs; with windows larger than about 5 × 5 the image acquires an artificial, "painted," effect (Figure 3.3-3).

In a manner similar to the median, minimum, and maximum filter, order filters can be defined to select a specific pixel rank within the ordered set. For example, we may find for certain types of pepper noise that selecting the second highest value works better than selecting the maximum value. This type of ordered selection is very

Figure 3.3-3 Various Window Sizes for Maximum and Minimum Filters

a. Result of minimum filtering Figure 3.3-2a; mask size = 5 × 5.

b. Result of minimum filtering Figure 3.3-2a; mask size = 9 × 9.

c. Result of maximum filtering Figure 3.3-2c; mask size = 5 × 5.

d. Result of maximum filtering Figure 3.3-2c; mask size = 9 × 9.

sensitive to the type of images and their use—it is application specific. Another example might be selecting the third value from the lowest, and using a larger window, for specific types of salt noise. It should be noted that, in general, a minimum or low rank filter will tend to darken an image and a maximum or high rank filter will tend to brighten an image—this effect is especially noticeable in areas of fine detail and high contrast.

The final two order filters are the midpoint and alpha-trimmed mean filters. They are actually both order and mean filters because they rely on ordering the pixel values, but they are then calculated by an averaging process. The *midpoint filter* is the average of the maximum and minimum within the window, as follows:

$$\text{ordered set} \rightarrow I_1 \leq I_2 \leq I_3 \leq \ldots \leq I_{N^2}$$

$$\text{Midpoint} = \frac{I_1 + I_{N^2}}{2}$$

The midpoint filter is most useful for gaussian and uniform noise, as illustrated in Figure 3.3-4.

The *alpha-trimmed mean* is the average of the pixel values within the window, but with some of the endpoint-ranked values excluded. It is defined as follows:

$$\text{ordered set} \rightarrow I_1 \leq I_2 \leq \ldots \leq I_{N^2}$$

$$\text{Alpha-trimmed mean} = \frac{1}{N^2 - 2T} \sum_{i=T+1}^{N^2-T} I_i$$

where T is the number of pixel values excluded at each end of the ordered set, and can range from 0 to $(N^2 - 1)$.

Figure 3.3-4 Midpoint Filter

a. Image with gaussian noise—variance = 300; mean = 0.

b. Result of midpoint filter; mask size = 3.

Figure 3.3-4 (Continued)

c. Image with uniform noise—variance = 300; mean = 0.

d. Result of midpoint filter; mask size = 3.

The alpha-trimmed mean filter ranges from a mean to median filter, depending on the value selected for the T parameter. For example, if $T = 0$, the equation reduces to finding the average gray-level value in the window, which is an arithmetic mean filter. If $T = (N^2 - 1)/2$, the equation becomes a median filter. This filter is useful for images containing multiple types of noise, such as gaussian and salt-and-pepper noise. In Figure 3.3-5 are the results of applying this filter to an image with both gaussian and salt-and-pepper noise.

Figure 3.3-5 Alpha-Trimmed Mean Filter

a. Image with gaussian and salt-and-pepper noises—variance = 200; mean = 0.

b. Result of alpha-trimmed mean filter; mask size = 3, trim size = 1.

Figure 3.3-5 (Continued)

c. Result of alpha-trimmed mean filter; mask size = 3, trim size = 4.

d. Result of alpha-trimmed mean filter; mask size = 5, trim size = 5.

3.3.2 Mean Filters

The mean filters function by finding some form of an average within the $N \times N$ window, using the sliding window concept to process the entire image. The most basic of these filters is the *arithmetic mean filter*, which finds the arithmetic average of the pixel values in the window, as follows:

Figure 3.3-6 Arithmetic Mean Filter

a. Image with gaussian noise—variance = 300; mean = 0.

b. Image with gamma noise—variance = 300; alpha = 1.

Figure 3.3-6 (Continued)

c. Result of arithmetic mean filter on image with gaussian noise; mask size = 3.

d. Result of arithmetic mean filter on image with gamma noise; mask size = 3.

e. Result of arithmetic mean filter on image with gaussian noise; mask size = 5.

f. Result of arithmetic mean filter on image with gamma noise; mask size = 5.

$$\text{Arithmetic Mean} = \frac{1}{N^2} \sum_{(r,c) \in W} d(r, c)$$

where N^2 = the number of pixels in the $N \times N$ window, W.

The arithmetic mean filter smooths out local variations within an image, so it is essentially a lowpass filter. It can be implemented with a convolution mask where all the mask coefficients are $1/N^2$. This filter will tend to blur an image while mitigating the noise effects. In Figure 3.3-6 are the results of an arithmetic mean applied to images with various types of noise. It can be seen that the larger the mask size, the more pronounced the blurring effect. This type of filter works best with gaussian and uniform noise. The blurring effect, which reduces image details, is undesirable, and the other mean filters are designed to minimize this loss of detail information.

The *contra-harmonic mean filter* works well for images containing salt OR pepper type noise, depending on the filter order R:

Figure 3.3-7 Contra-Harmonic Mean Filter

a. Image with salt noise—probability = .04.

b. Result of contra-harmonic mean filter; mask size = 3; order = -3.

c. Image with pepper noise—probability = .04.

d. Result of contra-harmonic mean filter; mask size = 3; order = +3.

$$\text{Contra-Harmonic Mean} = \frac{\sum_{(r,c)\in W} d(r,c)^{R+1}}{\sum_{(r,c)\in W} d(r,c)^{R}}$$

where W is the $N \times N$ window under consideration.

For negative values of R, it eliminates salt-type noise, whereas for positive values of R, it eliminates pepper-type noise. This is shown in Figure 3.3-7.

The *geometric mean filter* works best with gaussian noise and retains detail information better than an arithmetic mean filter. It is defined as the product of the pixel values within the window, raised to the $1/N^2$ power:

$$\text{Geometric Mean} = \prod_{(r,c)\in W} [d(r,c)]^{\frac{1}{N^2}}$$

In Figure 3.3-8 are the results of applying this filter to images with gaussian (a, b) and pepper noise (c, d). As shown in Figure 3.3-8d, this filter is ineffective in the presence of pepper noise—with zero (or very low) values present in the window, the equation returns a zero (or very small) number.

The *harmonic mean filter* also fails with pepper noise but works well for salt noise. It is defined as follows:

$$\text{Harmonic Mean} = \frac{N^2}{\sum_{(r,c)\in W} \frac{1}{d(r,c)}}$$

This filter also works with gaussian noise, retaining detail information better than the arithmetic mean filter. In Figure 3.3-9 are the results from applying the harmonic

Figure 3.3-8 Geometric Mean Filter

a. Image with gaussian noise—variance = 300; mean = 0.

b. Result of geometric mean filter on image with gaussian noise; mask size = 3.

Figure 3.3-8 (Continued)

c. Image with pepper noise—probability = .04.

d. Result of geometric mean filter on image with pepper noise; mask size = 3.

mean filter to an image with gaussian noise (a, b) and to an image corrupted with salt noise (c, d).

The Y_p *mean filter* is defined as follows:

$$Y_p \text{ Mean} = \left[\sum_{(r,c) \in W} \frac{d(r,c)^P}{N^2} \right]^{\frac{1}{P}}$$

Figure 3.3-9 Harmonic Mean Filter

a. Image with gaussian noise—variance = 300; mean = 0.

b. Result of harmonic mean filter on image with gaussian noise; mask size = 3.

Figure 3.3-9 (Continued)

c. Image with salt noise—probability = .04.

d. Result of harmonic mean filter on image with salt noise; mask size = 3.

This filter removes salt noise for negative values of P and pepper noise for positive values of P. Figure 3.3-10 illustrates the use of the Y_p filter.

3.3.3 Adaptive Filters—Minimum Mean-Square Error Filter

The previously described filters are adaptive in the sense that their output depends on the underlying pixel values. Some, such as the alpha-trimmed mean, can vary between a mean and median filter, but this change in filter behavior is fixed for a given value of the T parameter. However, an *adaptive filter* alters its basic behavior as the image is processed; it may act like a mean filter on some parts of the image and a median filter on other parts of the image. The typical criteria used to determine the filter behavior are the local image characteristics, usually measured by the local gray-level statistics. The minimum mean-square error (MMSE) filter is a good example of an adaptive filter, which exhibits varying behavior based on local image statistics. The MMSE filter works best with gaussian or uniform noise and is defined as follows:

$$\text{MMSE} = d(r,c) - \frac{\sigma_n^2}{\sigma_l^2}\left[d(r,c) - m_l(r,c)\right]$$

where

σ_n^2 = noise variance

σ_l^2 = local variance (in the window under consideration)

m_l = local mean (average in the window under consideration)

With no noise in the image, the noise variance equals zero, and this equation will return the original unfiltered image. In background regions of the image, areas of

Figure 3.3-10 Y_p Mean Filter

a. Image with salt noise—probability = .04.

b. Result of *Yp* mean filter on image with salt noise; mask size = 3, order = -3.

c. Image with pepper noise—probability = .04.

d. Result of *Yp* mean filter on image with pepper noise; mask size = 3, order = +3.

fairly constant value in the original uncorrupted image, the noise variance will equal the local variance, and the equation reduces to the mean filter. In areas of the image where the local variance is much greater then the noise variance, the filter returns a value close to the unfiltered image data. This is desired since high local variance implies high detail (edges), and an adaptive filter tries to preserve the original image detail. In general, the MMSE filter returns a value that consists of the unfiltered image data $d(r, c)$, with some of the original value subtracted out and some of the local mean added. The amount of the original and local mean used to modify the original are weighted by the noise to local variance ratio, σ_n^2/σ_i^2. As this ratio increases, implying primarily noise in the window, the filter returns primarily the local average. As this ratio goes down, implying high local detail, the filter returns more of the original unfiltered image. By operating in this manner, the MMSE filter adapts itself to the local image statistics, preserving image details while removing noise. Figure 3.3-11 illustrates the use of the MMSE filter on an image with added gaussian noise. Here we specify the window (kernel) size and the noise variance to be used. More information on adaptive filters can be found in the references.

Figure 3.3-11 MMSE Filter

a. Original image.

b. Image with gaussian noise—variance = 300; mean = 0.

c. Result of MMSE filter—kernel size = 3; noise variance = 300.

d. Result of MMSE filter—kernel size = 9; noise variance = 300.

3.4 Frequency Domain Filters

Frequency domain filtering operates by using the Fourier transform representation of images. This representation consists of information about the spatial frequency content of the image, also referred to as the spectrum of the image. In Figure 3.4-1 is the general model for frequency domain filtering. The Fourier transform is performed on three spatial domain functions: 1) the degraded image, $d(r, c)$, 2) the degradation function, $h(r, c)$, and 3) the noise model, $n(r, c)$. Next, the frequency domain filter is applied to the Fourier transform outputs, $N(u, v)$, $D(u, v)$, and $H(u, v)$. The output of the filter operation undergoes an inverse Fourier transform to give the restored image.

The specific models used for $h(r, c)$ and $n(r, c)$ depend on the application and, in practice, often must be estimated. In some cases, they may not be explicitly required (as in the previous section on spatial filters for noise removal, where the $h(r, c)$ was assumed unnecessary). The typical noise models are provided in Section 3.2. The degradation function can be experimentally determined in various ways.

Another name for the degradation function is the point spread function (PSF). The *point spread function* (the 2-D equivalent of the impulse response) describes what happens to a single point of light when it passes through a system. The PSF for a *linear, shift invariant system* completely describes the system. This makes it easy to find the degradation function for a system, *if the system is available and the conditions under which the image was acquired have not changed*—all we need to do is to

Figure 3.4-1 Frequency Domain Filtering

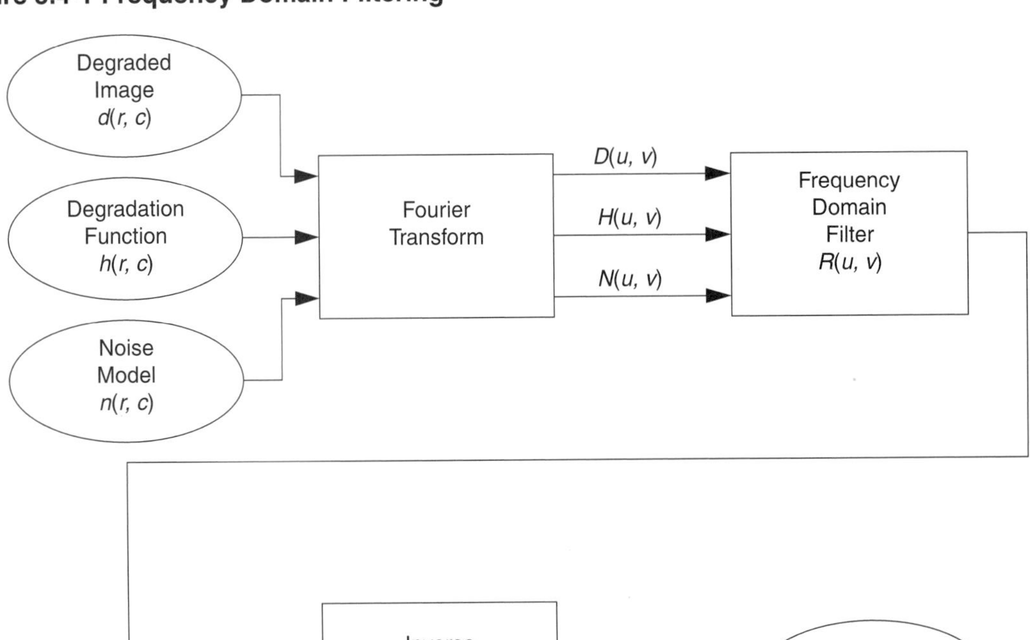

send a point of light through the system and see what comes out. The output is the PSF, in this case $h(r, c)$. However, it is not always practical to implement a point source of light, and a more reliable method is to use sinusoidal inputs at many different frequencies to determine the overall frequency response $H(u, v)$. For many applications the system that created the images may not be available, or the conditions under which the image was acquired are unknown. Often, in practice, the PSF must be estimated.

Most of the types of degradation that lend themselves to this type of analysis are some form of blurring. The blur can be linear in one direction only—horizontal, vertical, or diagonal; it can be circularly symmetric—the same blur in all directions. The blur may also be a combination of these different directional types of blur. The amount of blur can be experimentally determined by observation of a known point or line in an image. If a point can be found that we believe should be only one pixel wide (for example, a star in an astronomical image or a point found through the use of test charts), we can estimate the PSF by measuring the width of the point in the blurred image. We can find lines or edges that we believe should be only one pixel wide and measure their blurred width. This gives us some idea of how wide the PSF mask should be.

The PSF mask will be used to model $h(r, c)$; typical models are shown in Figure 3.4-2. The nonzero terms, designated by x's in the figure, have a uniform, gaussian, or centered-weighted distribution. In Figure 3.4-3 are shown representative values for the various types of blur filter masks. The optimal results are often found by a combination of experimentation and intuition—the image restoration process is, in general, an art not a science. After a blur filter mask has been selected, the mask must be padded with zeros up to the size of the image, before the Fourier transform of $h(r, c)$ is determined. After the blur mask has been zero-padded, this extended $h(r, c)$ needs to be shifted, with wrap-around, so that the center of the original blur mask is at the (0, 0) point in the image, that is the upper-left corner. In other words, the coefficients will appear in the corners, based on Fourier symmetry. If this is not done, phase shifts will occur in the output image.

The frequency domain filters incorporate information regarding the noise and the PSF into their model and are based on the mathematical model provided in Section 3.1.1:

$$
\begin{aligned}
D(u, v) &= H(u, v)I(u, v) + N(u, v) \\
\text{where } D(u, v) &= \text{Fourier transform of the degraded image} \\
H(u, v) &= \text{Fourier transform of the degradation function} \\
I(u, v) &= \text{Fourier transform of the original image} \\
N(u, v) &= \text{Fourier transform of the additive noise function}
\end{aligned}
$$

In order to obtain the restored image, the general form is as follows:

$$
\begin{aligned}
\hat{I}(r, c) &= F^{-1}[\hat{I}(u, v)] = F^{-1}\left[R_{\text{type}}(u, v)D(u,v)\right] \\
\text{where } \hat{I}(r, c) &= \text{the restored image, an approximation to } I(r, c) \\
F^{-1}[] &= \text{the inverse Fourier transform} \\
R_{\text{type}}(u, v) &= \text{the Restoration (frequency domain) filter,} \\
&\quad \text{the subscript defines the type of filter}
\end{aligned}
$$

The filters discussed here include the inverse filter, the classical Wiener filter, the parametric Wiener filter, the power spectrum equalization filter, the constrained least-squares filter, the geometric mean filter, and the notch filter. A general mathematical model using the geometric mean filter is provided, and from this model many of the other filters can be generated.

Figure 3.4-2 Blur Masks

X	X	X	X	X

a. Horizontal PSF mask; x = nonzero term.

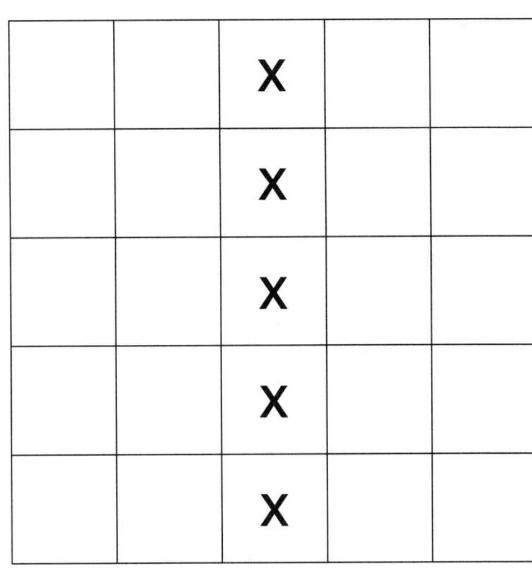

b. Vertical PSF mask; x = nonzero term.

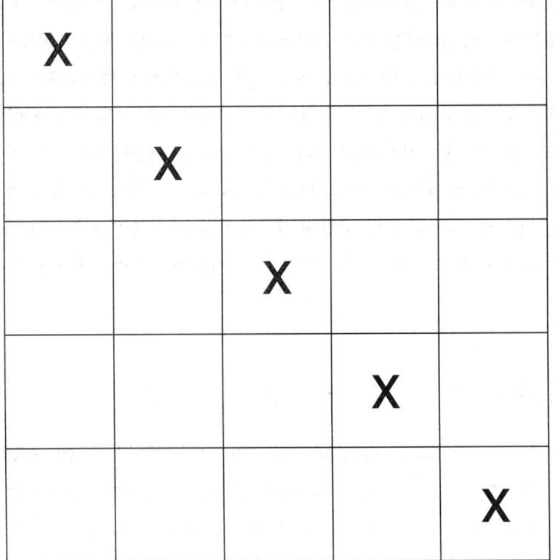

c. Diagonal PSF mask; x = nonzero term.

	X	X	X	
X	X	X	X	X
X	X	X	X	X
X	X	X	X	X
	X	X	X	

d. Circular PSF mask; x = nonzero term.

Figure 3.4-3 Typical Blur Mask Coefficients

0	0	0	0	0
0	0	0	0	0
1	1	1	1	1
0	0	0	0	0
0	0	0	0	0

a. Horizontal PSF mask with uniform blur.

0	0	1	0	0
0	0	1	0	0
0	0	4	0	0
0	0	1	0	0
0	0	1	0	0

b. Vertical PSF mask with center-weighting.

1	0	0	0	0
0	2	0	0	0
0	0	4	0	0
0	0	0	2	0
0	0	0	0	1

c. Diagonal PSF mask with gaussian distribution.

0	1	1	1	0
1	2	4	2	1
1	4	8	4	1
1	2	4	2	1
0	1	1	1	0

d. Circular PSF mask with gaussian distribution.

3.4.1 Inverse Filter

The inverse filter uses the foregoing model, with the added assumption of no noise ($N(u, v) = 0$). If this is the case, the Fourier transform of the degraded image is

$$D(u, v) = H(u, v)I(u, v) + 0$$

So the Fourier transform of the original image can be found as follows:

$$I(u, v) = \frac{D(u, v)}{H(u, v)} = D(u, v)\frac{1}{H(u, v)}$$

To find the original image, we take the inverse Fourier transform of $I(u, v)$:

$$I(r, c) = F^{-1}[I(u, v)] = F^{-1}\left[\frac{D(u, v)}{H(u, v)}\right] = F^{-1}\left[D(u, v)\frac{1}{H(u, v)}\right]$$

where $F^{-1}[\ ]$ represents the inverse Fourier transform.

The equation implies that the original, undegraded image can be obtained by multiplying the Fourier transform of the degraded image $D(u, v)$ by $1/H(u, v)$ and then inverse Fourier transforming the result. Thus, the restoration filter applied is $1/H(u, v)$, the *inverse filter*. Note that this inversion is a point-by-point inversion, *not* a matrix inversion.

EXAMPLE 3 - 2

$$H(u, v) = \begin{bmatrix} 50 & 50 & 25 \\ 20 & 20 & 20 \\ 20 & 35 & 22 \end{bmatrix} \qquad \frac{1}{H(u, v)} = \begin{bmatrix} \frac{1}{50} & \frac{1}{50} & \frac{1}{25} \\ \frac{1}{20} & \frac{1}{20} & \frac{1}{20} \\ \frac{1}{20} & \frac{1}{35} & \frac{1}{22} \end{bmatrix}$$

To find $1/H(u, v)$, we take each term separately and divide it into 1.

Unfortunately, in practice, complications arise when this technique is applied. If any points in $H(u, v)$ are zero, we face a mathematical dilemma—division by zero. If the assumption of no noise is correct, then the degraded image transform $D(u, v)$ will also have corresponding zeros, and we are left with an indeterminate ratio, 0/0. If the assumption is incorrect, and the image has been corrupted by additive noise, then the zeros will not coincide, and the image restored by the inverse filter will be obscured by the contribution of the noise terms. This can be seen by considering the following equation:

$$D(u, v) = H(u, v)I(u, v) + N(u, v)$$

Then, when we apply the inverse filter, we obtain

$$\hat{I}(u, v) = \frac{D(u, v)}{H(u, v)} = \frac{H(u, v)I(u, v)}{H(u, v)} + \frac{N(u, v)}{H(u, v)}$$
$$= I(u, v) + \frac{N(u, v)}{H(u, v)}$$

As the values in $H(u, v)$ become very small, the second term becomes very large, and it overshadows the $I(u, v)$ term, which corresponds to the original image we are trying to recover.

One method to deal with this problem is to limit the restoration to a specific radius about the origin in the spectrum, called the restoration cutoff frequency. For spectral components beyond this radius, we can set the filter gain to 0 ($\hat{I}(u, v) = 0$). This is the equivalent of an ideal lowpass filter, which may result in blurring and ringing. In practice, the selection of the cutoff frequency must be experimentally determined and is highly application specific. In Figure 3.4-4 we see the result of applying the inverse filter to an image blurred by an 11×11 gaussian convolution mask. Here

Figure 3.4-4 Inverse Filter

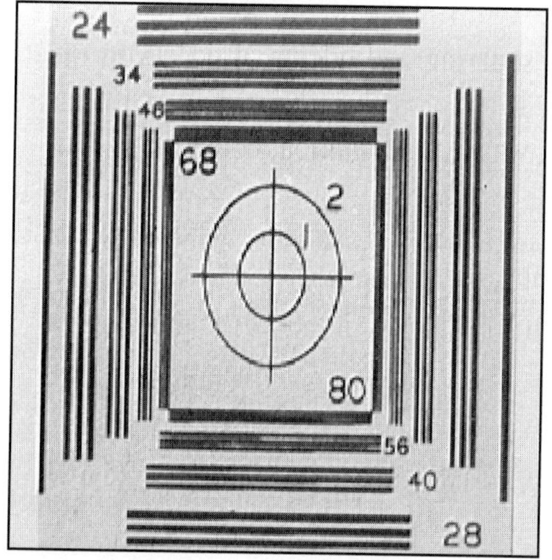

a. Original image.

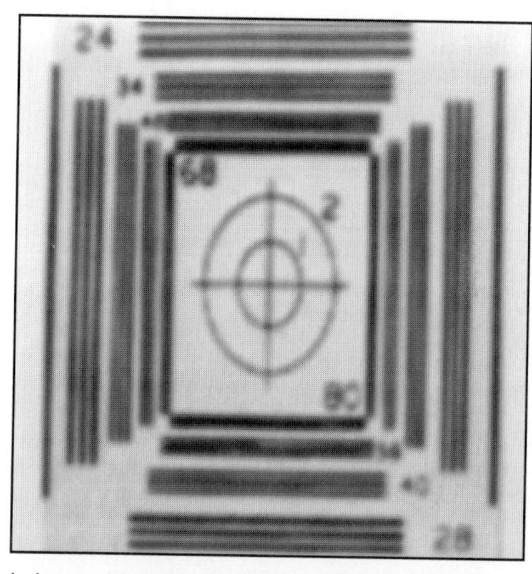

b. Image blurred with an 11×11 gaussian convolution mask.

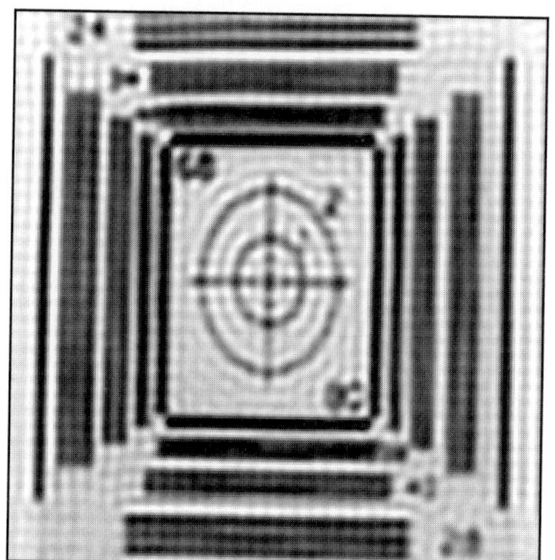

c. Inverse filter, with cutoff frequency = 40, histogram stretched with 3% low and high clipping to show detail.

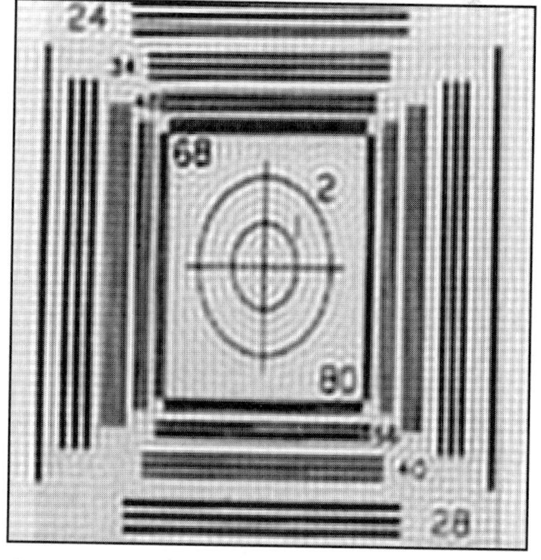

d. Inverse filter, with cutoff frequency = 60, histogram stretched.

Figure 3.4-4 (Continued)

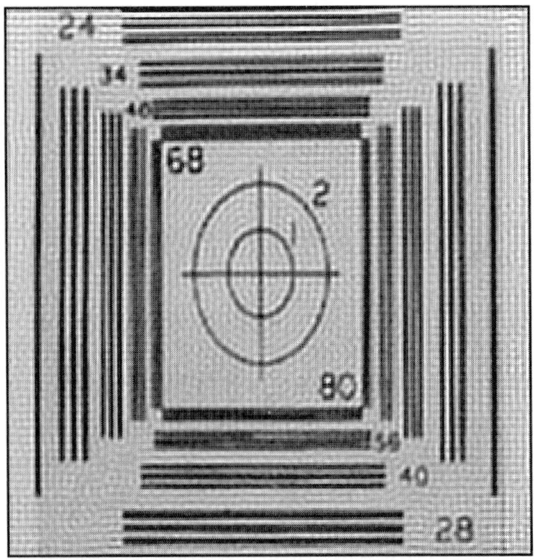

e. Inverse filter, with cutoff frequency = 80, histogram stretched.

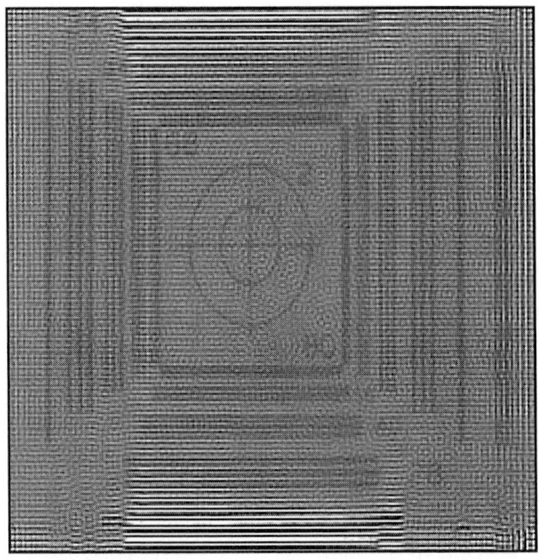

f. Inverse filter, with cutoff frequency = 100, histogram stretched.

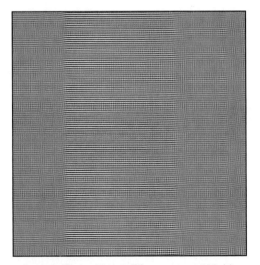

g. Inverse filter, with cutoff frequency = 120, histogram stretched.

we see that selection of a cutoff frequency that is too low may provide poor results, and with a cutoff frequency too high the resulting image is overwhelmed by noise effects.

With some types of degradation, the function $H(u, v)$ falls off quickly as we move away from the origin in the spectrum. In this case we may want to set the filter gain to 1 for frequencies beyond the restoration cutoff. Another possibility is to model a Butterworth filter, or something between the extremes of setting the gain to 0 or 1. In practice a similar result can be achieved by limiting the gain of the filter to some maximum value.

3.4.2 Wiener Filter

The Wiener filter, also called a minimum mean-square estimator (developed by Norbert Wiener in 1942), alleviates some of the difficulties inherent in inverse filtering by attempting to model the error in the restored image through the use of statistical methods. After the error is modeled, the average error is mathematically minimized, thus the term *minimum mean-square estimator*. The resulting equation is the Wiener filter:

$$R_W(u,v) = \frac{H^*(u,v)}{|H(u,v)|^2 + \left[\dfrac{S_n(u,v)}{S_I(u,v)}\right]}$$

$$\text{where } H^*(u,v) = \text{complex conjugate of } H(u,v)$$

$$S_n(u,v) = |N(u,v)|^2 = \text{power spectrum of the noise}$$

$$S_I(u,v) = |I(u,v)|^2 = \text{power spectrum of the original image}$$

This equation assumes a square image of size $N \times N$. The complex conjugate can be found by negating the imaginary part of a complex number. Other practical considerations are discussed in Section 3.4.6. Examining this equation will provide us with some understanding of how it works.

If we assume that the noise term $S_n(u, v)$ is zero, this equation reduces to an inverse filter since $|H(u, v)|^2 = H^*(u, v)H(u, v)$. As the noise term increases, the denominator of the Wiener filter increases, thus decreasing the value of $R_W(u, v)$. Thus, as the contribution of the noise increases, the filter gain decreases. This seems reasonable—in portions of the spectrum uncontaminated by noise we have an inverse filter, whereas in portions of the spectrum heavily corrupted by noise, the filter attenuates the signal, with the amount of attenuation being determined by the ratio of the noise spectrum to the uncorrupted image spectrum.

The Wiener filter is applied by multiplying it by the Fourier transform of the degraded image, and the restored image is obtained by taking the inverse Fourier transform of the result, as follows:

$$\hat{I}(r,c) = F^{-1}\left[\hat{I}(u,v)\right] = F^{-1}\left[R_W(u,v)D(u,v)\right]$$

Figure 3.4-5 compares the inverse filter and the Wiener filter. The filters are applied to images that have been blurred and then had various amounts of gaussian noise added. With small amounts of noise, the inverse filter works adequately, but when the noise level is increased, the Wiener filter results are obviously superior.

In practical applications the original, uncorrupted image is not typically available, so the power spectrum ratio is replaced by a parameter K whose optimal value must be experimentally determined:

$$R_W(u,v) = \frac{H^*(u,v)}{|H(u,v)|^2 + K}$$

Making the K parameter a function of the frequency domain variables (u, v) may also provide some added benefits. Because the noise typically dominates at high frequencies, it seems reasonable to have the value of K increase as the frequency increases, which will cause the filter to attenuate the signal at high frequencies. The following filter applies this idea.

3.4.3 Constrained Least-Squares Filter

The constrained least-squares filter provides a filter that can eliminate some of the artifacts caused by other frequency domain filters. This is done by including a

Figure 3.4-5 Wiener Filter

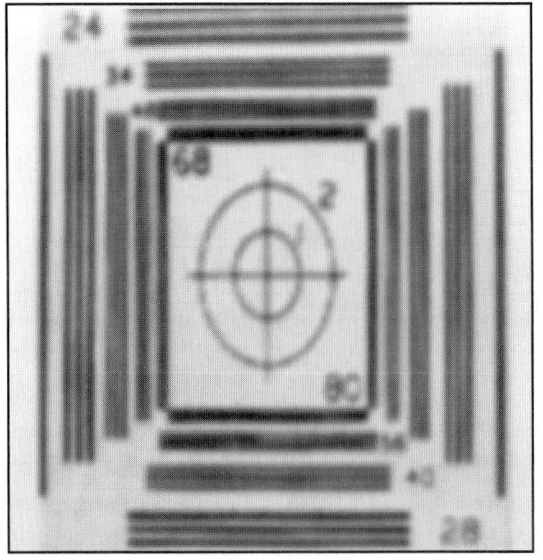

a. Image blurred with an 11×11 gaussian convolution mask.

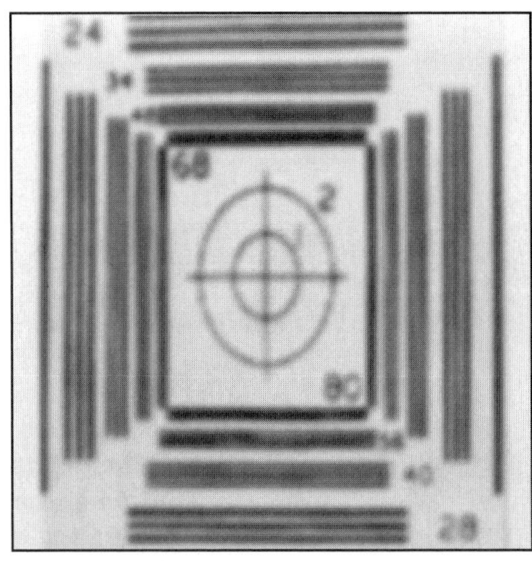

b. Image with gaussian noise—variance = 5; mean = 0.

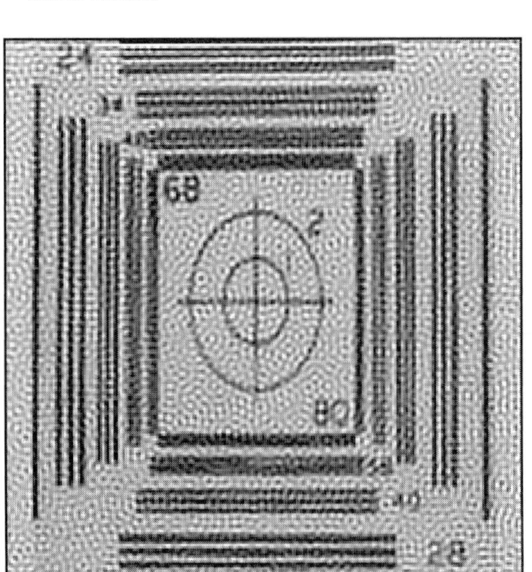

c. Inverse filter, with cutoff frequency = 80, histogram stretched with 3% low and high clipping to show detail.

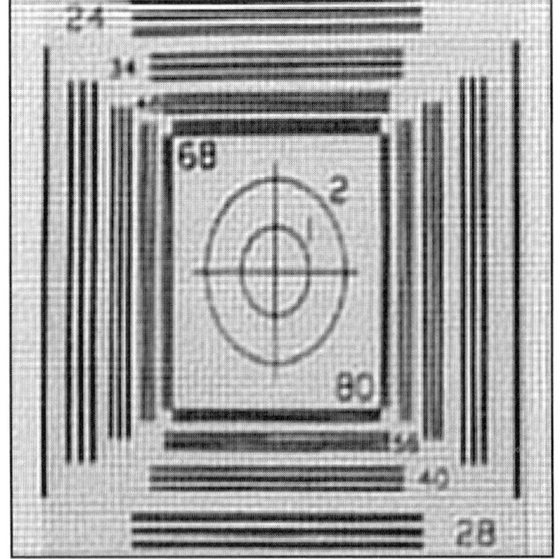

d. Wiener filter, with cutoff frequency = 80, histogram stretched.

Figure 3.4-5 (Continued)

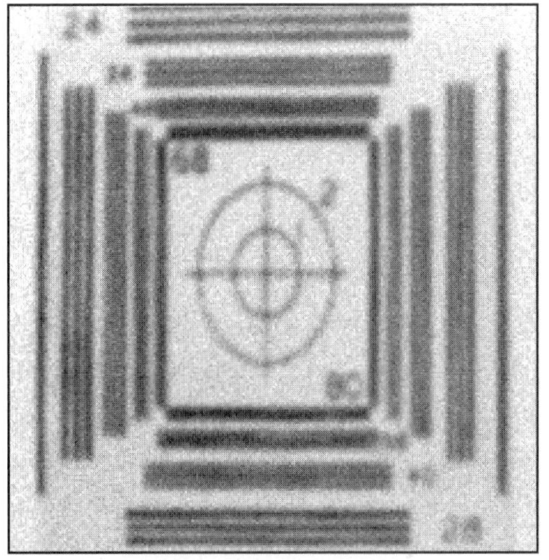

e. Image with gaussian noise—variance = 200; mean = 0.

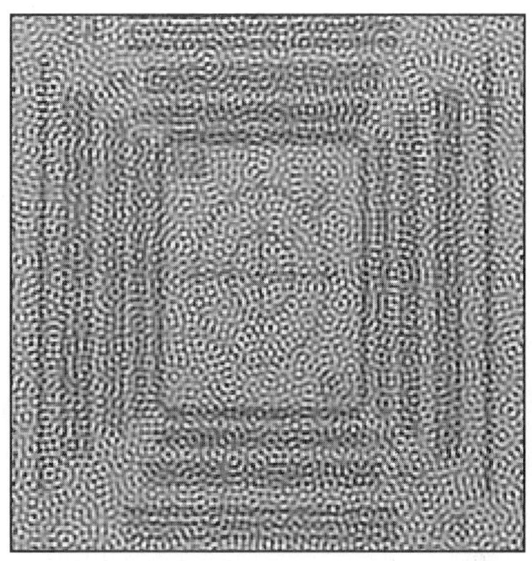

f. Inverse filter, with cutoff frequency = 80, histogram stretched.

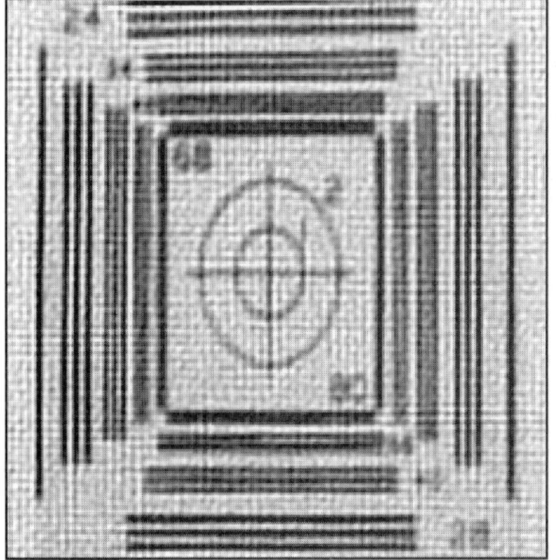

g. Wiener filter, with cutoff frequency = 80, histogram stretched.

smoothing criterion in the filter derivation, so that the result will not have undesirable oscillations (these appear as "waves" in the image), as sometimes occurs with other frequency domain filters. The constrained least-squares filter is given by

$$R_{\text{CLS}}(u, v) = \frac{H^*(u, v)}{|H(u, v)|^2 + \gamma |P(u, v)|^2}$$

where γ = adjustment factor

$P(u, v)$ = the Fourier transform of smoothness criterion function

The adjustment factor's value is experimentally determined and is application dependent. A standard function to use for $p(r, c)$ (the inverse Fourier transform of $P(u, v)$) is the laplacian filter mask, as follows:

$$p(r, c) = \begin{bmatrix} 0 & -1 & 0 \\ -1 & 4 & -1 \\ 0 & -1 & 0 \end{bmatrix}$$

However, before $P(u, v)$ is calculated, the $p(r, c)$ function must be extended with zeros (zero-padded) to the same size as the image. Figure 3.4-6 shows the results of applying this filter.

The constrained least-squares filter is applied by multiplying it by the Fourier transform of the degraded image, and the restored image is obtained by taking the inverse Fourier transform of the result as follows:

$$\hat{I}(r, c) = F^{-1}\left[\hat{I}(u, v)\right] = F^{-1}\left[R_{\mathrm{CLS}}(u, v)D(u, v)\right]$$

3.4.4 Geometric Mean Filters

The geometric mean filter equation provides us with a general form for many of the frequency domain restoration filters. It is defined as follows:

$$R_{\mathrm{GM}}(u, v) = \left[\frac{H^*(u, v)}{|H(u, v)|^2}\right]^{\alpha}\left[\frac{H^*(u, v)}{|H(u, v)|^2 + \gamma\left[\frac{S_n(u, v)}{S_I(u, v)}\right]}\right]^{1-\alpha}$$

The terms are as previously defined, with γ and α being positive real constants. If $\alpha = 1/2$ and $\gamma = 1$, this filter is called a *power spectrum equalization filter* (also called a *homomorphic filter*). If $\alpha = 1/2$, then this filter is an average between the inverse filter and the Wiener filter, hence the term geometric mean, although it is standard to refer to the general form of the equation as *geometric mean filter(s)*.

The geometric mean filter is applied by multiplying it by the Fourier transform of the degraded image, and the restored image is obtained by taking the inverse Fourier transform of the result, as follows:

$$\hat{I}(r, c) = F^{-1}\left[\hat{I}(u, v)\right] = F^{-1}\left[R_{\mathrm{GM}}(u, v)D(u, v)\right]$$

If $\alpha = 0$, this filter is called a *parametric Wiener filter*. The equation reduces to the Wiener filter equation, but with γ included as an adjustment parameter:

$$R_{\mathrm{PW}}(u, v) = \frac{H^*(u, v)}{|H(u, v)|^2 + \gamma\left[\frac{S_n(u, v)}{S_I(u, v)}\right]}$$

When $\gamma = 1$, this filter becomes a standard Wiener filter, and when $\gamma = 0$, this filter becomes the inverse filter. As γ is adjusted, the results vary between these two filters, with larger values providing more of the Wiener filtering effect.

Figure 3.4-6 Constrained Least-Squares Filter

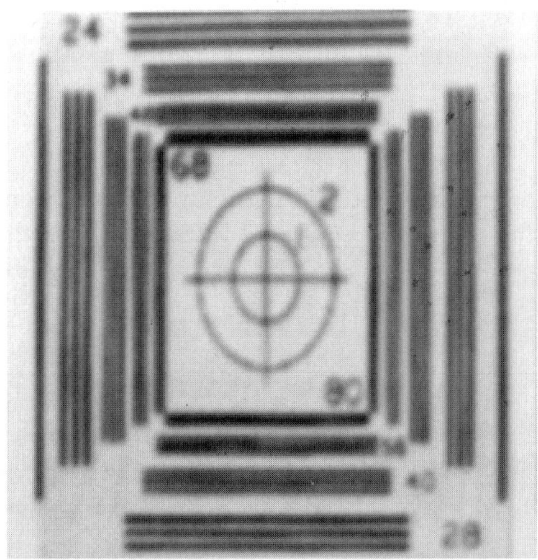

a. Blurred image with added gaussian noise, mean = 0, variance = 5.

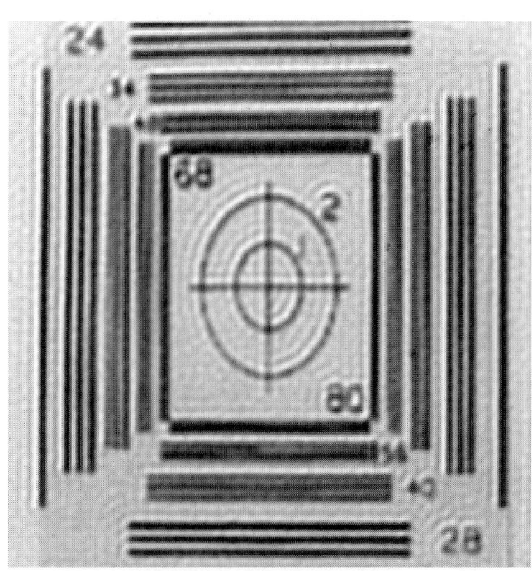

b. Result of CLS filter on a.

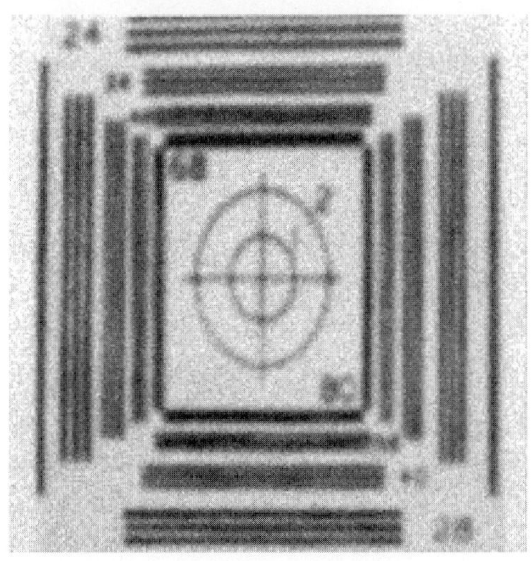

c. Blurred image with added gaussian noise, mean = 0, variance = 200.

d. Result of CLS filter on c.

The parametric Wiener filter is applied by multiplying it by the Fourier transform of the degraded image, and the restored image is obtained by taking the inverse Fourier transform of the result as follows:

$$\hat{I}(r, c) = F^{-1}\left[\hat{I}(u, v)\right] = F^{-1}\left[R_{\mathrm{PW}}(u, v)D(u, v)\right]$$

In general, the frequency domain filters work well for small amounts of blurring and moderate amounts of additive noise. The inverse filter is inadequate with too much noise, and the Wiener filter has the tendency to cause undesirable artifacts in the resultant image. The constrained least-squares filter helps to minimize the Wiener-type artifact, and the parametric Wiener and the geometric mean provide additional parameters, which can be adjusted for application-specific needs.

3.4.5 Notch Filter

The *notch filter* is a special form of a bandreject filter; instead of eliminating an entire ring of frequencies in the spectrum, it only "notches" out selected frequencies.

Figure 3.4-7 Notch Filter

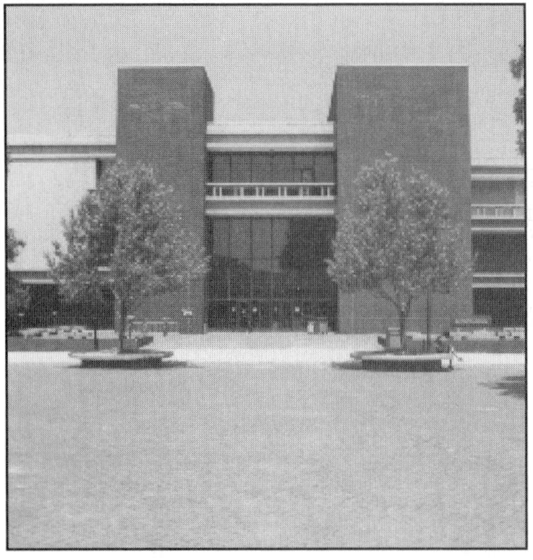

a. Original image.

b. Spectrum of original image.

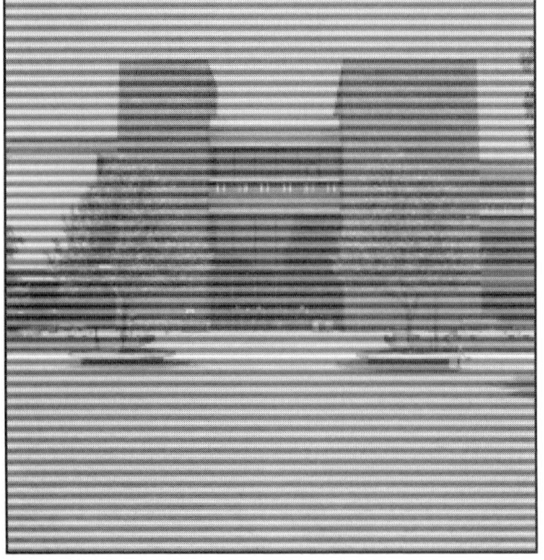

c. Image corrupted with sinusoidal noise.

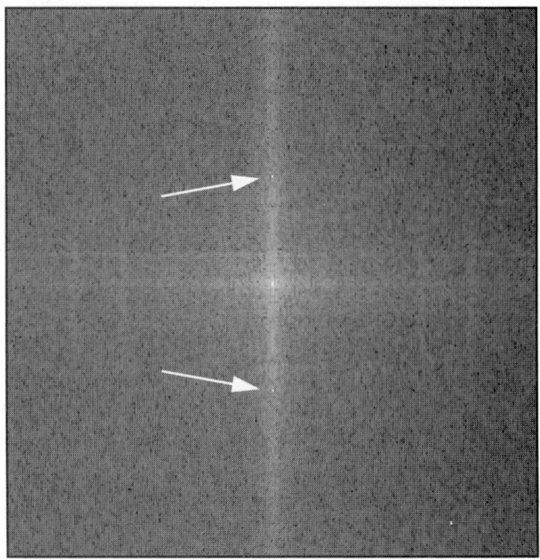

d. Spectrum of corrupted image; arrows point to contribution from interference.

Figure 3.4-7 (Continued)

e. Image restored by notch filtering.

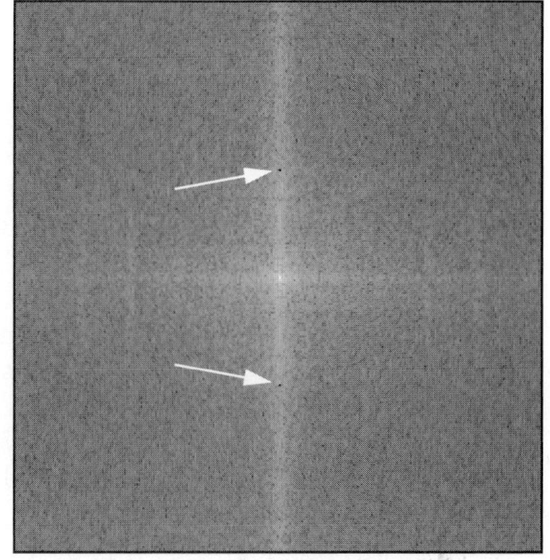

f. Spectrum of filtered image; arrows point to masked sinusoidal contribution.

g. Image further enhanced with histogram techniques.

This type of filter is most useful for an image that has been corrupted with a sinusoidal interference pattern. This type of image degradation is often seen in poor broadcast television images and is also a common artifact in images that have been obtained where the imaging device resides on some type of vibrating mechanical system such as a ship or a satellite.

For this type of image degradation, the spectrum will reveal the problem. Figures 3.4-7b, d show the type of spectrum that results from the sinusoidal interference. Bright spots in the spectrum corresponding to the interference can be seen. In

Figures 3.4-7e, f the restored image and the spectrum are shown. The portions of the spectrum that were causing the interference have been removed, effectively eliminating the interference pattern and noticeably improving the appearance of the image.

3.4.6 Practical Considerations

Using the Fourier transform as defined in Chapter 2, care must be taken when implementing the frequency domain filters. It is common practice to define the 2-D Fourier transform with a constant, $1/N$, in both the forward and inverse directions (as in Chapter 2), when it actually has a $1/N^2$ term in the forward direction only. This is done for symmetry, and it has no adverse effect on the Fourier transform pair since this is a linear process. However, it may affect the outcome of the frequency domain filters.

To avoid problems, the simplest method is to multiply each Fourier-transformed image by $1/N$, perform the filter calculations, multiply by the degraded image, and then multiply the result by N before passing it to the inverse Fourier transform. Note that for the power spectral density ratios, the division by N is not required because any constant multipliers will cancel when the ratio is taken. There are many different ways to deal with this problem, but it must be considered and dealt with appropriately or the results can be incorrect.

Care must be taken so that the degradation image $h(r, c)$ and the noise image $n(r, c)$ model the degradation process correctly. For example, most images are of type byte and thus have a range of 0 to 255. Typically the degradation image magnitude should be normalized, and the noise image should be mapped to the range of the added noise. Although we have discussed dealing with zeros in $H(u, v)$, it is also helpful to limit the gain of the restoration filter so that very small values in the denominator will not overwhelm the resulting image. When the image has been restored, simple postprocessing image enhancement methods, such as histogram equalization or a histogram stretch, can dramatically improve the visual results (compare Figures 3.4-6e, g).

Other methods of image restoration are explored in the references, including advanced preprocessing and postprocessing techniques to improve the results of the filters given here. Although many of these other methods are more complex, the improvements achieved are often minimal, or only applicable to a limited domain. Often image restoration requires a combination of techniques and, as with many computer imaging tasks, requires application-domain specific information.

3.5 GEOMETRIC TRANSFORMS

Geometric transforms are used to modify the location of pixel values within an image, typically to correct images that have been spatially distorted. These methods are often referred to as *rubber-sheet transforms* because the image is modeled as a sheet of rubber and stretched and shrunk, or otherwise manipulated, as required to correct for any spatial distortion. This type of distortion can be caused by defective optics in an image acquisition system, distortion in image display devices, or 2-D imaging of 3-D surfaces. The methods are used in mapmaking, image registration, image morphing, and other applications requiring spatial modification. It should be noted that the geometric transforms can also be used in image warping where the goal is to take a "good" image and distort it spatially.

The simplest geometric transforms—translate, rotate, zoom, and shrink—have already been discussed in Chapter 2. These transforms are limited to moving the pixels within an image in a fixed, regular manner and do not really distort the image but merely move pixel values. The more sophisticated geometric transforms, such as those discussed here, require two steps: 1) spatial transform and 2) gray-level interpolation. The model used for the geometric transforms is seen in Figure 3.5-1. The spatial transform provides the location of the output pixel, and the gray-level interpolation is necessary because pixel row and column coordinates provided by the spatial transform are not necessarily integers. The image is processed one pixel at a time, until the entire image has been transformed.

Figure 3.5-1 Geometric Transforms

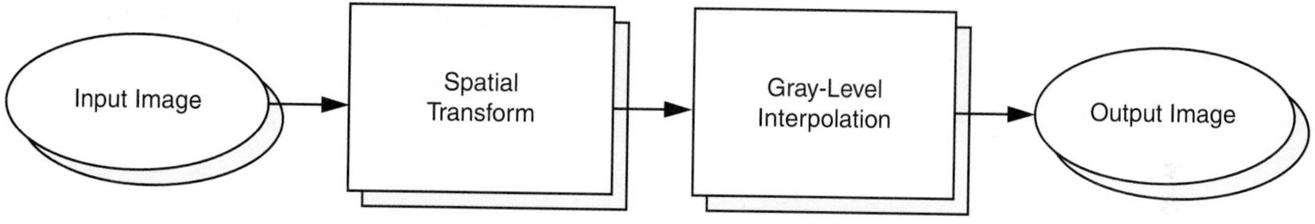

3.5.1 Spatial Transforms

Spatial transforms are used to map the input image location to a location in the output image; it defines how the pixel values in the output image are to be arranged. This process can be modeled as in Figure 3.5-2, where the original (undistorted) image is $I(r, c)$ and the distorted (or degraded) image is $d(\hat{r}, \hat{c})$. The distorted image coordinates can be defined by the two equations:

$\hat{r} = R(r, c)$ defines the row coordinate for the distorted image

$\hat{c} = C(r, c)$ defines the column coordinate for the distorted image

The primary idea presented here is to find a mathematical model for the geometric distortion process, specifically the two equations $R(r, c)$ and $C(r, c)$, and then apply the inverse process to find the restored image.

The type of distortion that occurs may vary across the image, so different equations for different portions of the image are often required. To determine the necessary equations, we need to identify a set of points in the original image that matches points

Figure 3.5-2 Spatial Transforms

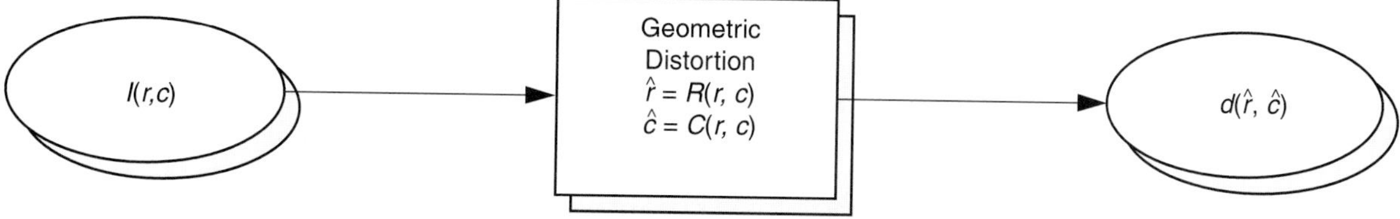

in the distorted image. These sets of points are called *tiepoints* and are used to define the equations $R(r, c)$ and $C(r, c)$. The form of these equations is typically bilinear, although higher-order polynomials can be used. The higher-order polynomials are much more computationally intensive, and there is no guarantee of better results—in some cases, the results may be worse (although it is wise to remember that we are dealing with subjective analysis regarding better or worse and that image restoration is more of an art than a science).

The method to restore a geometrically distorted image consists of three steps: 1) define quadrilaterals (four-sided polygons) with known or best-guessed tiepoints for the entire image, 2) find the equations $R(r, c)$ and $C(r, c)$ for each set of tiepoints, and 3) remap all the pixels within each quadrilateral subimage using the equations corresponding to those tiepoints.

Figure 3.5-3 illustrates step 1. The two images are divided into subimages, defined by the tiepoints (Figure 3.5-3a). Figure 3.5-3b shows not only the center sub-

Figure 3.5-3 Restoring Geometric Distortion

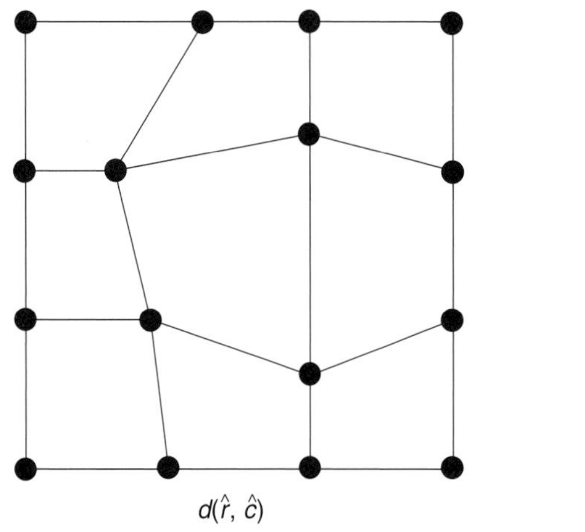

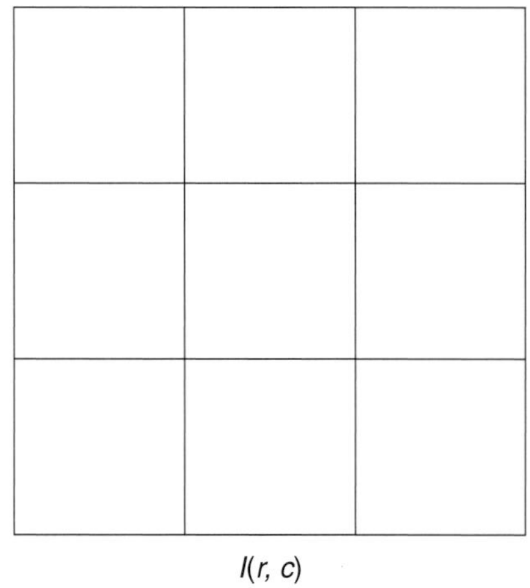

a. Images divided into quadrilateral subimages.

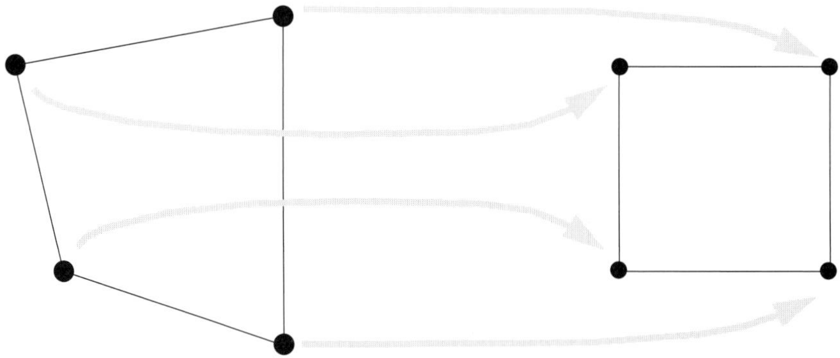

b. Center subimage showing corresponding points, also known as tiepoints.

image from both the distorted and the original images but also the corresponding tiepoints. The four corners are the tiepoints for this subimage and provide us with four pixels whose location is known in both images.

In step 2, using a bilinear model for the mapping equations, these four points are used to generate the equations:

$$R(r, c) = k_1 r + k_2 c + k_3 rc + k_4 = \hat{r}$$
$$C(r, c) = k_5 r + k_6 c + k_7 rc + k_8 = \hat{c}$$

The k_i values are constants to be determined by solving the eight simultaneous equations. Because we have defined four tiepoints, we have eight equations where r, c, $\hat{r}$, $\hat{c}$ are known (two for each point, one mapping the row coordinate, and one mapping the column coordinate). Now we can solve the eight equations for the eight unknowns, and we have the necessary equations for the coordinate mapping in step 3.

Step 3 involves application of the mapping equations $R(r, c)$ and $C(r, c)$ to all the (r, c) pairs in $I(r, c)$.

EXAMPLE 3-3

Assume that we have found the following mapping equations:

$$R(r, c) = 5r + 3c + 3rc + 2 = \hat{r}$$
$$C(r, c) = 1r + 1c + 2rc + 0 = \hat{c}$$

To find $I(2, 3)$, substitute $(r, c) = (2, 3)$ into the preceding equations. We find

$$R(r, c) = 5(2) + 3(3) + 3(2)(3) + 2 = 39$$
$$C(r, c) = 1(2) + 1(3) + 2(2)(3) + 0 = 17$$

Now, we let $I(2, 3) = d(39, 17)$.

Assuming that all the pixel value mappings worked out as well as the example, we could recover our original image $I(r, c)$ exactly. However, in practice the k_i values are not likely to cooperate and be integers. The following example illustrates this.

EXAMPLE 3-4

Assume that we have found the following mapping equations:

$$R(r, c) = 4.5r + 3c + 3.5rc + 2.4 = \hat{r}$$
$$C(r, c) = 1.6r + 1c + 2.4rc + 0 = \hat{c}$$

To find $I(2, 3)$, substitute $(r, c) = (2, 3)$ into the preceding equations. We find

$$R(r, c) = 4.5(2) + 3(3) + 3.5(2)(3) + 2.4 = 41.4$$
$$C(r, c) = 1.6(2) + 1(3) + 2.4(2)(3) + 0 = 20.6$$

Now, we want to set $I(2, 3) = d(41.4, 20.6)$.

The difficulty in this example arises when we try to determine the value of $d(41.4, 20.6)$. Because the digital images are defined only at the integer values for (r, c), gray-level interpolation must be performed.

3.5.2 Gray-Level Interpolation

The simplest method of gray-level interpolation is the *nearest neighbor method*, where the pixel is assigned the value of the closest pixel in the distorted image. In the preceding example the value of $I(2, 3)$ is set to the value of $d(41, 21)$, the row and column values determined by rounding the $\hat{r}$ and $\hat{c}$ result. This method is similar to the zero-order hold described in Section 2.2.1 for image enlargement. This method does not necessarily provide optimal results but has the advantage of being easy to implement and computationally fast.

Alternately, we can use a more advanced method to interpolate the value. In general these methods will be more computationally intensive but will provide more visually pleasing results. Figure 3.5-4 illustrates how this is done. The four surrounding pixel values in the distorted image are used to estimate the desired value, and this estimated value is used in the restored image. This can be done in a variety of ways. The easiest method is to find a *neighborhood average*. This can be done one dimensionally using the adjacent rows or columns, or it can be done two dimensionally using all four neighbors. The selection is application specific, but in general the 2-D average of the four neighbors will give better results.

Figure 3.5-4 Gray-Level Interpolation

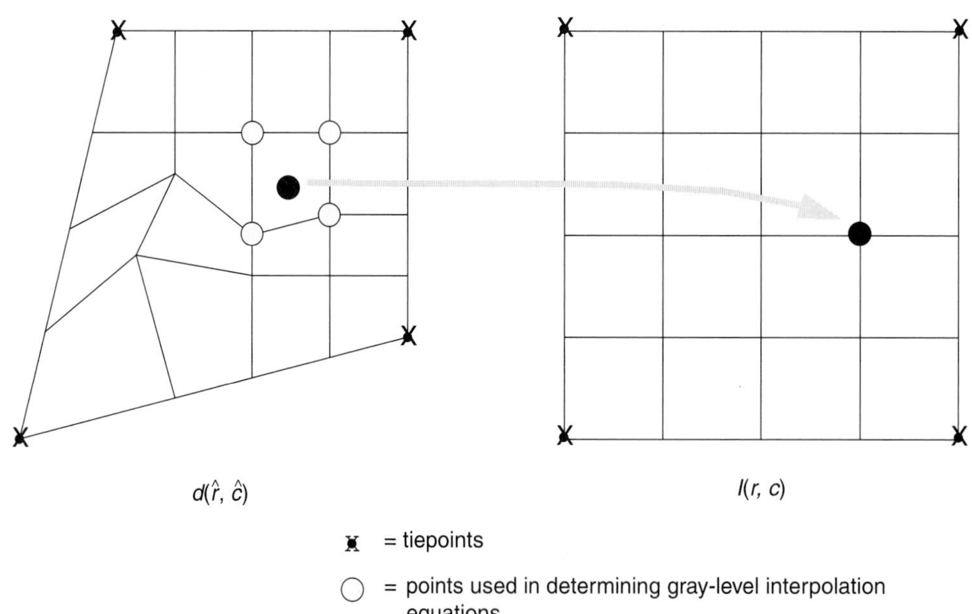

Also, a technique similar to the method used to find the spatial coordinates can be applied. This *bilinear interpolation* is done with the following equation:

$$g(\hat{r}, \hat{c}) = k_1\hat{r} + k_2\hat{c} + k_3\hat{r}\hat{c} + k_4$$

where $g(\hat{r}, \hat{c})$ = the gray-level interpolating equation

Note that these constants k_i are different than the constants used in the spatial mapping equations. The four unknown constants are found by using the four surrounding points shown in Figure 3.5-4. The values for row and column $(\hat{r}, \hat{c})$ and the gray-level values at each point are used.

EXAMPLE 3-5

Suppose that the four surrounding points are as follows:

$$d(\hat{r}, \hat{c}) \rightarrow d(1, 2) = 50,\ d(1, 3) = 55,\ d(2, 2) = 44,\ d(2, 3) = 48$$

Then we define the following four equations, and solve for the constants k_i:

$$50 = k_1(1) + k_2(2) + k_3(1)(2) + k_4$$
$$55 = k_1(1) + k_2(3) + k_3(1)(3) + k_4$$
$$44 = k_1(2) + k_2(2) + k_3(2)(2) + k_4$$
$$48 = k_1(2) + k_2(3) + k_3(2)(3) + k_4$$

Solving these equations simultaneously gives us:

$$k_1 = -4,\ k_2 = 6,\ k_3 = -1,\ k_4 = 44$$
$$\therefore g(\hat{r}, \hat{c}) = -4\hat{r} + 6\hat{c} - \hat{r}\hat{c} + 44$$

After the equation $g(\hat{r}, \hat{c})$ is found, the interpolated value can be determined. To do this we insert the noninteger values for row and column into the gray-level interpolating equation, and the resulting $g(\hat{r}, \hat{c})$ value is the interpolated gray-level value.

EXAMPLE 3-6

The preceding example assumes that the row and column coordinates are between rows 1 and 2 and columns 2 and 3; for example, $\hat{r} = 1.3$ and $\hat{c} = 2.6$. Applying these values to the preceding gray-level interpolating equation, we obtain

$$g(1.3, 2.6) = -4(1.3) + 6(2.6) - (1.3)(2.6) + 44 = 51.02 \approx 51$$

The gray-level value of 51 is then inserted into the restored image at the row and column location used to generate $\hat{r} = 1.3$ and $\hat{c} = 2.6$ from the mapping equations.

Figure 3.5-5 illustrates geometric restoration and compares the three gray-level interpolation methods discussed.

Figure 3.5-5 Geometric Restoration Example

a. Original image.

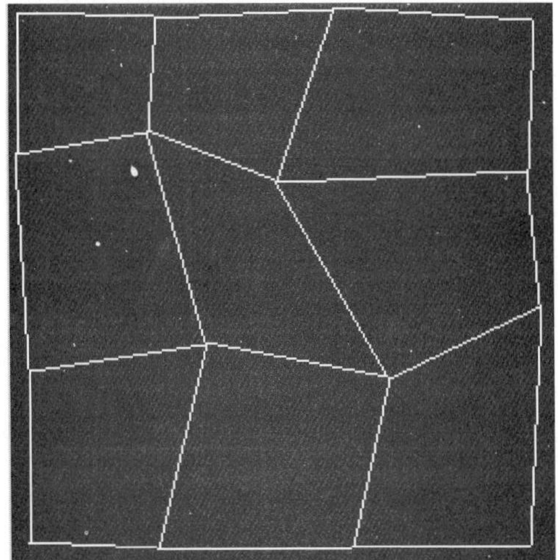

b. A mesh defined by 16 tiepoints will be used to first distort and then restore the image.

c. The original image has been distorted using the bilinear interpolation method.

d. Restoration by the nearest neighbor method shows the blocky effect that occurs at edges.

Figure 3.5-5: (Continued)

e. Restoration with neighborhood averaging interpolation provides smoother edges than with the nearest neighbor method, but it also blurs the image.

f. Restoration by bilinear interpolation provides optimal results. Note that some distortion occurs at the boundaries of the mesh quadrilaterals.

3.6 REFERENCES

The first complete text on image restoration is [Andrews/Hunt 77], and this book provides a solid foundation for the work that has been done since. For a survey of the field and a detailed look at some of the more recent research, see [Katsaggelos 91]. For more background and theory on image restoration, see [Bates/McDonnell 89] and [Sezan/Tekalp 96]. More information can also be found in the chapters on this topic in [Castleman 96], [Banks 90], [Jain 89], [Gonzalez/Woods 92], [Pratt 91], [Rosenfeld/Kak 82], [Lim 90], [Sid-Ahmed 95], and [Bracewell 95].

For the section on noise, the references [Andrews/Hunt 77], [Myler/Weeks 93], [Kennedy/Neville 86], and [Peebles 87] were consulted. More information regarding spatial filters can be found in [Myler/Weeks 93], [Tekalp 95], [Pitas/Venetsanopoulis 90], and [Haykin 91].

More information on frequency domain filters can be found in [Castleman 96], [Jain 89], [Gonzalez/Woods 92], [Pratt 91], [Rosenfeld/Kak 82], [Lim 90], [Sid-Ahmed 95], [Bates/McDonnell 89], [Jansson 97], and [Bracewell 95]. [Sid-Ahmed 95] and [Castleman 96] provide practical approaches to estimating the degradation function for image blurring. [Andrews/Hunt 77] provide more information on various PSFs, including spatially variant types, which are not discussed here. [Bates/McDonnell 89] provide advanced methods of preprocessing to improve the results of these filters. Much of the seminal work in communications theory, for example work on the Wiener filter, can be found in [Sloane/Wyner 93]. The geometric transforms information is in [Gonzalez/Woods 92], [Castleman 96], and [Pratt 91].

For general information on linear systems theory and digital signal processing theory, see [Oppenheim/Schafer 89], [Stanley/Dougherty/Dougherty 84], and [Tretter 76]. For more information on noise and estimation theory, see [VanTrees 68]. For more information on statistical or stochastic processes, see [Gray/Davisson 86] and [Peebles 87].

Andrews, H. C., and Hunt, B. R., *Digital Image Restoration*, Englewood Cliffs, NJ: Prentice Hall, 1977.

Banks, S., *Signal Processing, Image Processing and Pattern Recognition*, Englewood Cliffs, NJ: Prentice Hall, 1990.

Bates, R. H., and McDonnell, M. J., *Image Restoration and Reconstruction*, Oxford, UK: Oxford University Press, 1989.

Bracewell, R. N., *Two-Dimensional Imaging*, Englewood Cliffs, NJ: Prentice Hall, 1995.

Castleman, K. R., *Digital Image Processing*, Englewood Cliffs, NJ: Prentice Hall, 1996.

Gonzalez, R. C., and Woods, R. E., *Digital Image Processing*, Reading, MA: Addison-Wesley, 1992.

Gray, R. M., and Davisson, L. D., *Random Processes: A Mathematical Approach for Engineers*, Englewood Cliffs, NJ: Prentice Hall, 1986.

Haykin, S., *Adaptive Filter Theory*, Englewood Cliffs, NJ: Prentice Hall, 1991.

Jain, A. K., *Fundamentals of Digital Image Processing*, Englewood Cliffs, NJ: Prentice Hall, 1989.

Jansson, P. A. (Ed.), *Deconvolution of Images and Spectra*, 2nd Ed., New York: Academic Press, 1997.

Katsaggelos, A. K. (Ed.), *Digital Image Restoration*, New York: Springer-Verlag, 1991.

Kennedy, J. B., and Neville, A. M., *Basic Statistical Methods for Engineers and Scientists*, New York: Harper and Row, 1986.

Lim, J. S., *Two-Dimensional Signal and Image Processing*, Englewood Cliffs, NJ: Prentice Hall, 1990.

Myler, H. R., and Weeks, A. R., *Computer Imaging Recipes in C*, Englewood Cliffs, NJ: Prentice Hall, 1993.

Myler, H. R., and Weeks, A. R., *The Pocket Handbook of Image Processing Algorithms in C*, Englewood Cliffs, NJ: Prentice Hall, 1993.

Oppenheim, A. V., and Schafer, R. W., *Discrete-Time Signal Processing*, Englewood Cliffs, NJ: Prentice Hall, 1989.

Peebles, P. Z., *Probability, Random Variables, and Random Signal Principles*, New York: McGraw Hill, 1987.

Pitas, I., Venetsanopoulos, A. N., *Nonlinear Digital Filters*, Boston: Kluwer Academic, 1990.

Pratt, W. K., *Digital Image Processing*, New York: Wiley, 1991.

Rosenfeld, A., and Kak, A. C., *Digital Picture Processing*, San Diego, CA: Academic Press, 1982.

Sezan, I., and Tekalp, A. M., *Image Restoration*, Englewood Cliffs, NJ: Prentice Hall, 1996.

Sid-Ahmed, M. A., *Image Processing: Theory, Algorithms, and Architectures*, New York: McGraw Hill, 1995.

Sloane, N. J. A., and Wyner, A. D. (Eds.), *Claude Elwood Shannon, Collected Papers*, New York: IEEE Press, 1993.

Stanley, W. D., Dougherty, G. R., and Dougherty, R., *Digital Signal Processing*, Reston, VA: Reston Publishing, Prentice Hall, 1984.

Tekalp, A. M., *Digital Video Processing*, Englewood Cliffs, NJ: Prentice Hall, 1995.

Tretter, S. A., *Introduction to Discrete-Time Signal Processing*, New York: Wiley, 1976.

Van Trees, H. L., *Detection, Estimation, and Modulation Theory*, New York: Wiley, 1968.

CHAPTER 4

Image Enhancement

4.1 Introduction

Image enhancement techniques are used to emphasize and sharpen image features for display and analysis. *Image enhancement* is the process of applying these techniques to facilitate the development of a solution to a computer imaging problem. Consequently, the enhancement methods are application specific and are often developed empirically. Figure 4.1-1 illustrates the importance of the application by the feedback loop from the output image back to the start of the enhancement process and models the experimental nature of the development. In this figure we define the enhanced image as $E(r, c)$. The range of applications includes using enhancement techniques as preprocessing steps to ease the next processing step or as postprocessing steps to improve the visual perception of a processed image, or image enhancement may be an end in itself. Enhancement methods operate in the spatial domain by manipulating the pixel data or in the frequency domain by modifying the spectral components (Figure 4.1-2). Some enhancement algorithms use both the spatial and frequency domains.

The type of techniques includes *point operations*, where each pixel is modified according to a particular equation that is not dependent on other pixel values; *mask operations*, where each pixel is modified according to the values of the pixel's neighbors (using convolution masks); or *global operations*, where all the pixel values in the image (or subimage) are taken into consideration. Spatial domain processing methods include all three types, but frequency domain operations, by nature of the frequency (and sequency) transforms, are global operations. Of course, frequency domain operations can become "mask operations," based only on a local neighborhood, by performing the transform on small image blocks instead of the entire image.

Enhancement is used as a preprocessing step in some computer vision applications to ease the vision task, for example, to enhance the edges of an object to facilitate guidance of a robotic gripper. Enhancement is also used as a preprocessing step in

Figure 4.1-1 The Image Enhancement Process

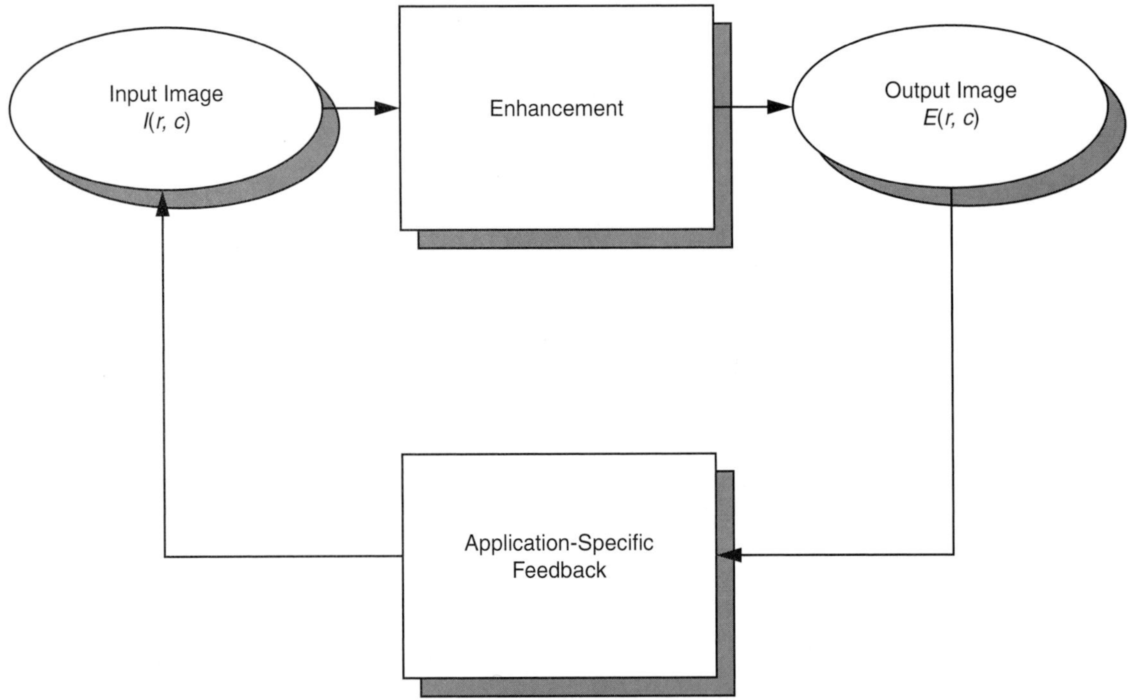

applications where human viewing of an image is required before further processing. For example, in one application, high-speed film images had to be correlated with a computer-simulated model of an aircraft. This process was labor intensive because the high-speed film generated many images per second and difficult because of the fact that the images were all dark. This task was made considerably easier by enhancing the images before correlating them to the model, enabling the technician to process many more images in one session.

Image enhancement is used for postprocessing to generate a visually desirable image. For instance, we may perform image restoration to eliminate image distortion and find that the output image has lost most of its contrast. Here, we can apply some basic image enhancement methods to restore the image contrast. Alternately, after a compressed image has been restored to its "original" state (decompressed), some postprocessing enhancement may significantly improve the look of the image. For example, the standard JPEG compression algorithm may generate an image with undesirable "blocky" artifacts, and postprocessing it with a smoothing filter (lowpass or mean) will improve the appearance.

Overall, image enhancement methods are used to make images look better. What works for one application may not be suitable for another application, so the development of enhancement methods require problem domain knowledge, as well as image enhancement expertise. Assessment of the success of an image enhancement algorithm is often "in the eye of the beholder," so image enhancement is as much an art as it is a science.

Figure 4.1-2 Image Enhancement

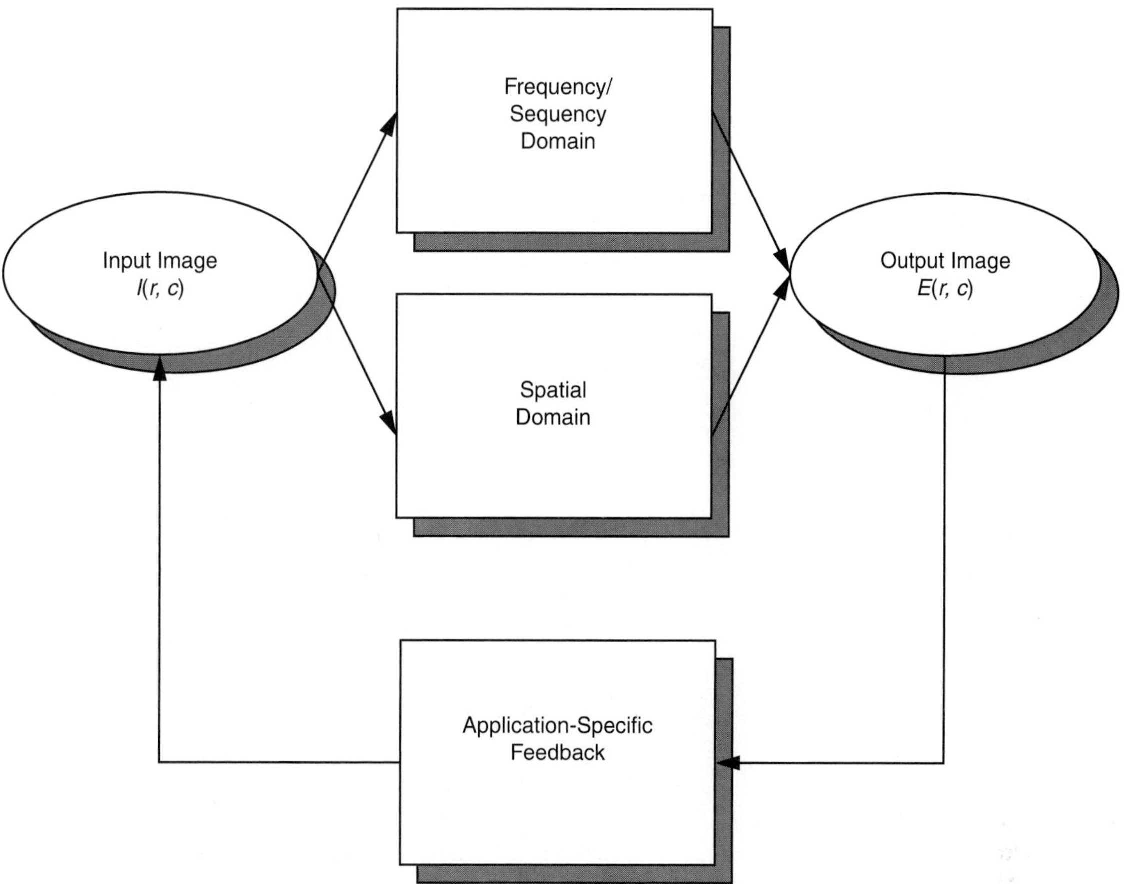

4.2 Gray-Scale Modification

Gray-scale modification (also called gray-level scaling) methods belong in the category of point operations and function by changing the pixel's (gray-level) values by a mapping equation. The *mapping equation* is typically linear (nonlinear equations can be modeled by piecewise linear models) and maps the original gray-level values to other, specified values. Typical applications include contrast enhancement and feature enhancement.

The primary operations applied to the gray scale of an image are to compress or stretch it. We typically compress gray-level ranges that are of little interest to us and stretch the gray-level ranges where we desire more information. This is illustrated in Figure 4.2-1a, where the original image data are shown on the horizontal axis and the modified values are shown on the vertical axis. The linear equations corresponding to the lines shown on the graph represent the mapping equations. If the slope of the line is between zero and one, this is called *gray-level compression*, whereas if the slope is greater than one, it is called *gray-level stretching*. In Figure 4.2-1a, the range of gray-level values from 28 to 75 is stretched, while the other gray values are left alone. The original and modified images are shown in Figures 4.2-1b, c, where we can see that stretching this range exposed previously hidden visual information. In some cases we may want to stretch a specific range of gray levels, while clipping the values at the low

Figure 4.2-1 Gray-Scale Modification

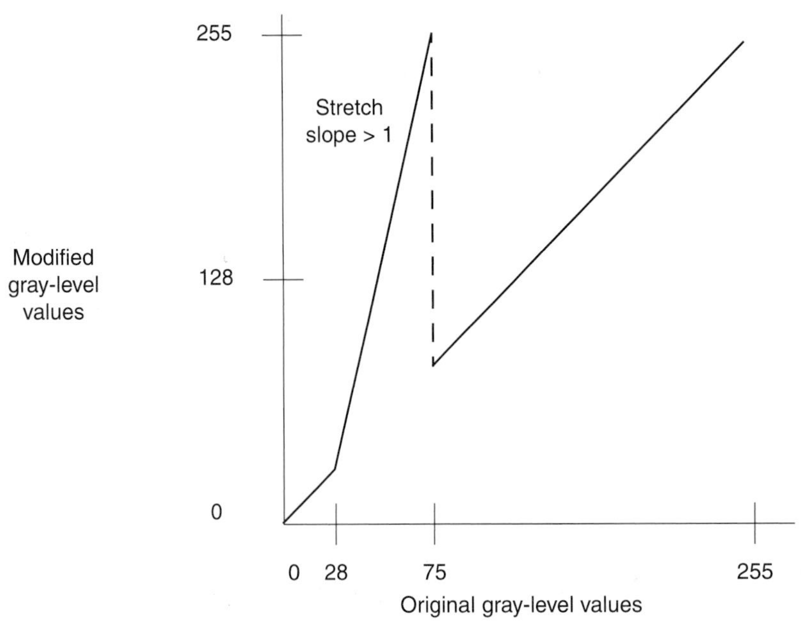

a. Gray-level stretching.

b. Original image.

c. Image after modification.

and high ends. Figure 4.2-1d illustrates a linear function to stretch the gray levels between 50 and 200, while clipping any values below 50 to 0 and any values above 200 to 255. The original and modified images are shown in Figures 4.2-1e, f, where we see the resulting enhanced image.

Figure 4.2-1 (Continued)

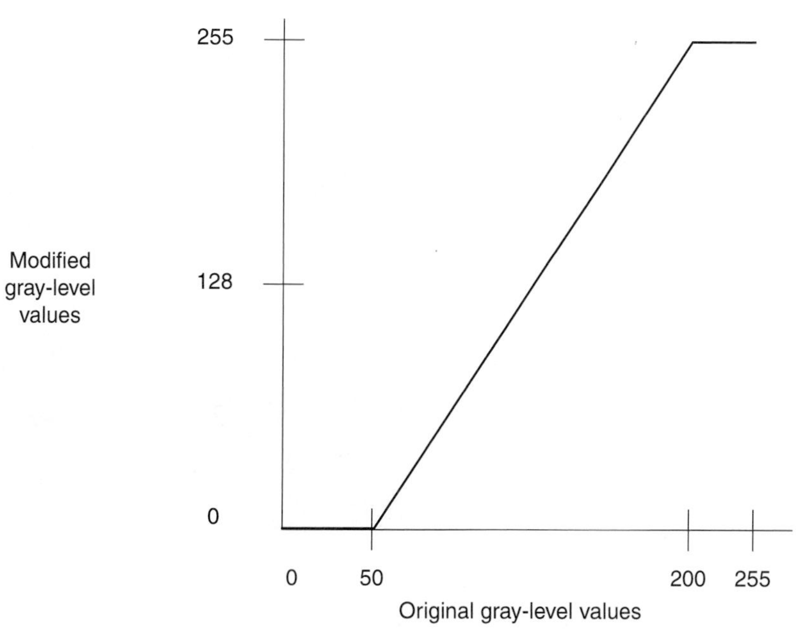

d. Gray-level stretching with clipping at ends.

e. Original image.

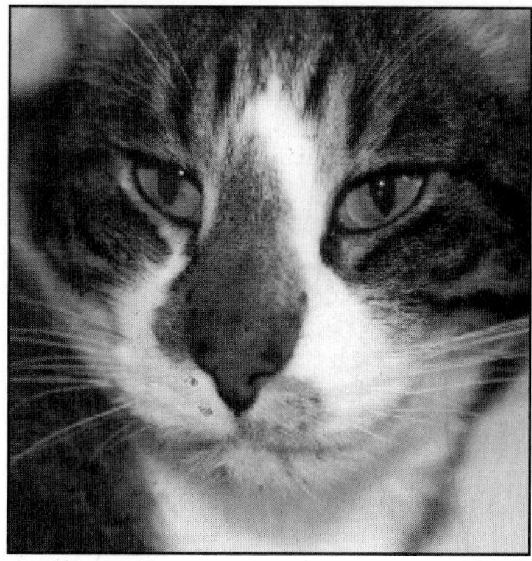

f. Image after modification.

Another type of mapping equation, used for feature extraction, is called intensity-level slicing. Here we are selecting specific gray-level values of interest and mapping them to a specified (typically high/bright) value. For example, we may have an application where it has been empirically determined that the objects of interest are in the gray-level range of 150 to 200. Using the mapping equations illustrated in Figures 4.2-2a, c, we can generate the resultant images in Figures 4.2-2b, d. With this type of operation, we can either leave the "background" gray-level values alone (Figure 4.2-2c) or turn them black (Figure 4.2-2e). Note that they do not need to be turned black; any gray-level value may be specified.

Figure 4.2-2 Intensity-Level Slicing

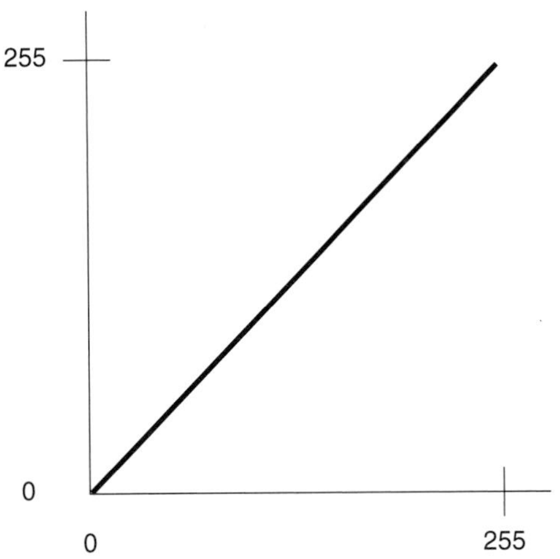

a. This operation returns the original gray levels.

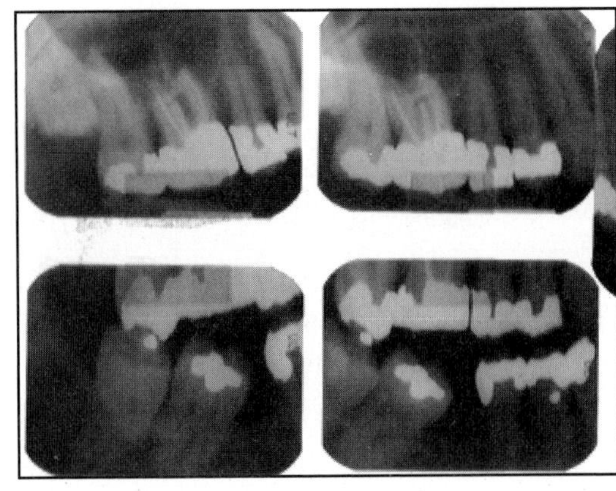

b. Original image.

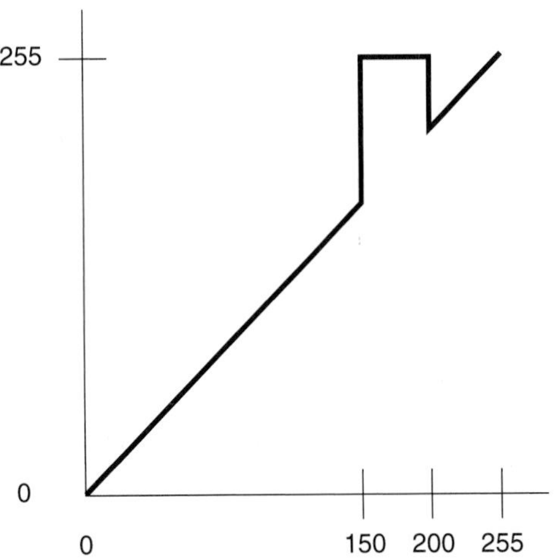

c. This operation intensifies the desired gray-level range, while not changing the other values.

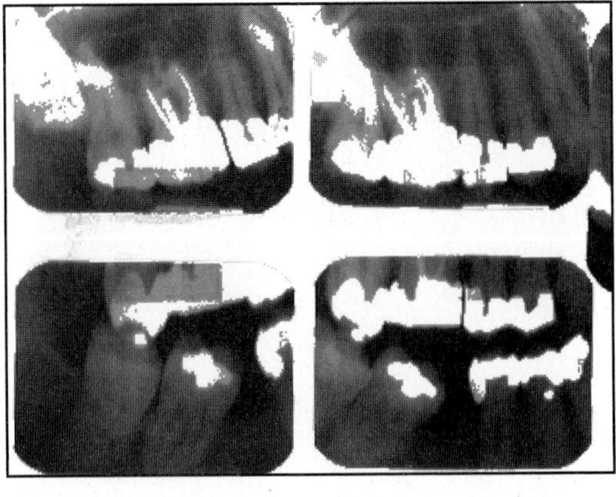

d. Image sliced to emphasize gray values from 150 to 200; background unchanged.

Figure 4.2-2 (Continued)

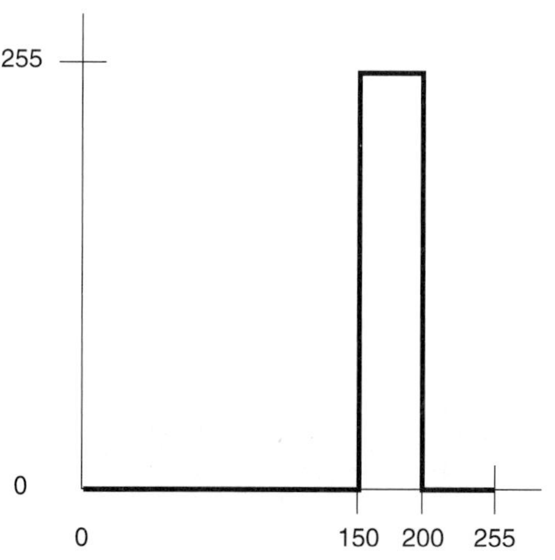

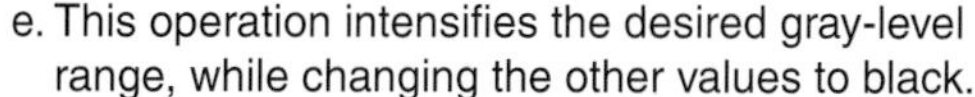

e. This operation intensifies the desired gray-level range, while changing the other values to black.

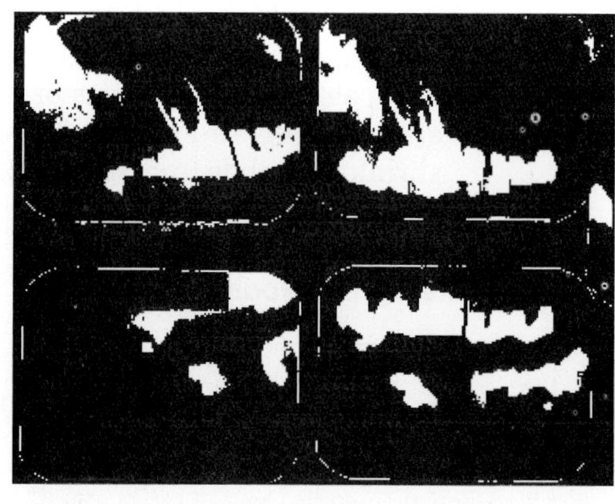

f. Image sliced to emphasize gray values from 150 to 200; background changed to black.

4.2.1 Histogram Modification

An alternate perspective to gray-level modification that performs a similar function is referred to as histogram modification. The *gray-level histogram* of an image is the distribution of the gray levels in an image. In Figure 4.2-3 we see an image and its corresponding histogram. In general, a histogram with a small spread has low contrast, and a histogram with a wide spread has high contrast, whereas an image with

Figure 4.2-3 Histogram

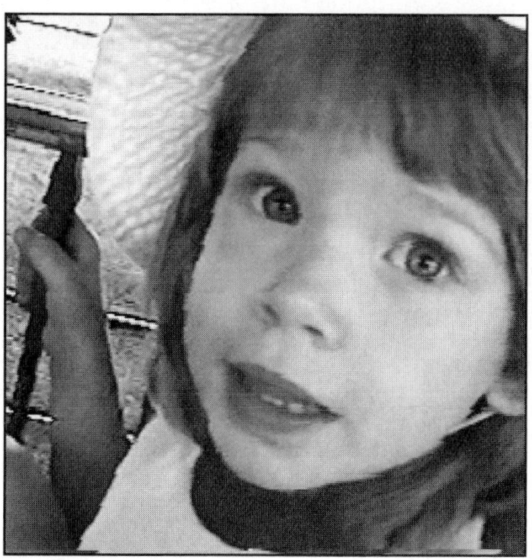

a. An 8-bit image.

Number of pixels

0 Gray value 255

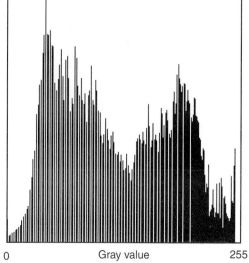

b. Histogram of the image.

its histogram clustered at the low end of the range is dark, and a histogram with the values clustered at the high end of the range corresponds to a bright image (see Figure 2.6-7). The histogram can also be modified by a mapping function, which will either stretch, shrink (compress), or slide the histogram. Histogram stretching and histogram shrinking are forms of gray-scale modification, sometimes referred to as *histogram scaling*. In Figure 4.2-4 we see a graphical representation of histogram stretch, shrink, and slide (see Section 6.3.2 and Image Processing exercise #6 in Chapter 8).

Figure 4.2-4 Histogram Modification

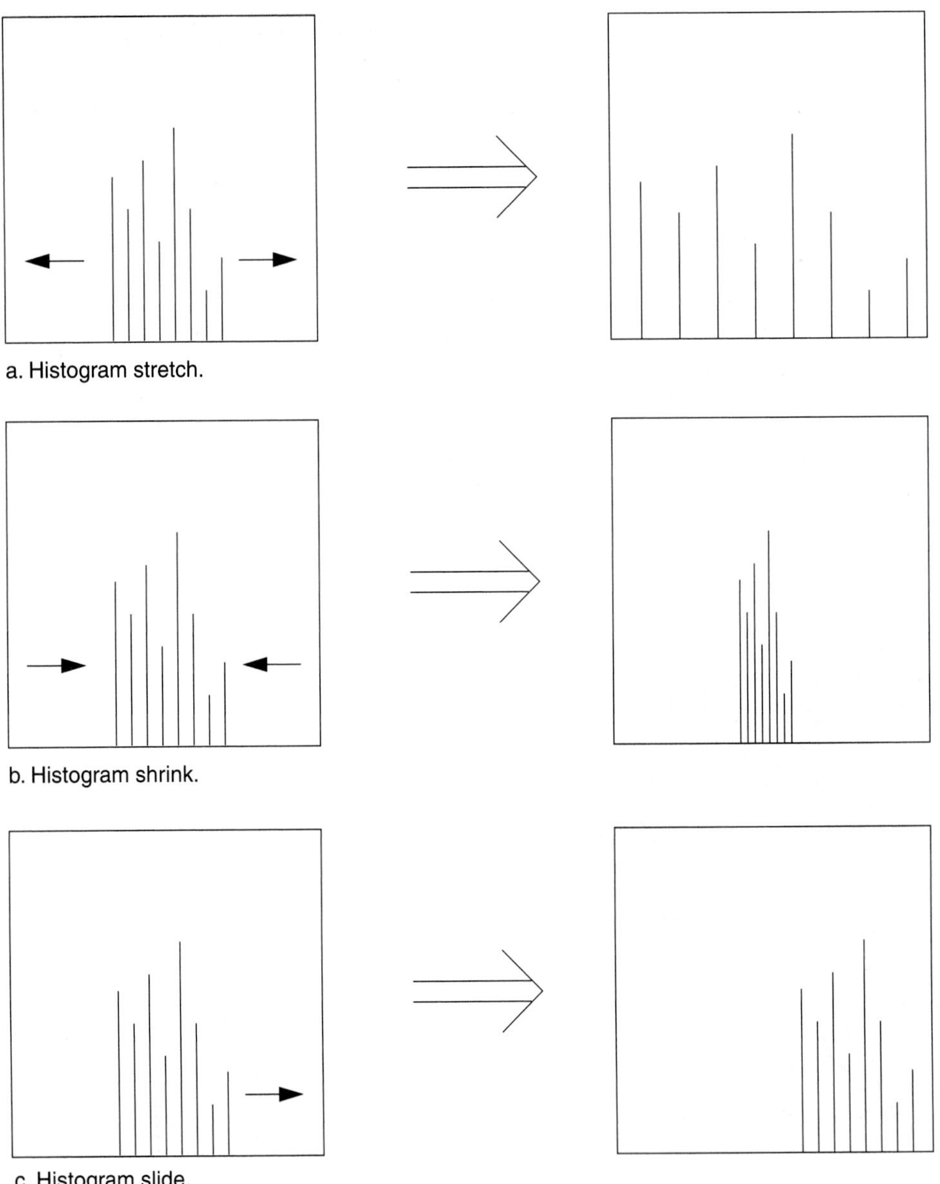

a. Histogram stretch.

b. Histogram shrink.

c. Histogram slide.

The mapping function for a histogram stretch can be found by the following equation:

$$\text{Stretch}(I(r, c)) = \left[\frac{I(r, c) - I(r, c)_{\text{MIN}}}{I(r, c)_{\text{MAX}} - I(r, c)_{\text{MIN}}}\right][\text{MAX} - \text{MIN}] + \text{MIN}$$

where

$I(r, c)_{\text{MAX}}$ is the largest gray-level value in the image $I(r, c)$

$I(r, c)_{\text{MIN}}$ is the smallest gray-level value in $I(r, c)$

MAX and MIN correspond to the maximum and minimum gray-level values possible (for an 8-bit image these are 0 and 255)

This equation will take an image and stretch the histogram across the entire gray-level range, which has the effect of increasing the contrast of a low contrast image (see Figure 4.2-5). If a stretch is desired over a smaller range, different MAX and MIN values can be specified.

If most of the pixel values in an image fall within a small range, but a few outliers force the histogram to span the entire range, a pure histogram stretch will not improve the image. In this case it is useful to allow a small percentage of the pixel values to be clipped at the low and high end of the range (for an 8-bit image this means truncating at 0 and 255). Figure 4.2-6 shows an example of this where we see a definite improvement with the stretched and clipped histogram compared to the pure histogram stretch.

The opposite of a histogram stretch is a histogram shrink, which will decrease image contrast by compressing the gray levels. The mapping function for a histogram shrink can be found by the following equation:

Figure 4.2-5 Histogram Stretching

a. Low-contrast image.

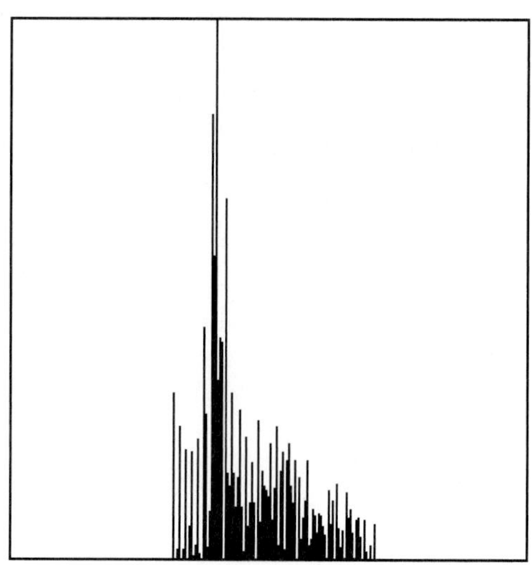

b. Histogram of low-contrast image.

$$\text{Shrink}(I(r, c)) = \left[\frac{\text{Shrink}_{\text{MAX}} - \text{Shrink}_{\text{MIN}}}{I(r, c)_{\text{MAX}} - I(r, c)_{\text{MIN}}}\right]\left[I(r, c) - I(r, c)_{\text{MIN}}\right] + \text{Shrink}_{\text{MIN}}$$

where

$I(r, c)_{\text{MAX}}$ is the largest gray-level value in the image $I(r, c)$

$I(r, c)_{\text{MIN}}$ is the smallest gray-level value in $I(r, c)$

Figure 4.2-5 (Continued)

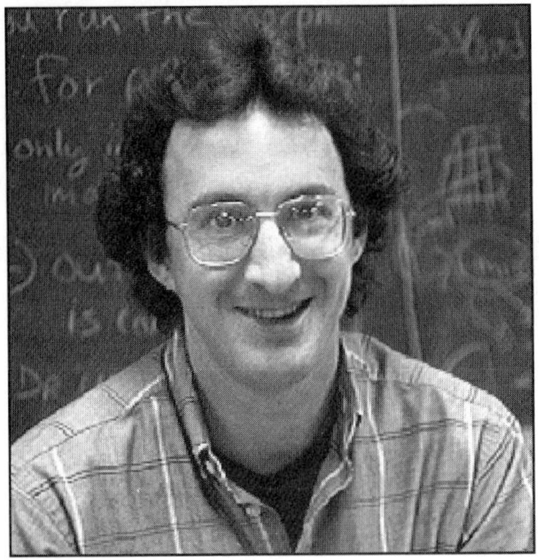

c. Image after histogram stretching.

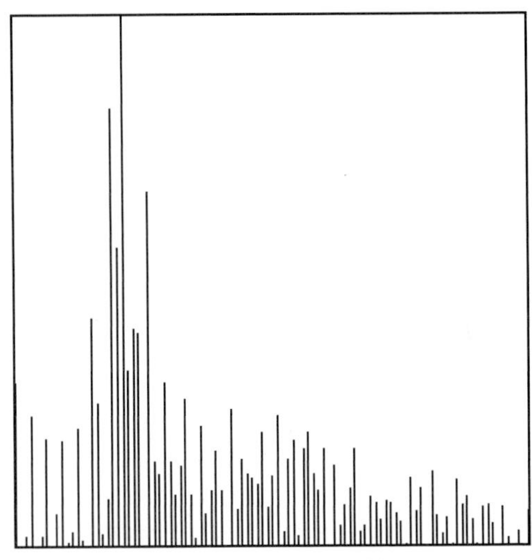

d. Histogram of image after stretching.

Figure 4.2-6 Histogram Stretching with Clipping

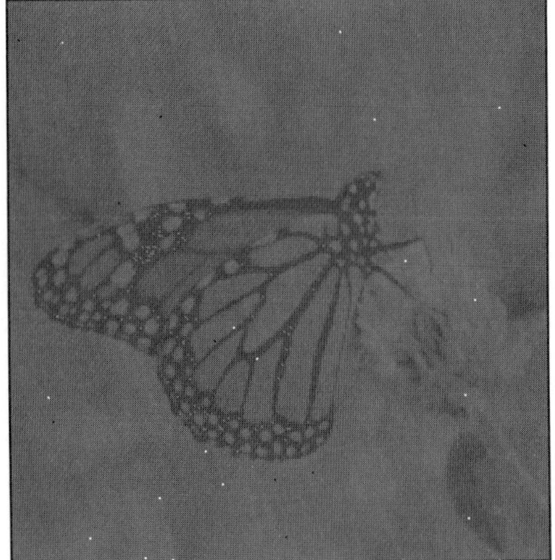

a. Original image.

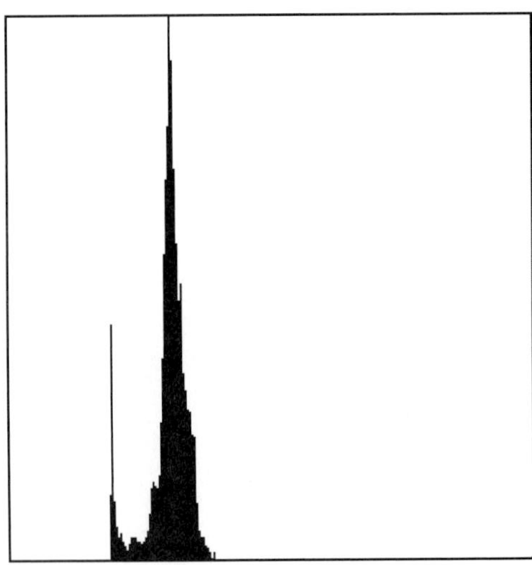

b. Histogram of original image.

$Shrink_{MAX}$ and $Shrink_{MIN}$ correspond to the maximum and minimum desired in the compressed histogram

Figure 4.2-7 illustrates a histogram shrink procedure. In Figures 4.2-7a, b we see an original image and its histogram, and Figures 4.2-7c, d show the result of the histogram shrink. In general, this process produces an image of reduced contrast and may not seem to be useful as an image enhancement tool. However, we will see (Section

Figure 4.2-6 (Continued)

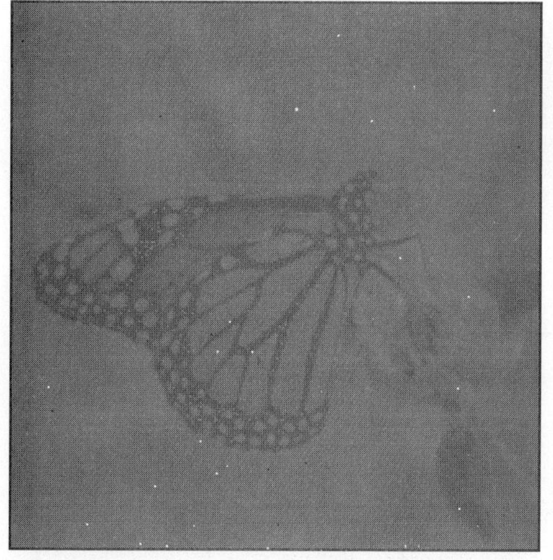

c. Image after histogram stretching without clipping.

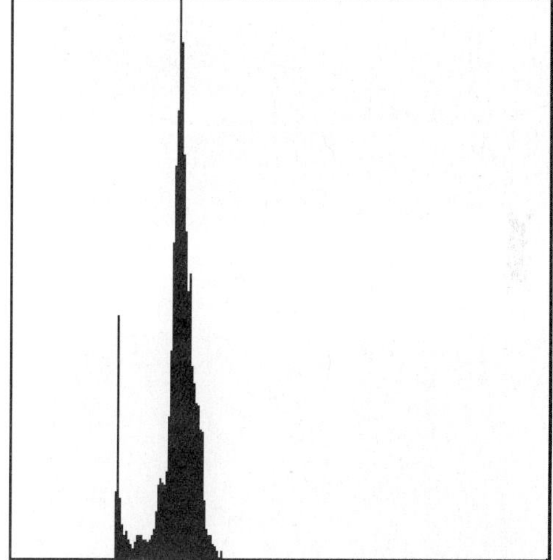

d. Histogram of image (c).

e. Image after histogram stretching with clipping 3% low and high values.

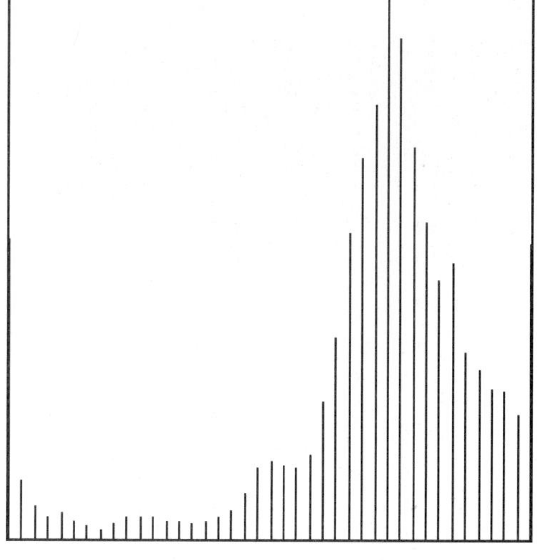

f. Histogram of image (e).

Figure 4.2-7 Histogram Shrinking

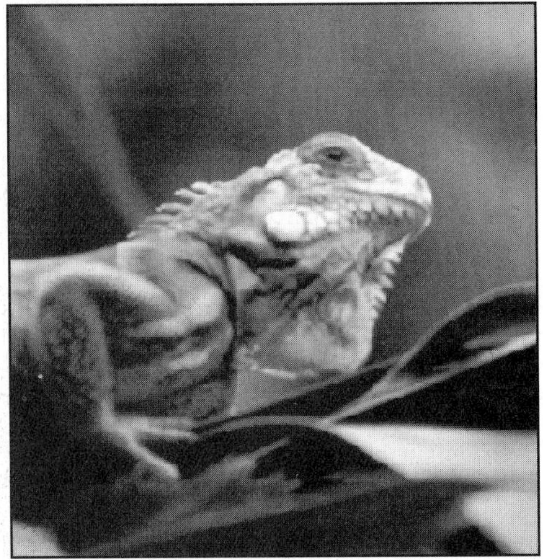

a. Original image.

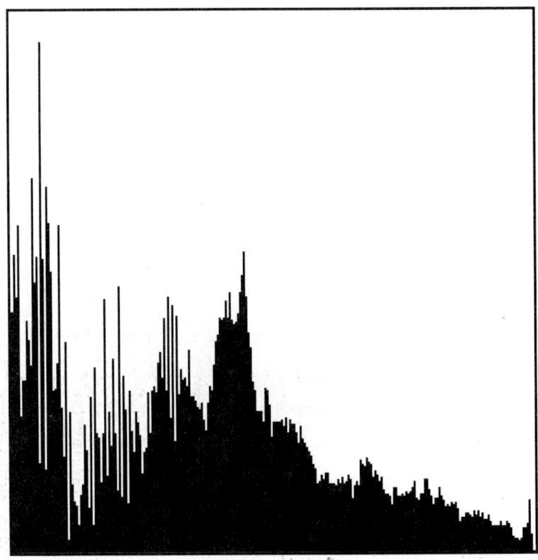

b. Histogram of image (a).

c. Image after shrinking the histogram to the range [75,175].

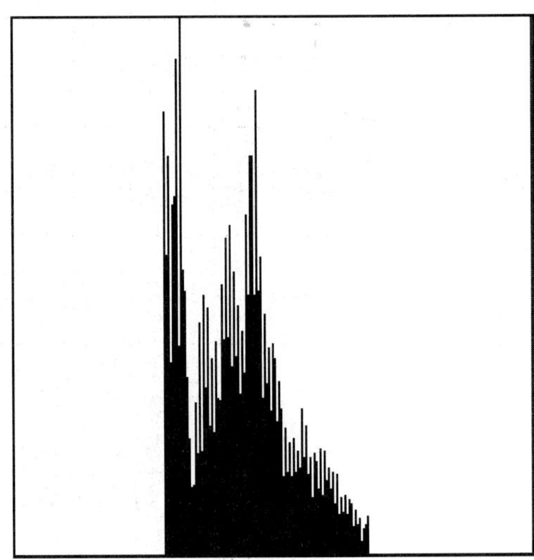

d. Histogram of image (c).

4.4) an image-sharpening algorithm that uses the histogram shrink process as part of an enhancement technique.

The histogram slide technique can be used to make an image either darker or lighter but retain the relationship between gray-level values. This can be accomplished by simply adding or subtracting a fixed number from all the gray-level values, as follows:

$$\text{Slide}(I(r, c)) = I(r, c) + \text{OFFSET}$$

where the OFFSET value is the amount to slide the histogram.

In this equation we assume that any values slid past the minimum and maximum values will be clipped to the respective minimum or maximum. A positive OFFSET value will increase the overall brightness, whereas a negative OFFSET will create a darker image. Figure 4.2-8 shows a dark image that has been brightened by a histogram slide with a positive OFFSET value.

Histogram equalization is a popular technique for improving the appearance of a poor image. Its function is similar to that of a histogram stretch but often provides more visually pleasing results across a wider range of images. *Histogram equalization* is a technique where the histogram of the resultant image is as flat as possible (with histogram stretching the overall shape of the histogram remains the same). The theoretical basis for histogram equalization involves probability theory, where we treat

Figure 4.2-8 Histogram Sliding

a. Original image.

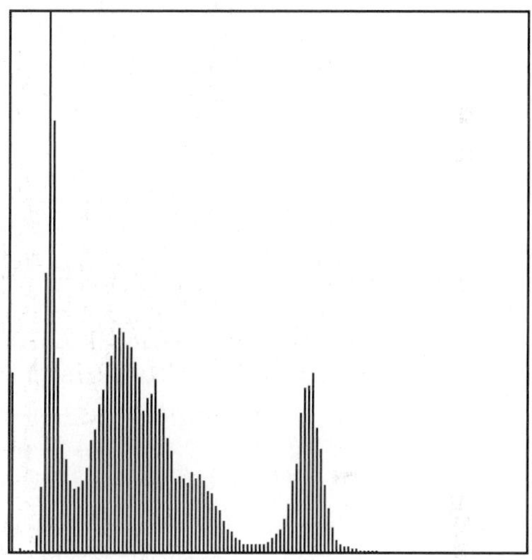

b. Histogram of original image.

c. Image after positive-value histogram sliding.

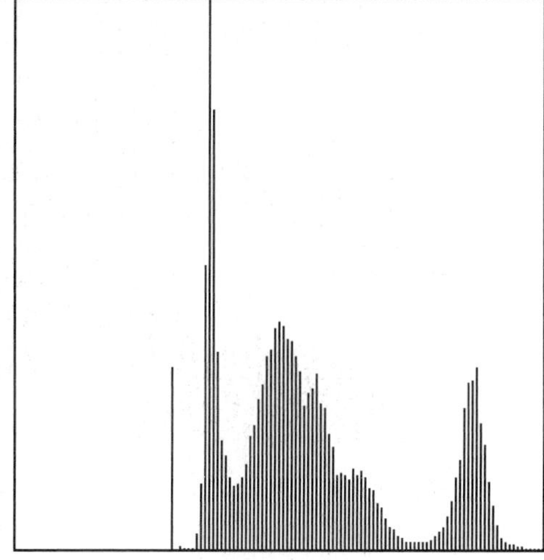

d. Histogram of image after sliding.

the histogram as the probability distribution of the gray levels. This is reasonable, since the histogram is the distribution of the gray levels for a particular image.

The histogram equalization process for digital images consists of four steps: 1) find the running sum of the histogram values, 2) normalize the values from step 1 by dividing by the total number of pixels, 3) multiply the values from step 2 by the maximum gray level value and round, and 4) map the gray-level values to the results from step 3 using a one-to-one correspondence. An example will help to clarify this process:

EXAMPLE 4-1

We have an image with 3 bits/pixel, so the possible range of values is 0 to 7. We have an image with the following histogram:

Gray-Level Value	*Number of Pixels (Histogram values)*
0	10
1	8
2	9
3	2
4	14
5	1
6	5
7	2

STEP 1: Create a running sum of the histogram values. This means that the first value is 10, the second is 10 + 8 = 18, next is 10 + 8 + 9 = 27, and so on. Here we get 10, 18, 27, 29, 43, 44, 49, 51.

STEP 2: Normalize by dividing by the total number of pixels. The total number of pixels is 10 + 8 + 9 + 2 + 14 + 1 + 5 + 0 = 51 (note this is the last number from step 1), so we get: 10/51, 18/51, 27/51, 29/51, 43/51, 44/51, 49/51, 51/51.

STEP 3: Multiply these values by the maximum gray-level values, in this case 7, and then round the result to the closest integer. After this is done we obtain 1, 2, 4, 4, 6, 6, 7, 7.

STEP 4: Map the original values to the results from step 3 by a one-to-one correspondence. This is done as follows:

Original Gray-Level Value	*Histogram Equalized Values*
0	1
1	2
2	4
3	4
4	6
5	6
6	7
7	7

All pixels in the original image with gray level 0 are set to 1, values of 1 are set to 2, 2 set to 4, 3 set to 4, and so on. In Figure 4.2-9 we see the original histogram and the resulting histogram-equalized histogram. Although the result is not flat, it is closer to being flat than the original histogram.

Figure 4.2-9 Histogram Equalization

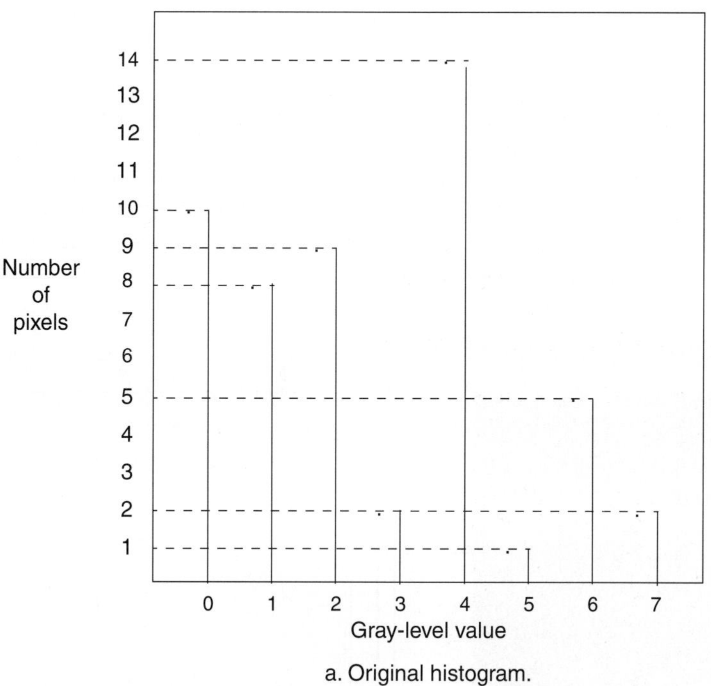

a. Original histogram.

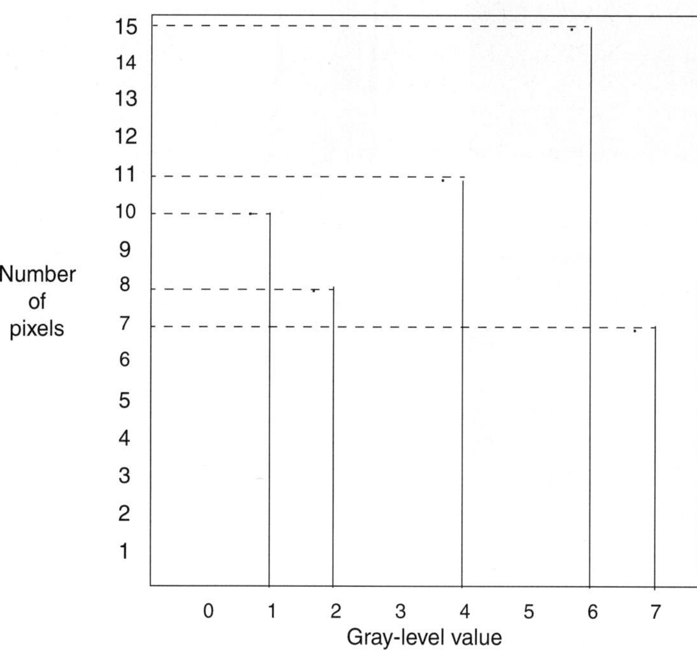

b. After histogram equalization.

Histogram equalization of a digital image will not typically provide a histogram that is perfectly flat, but it will make it as flat as possible. For the equalized histogram to be completely flat, the pixels at a given gray level might need to be redistributed across more than one gray level. This could be done but would greatly complicate the process, as some redistribution criteria would need to be defined. In most cases the visual gains achieved by doing this would be negligible and could in some cases be negative. In practice, it is not done.

Figure 4.2-10 shows the result of histogram equalizing two images with very poor contrast. In Figures 4.2-10a–d we see the results of applying histogram equalization to a dark image, and in Figures 4.2-10e–h we see results from a bright image. The results of this process are often very dramatic.

Figure 4.2-10 Histogram Equalization Example

a. Original dark image.

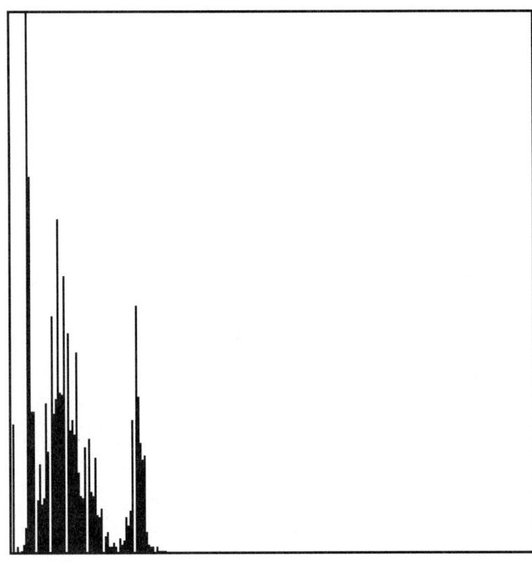

b. Histogram of image (a).

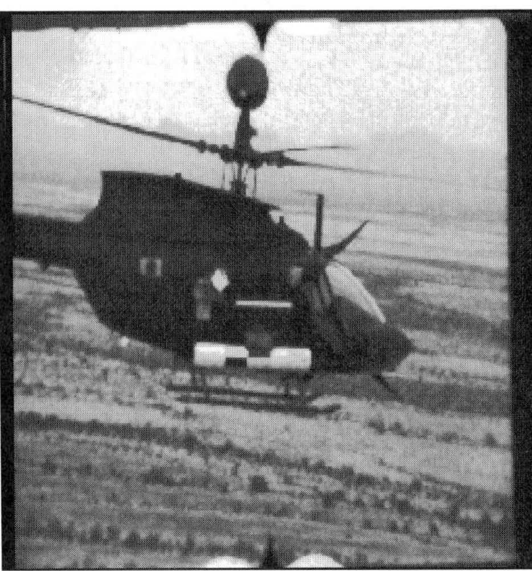

c. Dark image after histogram equalization.

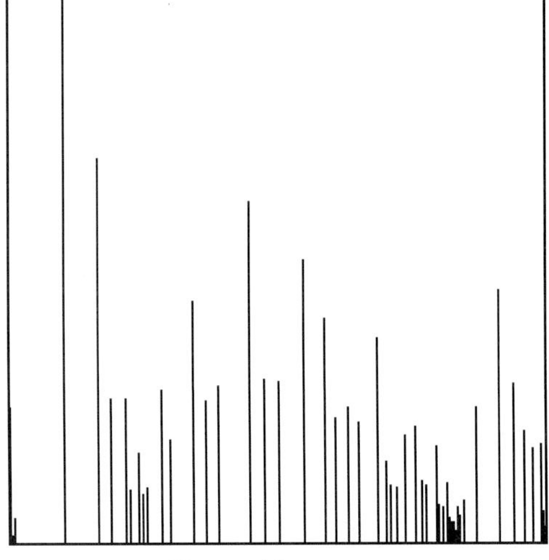

d. Histogram of image (c).

Figure 4.2-10 (Continued)

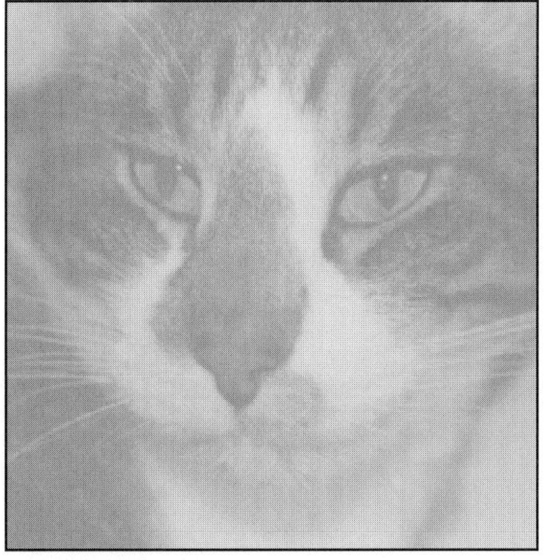

e. Original light image.

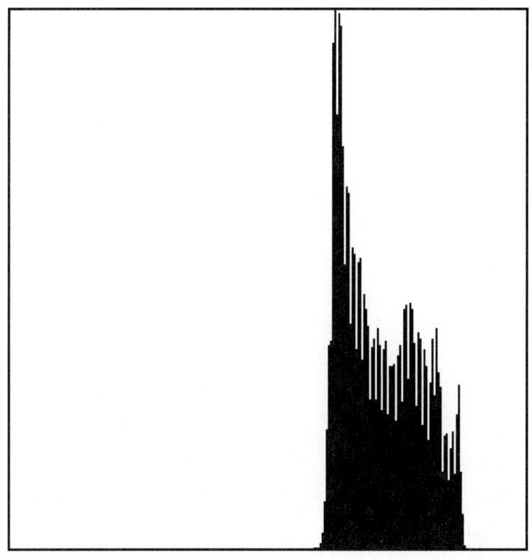

f. Histogram of image (e).

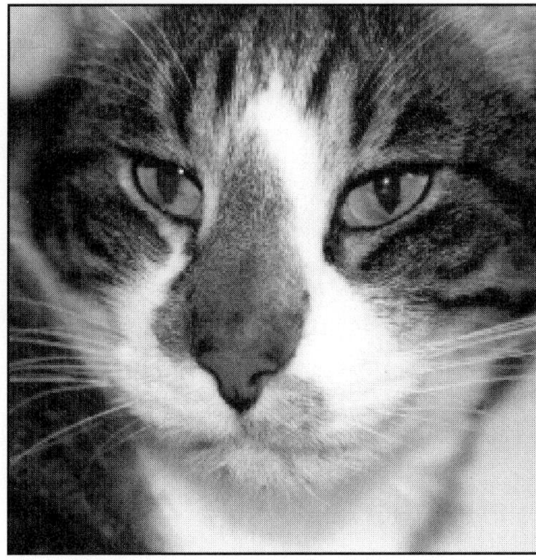

g. Light image after histogram equalization.

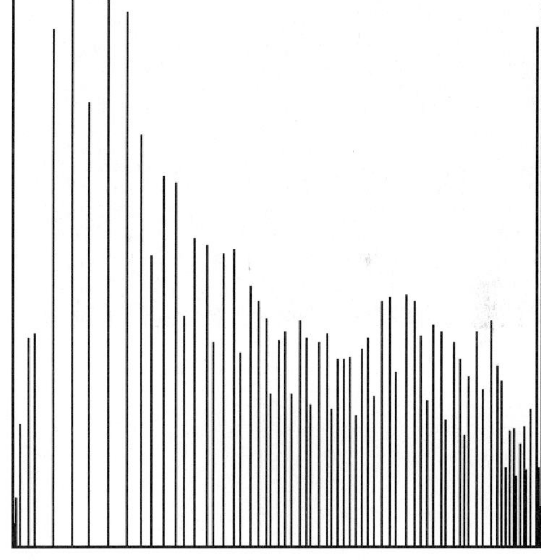

h. Histogram of image (g).

Histogram equalization may not always provide the desired effect because its goal is fixed—to distribute the gray-level values as evenly as possible. To allow for interactive histogram manipulation, the ability to specify the histogram is necessary. *Histogram specification* is the process of defining a histogram and modifying the histogram of the original image to match the histogram as specified. This process can be implemented by: 1) finding the mapping table to histogram-equalize the image, 2) specifying the desired histogram, 3) finding the mapping table to histogram-equalize the values of the desired histogram, and 4) mapping the original values to the values from step 3, by using the table from step 1. This process is best illustrated by example.

EXAMPLE 4-2

STEP 1: For this we will use the data from the previous example, where the histogram-equalization mapping table is given by:

Original Gray-Level Value—O	*Histogram Equalized Values—H*
0	1
1	2
2	4
3	4
4	6
5	6
6	7
7	7

STEP 2: Specify the desired histogram:

Gray-Level Value	*Number of Pixels in Desired Histogram*
0	1
1	5
2	10
3	15
4	20
5	0
6	0
7	0

STEP 3: Find the histogram equalization mapping table for the desired histogram:

Gray-Level Value	*Histogram Equalized Values—S*
0	round(1/51)*7 = 0
1	round(6/51)*7 = 1
2	round(16/51)*7 = 2
3	round(31/51)*7 = 4
4	round(51/51)*7 = 7
5	round(51/51)*7 = 7
6	round(51/51)*7 = 7
7	round(51/51)*7 = 7

STEP 4: Map the original values to values from step 3 by using the table from step 1. This is done by setting up a table created by combining the tables from steps 1 and 3. We will denote O for the original gray levels, H for the histogram-equalized levels, S for the specified and histogram-equalized values, and M for our final mapping, which will provide the desired histogram. The combined table consists of O and H from the step 1 table, S from the second column of the step 3 table, and M, which provides the resulting gray-level values and will be generated in this step. Here is the resulting table:

O	H	S	M
0	1	0	1
1	2	1	2
2	4	2	3
3	4	4	3
4	6	7	4
5	6	7	4
6	7	7	4
7	7	7	4

The M column for this table is obtained by mapping the value in H to the closest value in S and then using the corresponding row in O for the entry in M. For example, the first entry in H is 1. We find the closest value in S, which is 1. This 1 from S appears in row 1, so we write a 1 for that entry in M. Another example, the third entry in H is 4. We find the closest value in S, which is 4. This 4 from S appears in row 3, so we write a 3 for that entry in M. If we consider the fifth entry in H, we see that 6 must map to 7 (the closest value), but the 7 appears on rows 4, 5, 6, 7. Which one do we select? It depends on what we want; picking the largest value will provide maximum contrast, but picking the smallest (closest) value will produce a more gradually changing image. Typically, the smallest is chosen because we can always perform a histogram stretch or equalization on the output image, if we desire to maximize contrast.

In practice, the desired histogram is often specified by a continuous (possibly nonlinear) function, for example, a sine or a log function. To obtain the numbers for the specified histogram, the function is sampled, and the values are normalized to 1 and then multiplied by the total number of pixels in the image. Using a hyperbolic function for the specified histogram is called *histogram hyperbolization* and is based on a mathematical model of the response of the cones in the human visual system. Histogram hyperbolization is an attempt to make the perceived brightness levels equally likely—sort of a "histogram equalization" based on a model of the visual system.

4.2.2 Adaptive Contrast Enhancement

Adaptive contrast enhancement refers to modification of the gray-level values within an image based on some criterion that adjusts its parameters as local image characteristics change. The most straightforward method of doing this is to perform a histogram modification technique, applying it to the image on a block-by-block basis instead of doing it globally (on the entire image). In Figure 4.2-11 are the results of applying a histogram equalization to various block sizes within an image, and these are contrasted with a global histogram equalization. This technique is also called *local enhancement*.

The *adaptive contrast enhancement* (ACE) filter is used with an image that appears to have uneven contrast, where we want to adjust the contrast differently in different regions of the image. It works by using both local and global image statistics to determine the amount of contrast adjustment required. This filter is adaptive in the sense that its behavior changes based on local image statistics, unlike the histogram

Figure 4.2-11 Local Histogram Equalization

a. Original image.

b. Image after global histogram equalization.

c. Image after local histogram equalization, window = 7×7.

d. Image after local histogram equalization, window = 15×15.

modification techniques, which use only global parameters and result in fixed gray-level transformations. The image is processed using the sliding window concept (see Figure. 3.3-1), the local image statistics are found by considering only the current window (subimage), and the global parameters are found by considering the entire image. It is defined as follows:

$$\text{ACE} = k_1 \left[\frac{m_{I(r,c)}}{\sigma_l(r,c)} \right] \left[I(r,c) - m_l(r,c) \right] + k_2 \; m_l(r,c)$$

where $m_{I(r,c)}$ = is the mean for the entire image $I(r, c)$

σ_l = local standard deviation (in the window under consideration)

m_l = local mean (average in the window under consideration)

k_1, k_2 = constants, vary between 0 and 1

From the equation we can see that this filter subtracts the local mean from the original data and weights the result by the local gain factor, $k_1 \, [m_{I(r,c)}/\sigma_l(r, c)]$. This has the effect of intensifying local variations and can be controlled by the constant, k_1. Areas of low contrast (low values of $\sigma_l(r, c)$) are boosted. The mean is then added back to the result, weighted by k_2, to restore the local average brightness.

In practice it is often helpful to shrink the histogram of the image before applying this filter and to limit the gain from the local gain factor. Figure 4.2-12 illustrates

Figure 4.2-12 Adaptive Contrast Enhancement Filter

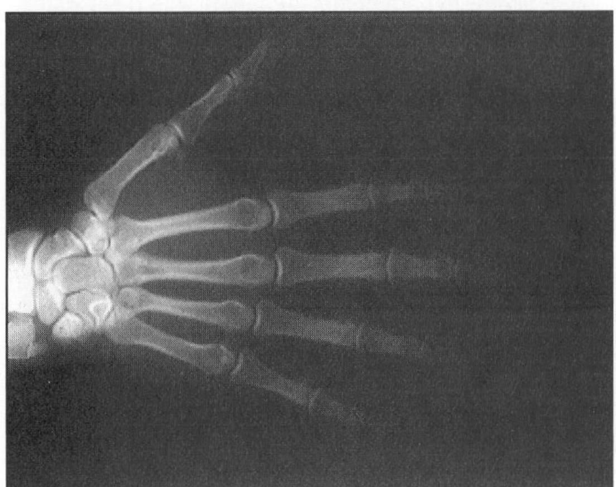

a. Original image.

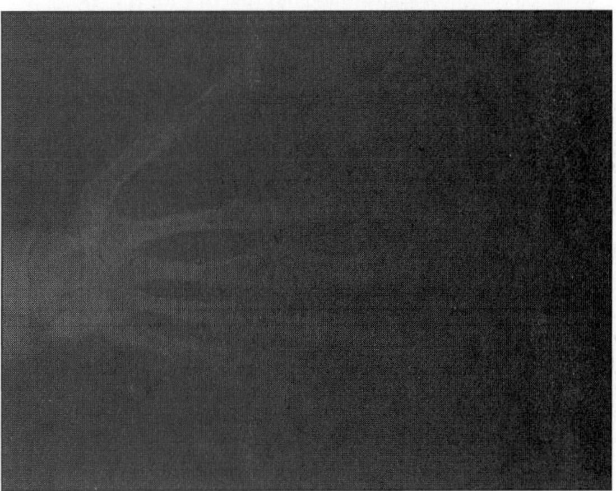

b. Histogram shrink version of (a)—Range [25,60].

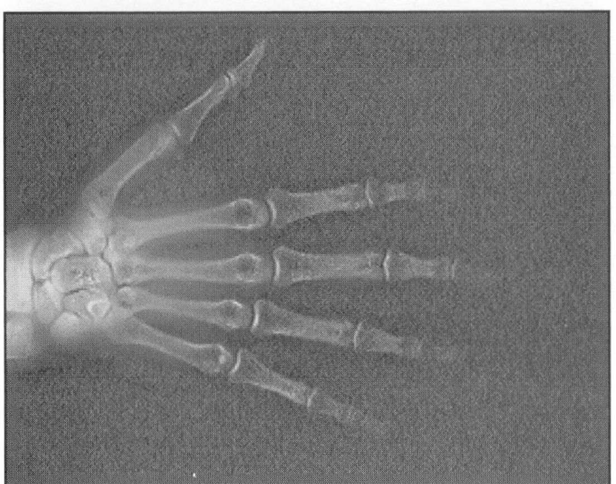

c. ACE filter with parameters—$k_1 = 0.9$; $k_2 = 1.0$; local gain max = 5.

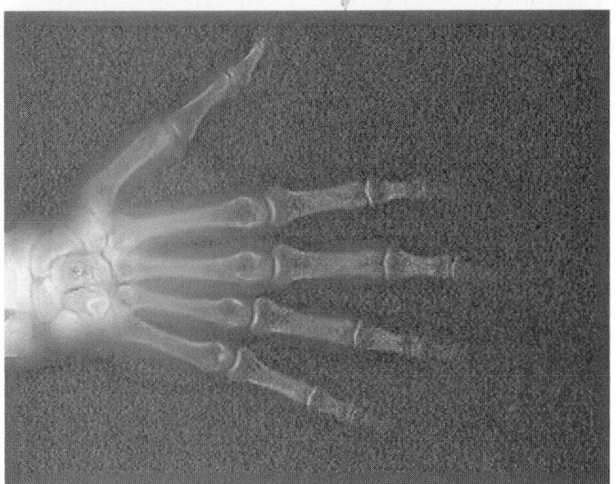

d. ACE filter with parameters—$k_1 = 0.5$; $k_2 = 1.0$; local gain max = 5.

Figure 4.2-12 (Continued)

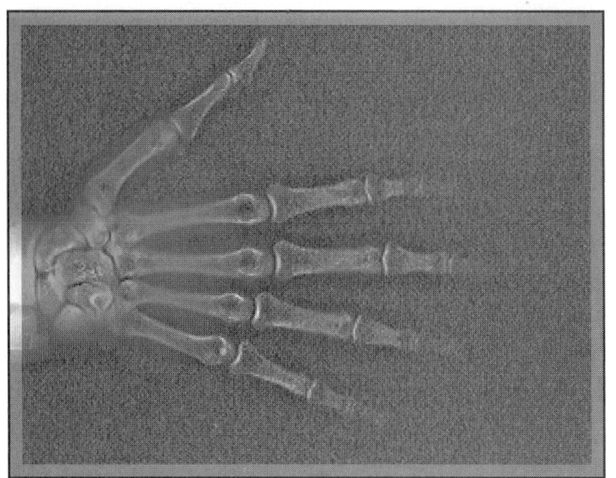

e. ACE filter with parameters—k_1 = 0.9; k_2 = 0.5; local gain max = 5.

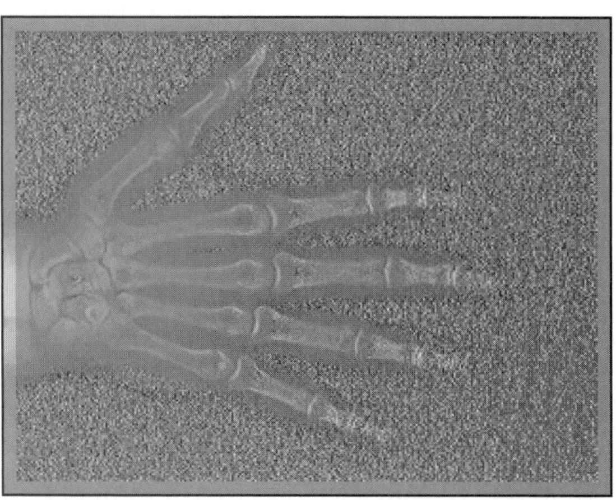

f. ACE filter with parameters—k_1 = 0.9; k_2 = 0.5; local gain max = 25.

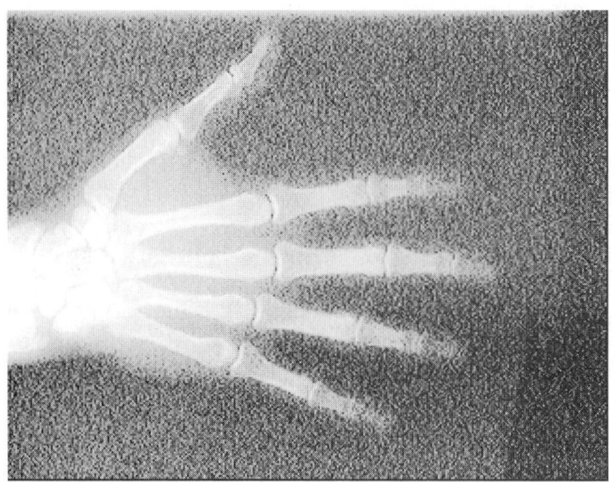

g. Histogram equalized version of original image (a).

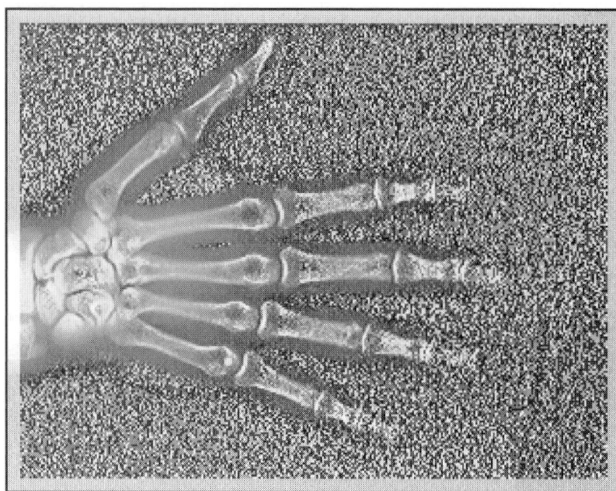

h. Histogram equalized version of image (f).

results from this filter with a local window size of 15 and by varying the other parameters. Figures 4.2-12c, d illustrate that by lowering k_1, which controls the local gain factor, bright areas (wrist area) in the original image tend to get washed out. In Figures 4.2-12-e, f we see that increasing the local gain factor maximum reduces contrast in the output image. Figures 4.2-12g, h shows the result of histogram equalization applied to the original compared to applying histogram equalization to one of the adaptive contrast filtered images. Here we see that we retain more detail in both the bright and dark areas of the image, but note that most of the detail in the dark, background areas appears to be attributable to noise, suggesting that some noise removal could improve the results. This illustrates the experimental nature of developing image enhancement algorithms and the fact that the algorithms tend to be application dependent.

4.2.3 Color

The human visual system can perceive thousands of colors in a small spatial area but only about 100 gray levels. An enhancement technique called pseudo-color takes advantage of this aspect of our visual perception to enhance gray-level images. *Pseudo-color* involves mapping the gray-level values of a monochrome image to red, green, and blue values, creating a color image. The pseudo-color techniques can be applied in both the spatial and frequency domains. Pseudo-color is often applied to images where the relative values are important, but the specific representation is not, for example, X-ray images.

In the spatial domain a gray level to color transform is defined, which has three different mapping equations for each of the red, green, and blue color bands. The equations selected are application specific and are functions of the gray levels in the image, $I(r, c)$. So we have three equations, as follows:

$$\begin{aligned} I_R(r, c) &= R\,[I(r, c)] \\ I_G(r,c) &= G\,[I(r, c)] \\ I_B(r, c) &= B\,[I(r, c)] \end{aligned}$$

where $R[\ ]$, $G[\ ]$, and $B[\ ]$ are the mapping equations to map the gray levels to the red, green, and blue components. These equations can be linear or nonlinear. A simple example, called *intensity slicing*, splits the range of gray levels into separate colors. For this, the gray levels that fall within a specified range are mapped to fixed RGB values (colors). Figure 4.2-13 illustrates the intensity slicing method for pseudo-color. Figure 4.2-13a shows the gray-scale range evenly divided into four different colors. The colors in the first range, 0 to MAX/4, are mapped to Color_1; the colors in the second range, MAX/4 to MAX/2, are mapped to Color_2, and so on. If we define Color_i as (R_i, G_i, B_i), we obtain the mapping equations given in Figure 4.2-13b. In this case the equations are constants over specified ranges; however, they can be any type of equations.

Alternately, we can perform pseudo-color in the frequency domain. This is typically done by performing a Fourier transform on the image and then applying a lowpass, bandpass, and highpass filter to the transformed data. These three filtered outputs are then inverse transformed, and the individual outputs are used as the RGB components of the color image. A block diagram of the process is illustrated in Figure 4.2-14a. Typical postprocessing includes histogram equalization, but it is application dependent. Although these filters may be of any type, they are often chosen to cover the entire frequency domain by dividing it into three separate bands, corresponding to lowpass, bandpass, and highpass filters (Figure 4.2-14b).

In Figure 4.2-15 a gray-level-mapping pseudo-color technique has been applied to a dental X-ray image. The original image is shown in Figure 4.2-15a, where it is difficult to discern the image features. In Figure 4.2-15b the image has undergone the pseudo-color mapping, followed by a luminance transform. Here we can observe the improved visual information (the pseudo-color version of this image is on CD-ROM, called *toothcol*).

The pseudo-color techniques provide us with methods to change a gray-scale image into a color image. We may wish to apply some of the techniques, such as histogram modification, to color images. One method for doing this is to treat each of the three bands in color images as separate gray-scale images. Thus, we can apply any

Figure 4.2-13 Pseudo-Color in the Spatial Domain

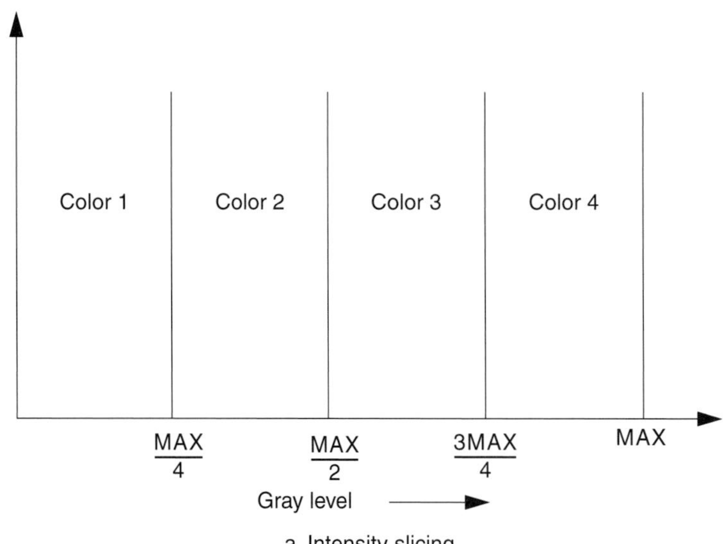

a. Intensity slicing.

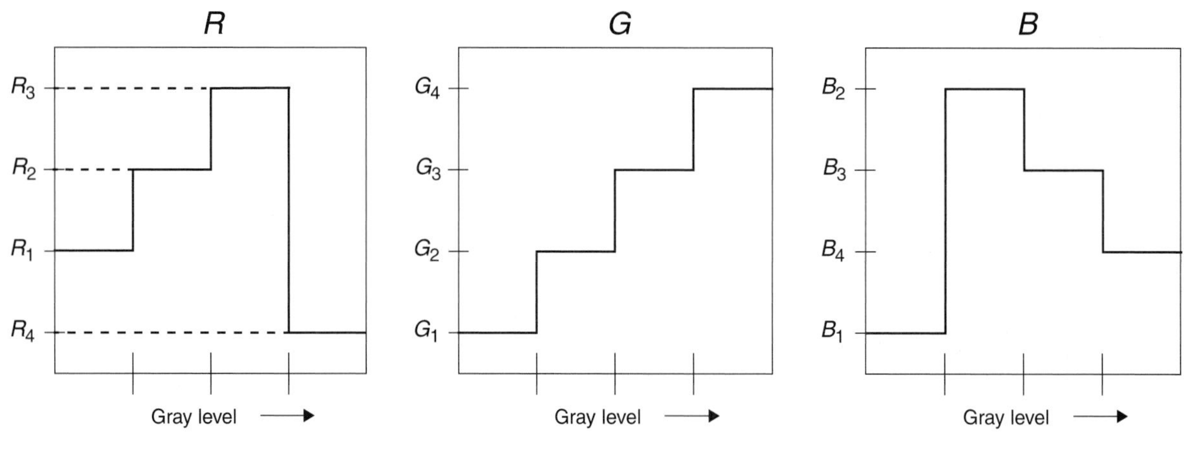

b. Mapping equations.

and all of the gray-scale modification techniques, including histogram modification, to color images by applying the method to each color band separately. The problem with this approach is that the colors will change, which is typically not the desired effect. We need to retain the relative color (the ratios between red, green, and blue for each pixel) in order to avoid color shifts. The relative color can be retained by applying the gray-scale modification technique to one of the color bands (red, green, or blue) and then using the ratios from the original image to find the other values. Typically the most important color band is selected, and this choice is very much application specific. This technique will not always provide us with the desired result, either. Often, we really want to apply the gray-scale modification method to the image brightness

Figure 4.2-14 Pseudo-Color in the Frequency Domain

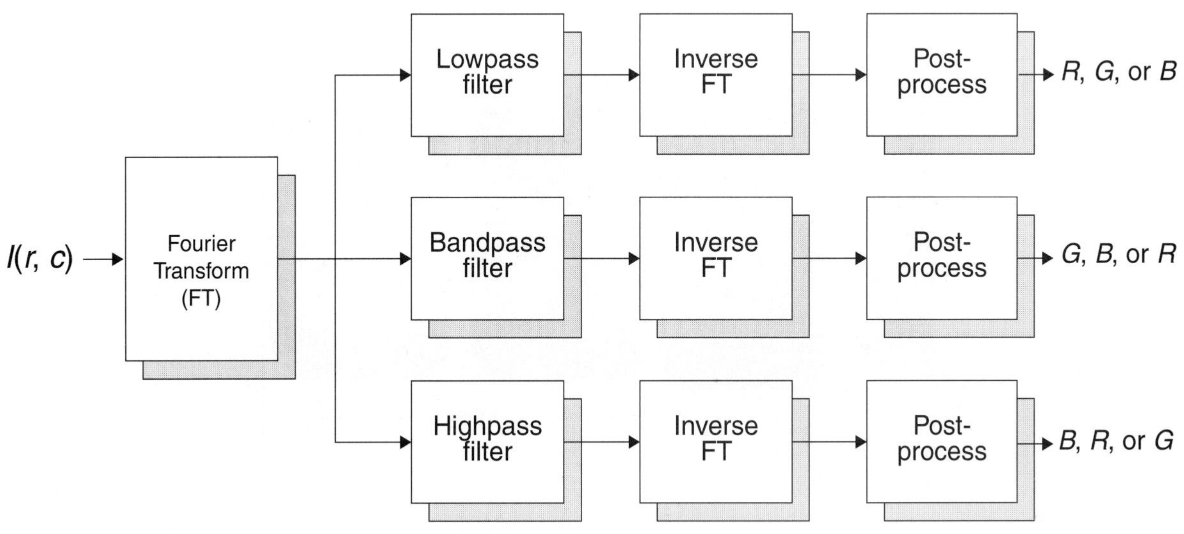

a. Block diagram of process.

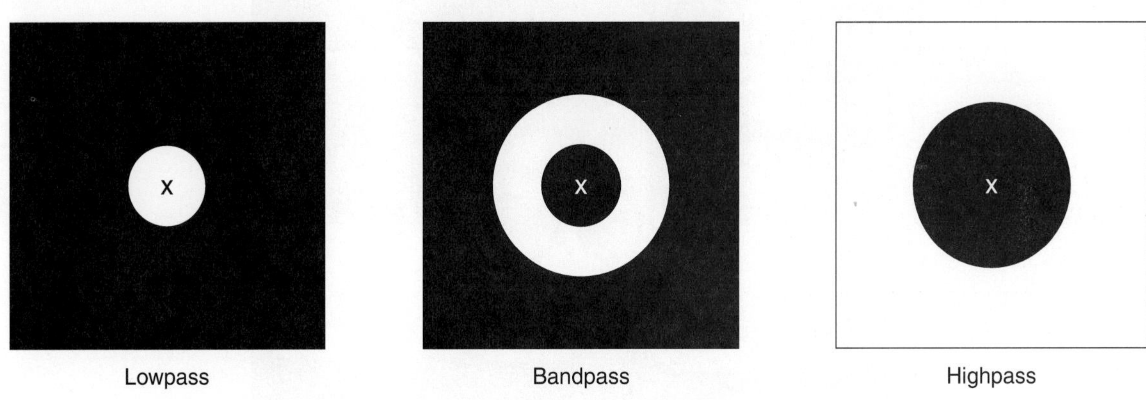

b. Fourier filters (x = origin).

only, even with color images. This is done by using the HSL transform, applying the gray-scale modification technique to the lightness band only, and then performing the inverse HSL transform. This effect will be similar to application to gray-scale images.

4.3 IMAGE SHARPENING

Image sharpening deals with enhancing detail information in an image. The detail information is typically contained in the high spatial frequency components of the image, so most of the techniques contain some form of highpass filtering. The detail

Figure 4.2-15 Pseudo-Color Enhancement

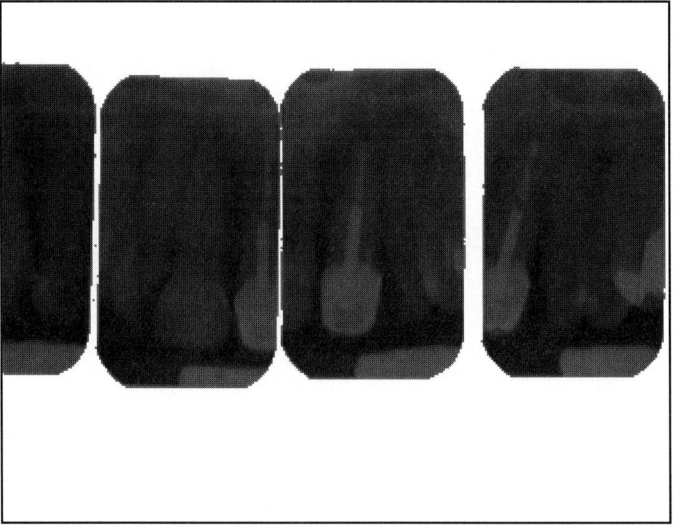

a. Gray-scale *X*-ray image.

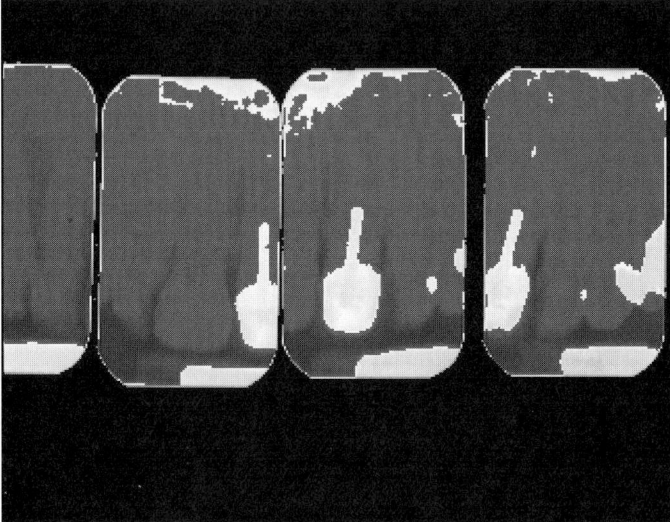

b. Pseudo-color-enhanced image, gray-level mapping technique.

information includes edges and, in general, corresponds to image features that are small spatially. This information is visually important because it delineates object and feature boundaries and is important for textures in objects.

The image sharpening discussion here is not meant to be comprehensive, but representative of the algorithms and techniques currently available. Mask operations in the spatial domain and their equivalent global operations in the frequency domain are considered.

4.3.1 Highpass Filtering

Highpass filters were introduced in Chapter 2. Here we will consider techniques specifically for image sharpening. Highpass filtering for image enhancement typically requires some form of postprocessing, such as histogram equalization, to create an acceptable image. Additionally, highpass filtering alone is seldom used for enhancement but is often part of a more complex enhancement algorithm. Highpass filtering, in the form of edge detection, is often used in computer vision applications to delineate object outlines.

Edge detectors are spatial domain convolution mask approximations to the equivalent frequency domain filter. One method to find an approximate spatial convolution mask that minimizes mean-square error is to use the Moore-Penrose generalized inverse matrix. This technique is beyond the scope of the discussion here, but more information can be found in the references. *Phase contrast filtering*, also discussed in the references, is similar to highpass filtering but is based on the assumption that most visual information is in the phase.

4.3.2 High-Frequency Emphasis

As we saw in Chapter 2, high-frequency emphasis can be used to enhance details in an image. The highpass filter alone will accentuate edges in the image but loses a large portion of the visual information by filtering out the low spatial frequency components. This problem is solved with the high-frequency emphasis filter, which retains some of the low-frequency information (see Figures 2.5-22 and 2.5-23) by adding an offset to the filter function. When this is done, care must be taken to avoid overflow in the resulting image. The results from overflow will appear as noise, typically white and black points (depending on how the data conversion is handled). This problem can be avoided by making certain that any intermediate values are promoted to data types that will handle the extended range and then remapping the data before displaying it.

A similar result can be obtained in the spatial domain by using a high boost spatial filter (see Image Processing exercise #4 in Chapter 8). The high boost spatial filter mask is of the following form:

$$\begin{bmatrix} -1 & -1 & -1 \\ -1 & x & -1 \\ -1 & -1 & -1 \end{bmatrix}$$

This mask is convolved with the image, and the value of x determines the amount of low-frequency information retained in the resulting image. A value of 8 will result in a highpass filter (the output image will contain only the edges), while larger values will retain more of the original image. If values of less than 8 are used for x, the resulting image will appear as a negative of the original. Figure 4.3-1 shows the results from using various values of x for high boost spatial filtering. These images have been histogram equalized as a postprocessing step for further enhancement. As was done with the edge detection spatial masks, this mask can be extended with –1's and a corresponding increase in the value of x. Larger masks will emphasize the edges more (make them wider) and help to mitigate the effects of any noise in the original image. For example, a 5×5 version of this mask is

$$\begin{bmatrix} -1 & -1 & -1 & -1 & -1 \\ -1 & -1 & -1 & -1 & -1 \\ -1 & -1 & x & -1 & -1 \\ -1 & -1 & -1 & -1 & -1 \\ -1 & -1 & -1 & -1 & -1 \end{bmatrix}$$

If we create an $N \times N$ mask, the value for x for a highpass filter is $N \times N - 1$, in this case 24 $(5 \times 5 - 1)$.

4.3.3 Homomorphic Filtering

The digital images we process are created from optical images. Optical images consist of two primary components, the lighting component and the reflectance component. The lighting component results from the lighting conditions present when the

Figure 4.3-1 High Boost Spatial Filter

a. Original image.

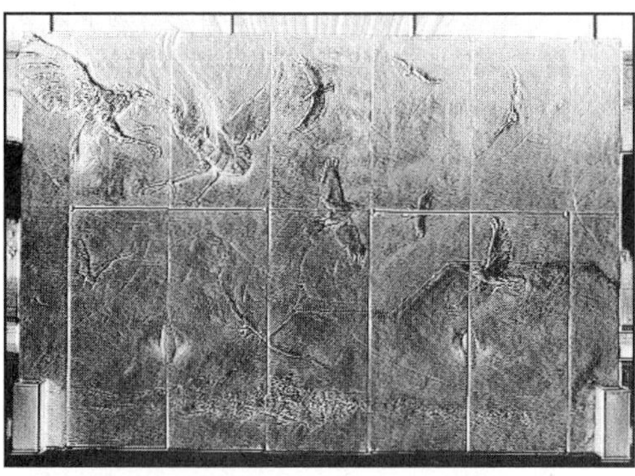

b. High boost filter, 3×3 mask. A small center value, $x = 5$, results in a negative of the original.

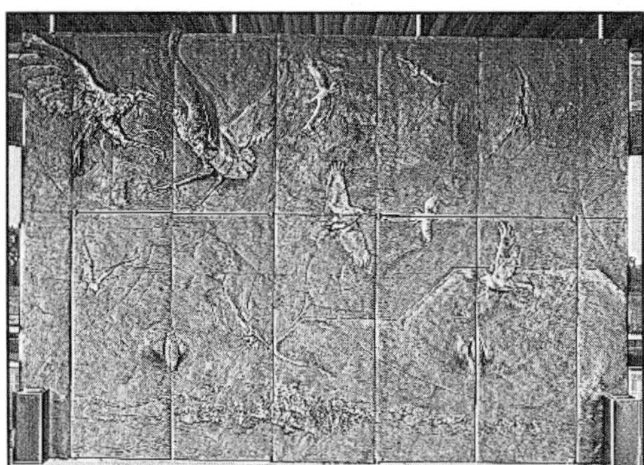

c. High boost filter, 3×3 mask. A center value of $x = 8$ results in enhanced details.

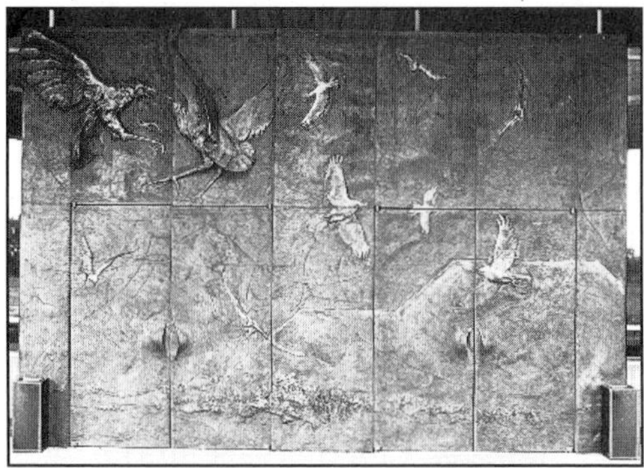

d. High boost filter, 3×3 mask. A large center value, $x = 15$, retains more of the original image.

image is captured and can change as the lighting conditions change. The reflectance component results from the way the objects in the image reflect light and are determined by the intrinsic properties of the object itself, which (normally) do not change. In many applications it is useful to enhance the reflectance component, while reducing the contribution from the lighting component. *Homomorphic filtering* is a frequency domain filtering process that compresses the brightness (from the lighting conditions) while enhancing the contrast (from the reflectance properties of the objects).

The image model for homomorphic filters is as follows:

$$I(r, c) = L(r, c)R(r, c)$$

where $L(r, c)$ represents contribution of the lighting conditions
$R(r, c)$ represents contribution of the reflectance properties of the objects

The homomorphic filtering process assumes that $L(r, c)$ consists of primarily slow spatial changes (low spatial frequencies) and is responsible for the overall range of the brightness in the image. The assumptions for $R(r, c)$ are that it consists primarily of high spatial frequency information, which is especially true at object boundaries, and that it is responsible for the local contrast (the spread within a small spatial area). These simplifying assumptions are valid for many types of real images.

The homomorphic filtering process consists of five steps: 1) a natural log transform (base e), 2) the Fourier transform, 3) filtering, 4) the inverse Fourier transform, and 5) the inverse log function—the exponential. This process is illustrated in a block diagram in Figure 4.3-2. The first step allows us to decouple the $L(r, c)$ and $R(r, c)$ components because the logarithm function changes a product into a sum. Step 2 puts

Figure 4.3-2 The Homomorphic Filtering Process

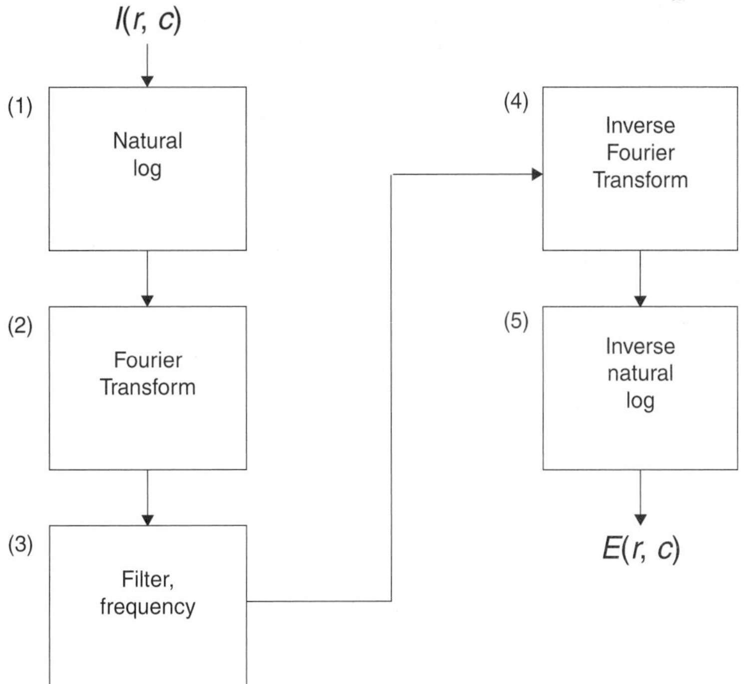

the image into the frequency domain so that we can perform the filtering in step 3. Next, steps 4 and 5 do the inverse transforms from steps 1 and 2 to get our image data back into the spatial domain. The only factor left to be considered is the filter function, $H(u, v)$.

The typical filter for the homomorphic filtering process is shown in Figure 4.3-3. Here we see that we can specify three parameters—the high-frequency gain, the low-frequency gain, and the cutoff frequency. Typically the high-frequency gain is greater than 1, and the low-frequency gain is less than 1. This provides us with the desired effect of boosting the $R(r, c)$ components, while reducing the $L(r, c)$ components. The selection of the cutoff frequency is highly application specific and needs to be chosen

Figure 4.3-3 Homomorphic Filtering

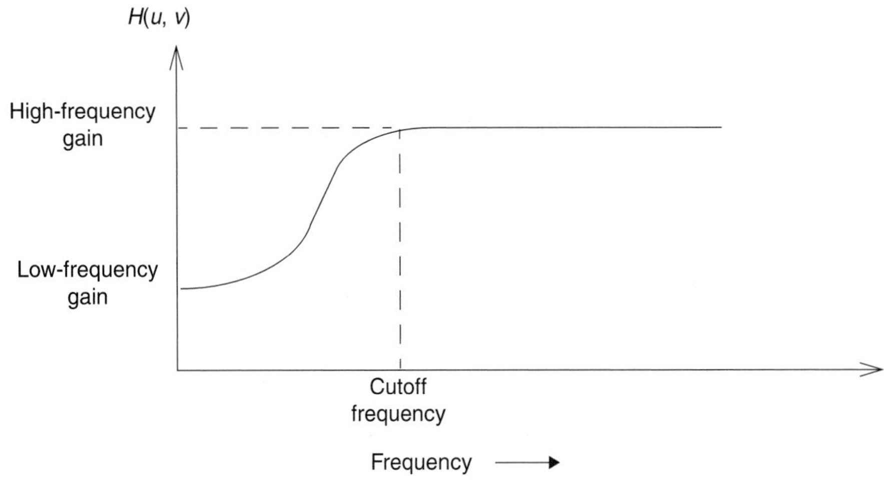

a. Cross-section of homomorphic filter, $H(u, v)$.

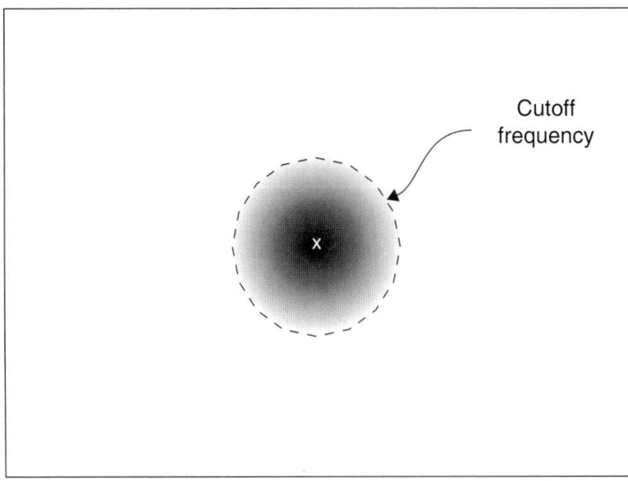

b. 2-D filter diagram (x = origin).

so that no important information is lost. In practice the values for all three parameters are often determined empirically.

Figure 4.3-4 shows results from application of homomorphic filtering to a poor image. In this case, the homomorphic filter returns an image of low contrast, so the contrast is enhanced by a histogram stretch procedure. A comparison is made between the homomorphic filter followed by a histogram stretch (Figure 4.3-4b) and simply stretching the original image's histogram (Figure 4.3-4c). We see that the homomorphic filter provides an image with greater visual detail.

Figure 4.3-4 Homomorphic Filtering Example

a. Original image.

b. Result of homomorphic filter followed by histogram stretching—upper gain = 1.2; lower gain = 0.5; cutoff frequency = 16.

c. Result of histogram stretch without the use of homomorphic filtering.

4.3.4 Unsharp Masking

The unsharp masking enhancement algorithm is representative of practical image sharpening methods. It combines many of the operations discussed, including filtering and histogram modification. A flowchart for this process is shown in Figure 4.3-5. Here we see that the original image is lowpass filtered, followed by a histogram shrink to the lowpass-filtered image. The resultant image from these two operations is

Figure 4.3-5 Unsharp Masking Enhancement

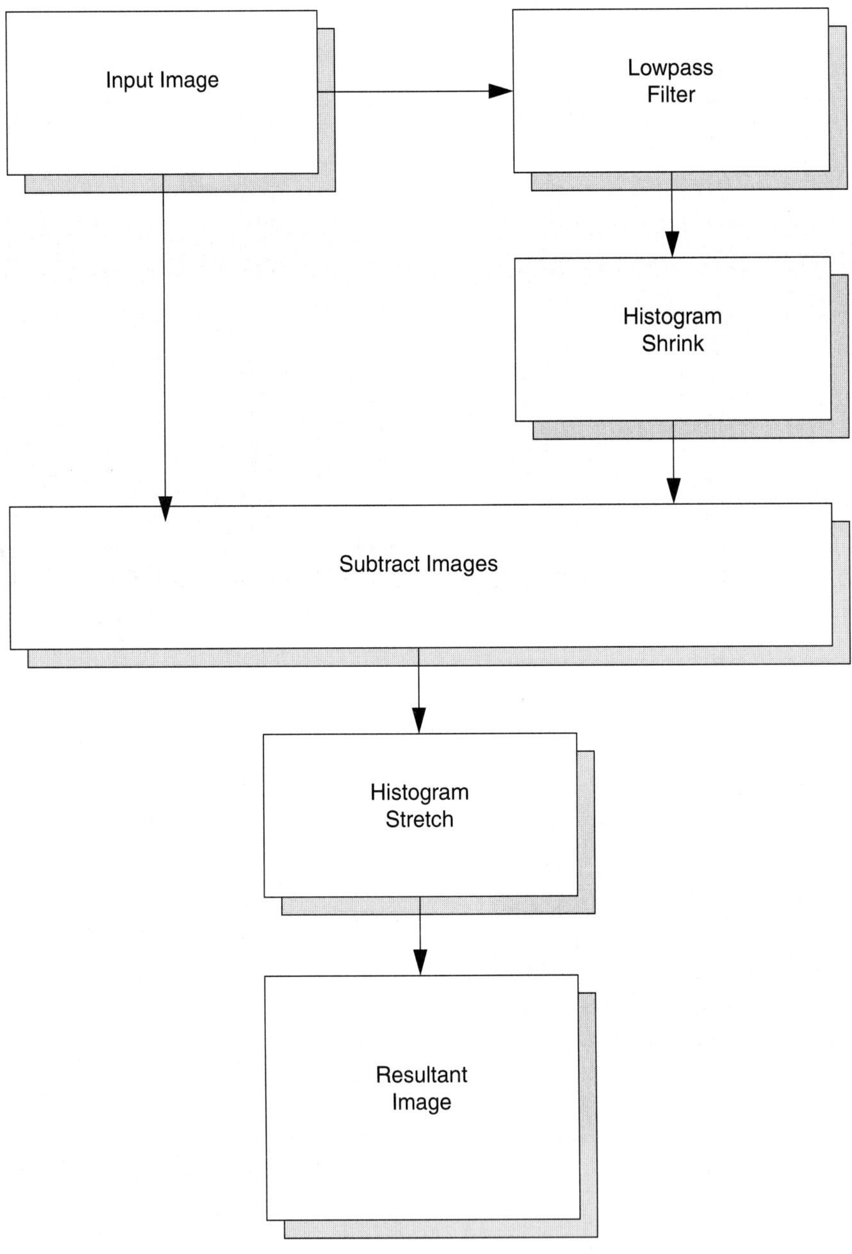

then subtracted from the original image, and the result of this operation undergoes a histogram stretch to restore the image contrast. This process works because subtracting a slowly changing edge (the lowpass-filtered image) from faster changing edges (in the original) has the visual effect of causing overshoot and undershoot at the edges, which has the effect of emphasizing the edges. By scaling the lowpassed image with a histogram shrink, we can control the amount of edge emphasis desired (see Image Processing exercise #7 in Chapter 8). In Figure 4.3-6 are the results of applying the unsharp masking algorithm with different ranges for the histogram shrink process. Here we see that as the range for the histogram shrink is increased, the resulting image has a greater edge emphasis.

Figure 4.3-6 Unsharp Masking

a. Original image.

b. Unsharp masking with lower limit = 0, upper = 100, with 2% low and high clipping.

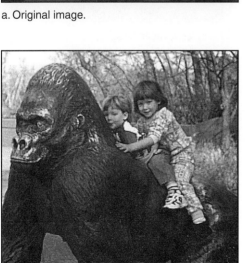

c. Unsharp masking with lower limit = 0, upper = 150, with 2% low and high clipping.

d. Unsharp masking with lower limit = 0, upper = 200, with 2% low and high clipping.

4.4 Image Smoothing

Image smoothing is used for two primary purposes: to give an image a softer or special effect or to eliminate noise. In the previous chapter we discussed filtering to eliminate noise, so here we will focus on creating a softer effect (see Image Processing exercise #4 in Chapter 8). Image smoothing is accomplished in the spatial domain by considering a pixel and its neighbors and eliminating any extreme values in this group. This is done by various types of mean and median filters (Chapter 3). In the frequency domain, image smoothing is accomplished by some form of lowpass filtering. Because the high spatial frequencies contain the detail, including edge, information, the elimination of this information via lowpass filtering will provide a smoother image. Any

Figure 4.4-1 Mean Filters (3 × 3)

a. Original image.

b. Aritmetic mean filter.

c. Alpha-trimmed mean filter—trim size = 0.

d. Contra-harmonic mean filter—order = +1.

Figure 4.4-1 (Continued)

e. Geometric mean filter.

f. Harmonic mean filter.

g. Y_p mean filter—order = +1.

h. Midpoint filter.

fast or sharp transitions in the image brightness will be filtered out, thus providing the desired effect.

4.4.1 Mean and Median Filtering

The use of mean filters to eliminate noise in images was discussed in Chapter 3, and we have seen that the various types of mean filters are most effective with different types of noise. For image smoothing, the results from most of the spatial mean filters are visually similar—the image is smoothed out providing a softer visual effect (see Figure 4.4-1). We can use a larger mask size for a greater smoothing effect. In Figure 4.4-2 we see the results of using an arithmetic filter and various mask sizes. We

Figure 4.4-2 Image Smoothing with an Arithmetic Mean Filter

a. Original image.

b. 3 × 3 arithmetic mean filter.

c. 5 × 5 arithmetic mean filter.

d. 7 × 7 arithmetic mean filter.

can see that as the mask size increases, the amount of smoothing increases, and at some point the smoothing becomes a noticeable blurring.

A median filter can also be used to create a similar smoothing effect, but with large mask sizes it creates an almost painted (and blurred) look. In Figure 4.4-3 we see the results from applying a median filter with various mask sizes. We can see that details smaller than the mask size are eliminated, and as the mask gets larger the image begins to take on an artificial look. With a large mask the median filter will take a long time to process, so in practice a fast algorithm or a pseudo-median filter is used. Most of the fast algorithms operate by efficiently maintaining the sorting of the

Figure 4.4-3 Image Smoothing with a Median Filter

a. Original image.

b. 3 × 3 median filter.

c. 5 × 5 median filter.

d. 7 × 7 median filter.

data as we move across the image. A pseudo-median filter is explored in the skin tumor application discussed in Chapter 7.

4.4.2 Lowpass Filtering

In the frequency domain, lowpass filtering can be done as discussed in Chapter 2. An ideal filter can cause undesirable artifacts, whereas the Butterworth filter does not. Lowpass filtering creates an image with a smooth appearance because it suppresses any rapidly changing brightness values in the original image. The lowpass filters do this by attenuating high spatial frequency information, which corresponds to the rapid changes (edges). The amount of information suppressed is determined by the cutoff frequency of the filter.

In the spatial domain, we can define lowpass filter masks to perform filtering approximately equivalent to specified frequency domain filters. These approximations can be obtained with the Moore-Penrose generalized inverse (see references) and are typically of the form of average (mean) filters. The coefficients for these filter masks are all positive, unlike the highpass filters where the center is surrounded by negative coefficients. Here are some common spatial convolution masks for lowpass filtering:

$$\begin{bmatrix} 1 & 1 & 1 \\ 1 & 1 & 1 \\ 1 & 1 & 1 \end{bmatrix} \quad \begin{bmatrix} 1 & 1 & 1 \\ 1 & 2 & 1 \\ 1 & 1 & 1 \end{bmatrix} \quad \begin{bmatrix} 2 & 1 & 2 \\ 1 & 4 & 1 \\ 2 & 1 & 2 \end{bmatrix} \quad \begin{bmatrix} 1 & 2 & 1 \\ 2 & 4 & 2 \\ 1 & 2 & 1 \end{bmatrix}$$

As is seen, the coefficients in the mask may be biased, that is, they may not all be 1's. This is typically for application-specific reasons. For example, we may want to weight the center pixel, or the diagonal pixels, more heavily than the other pixels. Note also that these types of masks are often multiplied by $1/N$, where N is the sum of the mask coefficients. As examples, the first mask is multiplied by 1/9, the second by 1/10, and so on. This is the equivalent of linearly remapping the image data (typically to BYTE) after the convolution.

4.5 References

References that contain major chapters on image enhancement include [Pratt 91], [Gonzalez/Woods 92], [Baxes 94], [Lim 90], and [Jain 89]. Gray-scale modification is discussed in [Gonzalez/Woods 92], [Rosenfeld/Kak 82], and [Jain 89]. A more complete theoretical treatment of histogram modification is given in [Banks 90], [Jain 89], [Gonzalez/Woods 92], [Rosenfeld/Kak 82], [Castleman 96], and [Pratt 91]. Specifically, more details on histogram hyperbolization can be found in [Pratt 91] and [Banks 90]. A useful adaptive histogram modification technique is discussed in [Pratt 91]. A conceptual perspective to gray-level transforms and histogram modification is provided in [Baxes 94], whereas [Jain/Kasturi/Schnuck 95], [Myler/Weeks 93], and [Sid-Ahmed 95] provide a practical treatment. The adaptive contrast enhancement filter is discussed in [Gonzalez/Woods 92]. Pseudo-color is discussed in [Pratt 91], [Gonzalez/Woods 92], and [Jain 89].

[Lim 90], [Schalkoff 89], and [Gonzalez/Woods 92] provide different perspectives to unsharp masking. Phase contrast filtering is discussed in [Sid-Ahmed 95]. Details on the use of the Moore-Penrose matrix for generating convolution masks based on frequency domain filter models can be found in [Gonzalez/Woods 92]. [Galbiati 90] provides many edge enhancement convolution masks, and [Bracewell 95] discusses convolution masks and their relationship to digital filter theory. Various image sharpening and smoothing methods are discussed in all the following references.

Banks, S., *Signal Processing, Image Processing and Pattern Recognition*, Englewood Cliffs, NJ: Prentice Hall, 1990.

Baxes, G. A., *Digital Image Processing: Principles and Applications*, New York: Wiley, 1994.

Bracewell, R. N., *Two-Dimensional Imaging*, Englewood Cliffs, NJ: Prentice Hall, 1995.

Castleman, K. R., *Digital Image Processing*, Englewood Cliffs, NJ: Prentice Hall, 1996.

Galbiati, L. J., *Machine Vision and Digital Image Processing Fundamentals*, Englewood Cliffs, NJ: Prentice Hall, 1990.

Gonzalez, R. C., and Woods, R. E., *Digital Image Processing*, Reading, MA: Addison-Wesley, 1992.

Jain, A. K., *Fundamentals of Digital Image Processing*, Englewood Cliffs, NJ: Prentice Hall, 1989.

Jain, R., Kasturi, R., and Schnuck, B. G., *Machine Vision*, New York: McGraw Hill, 1995.

Lim, J. S., *Two-Dimensional Signal and Image Processing*, Englewood Cliffs, NJ: Prentice Hall, 1990.

Myler, H. R., and Weeks, A.R., *Computer Imaging Recipes in C*, Englewood Cliffs, NJ: Prentice Hall, 1993.

Pratt, W. K., *Digital Image Processing*, New York: Wiley, 1991.

Rosenfeld, A., and Kak, A. C., *Digital Picture Processing*, San Diego, CA: Academic Press, 1982.

Russ, J. C., *The Image Processing Handbook*, Boca Raton, FL: CRC Press, 1992.

Schalkoff, R. J., *Digital Image Processing and Computer Vision*, New York: Wiley, 1989.

Sid-Ahmed, M. A., *Image Processing: Theory, Algorithms, and Architectures*, Englewood Cliffs, NJ: Prentice Hall, 1995.

CHAPTER 5

Image Compression

5.1 Introduction

Image compression has been pushed to the forefront of the image processing field. This is largely a result of the rapid growth in computer power, the corresponding growth in the multimedia market, and the advent of the World Wide Web, which makes the Internet easily accessible for everyone. Additionally, the advances in video technology, including high-definition television, are creating a demand for new, better, and faster image compression algorithms. Compression algorithm development starts with applications to two-dimensional (2-D) still images. Because video and television signals consist of consecutive frames of 2-D image data, the development of compression methods for 2-D still data is of paramount importance. After these are developed, they are often extended to video (motion imaging). Here, we will focus on image compression of single frames of image data.

What is image compression? *Image compression* involves reducing the size of image data files, while retaining necessary information. The reduced file is called the *compressed file* and is used to reconstruct the image, resulting in the *decompressed image*. The original image, before any compression is performed, is called the *uncompressed* image file. The ratio of the original, uncompressed image file and the compressed file is referred to as the *compression ratio*. The compression ratio is denoted by:

$$\text{Compression Ratio} = \frac{\text{Uncompressed File Size}}{\text{Compressed File Size}} = \frac{\text{SIZE}_U}{\text{SIZE}_C}$$

It is often written as $\text{SIZE}_U{:}\text{SIZE}_C$.

EXAMPLE 5-1

The original image is 256×256 pixels, single-band (gray-scale), 8 bits per pixel. This file is 65,536 bytes (64k). After compression the image file is 6,554 bytes. The compression ratio is: $\text{SIZE}_U/\text{SIZE}_C = 65536/6554 = 9.999 \approx 10$. This can also be written as 10:1.

This is called a "10 to 1 compression" or a "10 times compression," or it can be stated as "compressing the image to 1/10 its original size." Another way to state the compression is to use the terminology of *bits per pixel*. For an $N \times N$ image:

$$\text{Bits per Pixel} = \frac{\text{Number of Bits}}{\text{Number of Pixels}} = \frac{(8)(\text{Number of Bytes})}{N \times N}$$

EXAMPLE 5-2

Using the preceding example, with a compression ratio of 65,536/6,554 bytes, we want to express this as bits per pixel. This is done by first finding the number of pixels in the image: 256 × 256 = 65,536 pixels. We then find the number of bits in the compressed image file: (6,554 bytes)(8 bits/byte) = 52,432 bits. Now we can find the bits per pixel by taking the ratio: 52,432/ 65,536 = 0.8 bits/pixel.

The reduction in file size is necessary to meet the bandwidth requirements for many transmission systems, as well as the storage requirements in computer databases. The amount of data required for digital images is enormous. For example, a single 512 × 512, 8-bit image requires 2,097,152 bits for storage. If we wanted to transmit this image over the World Wide Web, it would probably take minutes for transmission—too long for most people to wait.

EXAMPLE 5-3

To transmit an RGB (color) 512 × 512, 24-bit (8 bits/pixel/color) image via modem at 28.8 kbaud (kilobits/second), it would take about

$$\frac{(512 \times 512 \text{ pixels})(24 \text{ bits/pixel})}{(28.8 \times 1024 \text{ bits/second})} \approx 213 \text{ seconds} \approx 3.6 \text{ minutes}$$

EXAMPLE 5-4

To transmit a digitized color 35mm slide scanned at 3,000 × 2,000 pixels, and 24 bits, at 28.8 kbaud would take about:

$$\frac{(3000 \times 2000 \text{ pixels})(24 \text{ bits/pixel})}{(28.8 \times 1024 \text{ bits/second})} \approx 4883 \text{ seconds} \approx 81 \text{ minutes}$$

Couple this result with transmitting multiple images, or motion images, and the necessity of image compression can be appreciated.

The key to a successful compression scheme comes with the second part of the definition—*retaining necessary information*. To understand this we must differentiate between data and information. For digital images, *data* refers to the pixel gray-level values that correspond to the brightness of a pixel at a point in space. *Information* is an interpretation of the data in a meaningful way. Data are used to convey information, much like the way the alphabet is used to convey information via words. Information is an elusive concept; it can be application specific. For example, in a binary image that contains text only, the necessary information may only involve the text being readable, whereas for a medical image the necessary information may be every minute detail in the original image.

There are two primary types of image compression methods—those that preserve the data and those that allow some loss of data. The first type are called *lossless* methods because no data are lost, and the original image can be recreated exactly from the compressed data. For complex images these methods are limited to compressing the image file to about one half to one third its original size (2:1 to 3:1); often the achievable compression is much less. For simple images such as text-only images, lossless methods may achieve much higher compression. The second type of compression methods are called *lossy* because they allow a loss in the actual image data, so the original uncompressed image can*not* be created *exactly* from the compressed file. For complex images these techniques can achieve compression ratios of 10 or 20 and still retain high-quality visual information. For simple images or lower-quality results, compression ratios as high as 100 to 200 can be attained.

Compression algorithms are developed by taking advantage of the redundancy that is inherent in image data. Three primary types of redundancy can be found in images: 1) coding, 2) interpixel, and 3) psychovisual redundancy. *Coding redundancy* occurs when the data used to represent the image are not utilized in an optimal manner. For example, if we have an 8 bits/pixel image that allows 256 gray-level values, but the actual image contains only 16 gray-level values, this is a suboptimal coding—only 4 bits/pixel are actually needed. *Interpixel redundancy* occurs because adjacent pixels tend to be highly correlated. This is a result of the fact that in most images the brightness levels do not change rapidly, but change gradually, so that adjacent pixel values tend to be relatively close to each other in value (for video, or motion images, this concept can be extended to include *interframe redundancy*, redundancy between frames of image data). The third type, *psychovisual redundancy*, refers to the fact that some information is more important to the human visual system than other types of information. For example, we can only perceive spatial frequencies below about 50 cycles per degree (see Section 1.6) so that any higher-frequency information is of little interest to us.

The key in image compression algorithm development is to determine the minimal data required to retain the necessary information. This is achieved by taking advantage of the redundancy that exists in images. To determine exactly what information is important and to be able to measure image fidelity, we need to define an image fidelity criterion. Note that the information required is application specific, and that, with lossless schemes, there is no need for a fidelity criterion.

5.1.1 Fidelity Criteria

Fidelity criteria can be divided into two classes: 1) objective fidelity criteria and 2) subjective fidelity criteria. The *objective fidelity criteria* are borrowed from digital signal processing and information theory and provide us with equations that can be used to measure the amount of error in the reconstructed (decompressed) image. *Subjective fidelity criteria* require the definition of a qualitative scale to assess image quality. This scale can then be used by human test subjects to determine image fidelity. In order to provide unbiased results, evaluation with subjective measures requires careful selection of the test subjects and carefully designed evaluation experiments. The objective criteria, although widely used, are not necessarily correlated with our perception of image quality. However, they are useful as a relative measure in comparing different versions of the same image (see Figure 5.1-1).

Figure 5.1-1 Objective Fidelity Measures

a. Original image. The peak SNR of an image with itself is theoretically infinite, so a high SNR implies a good image and a lower SNR implies an inferior image.

b. Original image quantized to 16 gray levels using IGS. The peak SNR of it and original image is 35.01.

c. Original image with gaussian noise added with a variance of 200 and mean 0. Peak SNR of it and original image is 28.14.

d. Original image with gaussian noise added with a variance of 800 and mean 0. Peak SNR of it and original image is 22.73.

Commonly used objective measures are the root-mean-square error e_{RMS}, the root-mean-square signal-to-noise ratio SNR_{RMS}, and the peak signal-to-noise ratio SNR_{PEAK}. We can define the error between an original, uncompressed pixel value and the reconstructed (decompressed) pixel value as

$$\begin{aligned} \text{error}(r,c) &= \hat{I}(r,c) - I(r,c) \\ \text{where } I(r,c) &= \text{the original image} \\ \hat{I}(r,c) &= \text{the decompressed image} \end{aligned}$$

Next, we can define the total error in an $N \times N$ decompressed image as

$$\text{Total error} = \sum_{r=0}^{N-1} \sum_{c=0}^{N-1} [\hat{I}(r,c) - I(r,c)]$$

The *root-mean-square error* is found by taking the square root ("root") of the error squared ("square") divided by the total number of pixels in the image ("mean"):

$$e_{RMS} = \sqrt{\frac{1}{N^2} \sum_{r=0}^{N-1} \sum_{c=0}^{N-1} [\hat{I}(r,c) - I(r,c)]^2}$$

The smaller the value of the error metrics, the better the compressed image represents the original image. Alternately, with the signal-to-noise (SNR) metrics, a larger number implies a better image. The SNR metrics consider the decompressed image $\hat{I}(r,c)$ to be the "signal" and the error to be "noise." We can define the *root-mean-square signal-to-noise ratio* as

$$SNR_{RMS} = \sqrt{\frac{\sum_{r=0}^{N-1} \sum_{c=0}^{N-1} [\hat{I}(r,c)]^2}{\sum_{r=0}^{N-1} \sum_{c=0}^{N-1} [\hat{I}(r,c) - I(r,c)]^2}}$$

Another related metric, the *peak signal-to-noise ratio*, is defined as

$$SNR_{PEAK} = 10 \log_{10} \frac{(L-1)^2}{\frac{1}{N^2} \sum_{r=0}^{N-1} \sum_{c=0}^{N-1} [\hat{I}(r,c) - I(r,c)]^2}$$

where L = the number of gray levels (e.g., for 8 bits L = 256)

These objective measures are often used in research because they are easy to generate and seemingly unbiased, but remember that these metrics are not necessarily correlated to our perception of an image. The subjective measures are a better method for comparison of compression algorithms, if the goal is to achieve high-quality images as defined by our visual perception.

Subjective testing is performed by creating a database of images to be tested, gathering a group of people that are representative of the desired population, and then having all the test subjects evaluate the images according to a predefined scoring criterion. The results are then analyzed statistically, typically using the averages and standard deviations as metrics. Subjective fidelity measures can be classified into three categories. The first type are referred to as *impairment tests*, where the test subjects score the images in terms of how bad they are. The second type are *quality tests*, where the test subjects rate the images in terms of how good they are. The third type are called *comparison tests*, where the images are evaluated on a side-by-side basis. The comparison type tests are considered to provide the most useful results, as they provide a relative measure, which is the easiest metric for most people to determine. Impairment and quality tests require an absolute measure, which is more difficult to determine in an unbiased fashion. In Table 5.1-1 are examples of internationally accepted scoring scales for these three types of subjective fidelity measures.

Table 5.1-1 Subjective Fidelity Scoring Scales

Impairment	Quality	Comparison
5—Imperceptible	A—Excellent	+2 much better
4—Perceptible, not annoying	B—Good	+1 better
3—Somewhat annoying	C—Fair	0 the same
2—Severely annoying	D—Poor	–1 worse
1—Unusable	E—Bad	–2 much worse

5.1.2 Compression System Model

The compression system model consists of two parts: the compressor and the decompressor. The *compressor* consists of a preprocessing stage and encoding stage, whereas the *decompressor* consists of a decoding stage followed by a postprocessing stage (Figure 5.1-2). Before encoding, preprocessing is performed to prepare the image

Figure 5.1-2 Compression System Model

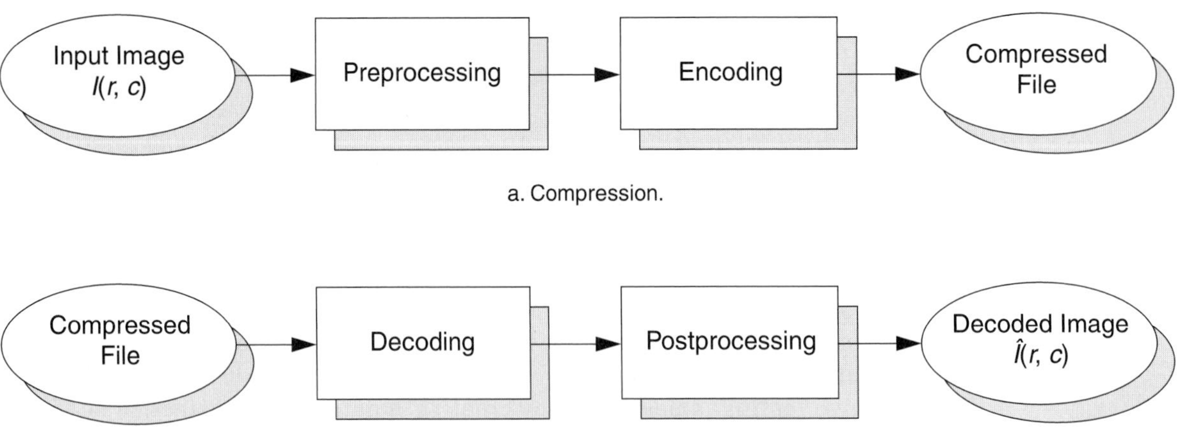

a. Compression.

b. Decompression.

for the encoding process, and consists of any number of operations that are application specific. After the compressed file has been decoded, postprocessing can be performed to eliminate some of the potentially undesirable artifacts brought about by the compression process. Often, many practical compression algorithms are a combination of a number of different individual compression techniques.

The compressor can be further broken down into stages as illustrated in Figure 5.1-3. The first stage in preprocessing is data reduction. Here, the image data can be reduced by gray-level and/or spatial quantization, or they can undergo any desired image enhancement (for example, noise removal) process. The second step in preprocessing is the mapping process, which maps the original image data into another mathematical space where it is easier to compress the data. Next, as part of the encoding process, is the quantization stage, which takes the potentially continuous data from the mapping stage and puts it in discrete form. The final stage of encoding involves coding the resulting data, which maps the discrete data from the quantizer onto a code in an optimal manner. A compression algorithm may consist of all the stages, or it may consist of only one or two of the stages.

The decompressor can be further broken down into the stages shown in Figure 5.1-4. Here the decoding process is divided into two stages. The first, the decoding stage, takes the compressed file and reverses the original coding by mapping the codes

Figure 5.1-3 The Compressor

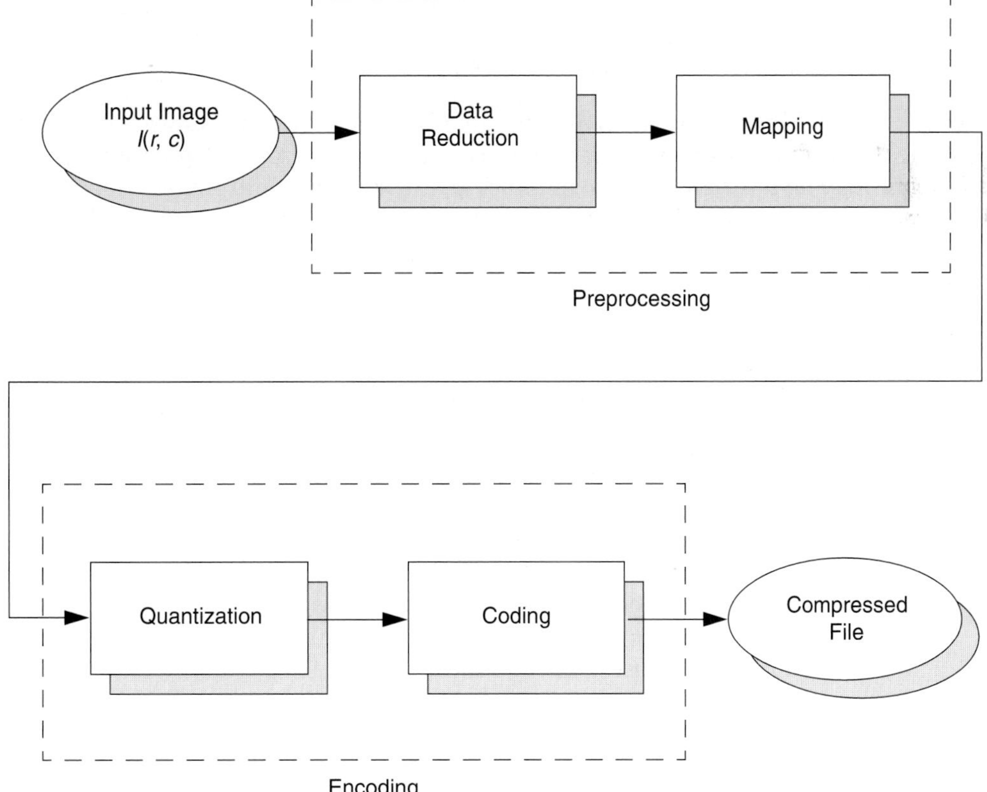

Figure 5.1-4 The Decompressor

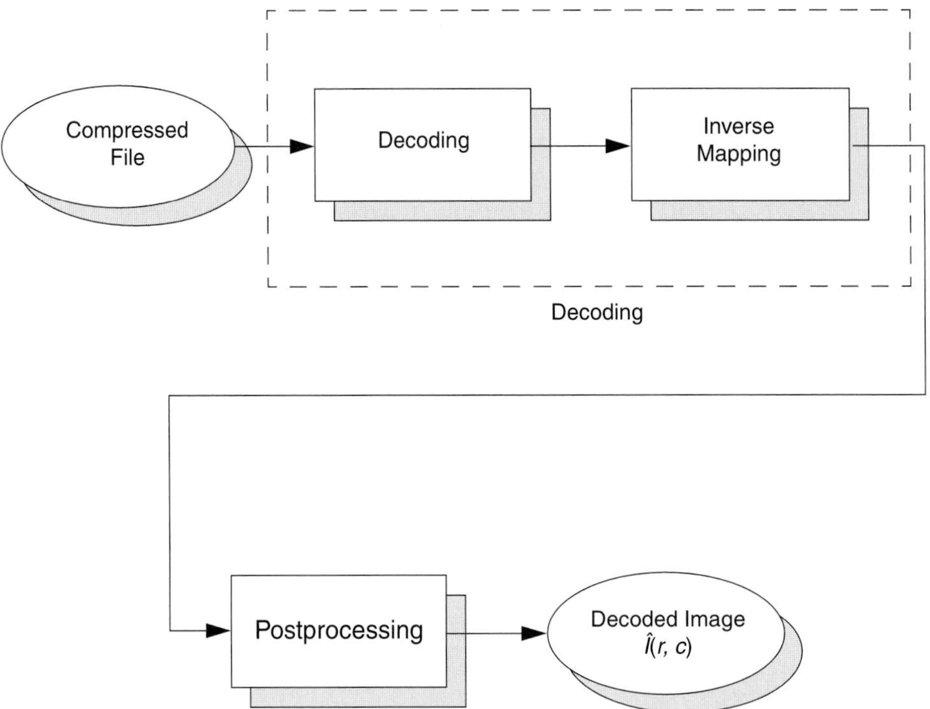

to the original, quantized values. Next, these values are processed by a stage that performs an inverse mapping to reverse the original mapping process. Finally, the image may be postprocessed to enhance the look of the final image. In some cases this may be done to reverse any preprocessing, for example, enlarging an image that was shrunk in the data reduction process. In other cases the postprocessing may simply enhance the image to ameliorate any artifacts from the compression process itself.

The development of a compression algorithm is highly application specific. During the preprocessing stage of compression, processes such as enhancement, noise removal, or quantization are applied. The goal of preprocessing is to prepare the image for the encoding process by eliminating any irrelevant information, where *irrelevant* is defined by the application. For example, many images that are for viewing purposes only can be preprocessed by eliminating the lower bit planes, without losing any useful information. In Figure 5.1-5 are shown the eight bit planes corresponding to an 8-bit image. Each bit plane is shown as an image by using white if the corresponding bit is a 1 and black if the bit is a 0. Here we see that the lower bit planes contain little information and can be eliminated with no significant information loss.

The mapping process is important because image data tend to be highly correlated. What this means is that there is a lot of redundant information in the data itself. Specifically, if the value of one pixel is known, it is highly likely that the adjacent pixel value is similar. By finding a mapping equation that decorrelates the data, this type of data redundancy can be removed. One method to do this is to find the difference between adjacent pixels and encode these values; this is called *differential*

Figure 5.1-5 Bit-Plane Images

a. Original, 8-bit image.

b. Bit-plane 7, the most significant bit.

c. Bit-plane 6.

d. Bit-plane 5.

e. Bit-plane 4.

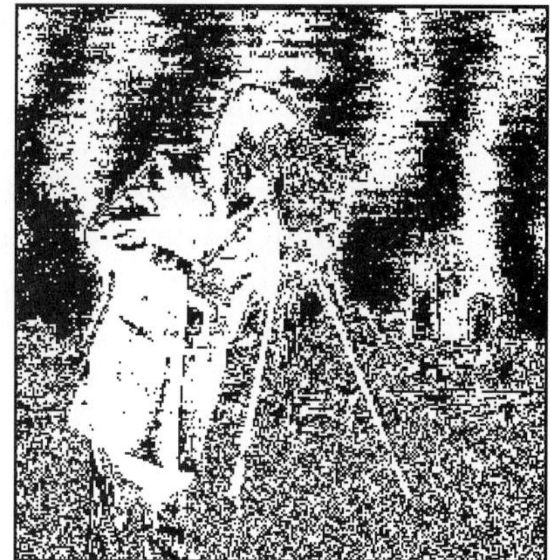

f. Bit-plane 3.

Figure 5.1-5 (Continued)

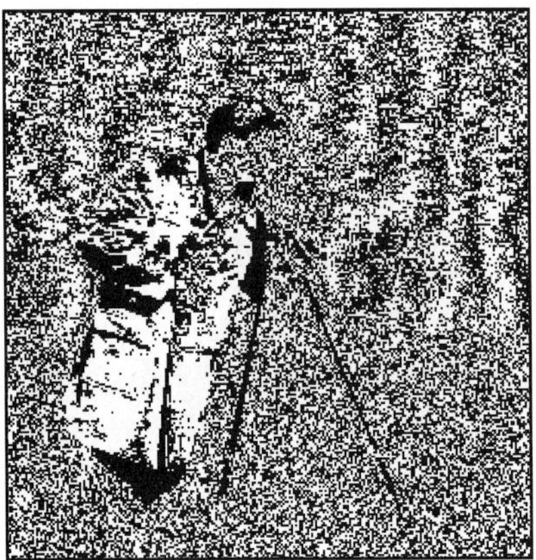

g. Bit-plane 2.

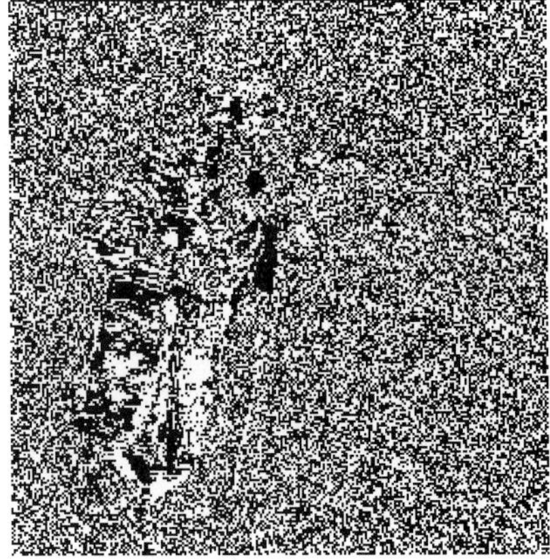

h. Bit-plane 1.

i. Bit-plane 0, the least significant bit.

coding. Secondly, the principal components transform (PCT, see Sections 1.7.3 and 2.4.3) can be used, which provides a theoretically optimal decorrelation. Also, the spectral domain is used for image compression, so this first stage may include mapping into the frequency or sequency domain. These methods are all *reversible*, that is information preserving, although all mapping methods are not reversible. The concept of reversibility is important to a compression method.

Depending on the mapping equation used, quantization may be necessary to convert the data into digital form. There are two ways to do this—uniform quantization or nonuniform quantization. In *uniform quantization* all the quanta, or subdivisions into which the range is divided, are of equal width. In *nonuniform quantization* these quantization bins are not all of equal width (see Figures 2.2-15 and 2.2-16). Often, nonuniform quantization bins are designed to take advantage of the response of the human visual system. For example, very high brightness levels appear the same—white—so wider quantization bins may be used over this range. In the spectral domain, the higher frequencies may also be quantized with wider bins because we are more sensitive to lower and midrange spatial frequencies. This concept of nonuniform quantization bin sizes can also be described as a *variable bit rate* because the wider quantization bins imply fewer bits to encode, whereas the smaller bins need more bits. Note that the quantization process is not reversible, so some information may be lost during quantization. Additionally, because it is not a reversible process, the inverse process does not exist, so it does not appear in the decompression model (Figure 5.1-4).

The coding stage of any image compression algorithm is very important. The coder provides a one-to-one mapping, each input is mapped to a unique output by the coder, so it is a reversible process. The code can be an *equal length code*, where all the code words are the same size, or an *unequal length code* with variable length code words. In most cases, an unequal length code is the most efficient for data compression but requires more overhead in the coding and decoding stages. Many of the lossless methods described here are primarily efficient coding techniques.

5.2 Lossless Compression Methods

Lossless compression methods are necessary in some imaging applications. For example, with medical images, the law requires that any archived medical images are stored without any data loss. Many of the lossless techniques were developed for non-image data and, consequently, are not optimal for image compression. In general, the lossless techniques alone provide marginal compression of complex image data, often in the range of only a 10% reduction in file size. However, lossless compression techniques may be used for both preprocessing and postprocessing in image compression algorithms. Additionally, for simple images the lossless techniques can provide substantial compression.

The underlying theory for lossless compression (also called *data compaction*) comes from the area of communications and information theory, with a mathematical basis in probability theory. One of the most important concepts used here is the idea of information content and randomness in data. Using information theory, an event that is less likely to occur is said to contain more information than an event that is more likely to occur. For example, consider the following statements:

1. The sun will come up tomorrow.
2. It will rain tomorrow.
3. A time machine will be invented in the next 10 years.

Which statement, in the sense stated previously, has the most information? Statement #1 contains relatively little information because this is an event that we all already know will occur—it has a 100% probability. Statement #2 contains more information, because the event "raining tomorrow" has a probability less than 100%. Statement #3 contains the most information because it is a highly unlikely event. This perspective on information is the *information theoretic definition* and should not be confused with our working definition that requires information to be useful, not simply novel. This background is provided to help explain some of the following concepts.

An important concept here is the idea of measuring the average information in an image, referred to as the *entropy*. The entropy for an $N \times N$ image can be calculated by this equation:

$$\text{Entropy} = -\sum_{i=0}^{L-1} p_i \log_2(p_i) \quad \text{(in bits per pixel)}$$

$$\text{where } p_i = \text{the probability of the } i\text{th gray level} = \frac{n_k}{N^2}$$

$$n_k = \text{the total number of pixels with gray value } k$$

$$L = \text{the total number of gray levels (e.g., 256 for 8 bits)}$$

This measure provides us with a theoretical minimum for the average number of bits per pixel that could be used to encode the image. This number is theoretically optimal and can be used as a metric for judging the success of a coding scheme.

EXAMPLE 5 – 5

Let $L = 8$, meaning that there are 3 bits/pixel in the original image. Now, lets say that the number of pixels at each gray-level value is equal (they have the same probability), that is:

$$p_0 = p_1 = \ldots = p_7 = \frac{1}{8}$$

Now, we can calculate the entropy as follows:

$$\text{Entropy} = -\sum_{i=0}^{7} p_i \log_2(p_i) = -\sum_{i=0}^{7} \frac{1}{8} \log_2\left(\frac{1}{8}\right) = 3$$

This tells us that the theoretical minimum for lossless coding for this image is 3 bits/pixel. In other words, there is no code that will provide better results than the one currently used (called the natural code, since $000_2 = 0$, $001_2 = 1$, $010_2 = 2$, ..., $111_2 = 7$). This example illustrates that the image with the most random distribution of gray levels, a uniform distribution, has the highest entropy.

EXAMPLE 5 – 6

Let $L = 8$, thus we have a natural code with 3 bits/pixel in the original image. Now let's say that the entire image has a gray level of 2, so

$$p_2 = 1, \text{ and } p_0 = p_1 = p_3 = p_4 = p_5 = p_6 = p_7 = 0$$

And the entropy is

$$\text{Entropy} = -\sum_{i=0}^{7} p_i \log_2(p_i) = -(1)\log_2(1) + 0 + \ldots + 0 = 0$$

This tells us that the theoretical minimum for coding this image is 0 bits/pixel. Why is this? Because the gray-level value is known to be 2. To code the entire image, we need only one value. This is called the certain event; it has a probability of 1.

The two preceding examples illustrate the range of the entropy:

$$0 \leq \text{Entropy} \leq \log_2(L)$$

The examples also illustrate the information theory perspective on information and randomness. The more randomness that exists in an image, the more evenly distributed the gray levels, and the more bits per pixel are required to represent the data. This also correlates to information—more randomness implies each individual value is less likely, which means more information is contained in each pixel value, so we need more bits to code each pixel value. This also provides us with one of the key concepts in coding theory: we want to assign a fewer number of bits to code more likely events. Intuitively, this makes sense. Given an image to code, a minimum overall file size will be achieved if a smaller number of bits is used to code the most frequent gray levels.

The entropy measure also provides us with a metric to evaluate coder performance. We can measure the average number of bits per pixel (*Length*) in a coder by the following:

$$L_{\text{ave}} = \sum_{i=0}^{L-1} l_i p_i$$

$$\text{where } l_i = \text{length in bits of the code for } i\text{th gray level}$$

$$p_i = \text{histogram-probability of } i\text{th gray level}$$

This can then be compared to the entropy, which provides the theoretical minimum. The closer L_{ave} is to the entropy, the better the coder.

5.2.1 Huffman Coding

The Huffman code, developed by D. Huffman in 1952, is a minimum length code. This means that given the statistical distribution of the gray levels (the histogram), the Huffman algorithm will generate a code that is as close as possible to the minimum bound, the entropy. This method results in a *variable length code*, where the code words are of unequal length. For complex images, Huffman coding alone will typically reduce the file by 10 to 50% (1.1:1 to 1.5:1), but this ratio can be improved to 2:1 or 3:1 by preprocessing for irrelevant information removal.

The Huffman algorithm can be described in five steps:

1. Find the gray-level probabilities for the image by finding the histogram.
2. Order the input probabilities (histogram magnitudes) from smallest to largest.
3. Combine the smallest two by addition.
4. GOTO step 2, until only two probabilities are left.
5. By working backward along the tree, generate code by alternating assignment of 0 and 1.

This procedure is best illustrated by example.

EXAMPLE 5-7

We have an image with 2 bits/pixel, giving four possible gray levels. The image is 10 rows by 10 columns. In Step 1 we find the histogram for the image. This is shown in Figure 5.2-1a, where we see that gray level 0 has 20 pixels, gray level 1 has 30 pixels, gray level 2 has 10 pixels, and

Figure 5.2-1 Huffman Coding Example

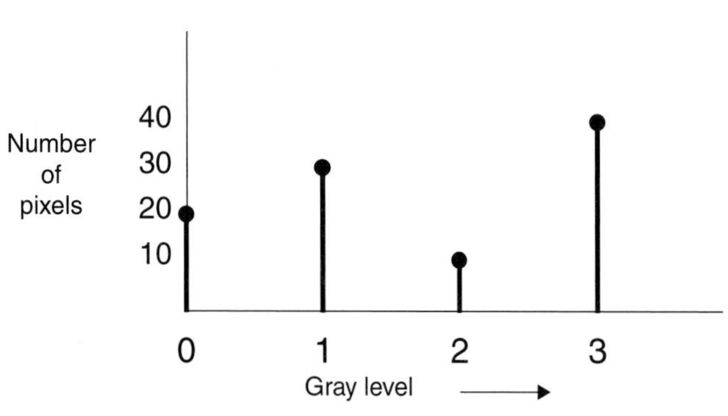

$$g_0 = \frac{20}{100} = 0.2$$

$$g_1 = \frac{30}{100} = 0.3$$

$$g_2 = \frac{10}{100} = 0.1$$

$$g_3 = \frac{40}{100} = 0.4$$

a. Step 1: Histogram.

$g_3 \rightarrow 0.4$

$g_1 \rightarrow 0.3$

$g_0 \rightarrow 0.2$

$g_2 \rightarrow 0.1$

b. Step 2: Order.

$0.4 \rightarrow 0.4$

$0.3 \rightarrow 0.3$

$0.2 \rightarrow 0.3$

0.1

c. Step 3: Add.

$0.4 \rightarrow 0.4 \rightarrow 0.4$

$0.3 \rightarrow 0.3 \rightarrow 0.6$

$0.2 \rightarrow 0.3$

0.1

$0.4 \rightarrow 0.4 \rightarrow 0.6$

$0.3 \rightarrow 0.3 \rightarrow 0.4$

$0.2 \rightarrow 0.3$

0.1

d. Step 4: Reorder and add until only two values remain.

gray level 3 has 40 pixels with the value. These are converted into probabilities by normalizing to the total number of pixels in the image. Next, in step 2, the probabilities are ordered as in Figure 5.2-1b. For step 3 we combine the smallest two by addition. Step 4 repeats steps 2 and 3, where we reorder (if necessary) and add the two smallest probabilities as in Figure 5.2-1d. This step is repeated until only two values remain. Because we have only two left in our example, we can continue to step 5 where the actual code assignment is made. The code assignment is shown in Figure 5.2-2. We start on the right-hand side of this tree and assign 0's and 1's, working our way back to the original probabilities. Figure 5.2-2a shows the first assignment of 0 and 1. A 0 is assigned to the 0.6 branch, and a 1, to the 0.4 branch. In Figure 5.2-2b the assigned 0 and 1 are brought back along the tree, and wherever a branch occurs the code is put on both branches. Now (Figure 5.2-2c) we assign the 0 and 1 to the branches labeled 0.3, appending to the existing code. Finally (Figure 5.2-2d), the codes are brought back one more level, and where the branch splits another assignment of 0 and 1 occurs (at the 0.1 and 0.2 branch). Now we have the Huffman code for this image as shown in Table 5.2-1.

Table 5.2-1 Huffman Code

Original Gray Level (Natural Code)	Probability	Huffman code
g_0: 00_2	0.2	010_2
g_1: 01_2	0.3	00_2
g_2: 10_2	0.1	011_2
g_3: 11_2	0.4	1_2

Note that two of the gray levels now have 3 bits assigned to represent them, but one gray level only has 1 bit assigned to represent it. The gray level represented by 1 bit, g_3, is the most likely to occur (40% of the time) and thus has the *least information in the information theoretic sense*. Remember that we learned from information theory that symbols with less information require fewer bits to represent them. The original image had an average of 2 bits/pixel, let us examine the entropy in bits per pixel and average bit length for the Huffman coded image file.

EXAMPLE 5-8

$$\text{Entropy} = -\sum_{i=0}^{3} p_i \log_2(p_i)$$

$$= -[(0.2)\log_2(0.2) + (0.3)\log_2(0.3) + (0.1)\log_2(0.1) + (0.4)\log_2(0.4)$$

$$\approx 1.846 \text{ bits/pixel}$$

(Note: $\log_2(x)$ can be found by taking $\log_{10}(x)$ and multiplying by 3.322)

$$L_{\text{ave}} = \sum_{i=0}^{L-1} l_i p_i$$

$$= 3(0.2) + 2(0.3) + 3(0.1) + 1(0.4)$$

$$= 1.9 \text{ bits/pixel} \quad \text{(Average length with Huffman code)}$$

Figure 5.2-2 Huffman Coding Example, Step 5

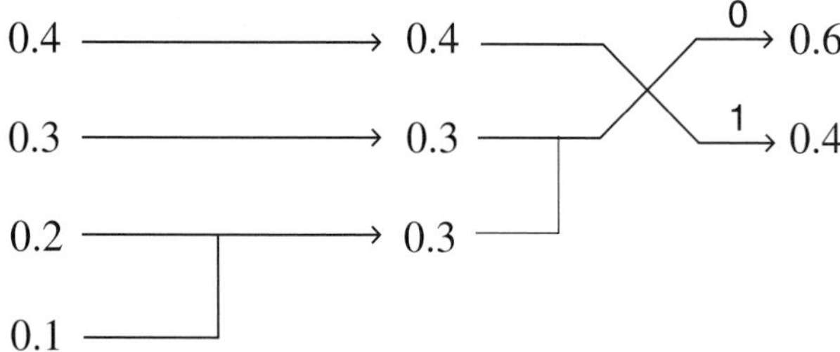

a. Assign 0 and 1 to the rightmost probabilities.

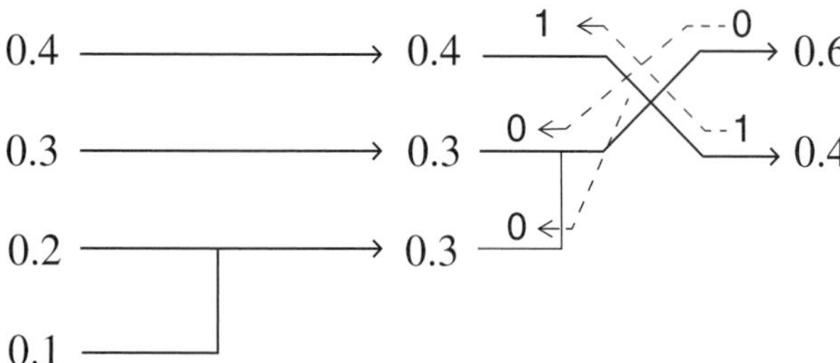

b. Bring 0 and 1 back along the tree.

c. Append 0 and 1 to previously added branches.

d. Repeat the process until the original branch is labeled.

In the example, we observe a 2.0:1.9 compression, which is about a 1.05 compression ratio, providing about 5% compression. From the example we can see that the Huffman code is highly dependent on the histogram, so any preprocessing to the histogram may help the compression ratio.

5.2.2 Run-Length Coding

Run-length coding (RLC) is an image compression method that works by counting the number of adjacent pixels with the same gray-level value. This count, called the *run length*, is then coded and stored. Here we will explore several methods of run-length coding—basic methods that are used primarily for binary (two-valued) images and extended versions for gray-scale images. We will also briefly discuss RLC standards.

Basic RLC is used primarily for binary images but can work with complex images that have been preprocessed by thresholding to reduce the number of gray levels to two. There are various ways to implement basic RLC, and the first step is to define the required parameters. We can either use horizontal RLC, counting along the rows, or vertical RLC, counting along the columns. In basic horizontal RLC the number of bits used for the coding depends on the number of pixels in a row. If the row has 2^n pixels, then the required number of bits is n, so that a run that is the length of the entire row can be coded.

EXAMPLE 5-9

A 256×256 image requires 8 bits, since $2^8 = 256$.

EXAMPLE 5-10

A 512×512 image requires 9 bits, since $2^9 = 512$.

The next step is to define a convention for the first RLC number in a row—does it represent a run of 0's or 1's? Defining the convention for the first RLC number to represent 0's, we can look at the following example.

EXAMPLE 5-11

The image is an 8×8 binary image, which requires 3 bits for each run-length coded word. In the actual image file are stored 1's and 0's, although upon display the 1's become 255 (white) and the 0's are 0 (black). To apply RLC to this image, using horizontal RLC:

$$\begin{bmatrix} 0 & 0 & 0 & 0 & 0 & 0 & 0 & 0 \\ 1 & 1 & 1 & 1 & 0 & 0 & 0 & 0 \\ 0 & 1 & 1 & 0 & 0 & 0 & 0 & 0 \\ 0 & 1 & 1 & 1 & 1 & 1 & 0 & 0 \\ 0 & 1 & 1 & 1 & 0 & 0 & 1 & 0 \\ 0 & 0 & 1 & 0 & 0 & 1 & 1 & 0 \\ 1 & 1 & 1 & 1 & 0 & 1 & 0 & 0 \\ 0 & 0 & 0 & 0 & 0 & 0 & 0 & 0 \end{bmatrix}$$

The RLC numbers are:

First row: 8

Second row: 0, 4, 4

Third row: 1, 2, 5

Fourth row: 1, 5, 2

Fifth row: 1, 3, 2, 1, 1

Sixth row: 2, 1, 2, 2, 1

Seventh row: 0, 4, 1, 1, 2

Eighth row: 8

Note that in the second and seventh rows, the first RLC number is 0, since we are using the convention that the first number corresponds to the number of zeros in a run.

This basic method can be extended to gray-level images by using a technique called bit-plane RLC. *Bit-plane RLC* works by applying basic RLC to each bit plane independently. In Figure 5.2-3 the idea of bit planes is illustrated. For each binary digit in the gray-level value, an image plane is created, and this image plane (a string of 0's and 1's) is then coded using RLC. Typical compression ratios of 0.5 to 1.2 are achieved with complex 8-bit monochrome images; so, without further processing, this is not a good compression technique for complex images. Bit-plane RLC is most useful for simple images, such as graphics files, where much higher compression ratios are achieved. The compression results using this method can be improved by preprocessing to reduce the number of gray levels, but then the compression is *not lossless*. In order for this method to be effective, the reduced image data (in *natural code*) needs to be mapped to a *Gray code* (named after Frank Gray), where adjacent numbers differ in only one bit. Because adjacent pixel values are highly correlated, adjacent pixel values tend to be relatively close in gray-level value, and this can be problematic for RLC.

EXAMPLE 5-12

In Figure 5.2-4 is shown the 4-bit Gray code and the natural binary code. The Gray code, by definition, only has one bit changing in adjacent codes. However, in, for example, the 7 to 8 transition with the natural code, all four bits change:

Natural Code	*Gray Code*
0 1 1 1	0 1 0 0
↓ ↓ ↓ ↓	↓ ↓ ↓ ↓
1 0 0 0	1 1 0 0

When a situation such as this example occurs, each bit plane experiences a transition, which adds a code for the run in each bit plane. However, with the Gray code, only one bit plane experiences the transition, so it only adds one extra code word.

Figure 5.2-3 Bit-Plane Run-Length Coding

b_3	b_2	b_1	b_0
0	0	0	0
0	0	0	1
0	0	1	0
⋮	⋮	⋮	⋮
0	1	1	0
1	1	1	1

a. 4 bits/pixel designation.

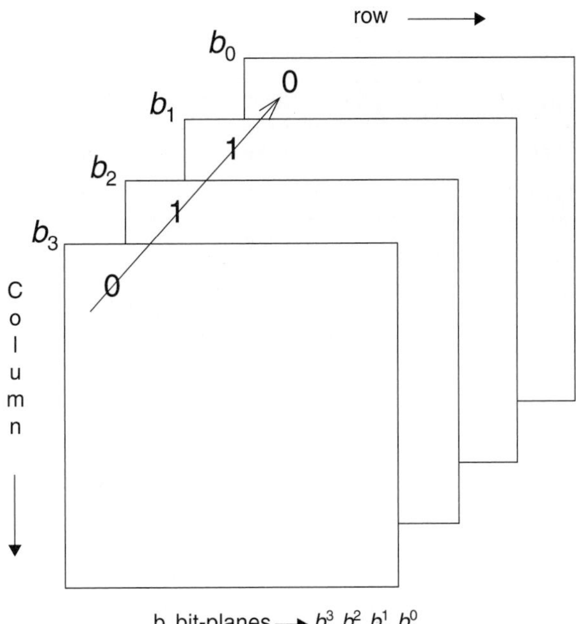

b. bit-planes → $b_0^3\ b_0^2\ b_0^1\ b_0^0$.

Another way to extend basic RLC to gray-level images is to include the gray level of a particular run as part of the code. Here, instead of a single value for a run, two parameters are used to characterize the run. The pair (G, L) correspond to the gray-level value G and the run-length L. This technique is only effective with images containing a small number of gray levels.

Figure 5.2-4 Gray Code

Decimal	4-bit Natural Code	4-bit Gray Code
0	0000	0000
1	0001	0001
2	0010	0011
3	0011	0010
4	0100	0110
5	0101	0111
6	0110	0101
7	0111	0100
8	1000	1100
9	1001	1101
10	1010	1111
11	1011	1110
12	1100	1010
13	1101	1011
14	1110	1001
15	1111	1000

a. Gray code versus natural code.

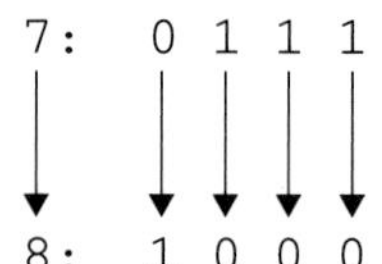

b. The natural code transition of 7 to 8 changes all four bits.

EXAMPLE 5-13

Given the following 8×8, 4-bit image:

$$\begin{bmatrix} 10 & 10 & 10 & 10 & 10 & 10 & 10 & 10 \\ 10 & 10 & 10 & 10 & 10 & 12 & 12 & 12 \\ 10 & 10 & 10 & 10 & 10 & 12 & 12 & 12 \\ 0 & 0 & 0 & 10 & 10 & 10 & 0 & 0 \\ 5 & 5 & 5 & 0 & 0 & 0 & 0 & 0 \\ 5 & 5 & 5 & 10 & 10 & 9 & 9 & 10 \\ 5 & 5 & 5 & 4 & 4 & 4 & 0 & 0 \\ 0 & 0 & 0 & 0 & 0 & 0 & 0 & 0 \end{bmatrix}$$

The corresponding gray-level pairs are as follows:

First row: 10,8

Second row: 10,5 12,3

Third row: 10,5 12,3

Fourth row: 0,3 10,3 0,3

Fifth row: 5,3 0,5

Sixth row: 5,3 10,2 9,2 10,1

Seventh row: 5,3 4,3 0,2

Eighth row: 0,8

These numbers are then stored in the RLC compressed file as:

10, 8, 10, 5, 12, 3, 10, 5, 12, 3, 0, 3, 10, 3, 0, 3, 5, 3, 0, 5, 5, 3, 10, 2, 9, 2, 10, 1, 5, 3, 4, 3, 0, 2, 0, 8

The decompression process requires the number of pixels in a row, and the type of coding used.

Standards for RLC have been defined by the International Telecommunications Union-Radio (ITU-R, previously CCIR). These standards, initially defined for use with FAX transmissions, have become popular for binary image compression. They use horizontal RLC but postprocess the resulting RLC with a Huffman encoding scheme. Newer versions of this standard also use a two-dimensional technique where the current line is coded based on a previous line. This additional processing helps to reduce the file size. These coding methods provide compression ratios of about 15 to 20 for typical documents.

5.2.3 Lempel-Ziv-Welch Coding

The *Lempel-Ziv-Welch* (LZW) coding algorithm works by coding strings of data. For images, these strings of data correspond to sequences of pixel values. It works by creating a string table that contains the strings and their corresponding codes. The string table is updated as the file is read, with new codes being inserted whenever a new string is encountered. If a string is encountered that is already in the table, the corresponding code for that string is put into the compressed file.

LZW coding uses code words with more bits than the original data. For example, with 8-bit image data, an LZW coding method could employ 10-bit words. The corresponding string table would then have 2^{10} = 1,024 entries. This table consists of the original 256 entries, corresponding to the original 8-bit data, and allows 768 other entries for string codes. The string codes are assigned during the compression process, but the actual string table is not stored with the compressed data. During decompression the information in the string table is extracted from the compressed data itself.

For the GIF (and TIFF) image file format the LZW algorithm is specified, but there has been some controversy over this because the algorithm is patented (by Unisys Corporation under patent #4,558,302). Because these image formats are widely used, other methods similar in nature to the LZW algorithm have been developed to be used with these, or similar, image file formats. Similar versions of this algorithm include the *adaptive Lempel-Ziv*, used in the UNIX compress function, and the *Lempel-Ziv 77* algorithm used in the UNIX gzip function.

5.2.4 Arithmetic Coding

In arithmetic coding there is not a direct correspondence between the code and the individual pixel values. *Arithmetic coding* transforms input data into a single floating point number between 0 and 1. As each input symbol (in this case, pixel value) is read, the precision required for this number becomes greater. Because images are very large and the precision of digital computers is finite, an entire image must be divided into small subimages to be encoded.

Arithmetic coding uses the probability distribution of the data (histogram), so it can theoretically achieve the maximum compression specified by the entropy. It works by successively subdividing the interval between 0 and 1, based on the placement of the current pixel value in the probability distribution. This is best illustrated by example.

EXAMPLE 5-14

Given a 16×16, 2-bit image with the histogram shown in Figure 5.2-5a, we can define an arithmetic coding probability table shown in Figure 5.2-5b. The probability values are the ratio of the specific gray-level value to the total number of pixels in the image (in this case $16 \times 16 = 256$). The initial subinterval specifies how the 0 to 1 interval is divided based on the distribution, where the width of the subinterval is equal to the probability, and the subinterval starts where the previous one stops. In Figure 5.2-5c the actual arithmetic coding process is illustrated, with an example pixel value sequence of 0, 0, 3, 1. Starting on the left, the initial 0 to 1 interval is subdivided, based on the probability distribution. Next, the first pixel value 0 is coded by extracting the subinterval corresponding to the 0 and subdividing it again, based on the same relative distribution. This process is repeated for each pixel value in the sequence until a final interval is determined, in this case from 58/1,024 to 62/1,024, or 0.056640625 to 0.060546875. Any value within this subinterval, such as 0.057 or 0.060, can be used to represent this sequence of gray-level values.

In practice, this technique may be used as part of an image compression scheme, but it is impractical to use alone.

5.3 Lossy Compression Methods

In order to achieve high compression ratios with complex images, lossy compression methods are required. Lossy compression provides tradeoffs between image quality and degree of compression, which allows the compression algorithm to be customized to the application. With some of the more advanced methods, images can be compressed 10 to 20 times with virtually no visible information loss, and 30 to 50 times with minimal degradation (see Figure 5.3-1). Image enhancement and restoration techniques can be combined with lossy compression schemes to improve the appearance of the decompressed image.

The lossy compression methods discussed are representative of the available tools for compression algorithm development and provide a wide variety of compression ratios and image quality. Many of the methods have adjustable parameters to allow the user to select the desired compression ratio and image fidelity. In general, a

Figure 5.2-5 Arithmetic Coding

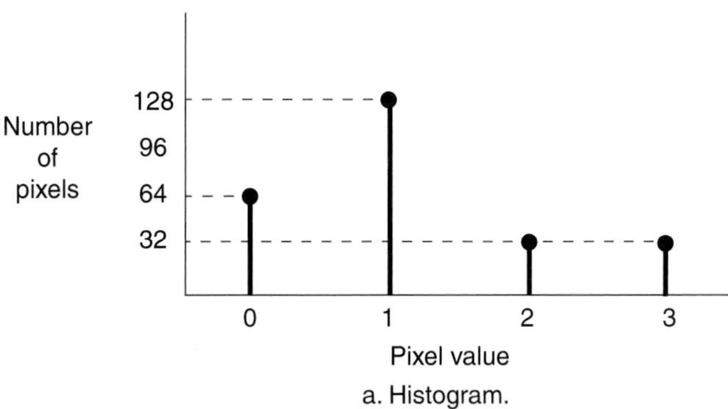

a. Histogram.

Pixel Value	Probability	Initial Subinterval
0	64/256 = 1/4	0 - 1/4
1	128/256 = 1/2	1/4 - 3/4
2	32/256 = 1/8	3/4 - 7/8
3	32/256 = 1/8	7/8 - 1

b. Probability table.

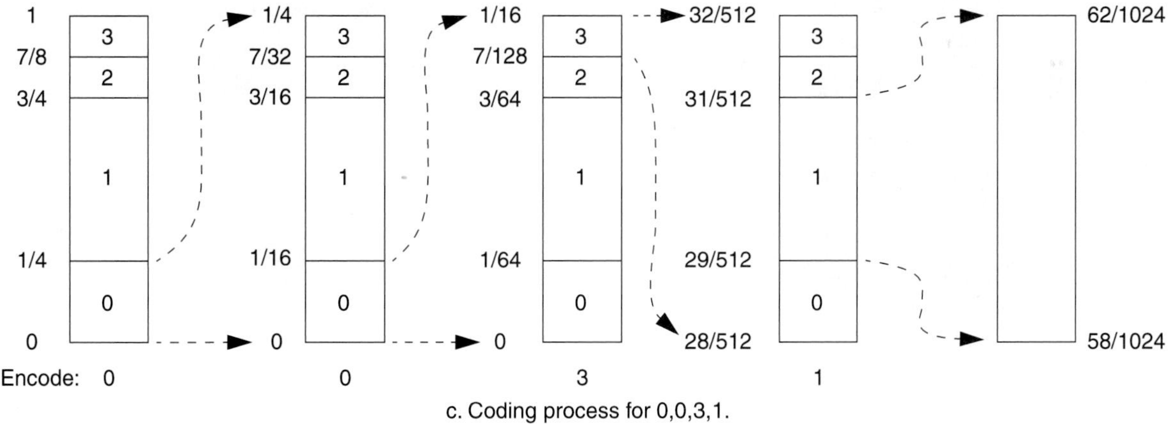

c. Coding process for 0,0,3,1.

higher compression ratio results in a poorer image, but the results are highly image dependent. A technique that works well for one application may not be suitable for another.

Lossy compression is performed in both the spatial and transform domains. We will explore methods that utilize each of these domains and some that use both. In the spatial domain we will discuss gray-level run-length coding (GLRLC), differential pre-

Figure 5.3-1 Lossy Image Compression

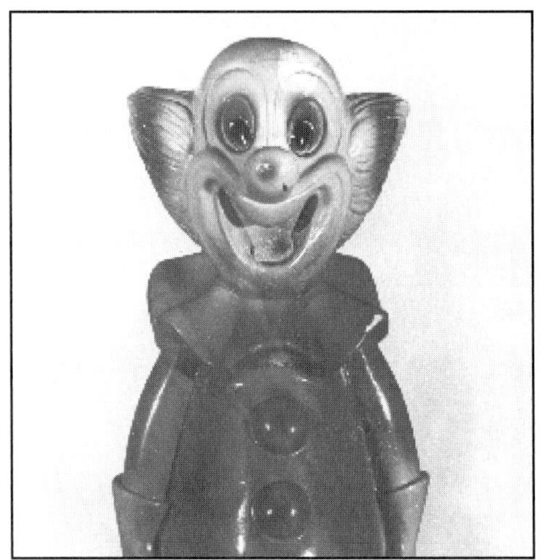

a. Original image.

b. JPEG compression, 10:1 ratio.

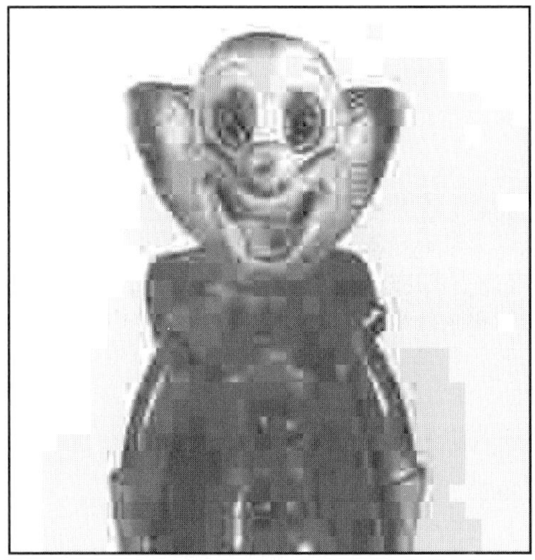

c. JPEG compression, 48:1 ratio.

d. Wavelet/vector quantization compression, 36:1 ratio.

dictive coding (DPC), block truncation coding (BTC), and vector quantization (VQ). In the transform domain we will discuss filtering, zonal coding, threshold coding, and the JPEG algorithm. We will also look at techniques for combining these methods into hybrid compression algorithms, which use both the spatial and transform domains.

5.3.1 Gray-Level Run-Length Coding

In Section 5.2.2 on lossless compression we discussed methods of extending basic run-length coding to gray-level images, by using bit-plane coding. The RLC technique can also be used for lossy image compression by reducing the number of gray levels and then applying standard RLC techniques. As with the lossless techniques, for this method to be effective, the reduced image data (in natural code) need to be mapped to a gray code, where adjacent numbers differ in only one bit.

A more sophisticated RLC algorithm for encoding gray-level images is called the *dynamic window-based RLC*. This algorithm relaxes the criterion of the runs being the same value and allows for the runs to fall within a gray-level range, called the *dynamic window range*. This range is dynamic because it starts out larger than the actual gray-level window range, and maximum and minimum values are narrowed down to the actual range as each pixel value is encountered. This process continues until a pixel is found out of the actual range. The image is encoded with two values, one for the run length and one to approximate the gray-level value of the run. This approximation can simply be the average of all the gray-level values in the run, or a more complex method may be used to calculate the representative value.

EXAMPLE 5-15

Assume the following pixel values in sequence:

65 67 66 64 63 68 70

and a window range of 5.

The first value is called the reference value (in this case 65). A dynamic window range is then defined that has

MINIMUM = reference – (window length – 1)

and

MAXIMUM = reference + (window length – 1)

In this case the dynamic window is [65 – (5 – 1)] to [65 + (5 – 1]) = 61 to 69. The next value encountered (67) is used to adjust this range. The range based on this value alone is from 63 to 71. The new dynamic range is based on the intersection of the range from this new value with the previous range, so the new range is 63 to 69. This process continues until the value of 68 is encountered. At this point the range has been narrowed down to 63 to 67, so 68 is out of range. This run is then encoded as

RUN LENGTH = 5

GRAY LEVEL = (65 + 67 + 66 + 64 + 63)/5 = 65.

In Figure 5.3-2 are results of the dynamic window-based RLC, where the average was used as the representative value. This particular algorithm also uses some preprocessing to allow for the run-length mapping to be coded so that a run can be any length and is not constrained by the length of a row (see reference for details).

5.3.2 Block Truncation Coding

Block truncation coding works by dividing the image into small subimages and then reducing the number of gray levels within each block. This reduction is performed by a quantizer that adapts to the local image statistics. The levels for the quantizer are chosen to minimize a specified error criterion, and then all the pixel values within each block are mapped to the quantized levels. The necessary information to decompress the image is then encoded and stored. Many different BTC algorithms have been defined by using various types of quantization and error criteria, as well as various preprocessing and postprocessing methods. The more sophisticated algorithms provide better results, but with a corresponding increase in computation time.

The basic form of BTC divides the image into 4×4 blocks and codes each block using a two-level quantizer. The two levels are selected so that the mean and variance of the gray levels within the block are preserved. Each pixel value within the block is then compared with a threshold, typically the mean, and then is assigned to one of the

Figure 5.3-2 Dynamic Window Range RLC

a. Original image.

b.Window length = 10; compression = 4:1.

c. Error image of (b), multiplied to show detail.

Figure 5.3-2 (Continued)

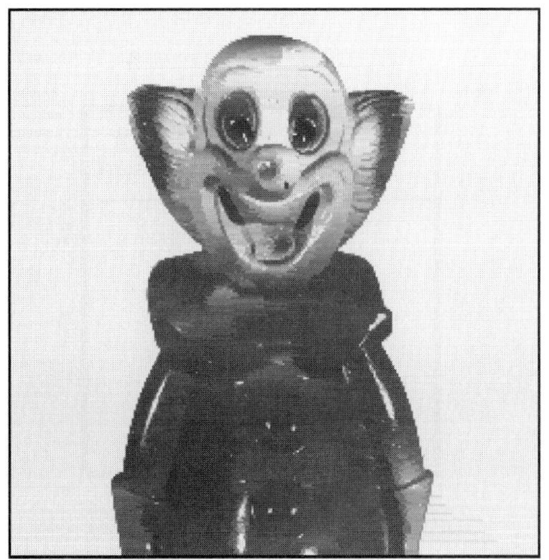

d. Window length = 35; compression = 8:1.

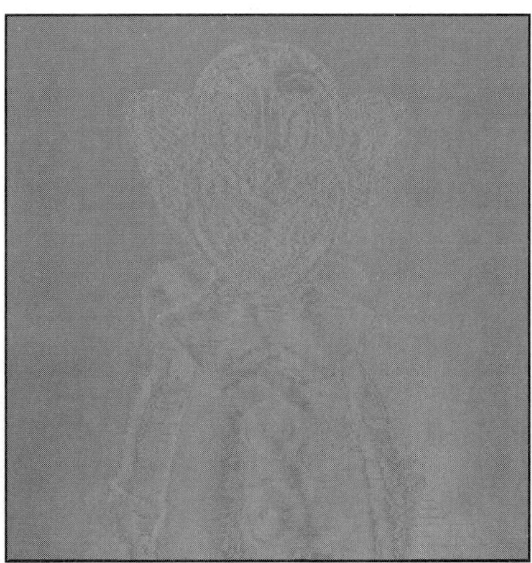

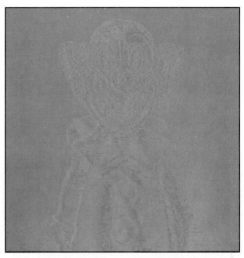

e. Error image of (d), multiplied to show detail.

f. Window length = 60; compression = 12.6:1.

g. Error image of (f), multiplied to show detail.

two levels. If it is above the mean, it is assigned the high-level code (1); if it is below the mean, it is assigned the low-level code (0).

After this is done, the 4×4 block is encoded with 4 bytes: 2 bytes to store the two levels and 2 bytes to store the bit string of 1's and 0's corresponding to the high and low codes for that particular block. This is illustrated in Figure 5.3-3, where we see that the bit string for the 4×4 block is packed into 2 bytes. Since the original 4×4 subimage has 16 bytes and the resulting code has 4 bytes (two for the high and low values, and two for the bit string), this provides a 16:4 or 4:1 compression. Although the results of this algorithm are image dependent, it tends to produce images with

Figure 5.3-3 Basic Block Truncation Coding

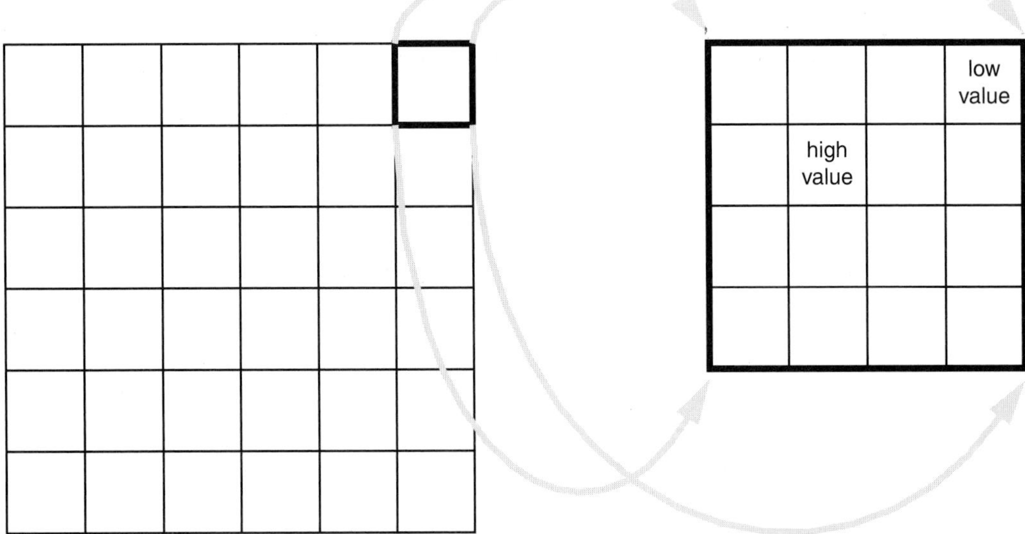

a. Divide image into 4 × 4 blocks.

b. Find high and low values for blocks.

1	1	0	1
0	0	0	0
1	1	1	1
1	1	0	1

c. Assign a 0 to each pixel less than the mean,1 to each pixel greater than the mean.

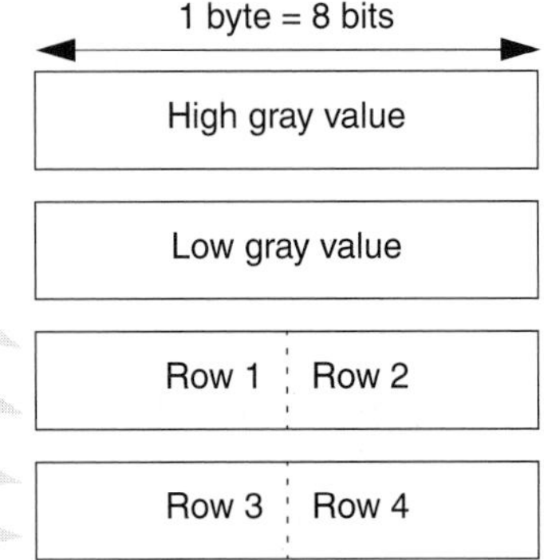

d. Encode 4 × 4 block with 4 bytes.

blocky effects as shown in Figure 5.3-4b. These artifacts can be smoothed by applying enhancement techniques such as median and average (lowpass) filters (Figures 5.3-4d, e). More advanced BTC algorithms can be explored in the references, and Figure 5.3-5 illustrates the multilevel BTC algorithm that uses a four-level quantizer. This algorithm allows for varying the block size, and a larger block size should provide higher compression, but with a corresponding decrease in image quality. With this particular implementation the compression ratio cannot be greater than 4, but we can see the decreasing image quality as the block size is increased.

5.3.3 Vector Quantization

Vector quantization is the process of mapping a vector that can have many values to a vector that has a smaller (quantized) number of values. For image compression, the vector corresponds to a small subimage, or block.

Figure 5.3-4 Block Truncation Coding

a. Original image.

b. Block truncation coded image, compression = 4:1.

c. Error image of (b), histogram stretched to show detail.

Figure 5.3-4 (Continued)

d. Image (b) postprocessed with a 3 × 3 median filter.

e. Image (b) postprocessed with a 3 × 3 averaging filter.

Figure 5.3-5 Multilevel Block Truncation Coding

a. Original image.

Figure 5.3-5 (Continued)

b. Multilevel BTC—block size = 8 × 8; compression = 3.99:1.

c. Error image of (b), multiplied to show detail.

d. Multilevel BTC—block size = 16 × 16; compression = 3.99:1

e. Error image of (d), multiplied to show detail.

Figure 5.3-5 (Continued)

f. Multilevel BTC—block size = 32 × 32; compression = 3.99:1.

g. Error image of (f), multiplied to show detail.

EXAMPLE 5-16

Given the following 4 × 4 subimage:

$$\begin{bmatrix} 65 & 70 & 71 & 75 \\ 71 & 70 & 71 & 81 \\ 81 & 80 & 81 & 82 \\ 90 & 90 & 91 & 92 \end{bmatrix}$$

this can be rearranged into a 1-D vector by putting the rows adjacent as follows:

[row1 row2 row3 row4] = [65 70 71 75 71 70 71 81 81 80 81 82 90 90 91 92]

The previous types of quantization deal with taking a single value and reducing the number of bits used to represent that value—this is called *scalar quantization* and is most easily achieved by rounding or truncation. Vector quantization treats the entire subimage (vector) as a single entity and quantizes it by reducing the total number of bits required to represent the subimage. This is done by utilizing a *codebook*, which stores a fixed set of vectors, and then coding the subimage by using the index (address) into the codebook.

EXAMPLE 5-17

Given an 8-bit, 256 × 256 image, we devise a vector quantization scheme that will encode each 4 × 4 block with one of the vectors in a codebook of 256 entries. Because the codebook has a length of 256, it requires the use of an 8-bit word for the index (address) into the codebook, which will be used to represent the vector. We determine that we want to encode a specific subimage with vector number 122 in the codebook. For this subimage we then store the number 122 as the index into the codebook. Then when the image is decompressed, the vector at the 122 address in the codebook is used for that particular subimage. This is illustrated in Figure 5.3-6. This will require 1 byte (8-bits) to be stored for each 4×4 block, providing a data reduction of 16 bytes for a 4×4 block to 1 byte, or 16:1.

In the example we achieved a 16:1 compression, but note that this assumes that the codebook is not stored with the compressed file. However, the codebook will need

Figure 5.3-6 Quantizing with a Codebook

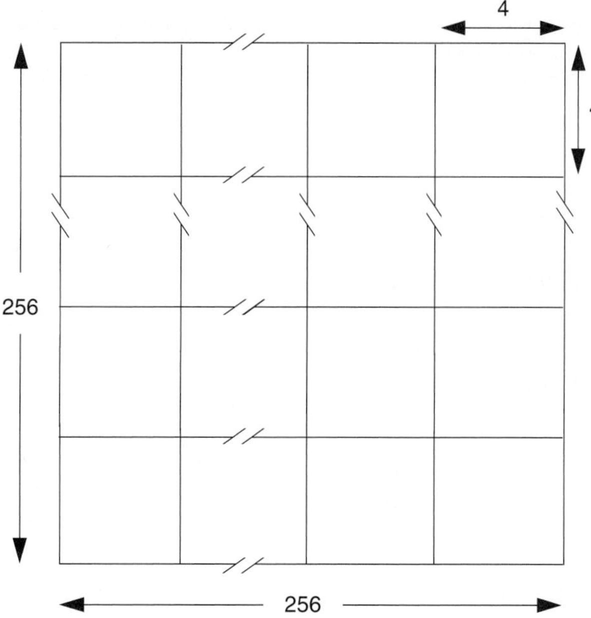

a. Original 256×256 image divided into 4×4 blocks.

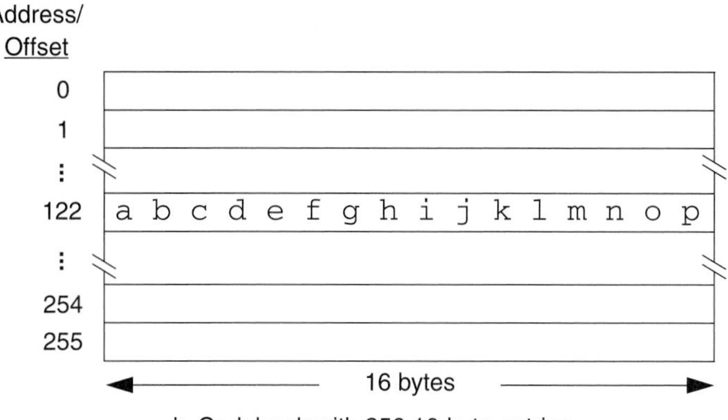

b. Codebook with 256 16-byte entries.

a	b	c	d
e	f	g	h
i	j	k	l
m	n	o	p

c. A subimage decompressed with vector #122.

to be stored unless a generic codebook that could be used for a particular type of image is devised—then we need only store the name of that particular codebook file. In the general case, better results will be obtained with a codebook that is designed for a particular image.

EXAMPLE 5-18

If we include the codebook in the compressed file from the previous example, the compression ratio will not be quite as good. For every 4×4 block we will have 1 byte. This gives us

$$\left(\frac{256 \text{ pixels}}{4 \text{ pixels/block}}\right)\left(\frac{256 \text{ pixels}}{4 \text{ pixels/block}}\right) = 4{,}096 \text{ blocks}$$

At 1 byte for each 4×4 block, this give us 4,096 bytes for the codebook addresses. Now we also include the size of the codebook, 256×16:

$$4{,}096 + (256)(16) = 8{,}192 \text{ bytes for the coded file}$$

The original 8-bit, 256×256 image contained

$$(256)(256) = 65{,}536 \text{ bytes}$$

Thus, we obtain a compression of

$$\frac{65{,}536}{8{,}192} = 8 \rightarrow 8{:}1 \text{ compression}$$

In this case, including the codebook cut the compression in half, from 16:1 to 8:1.

Now, how do we decide which vectors will be stored in the codebook? This is typically done by a training algorithm that finds a set of vectors that best represents the blocks in the image. This set of vectors is determined by optimizing some error criterion, where the error is defined as the sum of the distances between the original subimages and the resulting decompressed subimages. The standard method is to use the Linde-Buzo-Gray algorithm, which is described in Section 7.4. This technique can be applied in both the spatial and transform domains.

5.3.4 Differential Predictive Coding

Differential predictive coding works by predicting the next pixel value based on the previous values and encoding the difference between the predicted value and the actual value (for analog signals, this is also called differential pulse code modulation or DPCM). This technique takes advantage of the fact that adjacent pixels are highly correlated, which means that the difference between adjacent pixels is typically small. Because this difference is small, it will take only a small number of bits to represent it. The use of a predictor allows us to further reduce the amount of information to be encoded. By using a simple prediction equation, we can estimate the next pixel value and then encode only the difference between the estimate and the actual value. This error is then quantized, to further reduce the data and to optimize visual results, and can then be coded.

A block diagram of this process is shown in Figure 5.3-7, where we can see that the predictor must be in the feedback loop so that it matches the decompression system. The system must be initialized by retaining the first value(s) without any compression in order to calculate the first prediction. From the block diagram, we have the following:

$$\tilde{I} = \text{the predicted next pixel value}$$
$$\hat{I} = \text{the reconstructed pixel value}$$
$$e = I - \tilde{I} = \text{error}$$
$$\hat{e} = \hat{I} - \tilde{I} = \text{quantized error}$$

The prediction equation is typically a function of the previous pixel(s) and can also include global or application-specific information.

The theoretically optimum predictor, using only the previous value and based on minimizing mean-squared error between the original and the decompressed image, is given by

$$\tilde{I}(r, c+1) = \rho\hat{I}(r, c) + (1 - \rho)\,\overline{I}(r, c)$$
where $\overline{I}(r, c)$ = the average value for the image
ρ = the normalized correlation between pixel values

Figure 5.3-7 Differential Predictive Coding

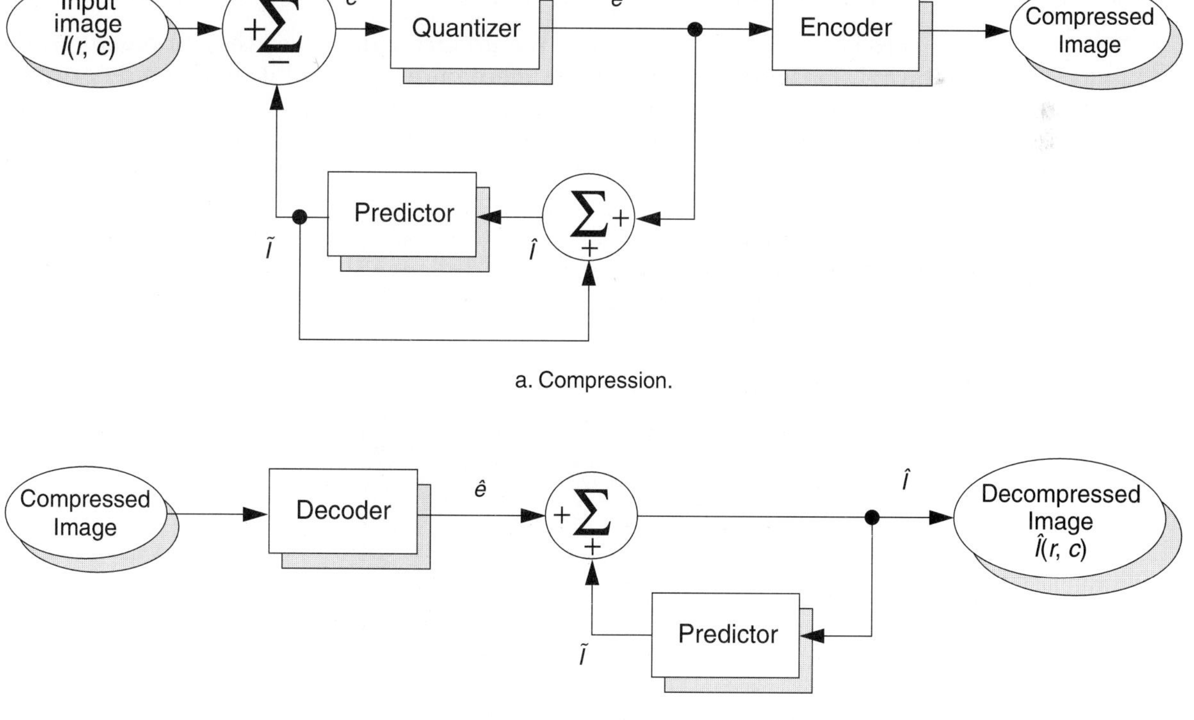

For most images ρ is between 0.85 and 0.95. When the next pixel value has been predicted, the error is calculated:

$$e(r, c+1) = I(r, c+1) - \tilde{I}(r, c+1)$$

This error signal is then quantized such that

$$\hat{e}(r, c+1) = \hat{I}(r, c+1) - \tilde{I}(r, c+1)$$

This quantized error can then be encoded using a lossless encoder, such as a Huffman coder. It should be noted that it is important that the predictor uses the same values during both compression and decompression, specifically the reconstructed values and not the original values (see Figure 5.3-7). In Figure 5.3-8 we see the results from using the original image values in the prediction, compared to using the reconstructed (decompressed) pixel values in the predictor. With these examples the quantization used was simply truncation ("clipping").

The prediction equation can be one-dimensional or two-dimensional, that is, it can be based on previous values in the current row only or on previous rows also (see Figure 5.3-9). The following prediction equations are typical examples of those used in practice, with the first being one-dimensional and the next two being two-dimensional:

$$\tilde{I}(r, c+1) = 0.97\hat{I}(r, c)$$
$$\tilde{I}(r, c+1) = 0.49\hat{I}(r, c) + 0.49\hat{I}(r-1, c+1)$$
$$\tilde{I}(r, c+1) = 0.74\hat{I}(r, c) + 0.74\hat{I}(r-1, c+1) - 0.49\hat{I}(r-1, c)$$

Using more of the previous values in the predictor increases the complexity of the computations for both compression and decompression, and it has been determined

Figure 5.3-8 DPC Example

a. Original image.

Figure 5.3-8 (Continued)

b. DPC using original values in predictor, clipping to the maximum, 5 bits/pixel, normalized correlation .90.

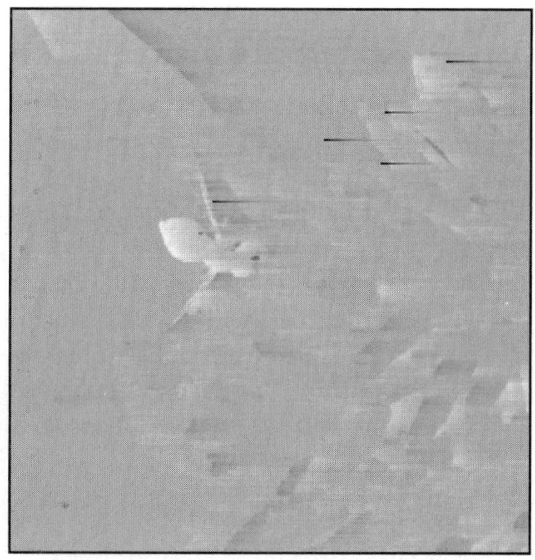

c. Error image of (b), multiplied to show detail.

d. DPC using reconstructed values in predictor, clipping to the maximum, 5 bits/pixel, normalized correlation .90.

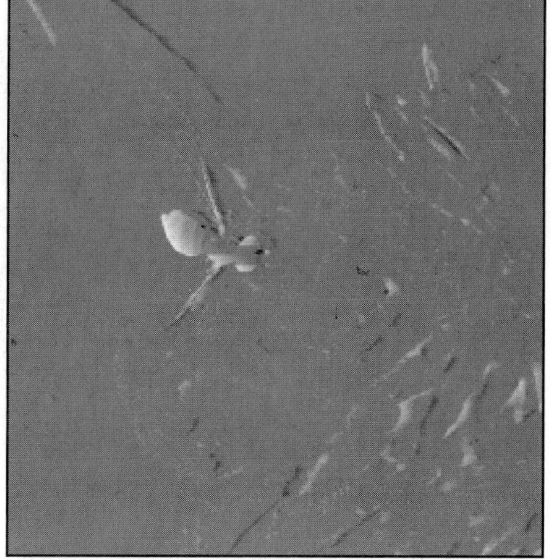

e. Error image of (d), multiplied to show detail.

that using more than three of the previous values provides no significant improvement in the resulting image.

The results of DPC can be further improved by using an optimal quantizer, such as the Lloyd-Max quantizer, instead of simply truncating the resulting error. The Lloyd-Max quantizer assumes a specific distribution for the prediction error. Assuming a 2-bit code for the error and a laplacian distribution for the error, the Lloyd-Max quantizer is defined as follows (see Figure 5.3-10):

ERROR RANGE		QUANTIZED VALUE
$0 \leq e < 1.102\sigma$	→	$+0.395\sigma$
$1.102\sigma \leq e < \infty$	→	$+1.81\sigma$
$-1.102\sigma \leq e < 0$	→	-0.395σ
$-1.102\sigma \leq e < -\infty$	→	-1.81σ

where σ = the standard deviation of the error distribution

Tables for the coefficients for n-bit codes can be found in the references. For most images, the standard deviation σ for the error signal is between 3 and 15. In Figure 5.3-11 is a comparison of using the Lloyd-Max quantizer and truncation as a quantization method. Figure 5.3-12 shows the error images and decompressed images using different bit rates for DPC compression (with Lloyd-Max quantization and a one-dimensional predictor).

Figure 5.3-9 DPC Predictor Dimensions

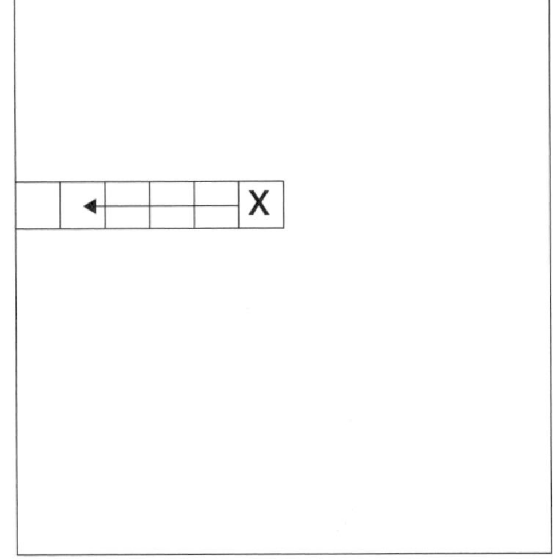

a. One-dimensional predictor, based on current row only; x = current pixel.

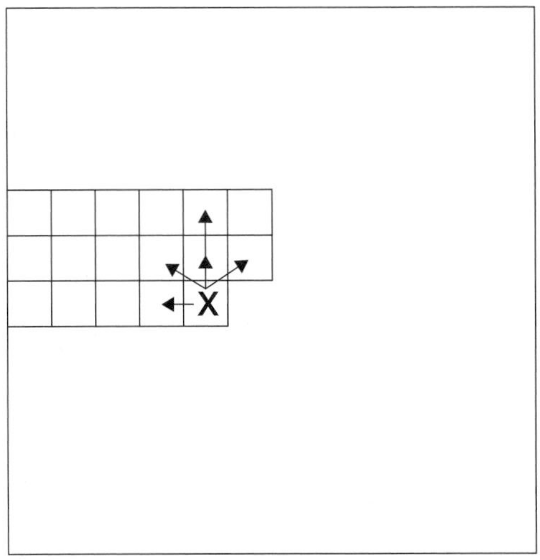

b. Two-dimensional predictor based on current and previous row or rows.

Figure 5.3-10 Lloyd-Max Quantizer

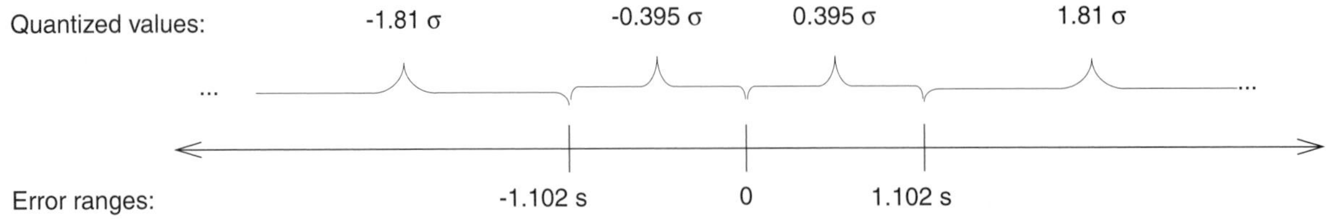

a. 2-bit Lloyd-Max quantizer with laplacian error distribution.

Figure 5.3-10 (Continued)

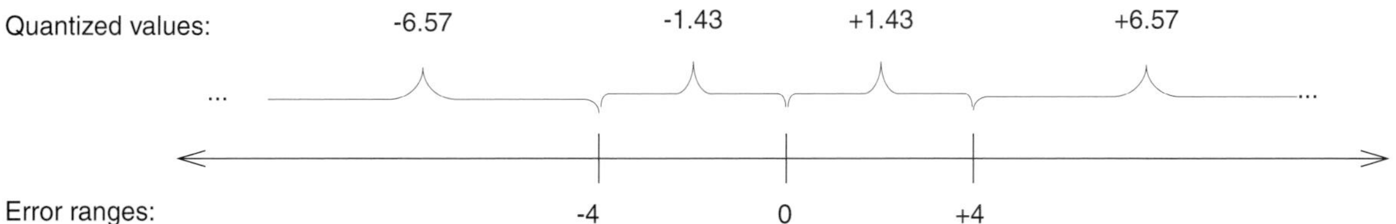

b. An example for $\sigma = 3.63$. The ±1.43 is typically rounded to ±1, and the ±6.57 is rounded to ±7.

Figure 5.3-11 DPC Quantization

a. Original image.

b. Lloyd-Max quantizer, using 2 bits/pixel, normalized correlation = 0.90, with standard deviation = 10.

c. Truncation quantizer, using 2 bits/pixel, normalized correlation = 0.90.

Figure 5.3-11 (Continued)

d. Lloyd-Max quantizer, using 4 bits/pixel, normalized correlation = 0.90, with standard deviation = 10.

e. Truncation quantizer, using 4 bits/pixel, normalized correlation = 0.90.

5.3.5 Transform Coding

Transform coding is a form of block coding done in the transform domain. The image is divided into blocks, or subimages, and the transform is calculated for each block. Any of the previously defined transforms can be used, frequency (e.g., Fourier) or sequency (e.g., Walsh), but it has been determined that the discrete cosine transform (DCT) is optimal for most images. After the transform has been calculated, the transform coefficients are quantized and coded. The primary reason this method is effective is because the frequency/sequency transform of images efficiently puts most of the information into relatively few coefficients so that many of the high-frequency coefficients can be quantized to 0 (eliminated completely). This type of transform is really just a special type of mapping that uses spatial frequency concepts as a basis for the mapping. Remember that for image compression the whole idea of mapping the original data into another mathematical space is to pack the information (or energy) into as few coefficients as possible.

The simplest form of transform coding is achieved by filtering; we can simply eliminate some of the high-frequency coefficients. This alone will not provide much compression because the transform data are typically floating point and thus 4 or 8 bytes/pixel (compared to the original pixel data at 1 byte/pixel), so quantization and coding are applied to the reduced data. The quantization process is partially performed by what is referred as bit allocation. *Bit allocation* is determining the number of bits to be used to code each coefficient. Typically, more bits are used for lower-frequency components.

Figure 5.3-12 DPC Quantization

a. Original image.

b. Lloyd-Max quantizer, using 1 bit/pixel, normalized correlation = 0.90, with standard deviation = 10.

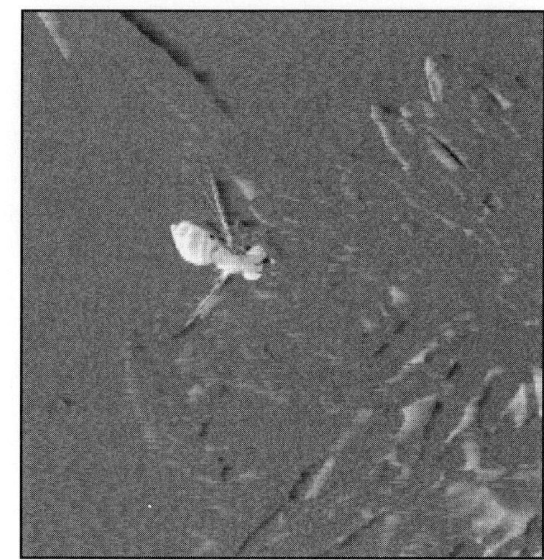

c. Error image for (b).

Figure 5.3-12 (Continued)

d. Lloyd-Max quantizer, using 2 bits/pixel, normalized correlation = 0.90, with standard deviation = 10.

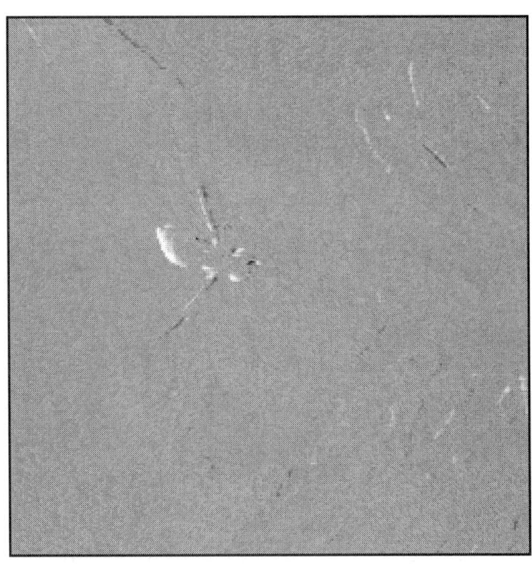

e. Error image for (d).

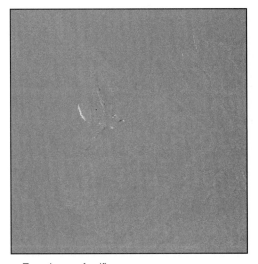

f. Lloyd-Max quantizer, using 3 bits/pixel, normalized correlation = 0.90, with standard deviation = 10.

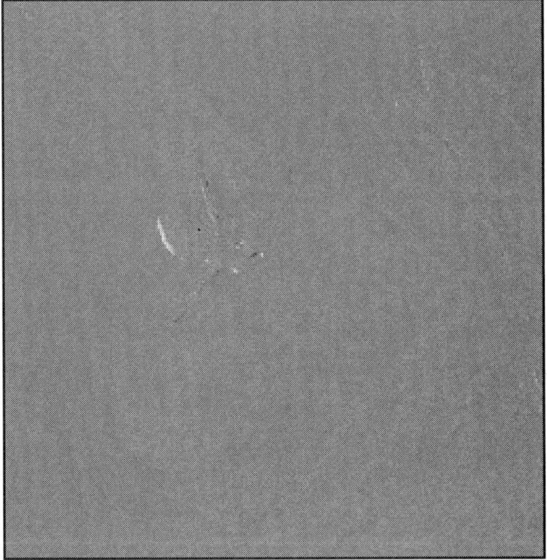

g. Error image for (f).

Figure 5.3-12 (Continued)

h. Lloyd-Max quantizer, using 4 bits/pixel, normalized correlation = 0.90, with standard deviation = 10.

i. Error image for (h).

j. Lloyd-Max quantizer, using 5 bits/pixel, normalized correlation = 0.90, with standard deviation = 10.

k. Error image for (j).

EXAMPLE 5-19

We have decided to use transform coding with a DCT on an image by dividing it into 4×4 blocks. The selected bit allocation can be represented by the following mask:

$$\begin{bmatrix} 8 & 6 & 4 & 1 \\ 6 & 4 & 1 & 0 \\ 4 & 1 & 0 & 0 \\ 1 & 0 & 0 & 0 \end{bmatrix}$$

where the numbers in the mask are the number of bits used to represent the corresponding transform coefficients (the upper-left corner corresponds to the zero-frequency coefficient, or average value, and the frequency increases to the right and down). This allows the lower frequencies less quantization (more resolution) because they have more bits to represent them.

Next a quantization scheme, such as Lloyd-Max quantization, is applied. Because the zero-frequency coefficient for real images contains a large portion of the energy in the image and is always positive, it is typically treated differently than the higher-frequency coefficients. After they have been quantized, the coefficients can be coded using, for example, a Huffman coding method.

In addition to simple filtering, two particular types of transform coding have been widely explored: zonal and threshold coding. These two vary in the method they use for selecting the transform coefficients to retain (using ideal filters for transform coding selects the coefficients based on their location in the transform domain). *Zonal coding* involves selecting specific coefficients based on maximal variance, whereas *threshold coding* selects the coefficients above a specific value. In zonal coding, a zonal mask is determined for the entire image by finding the variance for each frequency component. This variance is calculated by using each subimage within the image as a separate sample and then finding the variance within this group of subimages (see Figure 5.3-13). The *zonal mask* is a bitmap of 1's and 0's, where the 1's correspond to the coefficients to retain, and the 0's correspond to the ones to eliminate. In practice, the zonal mask is often predetermined because the low-frequency terms tend to contain the most information, and hence exhibit the most variance. In threshold coding a different threshold mask is required for each block, which increases file size as well as algorithmic complexity.

One of the most commonly used image compression standards is primarily a form of transform coding. The Joint Photographic Experts Group met initially in 1987 under the auspices of the International Standards Organization (ISO) to devise an optimal still image compression standard. The result was a family of image compression methods for still images. The JPEG standard uses the DCT and 8×8 pixel blocks as the basis for compression. Before computing the DCT, the pixel values are level-shifted so that they are centered at zero.

EXAMPLE 5-20

A typical 8-bit image has a range of gray levels from 0 to 255. Level-shifting this range to be centered at zero involves subtracting 128 from each pixel value, so the resulting range is from –128 to 127.

Figure 5.3-13 Zonal Coding

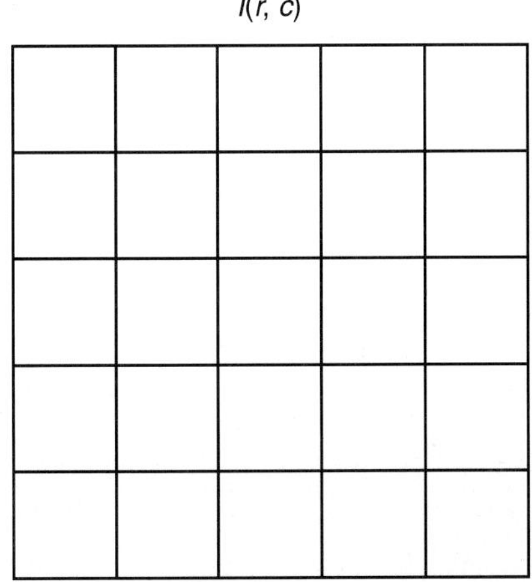

a. Divide the image into blocks.

$T(u, v)$

a	b	c	d	e
f	g ···			

b. Apply the transform to each block.

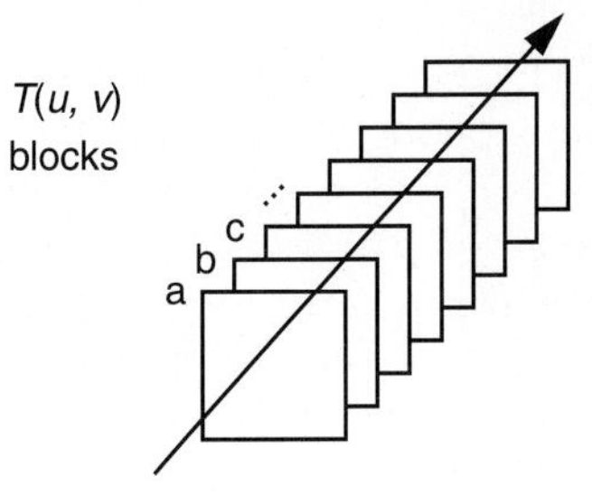

c. Treating each transform block from $T(u, v)$ as a separate sample, calculate the variance for each frequency component. Retain only the components with variance above a specified threshold.

1	1	1	1	0
1	1	1	0	0
1	1	0	0	0
1	0	0	0	0
0	0	0	0	0

d. Generate zonal masks—1 = retain; 0 = eliminate. A typical mask is shown.

After level-shifting, the DCT is computed. Next, the DCT coefficients are quantized by dividing by the values in a quantization table and then truncated. For color signals JPEG transforms the RGB components into the YCrCb color space, and subsamples the two color difference signals (Cr and Cb), since we perceive more detail in the luminance (brightness) than in the color information. After the coefficients are quantized, they are coded using a Huffman code. The zero-frequency coefficient is differentially encoded relative to the previous block. In Figure 5.3-14 are results of JPEG compression.

5.3.6 Hybrid Methods

Hybrid compression methods use both the spatial domain and the transform domain. For example, the original image (spatial domain) can be differentially mapped, and then this differential image can be transform coded. Alternately, a one-dimensional transform can be performed on the rows, and this transformed data can undergo differential predictive coding along the columns. These methods are often used for compression of analog video signals. For digital images these techniques can be applied to blocks (subimages), as well as rows or columns.

Figure 5.3-14 JPEG Compression

a. Original image.

b. JPEG compression = 10:1.

c. Error image for (b), multiplied by 8 to show detail.

Figure 5.3-14 (Continued)

d. JPEG compression = 20:1.

e. Error image for (d), multiplied by 8 to show detail.

f. JPEG compression = 30:1.

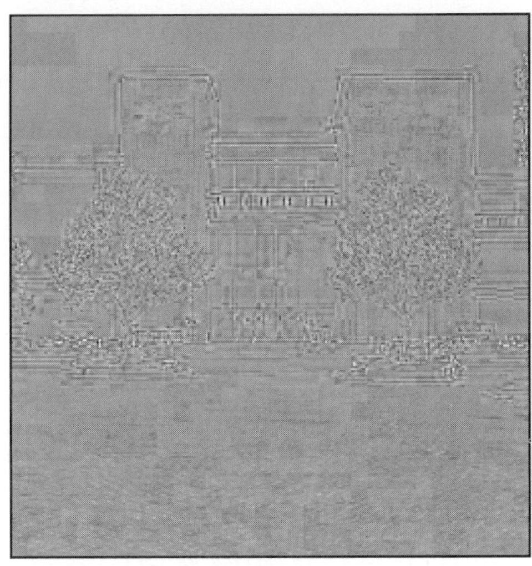

g. Error image for (f), multiplied by 8 to show detail.

Model-based image compression can be considered a hybrid method, although the transform used may be an object-based transform. *Model-based compression* works by finding models for objects within the image and using model parameters for the compressed file. The objects are often defined by lines or shapes (boundaries), so a Hough transform may be used, whereas the object interiors can be defined by statistical texture modeling. Methods have also been developed that use texture modeling in the wavelet domain. The model-based methods can achieve very high compression ratios, but the decompressed images often have an artificial look to them.

Wavelet-based compression shows much promise for the next generation of image compression methods. Because wavelets localize information in both the spatial and frequency domain, these are included under the hybrid method category. The wavelet transform combined with vector quantization has led to the development of compression algorithms with high compression ratios. The general method follows:

1. Perform the wavelet transform on the image by using convolution masks (described in Section 2.5.5).
2. Number the different wavelet bands from 0 to $N-1$, where N is the total number of wavelet bands and 0 is the lowest frequency (in both horizontal and vertical directions) band (see examples in Figure 5.3-15).
3. Scalar quantize the 0 band linearly to 8 bits.
4. Vector quantize the middle bands using a small block size (e.g., 2×2). Decrease the codebook size as the band number increases.
5. Eliminate the highest frequency bands.

Figure 5.3-16 shows results of this algorithm by varying the parameters to achieve different compression ratios. This wavelet/vector quantization (WVQ) compression algorithm is discussed in more detail in Section 7.4.

Figure 5.3-15 Wavelet/Vector Quantization Compression

a. Numbering for seven bands.

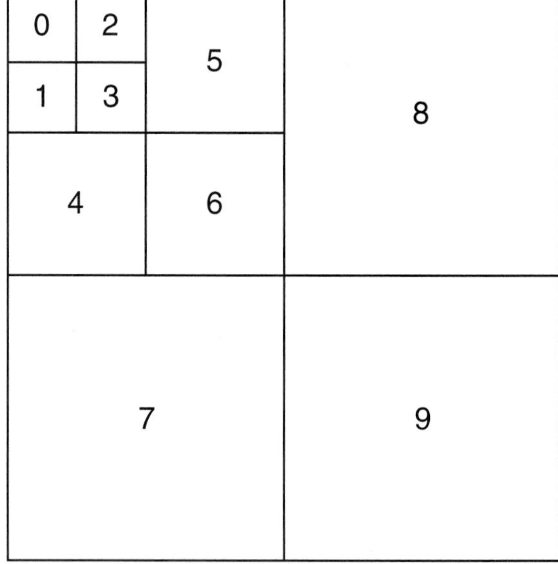

b. Numbering for ten bands.

Figure 5.3-16 Wavelet/Vector Quantization Compression Example

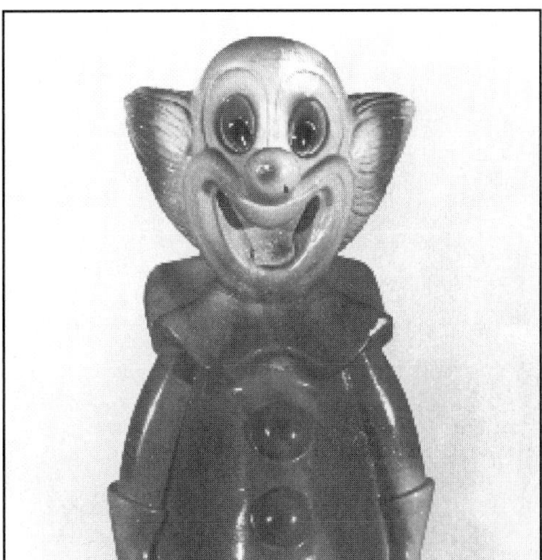

a. Original image.

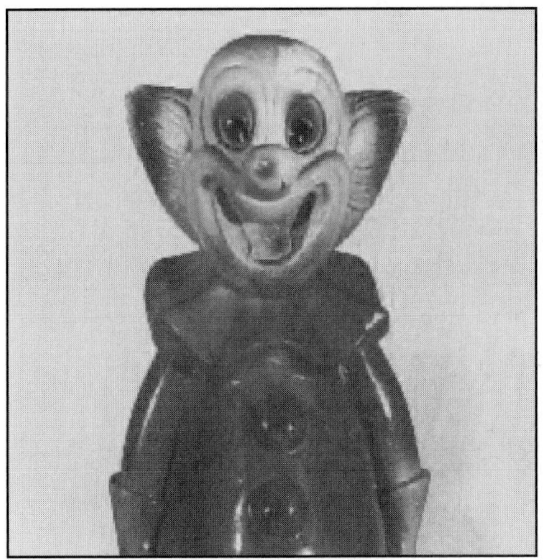

b. WVQ compression ratio 10:1.

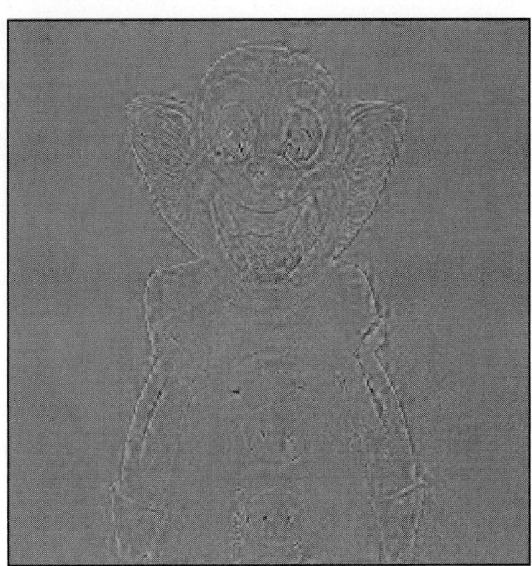

c. Error of image (b).

Figure 5.3-16 (Continued)

d. WVQ compression ratio 15:1.

e. Error of image (d).

f. WVQ compression ratio 33:1.

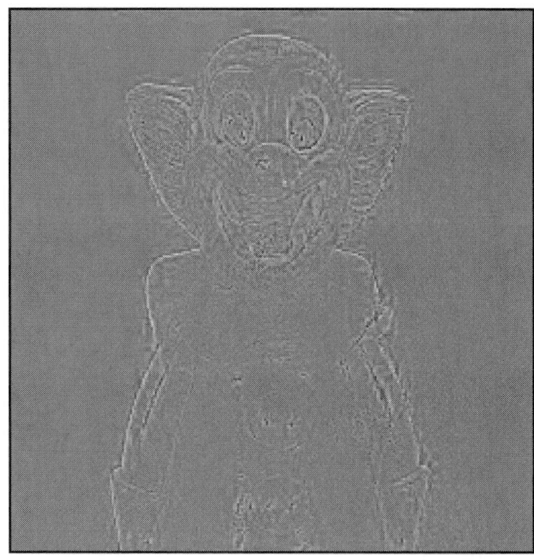

g. Error of image (f).

Figure 5.3-16 (Continued)

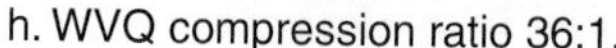

h. WVQ compression ratio 36:1.

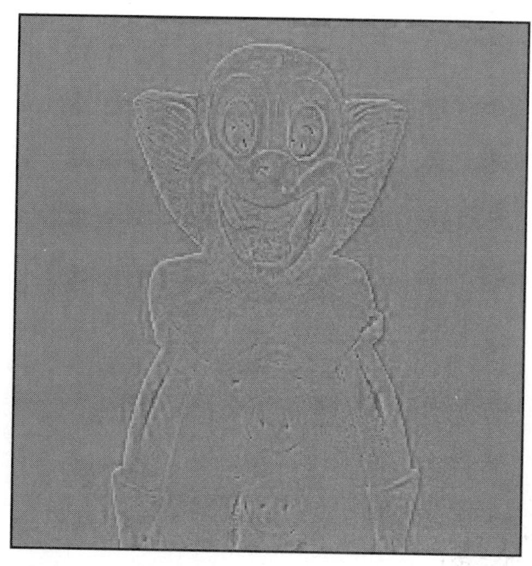

i. Error of image (h).

5.4 REFERENCES

More information on compression of video (motion) images can be found in [Tekalp 95], [Sid-Ahmed 95], [Clarke 95], [Kou 95], [Bhaskaran/Konstantinides 95], and [Netravali/Haskell 88]. An introduction to information theory is in [Tranter/Ziemer 85]. Information on coding redundancy can be found in [Gonzalez/Woods 92], and coding irrelevancy, in [Castleman 96]. The fidelity criteria are in [Gonzalez/Woods 92] and [Golding 78]. The compression system model is based on the model in [Gonzalez/Woods 92]. For a tutorial text on digital image compression, see [Rabbani/Jones 91].

The Huffman coding technique is found in [Huffman 52], [Gonzalez/Woods 92], [Netravali/Haskell 88], [Rosenfeld/Kak 82], [Jain 89], and [Sid-Ahmed 95]. More information on run-length coding, including two-dimensional methods, can be found in [Jain 89], [Tekalp 95], [Gonzalez/Woods 92], and [Hunter/Robinson 80]. Details on LZW coding are contained in [Ziv/Lempel 77] and [Welch 84]. The arithmetic coding method can be found in [Gonzalez/Woods 92].

The dynamic window-based RLC algorithmic details are in [Kumaran/Umbaugh 95]. Block truncation coding is explored in [Dasarathy 95], [Delp/Mitchel 79], [Rosenfeld/Kak 82] and [Wu/Coll 93]. Vector quantization is in [Tekalp 95], [Netravali/Haskell 88], and [Linde/Buzo/Gray 80]. Differential predictive techniques are explored in [Gonzalez/Woods 92], [Rosenfeld/Kak 82], [Jain 89], and [Netravali/Haskell 88]. More on transform coding can be found in [Gonzalez/Woods 92], [Netravali/Haskell 88], [Rosenfeld/Kak 82], [Jain 89], and [Sid-Ahmed 95].

Details of the image compression standards, including the RLC, JPEG, and MPEG, can be found in [Tekalp 95], [Kou 95], and [Bhaskaran/Konstantinides 95]. Information on the hardware and integrated circuits for these can be found in [Bhaskaran/Konstantinides 95].

A hybrid DPCM compression method for real-time video is described in [Whitehouse et al. 77]. Model-based compression using texture and the wavelet transform is described in [Ryan/Sanders/Fisher/Iverson 96]. More information on the wavelet/vector quantization compression algorithms can be found in [Kjoelen 95].

Bhaskaran, V., and Konstantinides, K., *Image and Video Compression Standards: Algorithms and Architectures*, Boston: Kluwer Academic Publishers, 1995.

Castleman, K. R., *Digital Image Processing*, Englewood Cliffs, NJ: Prentice Hall, 1996.

Clarke, R. J., *Digital Compression of Still Images and Video*, San Diego, CA: Academic Press, 1995.

Dasarathy, B. V., *Image Data Compression: Block Truncation Coding*, Los Alamitos, CA: IEEE Computer Society Press, 1995.

Delp, E. J., and Mitchell, O. R., "Image Compression Using Block Truncation Coding," *IEEE Transactions on Communications*, Vol. 27, No. 9, pp. 1335–1342, September 1979.

Golding, L. S., "Quality Assessment of Digital Television Signals," *SMPTE Journal*, Volume 87, pp. 153–157, March 1978.

Gonzalez, R. C., and Woods, R. E., *Digital Image Processing*, Reading, MA: Addison Wesley, 1992.

Huffman, D. A., "A Method for the Reconstruction of Minimum Redundancy Codes," *Proceedings of the IRE*, Vol. 40, No. 10, pp. 1098–1101, 1952.

Hunter, R., and Robinson, A. H., "International Digital Facsimile Coding Standards," *Proceedings of the IEEE*, Vol. 68, No. 7, pp. 854–867, 1980.

Jain, A. K., *Fundamentals of Digital Image Processing*, Englewood Cliffs, NJ: Prentice Hall, 1989.

Kjoelen, A., *Wavelet Based Compression of Skin Tumor Images*, Master's Thesis in Electrical Engineering, Southern Illinois University at Edwardsville, 1995.

Kou, W., *Digital Image Compression: Algorithms and Standards*, Boston: Kluwer Academic Publishers, 1995.

Kumaran, M., and Umbaugh, S. E, "A Dynamic Window-Based Runlength Coding Algorithm Applied to Gray-Level Images," *Graphical Models and Image Processing*, Vol. 57, No. 4, pp. 267–282, July 1995.

Linde, Y., Buzo, A., and Gray, R. M., "An Algorithm for Vector Quantizer Design," *IEEE Transactions on Communications*, Vol. 28, No. 1, pp. 84–89, January 1980.

Netravali, A. N., and Haskell, B. G., *Digital Pictures: Representation and Compression*, New York: Plenum Press, 1988.

Rabbani, M., and Jones, P. W., *Digital Image Compression Techniques*, Bellingham, WA: SPIE Optical Engineeering Press, 1991.

Rosenfeld, A., and Kak, A. C., *Digital Picture Processing*, San Diego, CA: Academic Press, 1982.

Ryan, T. W., Sanders L. D., Fisher, H. D., and Iverson, A. E., "Image Compression by Texture Modeling in the Wavelet Domain," *IEEE Transactions on Image Processing*, Vol. 5, No. 1, pp. 26–36, January 1996.

Sid-Ahmed, M. A., *Image Processing: Theory, Algorithms, and Architectures*, New York: McGraw Hill, 1995.

Tekalp, A. M., *Digital Video Processing*, Upper Saddle River, NJ: Prentice Hall, 1995.

Tranter, W. H, and Ziemer, R. E., *Principles of Communications: Systems, Modulation, and Noise*, Boston: Houghton Mifflin Company, 1985.

Welch, T. A., "A Technique for High-Performance Data Compression," *IEEE Computer*, Vol. 17, No. 6, pp. 8–19, 1984.

Whitehouse, H., Wrench, E., Weber, A., Claffie, G., Richards, J., Rudnick, J., Schaming, W., and Schanne, J., "A Digital Real Time Intraframe Video Bandwidth Compression System," *SPIE Application of Digital Image Processing IOCC 1977*, Vol. 119, pp. 64–78, 1977.

Wu, Y., and Coll, D. C., "Multilevel Block Truncation Coding Using a Minimax Error Criterion for High-Fidelity Compression of Digital Images," *IEEE Transactions on Communications*, Vol. 41, No. 8, August 1993.

Ziv, J., and Lempel, J., "A Universal Algorithm for Sequential Data Compression," *IEEE Transactions on Information Theory*, Vol. 24, No. 5, pp. 530–537, 1977.

II

CVIPtools

CHAPTER 6

Using CVIPtools

6.1 Introduction and Overview

CVIPtools is a software package developed at Southern Illinois University at Edwardsville, in conjunction with the writing of this book, to be used for computer imaging research and education. The first versions of the software were text menu-driven, and thus not very user-friendly to anyone but the developers. With the version 3.x series of CVIPtools, a graphical user interface (GUI) was added to make the program more readily accessible to more people.

CVIPtools is designed for the exploration of computer imaging by providing you with various standard, and state-of-the-art, algorithms which are organized by imaging task—analysis, restoration, enhancement, and compression. You can compare the results of applying the various algorithms by interactively modifying parameters and observing the output images. The software was initially designed to be run on high-end UNIX workstations, but since its inception in 1991, the personal computers (PC) and operating systems (Windows NT/95) have acquired the necessary power. As a result, CVIPtools has recently been ported to the PC/Windows environment.

CVIPtools differs from the commercially available imaging software in the objectives of its design. Because it is designed specifically for research and education, you have access to the many different parameters for the different algorithms and are not constrained by what the market has deemed "works best." In some cases, the algorithms may not work very well (for commercial applications) but have educational value. Some of these same algorithms that do not "work very well" may be useful to researchers for specific applications or may become part of a larger processing algorithm that does work well.

6.2 The Graphical User Interface

The GUI for CVIPtools allows for easy access to the CVIPtools functions. It consists of a main window, four application windows, a utilities window, an X-windows image viewer, and a help window. The application windows correspond to Chapters 2–5: Image Analysis, Image Restoration, Image Enhancement, and Image Compression. The main window allows access to the other windows and provides commonly needed information and utilities.

6.2.1 Main Window

When CVIPtools is invoked, the first window that appears is the CVIPtools main window, as shown in Figure 6.2-1. At the top is the selection bar, which allows you to select a category of operations. *File* is the leftmost option and is a drop-down menu that contains a list of file and file-related operations, as well as the *Exit* option. The next five selections—*Analysis, Restoration, Enhancement, Compression*, and *Utilities*—will invoke other primary windows that will be explored in later sections. The final entry on this selection bar is the *Help* selection, which allows you to obtain more detailed information about using the functions and, in some cases, information about the algorithms themselves. When you click the mouse on *Help*, the Help window appears as shown in Figure 6.2-2. Using the selection bar across the top of the Help window, which matches the selection bar on the main window, you may get help for any of the CVIPtools functions.

On the left side of the main window are buttons that can be used to display, save, delete, and rename images. These all relate to images that are stored in the image queue. The *image queue* stores images in random access memory (RAM), and you can

Figure 6.2-1 The Main Window

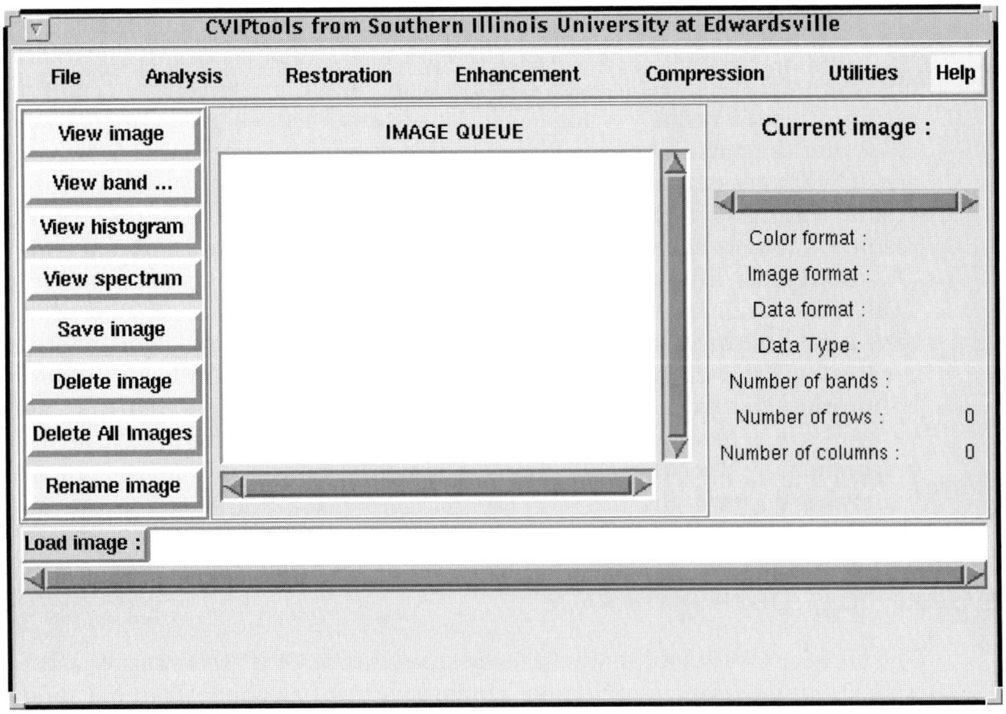

CVIPtools from Southern Illinois University at Edwardsville
File Analysis Restoration Enhancement Compression Utilities Help
View image
View band ...
View histogram
View spectrum
Save image
Delete image
Delete All Images
Rename image
IMAGE QUEUE
Current image :
Color format :
Image format :
Data format :
Data Type :
Number of bands :
Number of rows : 0
Number of columns : 0
Load image :

Figure 6.2-2 CVIPtools Help Window

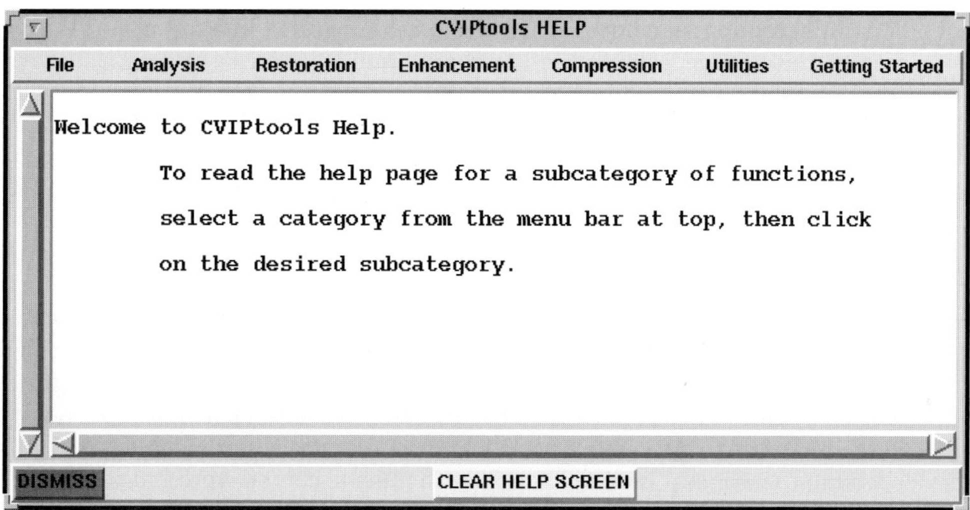

see the current contents in the center of the main window under the label IMAGE QUEUE. The image queue was implemented to facilitate fast processing—output images are automatically put into the image queue and are not written to disk files unless you explicitly save them with the *Save image* button.

In order for CVIPtools to process an image, it must first be loaded into the image queue. This is achieved by typing the image name into the Load image entry box at the bottom of the main window. When you enter an image name, the CVIPtools software automatically uses the environment variable CVIP_IMGPATH for image search paths. You can temporarily add image search paths with the *Add a temporary path* option listed under *File*. Alternately, you can click the mouse on the Load Image button, and a File Select Box window will appear. In this window you scroll through the directory structure searching for the image to be loaded (to select a directory double-click the left mouse button).

On the right side of the main window, information pertaining to the current image is displayed. The current image is highlighted in the queue and can be selected with a click of the left mouse button. At the bottom of the main window is a text area that is used for displaying messages, when needed.

6.2.2 CVIPtools Image Viewer

The CVIPtools image viewer is written in X-windows using the low-level Xlib functions for ease of portability. Because it is X-window-based, it typically requires a UNIX operating system, but CVIPtools is designed to allow Windows NT/95 users to use any of a number of freeware and shareware image viewers (see the README file on the CD for details). The CVIPtools image viewer displays images directly from memory, which greatly speeds display processing. It can be used as a stand-alone image viewer by typing *picture <image_name>*, but is normally used within the CVIPtools environment.

When an image is displayed, information about the image can be obtained by holding the middle mouse button (or right button, depending on your mouse setup). This information, as shown in Figure 6.2-3, includes image format, data type, data format, data range, color space, number of bands, and image height and width. Additionally, the row and column coordinates of the current mouse position are displayed at the bottom of the information window, and the corresponding pixel values are displayed under color bars whose heights represent those values. When the middle button is released, the row and column coordinates are passed back to the CVIPtools GUI and used, for example, in image geometry and feature extraction operations.

The image viewer allows you to perform standard image geometry operations, such as resizing, rotating, flipping, and changing the aspect ratio. It is important to note that *these operations affect only the image that is displayed, not the image in the CVIPtools image queue.* They are for viewing convenience only, and any changes to the image itself (in the queue) can be accomplished by using the CVIPtools geometry operations (available in the Utilities and Analysis→Geometry windows). Even if the image is resized within the viewer, the row and column coordinates displayed will still correspond to the original image. Therefore, the image can be enlarged to ease the selection of small image features, or the image can be shrunk to minimize screen use. The keyboard commands to perform these operations are listed in Table 6.2-1. To make it easy for those familiar with using XV, a commonly used X-windows viewer, we have used the same commands that XV uses. In addition to the keyboard commands, you can change the image size by grabbing the lower-right corner of the image with the left mouse button and dragging it.

The CVIPtools image viewer also allows you to select a specific portion of an image by drawing a box with a press and drag of the left mouse button. This informa-

Figure 6.2-3 Image Viewer Information Window

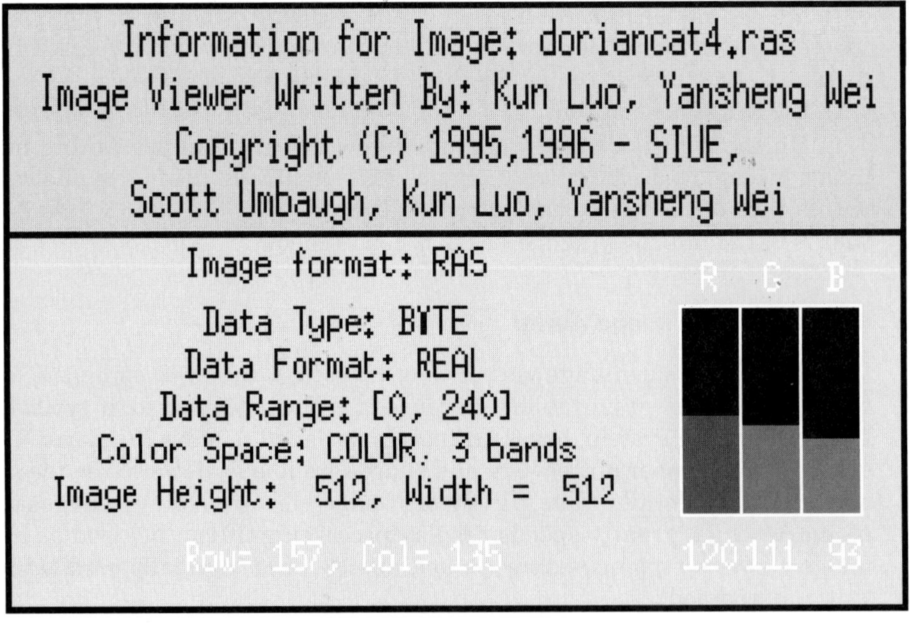

Table 6.2-1 Keyboard Commands for CVIPtools Image Viewer

RESIZE IMAGE	m	maximize image to entire screen
	M	maximize image while retaining aspect ratio
	>	double image size
	<	half image size
	.	enlarge image by 10%
	,	shrink image by 10%
CHANGE ASPECT RATIO	a	modify aspect ratio using shorter side
	A	modify aspect ratio using longer side
	4	modify aspect ratio to 4×3 using shorter side
	$	modify aspect ratio to 4×3 using longer side
ROTATE	t	turn 90 degrees clockwise
	T	turn 90 degrees counterclockwise
FLIP	h	horizontal flip
	v	vertical flip
OTHERS	n, <Enter>	change back to original image
	q, Q, Ctl-C	quit CVIPtools RamViewer
	e	histogram equalization

tion is automatically passed back to the CVIPtools GUI for use in, for example, the image crop function. After a select box has been drawn, you can move it by either clicking the left mouse button within the box and dragging it (hold down the left mouse button) or by using the arrow keys. Pressing shift on the keyboard combined with the arrow keys allows you to enlarge or shrink the select box. A new select box can be created at anytime and automatically destroys the first select box. After a select box has been drawn, it retains its relative position throughout any viewer-based image geometry operations.

6.2.3 Utilities Window

When you select Utilities from the main window, the Utilities window appears. Shown in Figure 6.2-4, the Utilities window has a selection bar at the top, which contains various categories of commonly used utilities: *Arithmetic / Logic, Convert, Create, Enhance, Filter, Size*, and *Miscellaneous*. Additionally, the selection bar has a *Help* button on the upper-right corner, which has a standard color of yellow (Caution! Help me!) and will bring up the Help window, and a *DISMISS* button on the lower-left corner, which has a standard color of red (Stop! Go away!) and will cause the window to

Figure 6.2-4 GUI Utilities Window and Menus

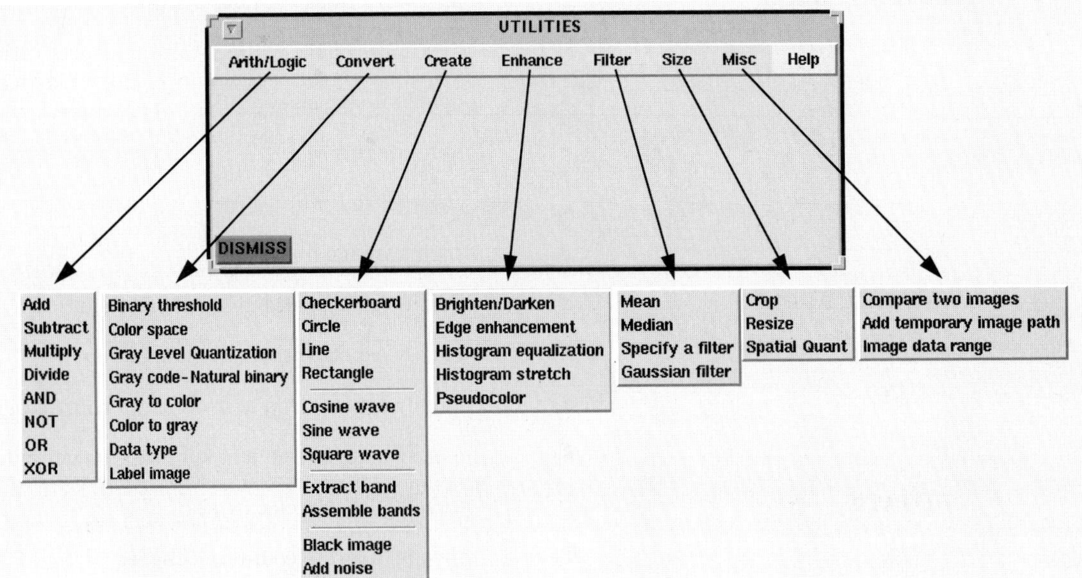

disappear. The figure also shows the drop-down menus that appear when selections are made from the selection bar.

When you make a selection from one of the drop-down menus, the necessary information appears in the Utilities window for that particular function, or class of functions. By limiting screen usage in this manner, the Utilities window is easily accessible when other primary windows are in use. The general philosophy guiding the design of the Utilities GUI is to maximize utility and usage, while minimizing use of screen space. In some cases, for example, with Utilities→Enhancement, only the most commonly used functions will appear in the Utilities window, and the choices for the various parameters may be limited. This allows Utilities to be used easily and quickly, and if you need more, you can select the main Enhancement window.

6.2.4 Analysis Window

When you first select Analysis from the main window, the Analysis selection bar appears (shown in Figure 6.2-5). The selection bar has selections across the top that pertain to Image Analysis: *Geometry, Edge / Line Detection, Segmentation, Transforms,* and *Features*. Additionally, the selection bar has a *Help* button on the upper-right cor-

Figure 6.2-5 GUI Analysis Selection Bar

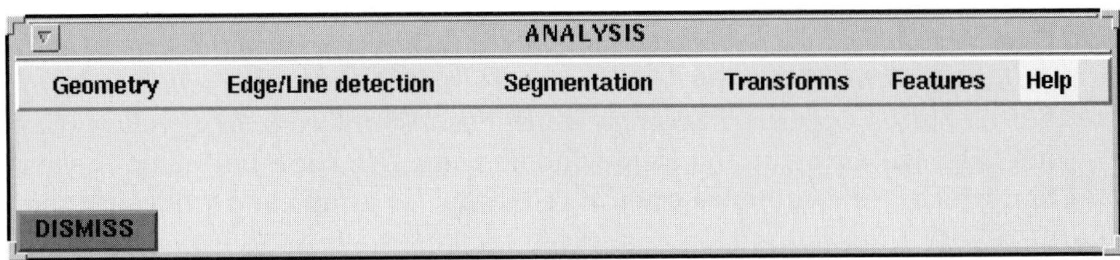

ner, and a *DISMISS* button on the lower-left corner. When you make a selection, by clicking one of the selection buttons with the left mouse button, the Analysis window expands and the CVIPtools functions available under that selection appear. Figure 6.2-6 shows the Analysis window after *Geometry* has been selected.

Notice that the heading on the window has been changed to Analysis→Geometry, and the window has been expanded to include the geometry functions. We also see that the *PERFORM* button has appeared in the lower-right corner. The *PERFORM* button performs a selected operation and has a standard color of green (Go! Do it!). Before we perform an operation, we need to select it by clicking on one of the radiobuttons on the left side of the window. A radiobutton looks like a square that is rotated 45 degrees. It has the background color when inactive (off) and changes color when activated (on). These are called radiobuttons because they function like the buttons on your car radio—only one can be active at a time. In this case, we can perform only one operation at a time.

After the operation has been selected, you can choose the necessary input parameters and enter the desired numbers in the entry box. An *entry box* has a white background and allows for input from the keyboard by placing the cursor over the entry box and clicking with the mouse. Note that these will initially contain default values, which allow for immediate use of an operation by simply selecting it via the radiobut-

Figure 6.2-6 GUI Analysis -> Geometry Window

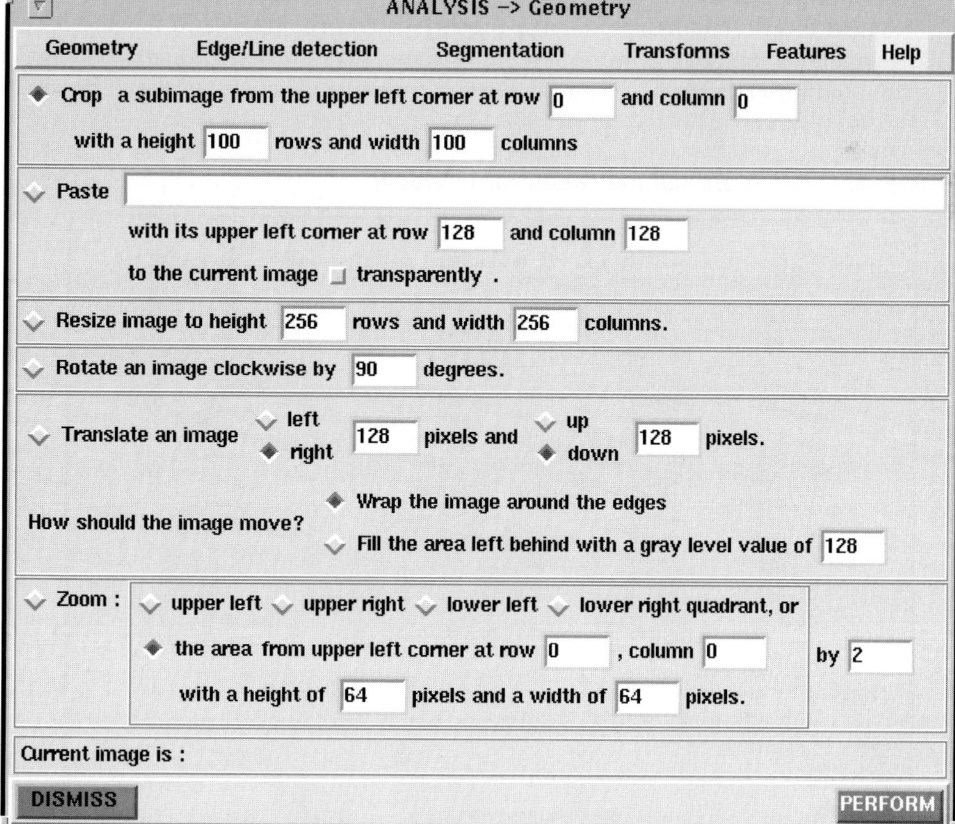

ton on the left and then clicking on the *PERFORM* button (assuming that an image has been selected).

Many of the selections in the Geometry window will use input from the CVIP-tools image viewer for row and column coordinates. For example, with the *Crop* and *Zoom* functions, if you draw a box on the image, those coordinates will be used in the GUI when you activate the *PERFORM* button, or you can enter the values directly in the entry box, if desired (note, however, that when using the mouse, the entry boxes are not updated until you press the *PERFORM* button).

Figure 6.2-7 shows the Analysis window after *Edge/Line detection* is selected. Here, on the left, we see the various edge detectors, which can be selected with radio-buttons. These are arranged in groups, based on the necessary input parameters. All the edge detectors can be preprocessed with various noise-removing filters and post-processed via a threshold operation. The prefiltering and postthresholding operations are selected with a special type of entry box that does not allow for keyboard input but does allow you to make a selection from a pop-up menu. Use the button on the right of the entry box that has a downward pointing arrow (↓), by clicking and dragging the mouse to the desired menu selection. The selection will appear in the entry box to provide feedback and to ease perusal of the setting of the input parameters.

Figure 6.2-8 shows the Analysis→Segmentation window. The top portion of the window contains segmentation algorithms that have only one or two input parameters. In the middle of this window are buttons for the *Multiresolution* and *Split and Merge* algorithms, which have a variety of homogeneity criteria from which to choose (for details see the on-line help pages). At the bottom of this window are the morphological filters. The morphological filters, which are often used for postprocessing on a segmented image to eliminate small objects, fill in holes, and smooth contours, can be divided into two groups. The first group contains *Closing, Opening, Erosion*, and *Dila-*

Figure 6.2-7 GUI Analysis -> Edge/Line Detection Window

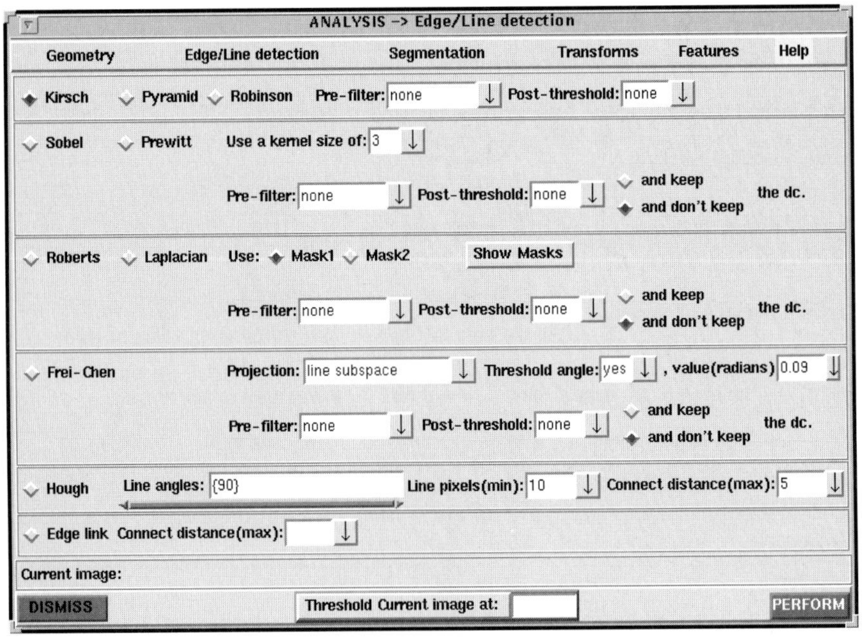

tion and will operate on any type of image. The second group, which consists of *Iterative Modification*, operates on binary images and allows for a variety of options as explained in Section 2.4.6. The GUI for the morphological filters is similar to the Utilities window in that the required input parameters appear only when you make the selection. Figure 6.2-8b shows the Analysis→Segmentation window after *Iterative Modification* is selected. Here we see that the input parameters required for iterative modification appear after it has been selected.

The Analysis→Transforms window is shown in Figure 6.2-9. The transforms that can be selected are in the upper portion of the window, while the filters can be seen in

Figure 6.2-8 GUI Analysis -> Segmentation Windows

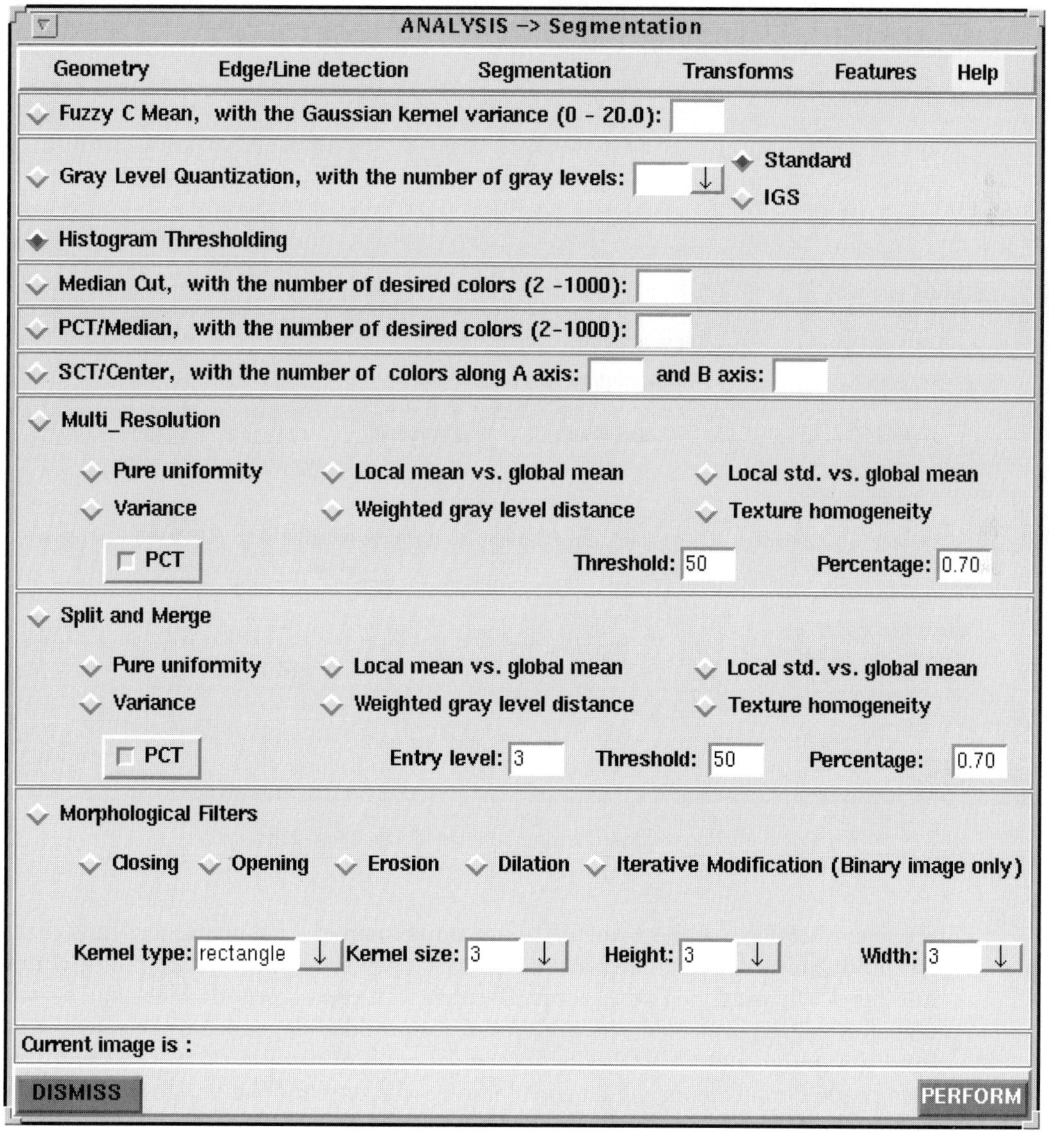

a. Default segmentation window.

Figure 6.2-8 (Continued)

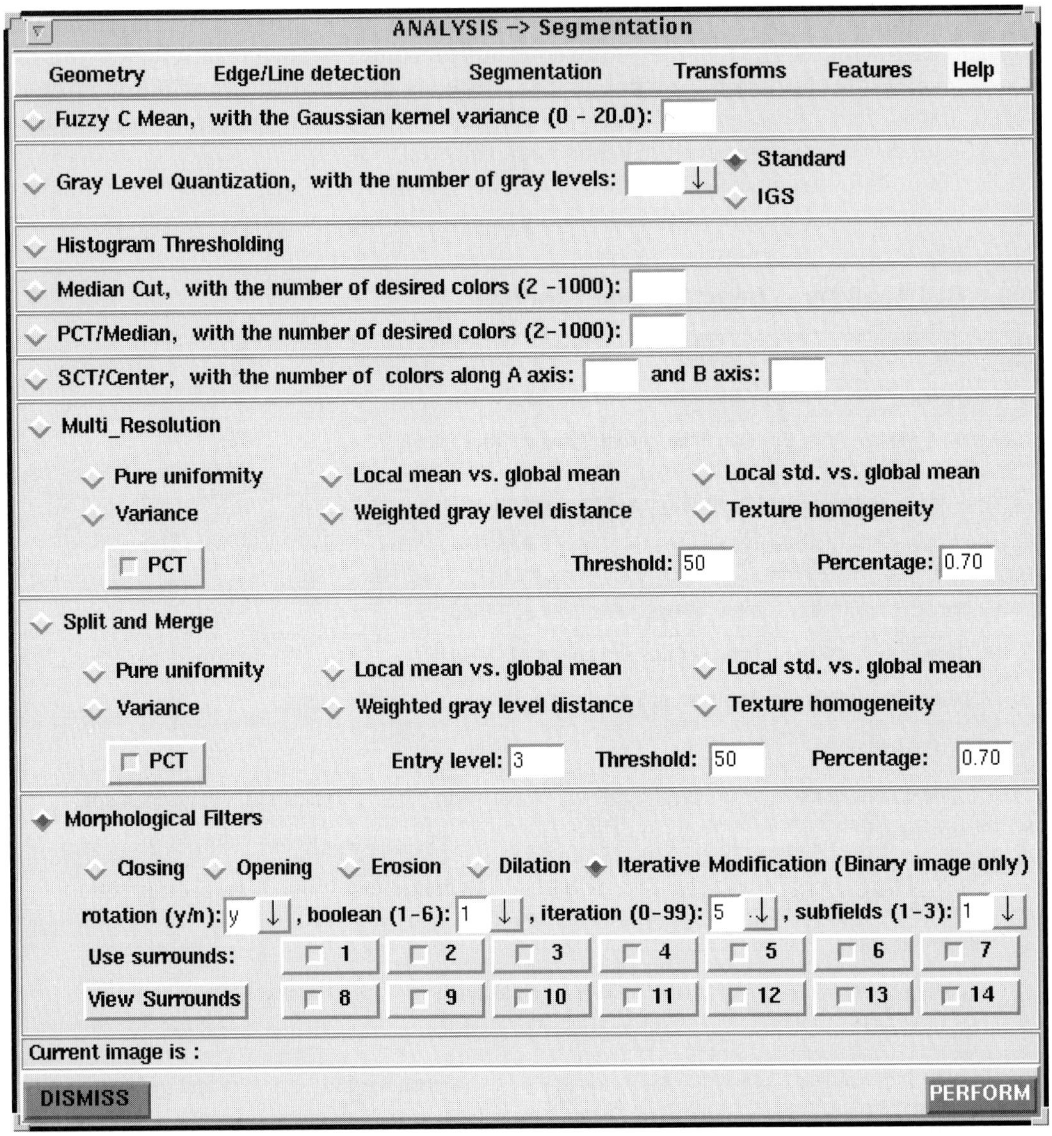

ANALYSIS -> Segmentation
Geometry · Edge/Line detection · Segmentation · Transforms · Features · Help
Fuzzy C Mean, with the Gaussian kernel variance (0 - 20.0):
Gray Level Quantization, with the number of gray levels: ↓ · Standard · IGS
Histogram Thresholding
Median Cut, with the number of desired colors (2 -1000):
PCT/Median, with the number of desired colors (2-1000):
SCT/Center, with the number of colors along A axis: and B axis:
Multi_Resolution
Pure uniformity · Local mean vs. global mean · Local std. vs. global mean
Variance · Weighted gray level distance · Texture homogeneity
PCT · Threshold: 50 · Percentage: 0.70
Split and Merge
Pure uniformity · Local mean vs. global mean · Local std. vs. global mean
Variance · Weighted gray level distance · Texture homogeneity
PCT · Entry level: 3 · Threshold: 50 · Percentage: 0.70
Morphological Filters
Closing · Opening · Erosion · Dilation · Iterative Modification (Binary image only)
rotation (y/n): y ↓, boolean (1-6): 1 ↓, iteration (0-99): 5 ↓, subfields (1-3): 1 ↓
Use surrounds: 1 · 2 · 3 · 4 · 5 · 6 · 7
View Surrounds · 8 · 9 · 10 · 11 · 12 · 13 · 14
Current image is :
DISMISS · PERFORM

b. Segmentation window showing Iterative Modification selection.

the lower portion. All the transforms are implemented with fast algorithms that require the input images to have dimensions that are powers of two, and, if not, the images are automatically extended with zeros. Because you can select the blocksize on which the transforms are performed, the number of zeros added depends on the selected blocksize (and the image size). CVIPtools determines the minimum number of zeros required along each dimension based on the blocksize and extends the image accordingly.

The discrete cosine transform (DCT), Fourier transform (FFT), and Haar, Walsh, and Hadamard transforms all require that you select the forward or inverse transform

Figure 6.2-9 GUI Analysis -> Transforms Window

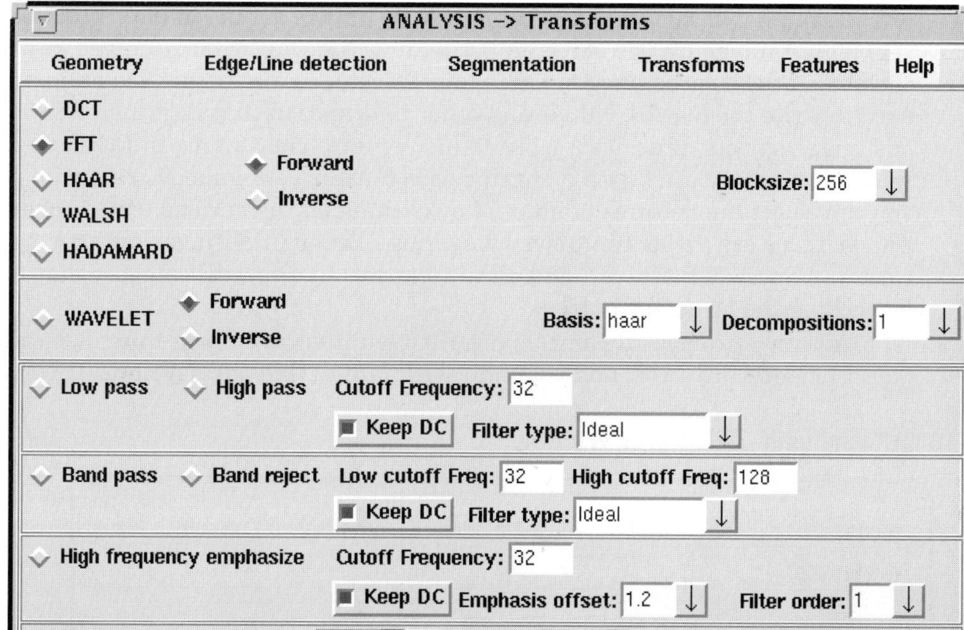

and the desired blocksize. When you select an inverse transform, the blocksize is automatically extracted from the image history that CVIPtools maintains for transform functions. With the wavelet transform, you can select the type of basis function desired, as well as the decomposition level. The wavelet algorithm is implemented as described in Chapter 2, and the decomposition level refers to the number of times that the image is decomposed with the wavelet transform. At each successive decomposition level, the wavelet transform operates on the lowpass-lowpass portion of the previous iteration of the transform. The output from any of the transforms is put in the image queue and then automatically displayed with a log remapping for optimal viewing.

Filtering can be performed on a transform output image, which is in the frequency, sequency, or wavelet domain. The filters available include *Lowpass, Highpass, Bandpass, Bandreject, High frequency emphasis*, and *Notch*. All these filters automatically perform the inverse transform after filtering to allow you to see the resulting image. The first four filters—lowpass, highpass, bandpass, and bandreject—allow you to select either an ideal or butterworth filter type and the cutoff frequency(s) and to retain or remove the component at the origin (DC). If you select a butterworth filter, you can also specify the desired order of the filter. Remember that the origin is in the upper-left corner for all transforms, except the FFT, where we put the origin in the center. This means that the cutoff frequency can range from 0 to the blocksize for most transforms, but with the FFT it ranges from 0 to one-half the blocksize.

The Analysis→Features window is shown in Figure 6.2-10. At the top of this window are the entry boxes for the necessary input image and file names. To select the original image, click on the image name in the image queue (in main window), and the segmented image can be selected by holding the Control key on the keyboard and clicking on the desired image in the queue. Next, enter a feature file name that will be used to write the feature values and other information that is gathered from the various image objects. If desired, you can also enter a class name in the Class entry box that will be associated with a specific object. After you make these initial selections, you can select the features of interest by clicking on the various feature checkbuttons. The features are divided into five categories—Binary, RST-invariant, Histogram, Texture, and Spectral features—and all the features in a specific category may be selected by clicking the *Select All* button.

After you select all the images, files, and features, you must select an image object by entering a row and column coordinate within the object or (if you are using

Figure 6.2-10 GUI Analysis -> Features Window

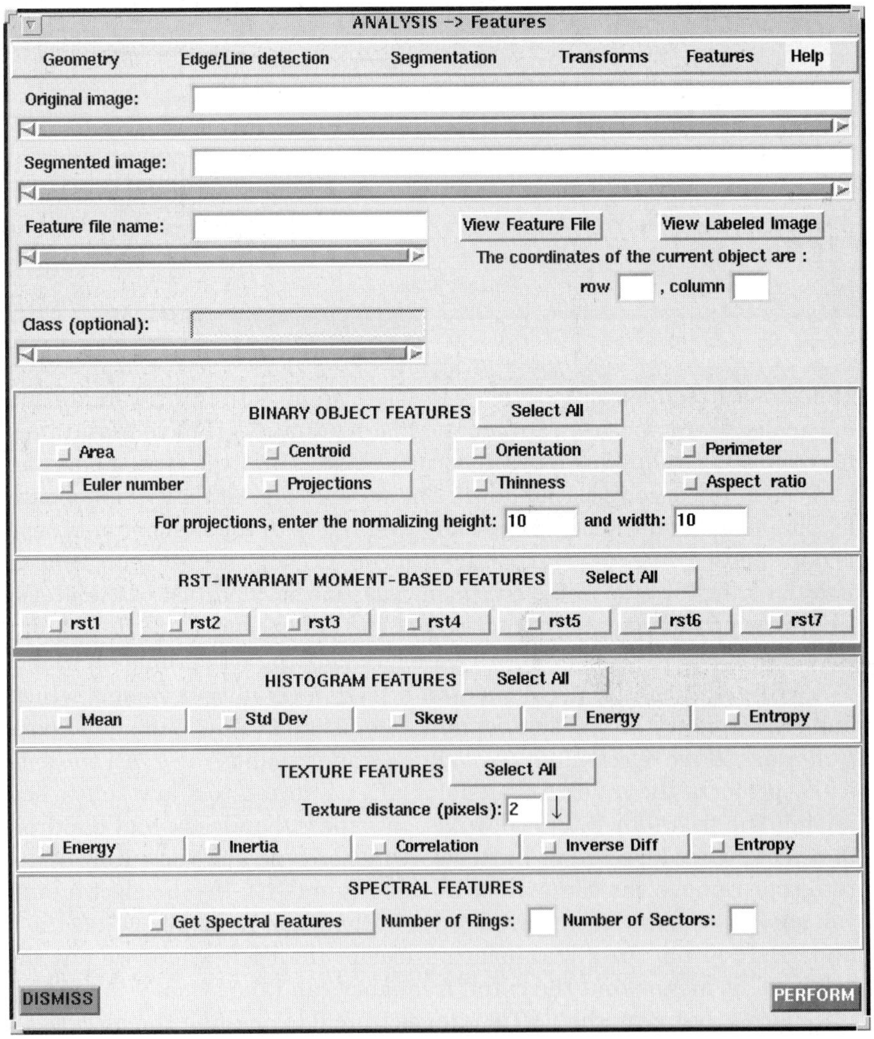

the CVIPtools viewer) by clicking the middle mouse button on the object. Now, the selected features are extracted and written to the feature file when you click the *PERFORM* button. This process can be repeated for as many images and objects as desired, by changing the image names and class, if necessary. Note that, if the selected features are modified, the program requires that you specify a different feature file name. This is necessary because, if the same file was used, it would lead to inconsistent feature files—different examples would have different features in the same places. A typical feature file is shown in Figure 6.2-11. Here we see that the file con-

Figure 6.2-11 Feature File

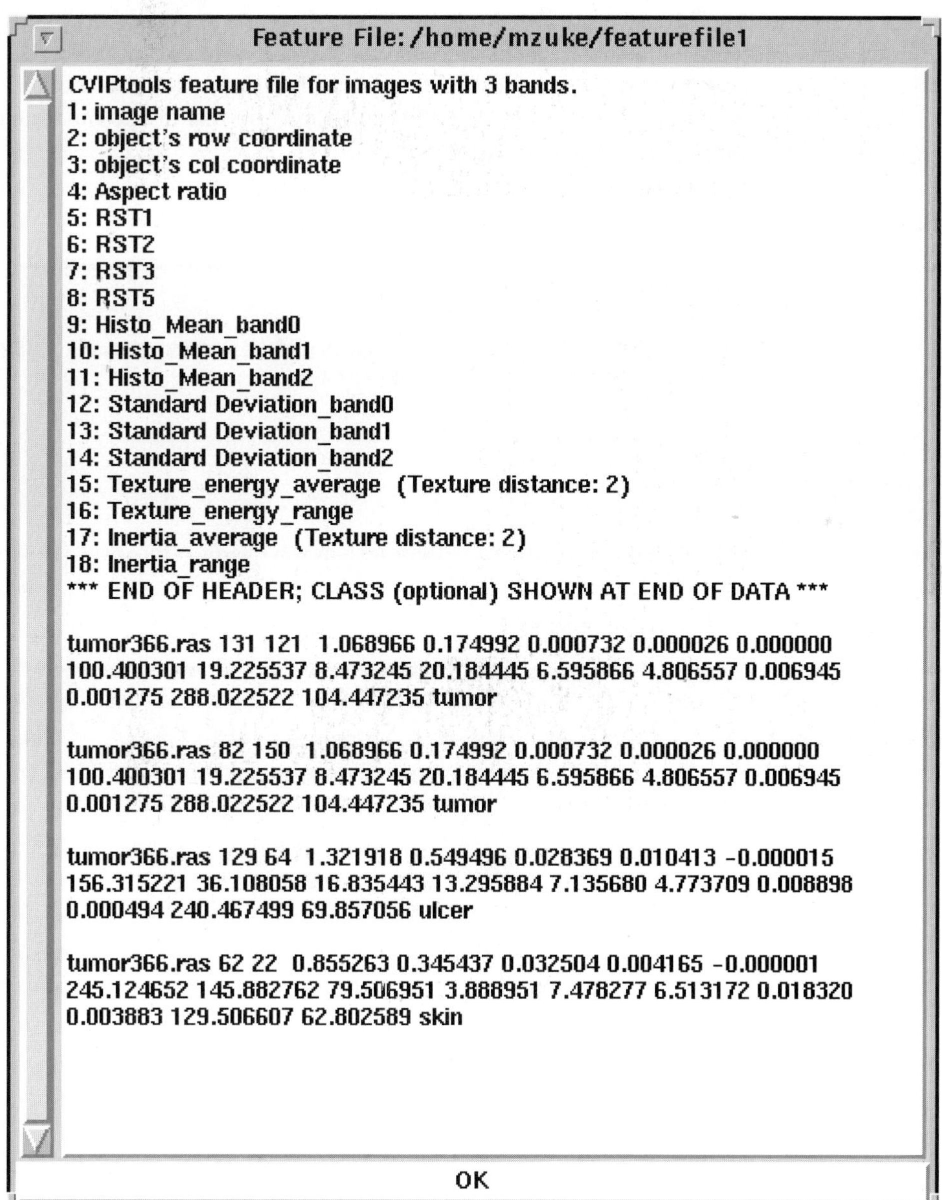

Feature File: /home/mzuke/featurefile1

CVIPtools feature file for images with 3 bands.
1: image name
2: object's row coordinate
3: object's col coordinate
4: Aspect ratio
5: RST1
6: RST2
7: RST3
8: RST5
9: Histo_Mean_band0
10: Histo_Mean_band1
11: Histo_Mean_band2
12: Standard Deviation_band0
13: Standard Deviation_band1
14: Standard Deviation_band2
15: Texture_energy_average (Texture distance: 2)
16: Texture_energy_range
17: Inertia_average (Texture distance: 2)
18: Inertia_range
*** END OF HEADER; CLASS (optional) SHOWN AT END OF DATA ***

tumor366.ras 131 121 1.068966 0.174992 0.000732 0.000026 0.000000
100.400301 19.225537 8.473245 20.184445 6.595866 4.806557 0.006945
0.001275 288.022522 104.447235 tumor

tumor366.ras 82 150 1.068966 0.174992 0.000732 0.000026 0.000000
100.400301 19.225537 8.473245 20.184445 6.595866 4.806557 0.006945
0.001275 288.022522 104.447235 tumor

tumor366.ras 129 64 1.321918 0.549496 0.028369 0.010413 -0.000015
156.315221 36.108058 16.835443 13.295884 7.135680 4.773709 0.008898
0.000494 240.467499 69.857056 ulcer

tumor366.ras 62 22 0.855263 0.345437 0.032504 0.004165 -0.000001
245.124652 145.882762 79.506951 3.888951 7.478277 6.513172 0.018320
0.003883 129.506607 62.802589 skin

OK

tains a header that specifies the content of each entry, followed by the feature data. The feature files, and feature extraction capability of CVIPtools, are designed so that the output can be easily imported into the many (public domain and commercial) pattern classification, neural network, and artificial intelligence programs for further analysis and development.

The feature extraction process works by labeling the segmented image, using the six-connectivity algorithm described in Section 2.6-2. The labeled image is then used in conjunction with the original image to extract the required features. For binary and RST-invariant features, the selected object is treated as a binary object. For histogram features, the labeled image is used as a mask on the original so that only object pixels are included in the calculations. For spectral features, all nonobject pixels are set to zero, a bounding box is found around the object, and the image is cropped before the Fourier spectrum is found. For the texture features, nonobject pixels are set to zero, but these are not included in the texture feature calculations. Note that both the labeled image and the feature file can be viewed via the use of the buttons directly to the right of the feature file name entry box.

Figure 6.2-12 GUI Restoration Selection Bar

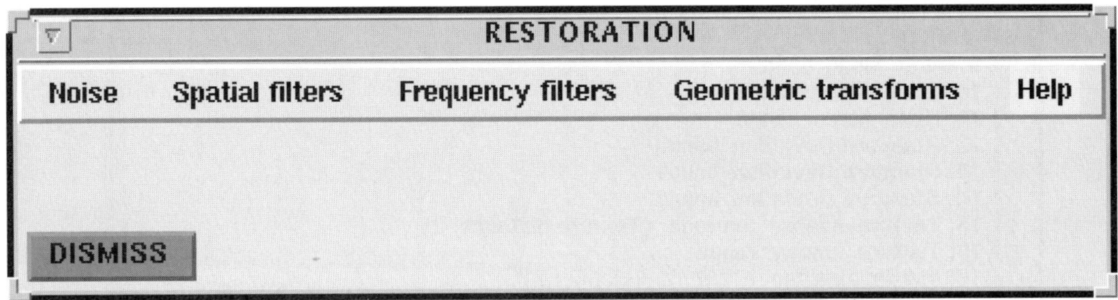

Figure 6.2-13 GUI Restoration -> Noise Window

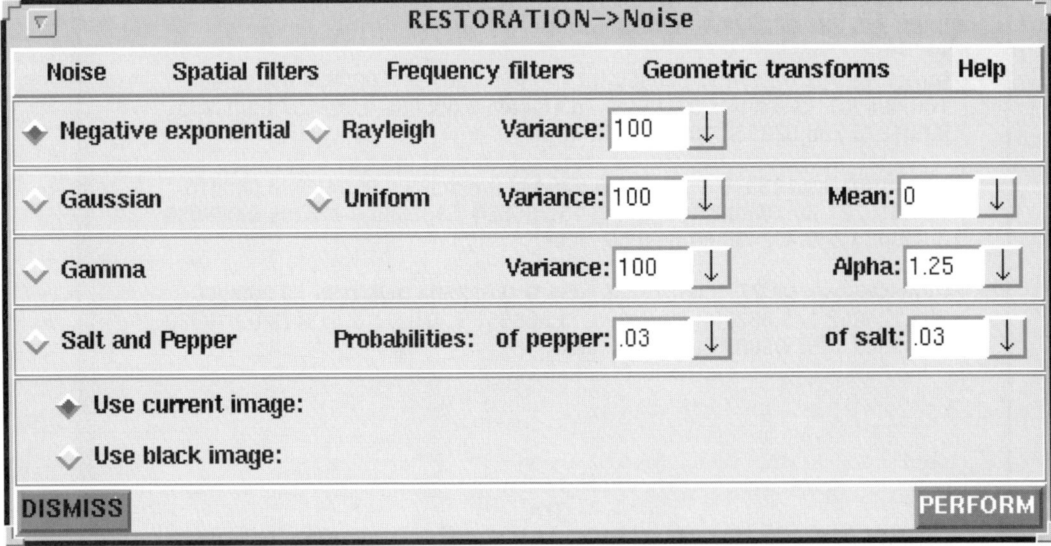

6.2.5 Restoration Window

The Restoration selection bar is shown in Figure 6.2-12. The selection bar has selections across the top that pertain to image restoration: *Noise, Spatial filters, Frequency filters*, and *Geometric transforms*. The *Noise* selection allows you to add to an image various types of noise, which can be used to help determine experimentally the type of noise that exists within a degraded image. When you select *Noise*, the Noise window, shown in Figure 6.2-13, appears. Here we see the six types of noise available; we can select the appropriate parameters for each different noise type. You can add noise to an existing image or create a noise-only image by selecting the *Use black image* option. The output image is put in the queue as a floating point image with the added noise, but it is remapped to byte for display purposes.

The second selection on the Restoration selection bar is *Spatial filters*; this window is shown in Figure 6.2-14. Spatial filters are divided into three categories: *Order,*

Figure 6.2-14 GUI Restoration -> Spatial Filter Window

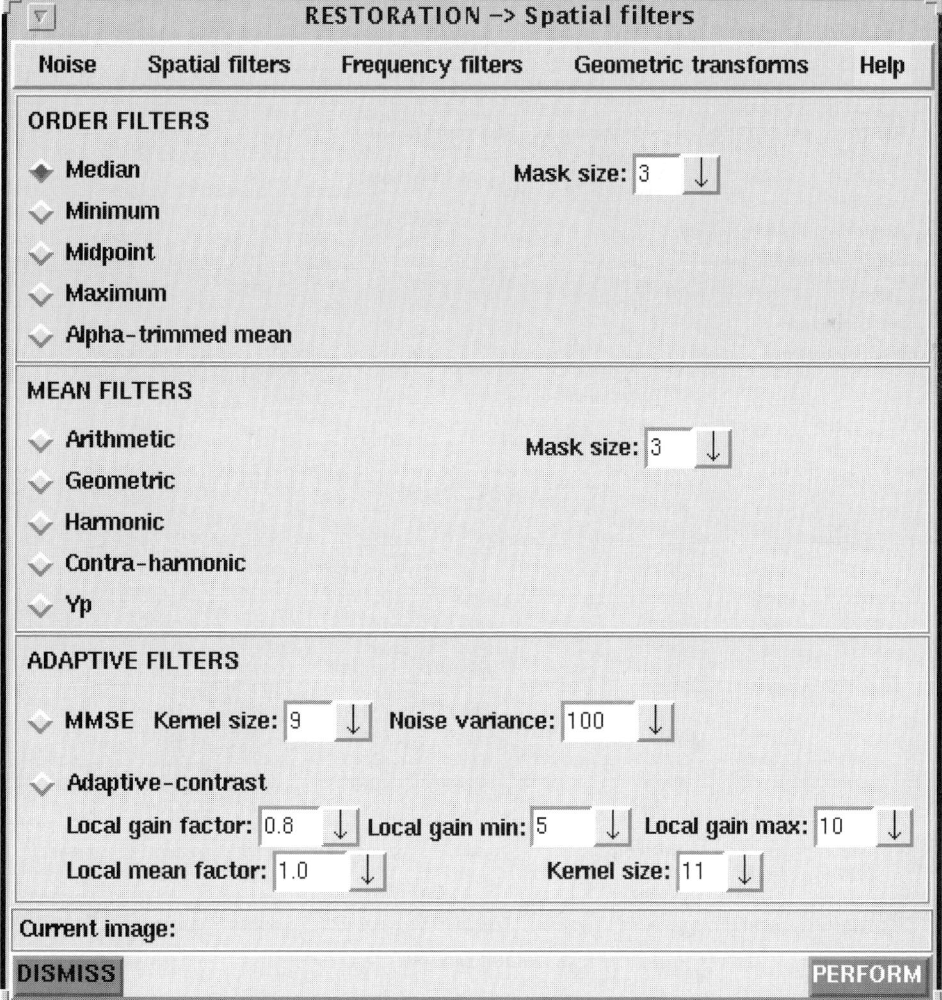

Mean, and *Adaptive* filters. These filters are used primarily for noise removal and are described in detail in Chapter 3. Because these are spatial filters, the mask size is required for all, but other parameters will be filter-specific. Any other necessary parameters have entry boxes that appear when the specific filter is selected. For example, the alpha-trimmed mean filter requires a *trim size*, while the contra-harmonic and *Yp* mean filters require the *filter order*.

The third selection on the Restoration selection bar is *Frequency filters*; the default window is shown in Figure 6.2-15a. The restoration filters described in Chapter 3 are available and can be selected by clicking one of the radiobuttons on the left. The necessary parameters and images are selected in the box to the right of the RESTORATION FILTERS box. Here, you can choose an image for the degradation function, $h(r, c)$, or specify the function by specifying the necessary parameters (see Figure 6.2-15b). Any other parameters necessary for the chosen filter will appear in the area under the degradation function selection area, for example, the noise function $n(r, c)$

Figure 6.2-15 GUI Restoration -> Frequency Filter Window

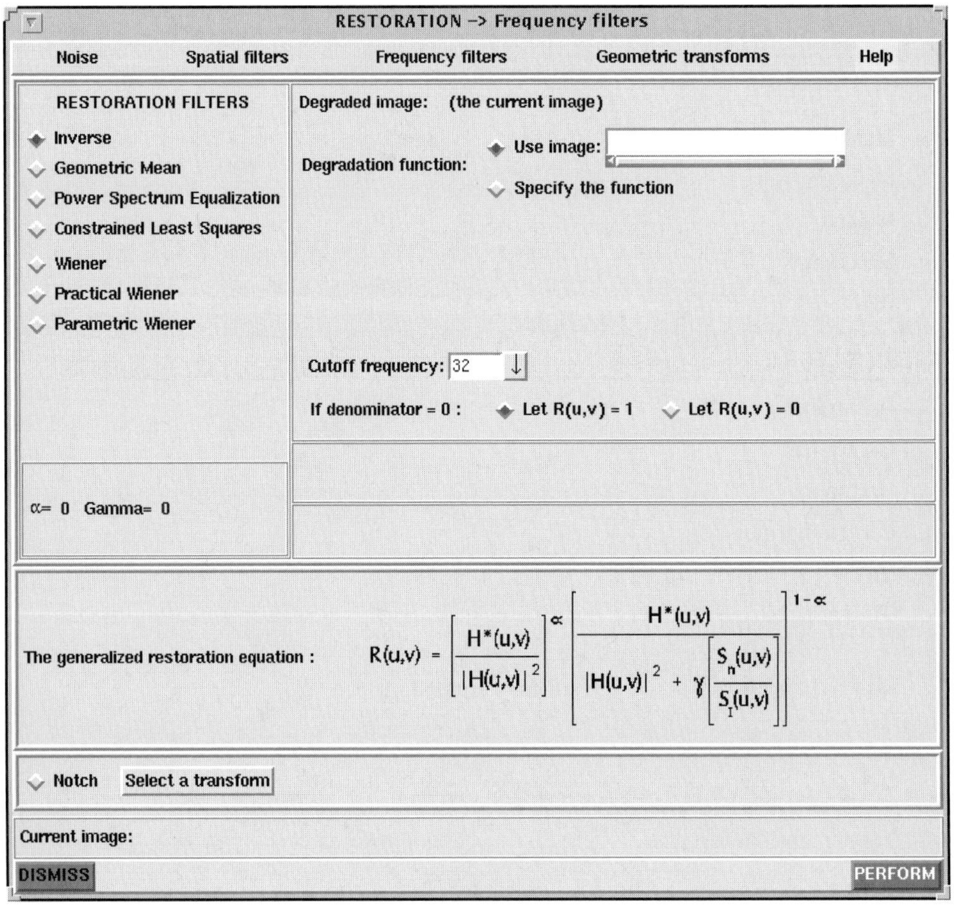

RESTORATION -> Frequency filters

Noise | Spatial filters | Frequency filters | Geometric transforms | Help

RESTORATION FILTERS

- Inverse
- Geometric Mean
- Power Spectrum Equalization
- Constrained Least Squares
- Wiener
- Practical Wiener
- Parametric Wiener

Degraded image: (the current image)

Degradation function: Use image: / Specify the function

Cutoff frequency: 32

If denominator = 0 : Let R(u,v) = 1 Let R(u,v) = 0

α= 0 Gamma= 0

The generalized restoration equation :

$$R(u,v) = \left[\frac{H^*(u,v)}{|H(u,v)|^2}\right]^{\alpha}\left[\frac{H^*(u,v)}{|H(u,v)|^2 + \gamma\left[\frac{S_n(u,v)}{S_I(u,v)}\right]}\right]^{1-\alpha}$$

Notch Select a transform

Current image:

DISMISS PERFORM

a. Default window.

Figure 6.2-15 (Continued)

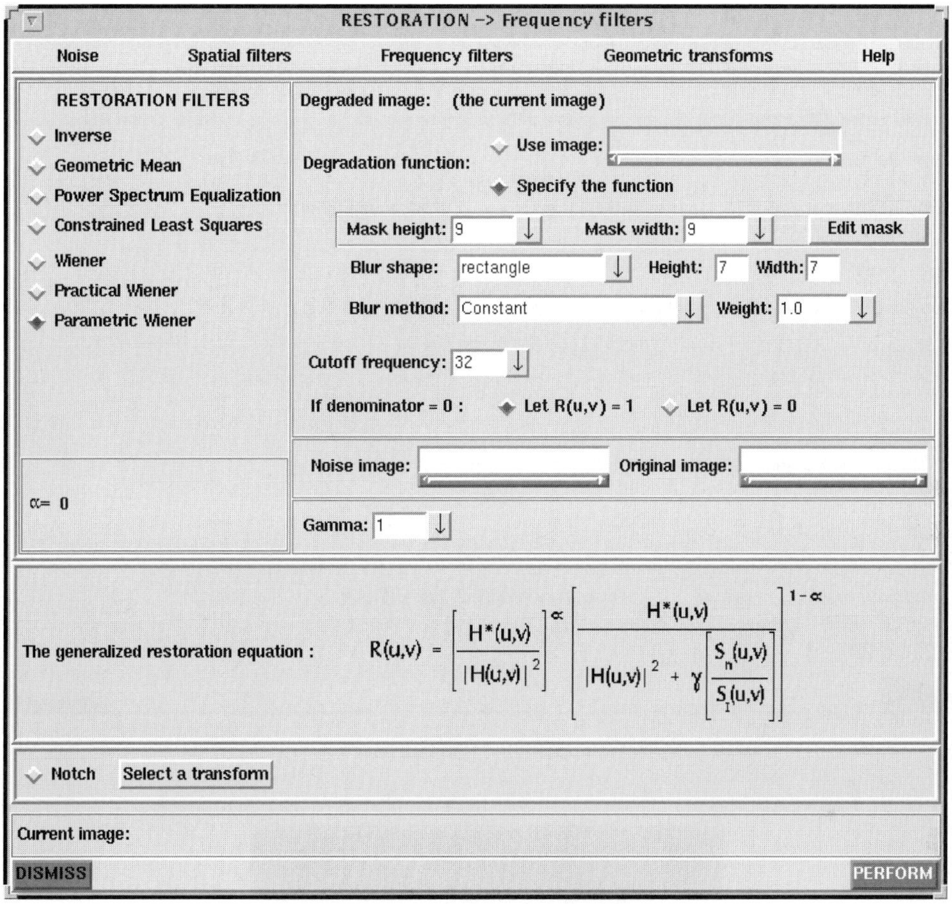

b. Specifying a degradation function.

or α or γ (see Figure 6.2-15b). At the bottom of this window, the notch filter is also provided, because it is useful to remove certain types of interference from images.

The fourth selection on the Restoration selection bar is *Geometric transforms*; this window is shown in Figure 6.2-16. Here we can apply the geometric restoration method described in Chapter 3, based on using quadrilaterals that define the tiepoints between the distorted image and the restored image. The file that contains the tiepoints is called the *mesh file*. In the upper portion of the window, you can enter a new mesh file, use an existing mesh file, or create a sine mesh. To enter a new mesh, you select the number of rows and columns and proceed to type in the (r, c) coordinates. With the CVIPtools image viewer, use the mouse to select the points desired. To use the mouse, first select a point in the image (with the middle mouse button) and then click on the CVIPtools SET button (with the left mouse button); repeat this sequence until all the points have been selected. Enter the mesh points in the order illustrated in the Restoration→Geometric transforms window. At the bottom of the

Figure 6.2-16 GUI Restoration -> Geometric Transform Window

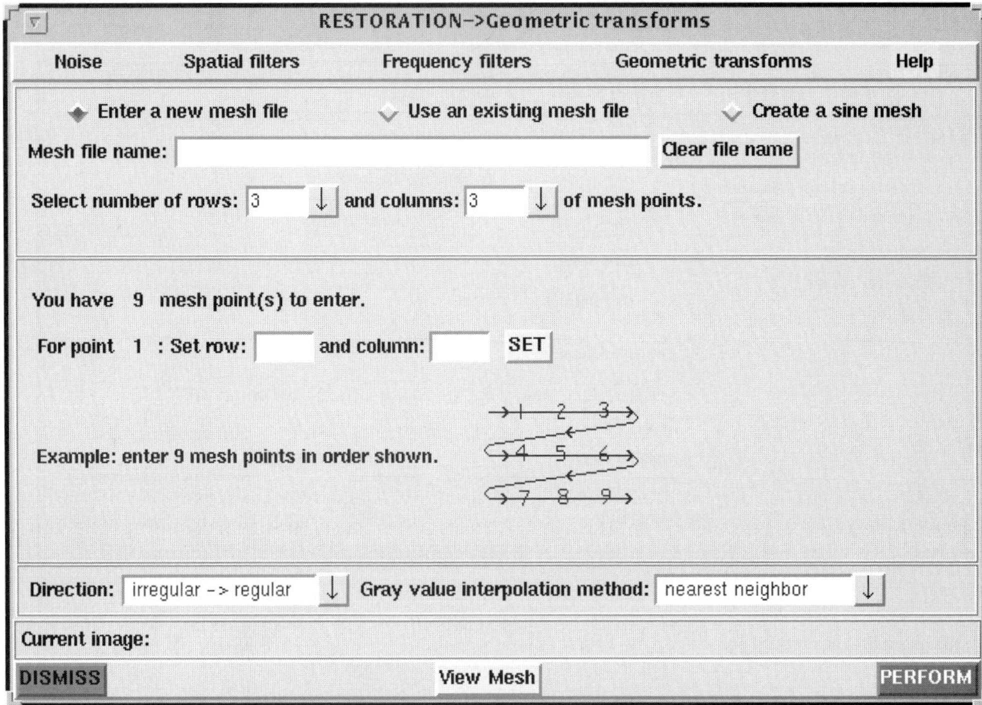

a. Default geometric transform window.

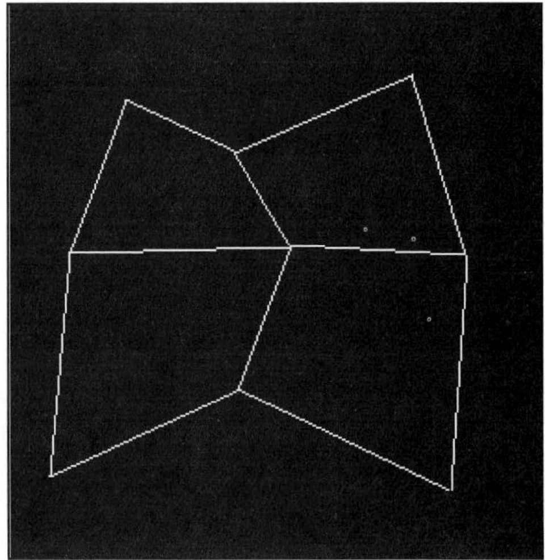

b. Sample mesh file.

window, select the direction of the geometric transform, using one of three different gray-value interpolation methods. The direction selected determines if the image is mapped from a regular grid to an irregular grid, *regular→irregular*, or from an irregular grid to a regular grid, *irregular→regular*. To restore a warped image, we normally select *irregular→regular*, and specify the irregular grid in the mesh file. You can warp an image by using the *regular→irregular* direction selection and restore it with the *irregular→regular* choice, using the same mesh file. Alternately, you can warp an image by using the *irregular→regular* direction selection and restore it with the *regular→irregular* choice. A mesh file can be viewed as an image by selecting the *View Mesh* button at the bottom of the window. A sample mesh file, illustrating an irregular grid with nine points, is displayed as an image in Figure 6.2-16b.

6.2.6 Enhancement Window

The Enhancement selection bar is shown in Figure 6.2-17. The selection bar has selections across the top that pertain to image enhancement: *Histograms, Pseudocolor, Sharpening*, and *Smoothing*. The *Histograms* selection brings up the window shown in Figure 6.2-18, which contains gray-level and histogram-based modification functions and has an extra button on the bottom for viewing the histogram of an image. All the histogram functions operate on images whose data type is byte, which ranges from 0 to 255. If an input image has a different data type, it is automatically converted to byte. The functions that modify the gray level directly, as opposed to modification by using the histogram, operate in a different manner. The *Linear modify* function will retain the input data type, remapping to byte for display only; whereas the *Gray level multiplication* function provides two choices. The first choice, *With BYTE clipping*, remaps any input data type to byte and then performs the multiplication and truncates any out of range data. The second choice, *Without BYTE clipping*, performs the multiplication and outputs an image of data type float, retaining all the information.

The second selection on the Enhancement selection bar is *Pseudocolor*; this window is shown in Figure 6.2-19. Here we see three principal pseudocolor methods: *Frequency domain mapping, Gray level mapping*, and *Intensity slicing*. In the frequency domain the Fourier transform and three filters are used. You specify two cutoff frequencies, which are used to generate the three filters. All frequencies below the low cutoff are used for the lowpass filter; the frequencies between the high and low are used for the bandpass filter; and the frequencies above the high cutoff are used for the

Figure 6.2-17 GUI Enhancement Selection Bar

ENHANCEMENT

Histograms Pseudocolor Sharpening Smoothing Help

DISMISS

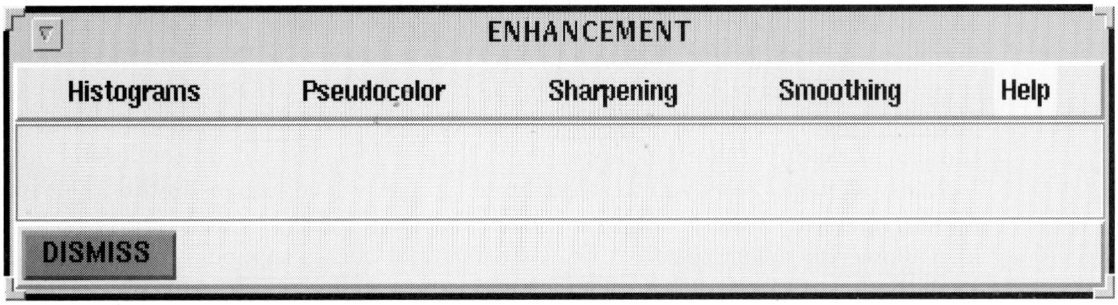

Figure 6.2-18 GUI Enhancement -> Histogram Window

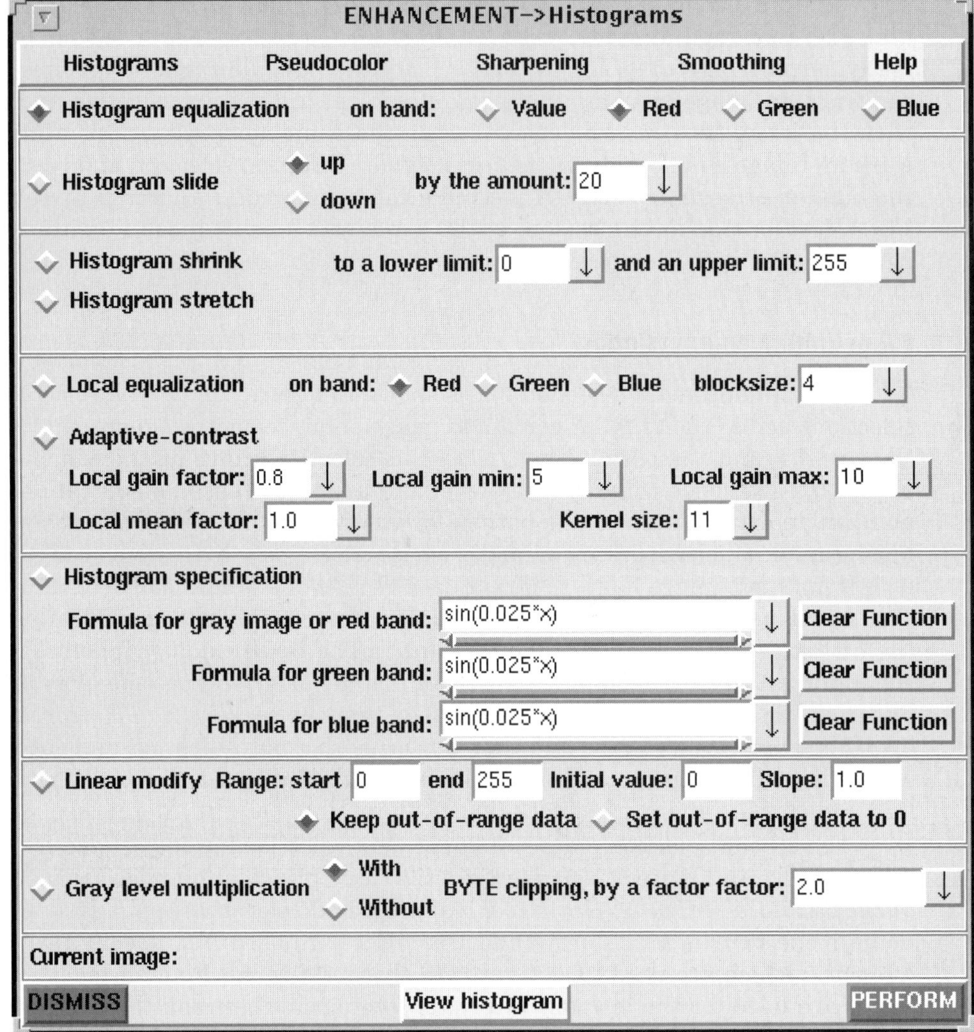

highpass filter. You can map the output from each filter to one of the red, green, and blue (RGB) color bands. For gray-level mapping you can select from a set of predefined functions, defined graphically in a pop-up menu, for each of the red, green, and blue color bands. If a more customized mapping is desired, you can use the *Assemble bands* and *Add* functions in Utilities→Create and Utilities→Arith/Logic, in conjunction with the *Linear modify* under Enhancement→Histograms. The third pseudocolor option, *Intensity slicing*, allows you to select up to four gray-level ranges in the input image and assign each of these ranges to a color specified by RGB values. *Intensity slicing* also allows you either to set values that have not been assigned a color to zero or to keep them as the gray-level values in the original image. With this option you can

Figure 6.2-19 GUI Enhancement -> Pseudocolor Window

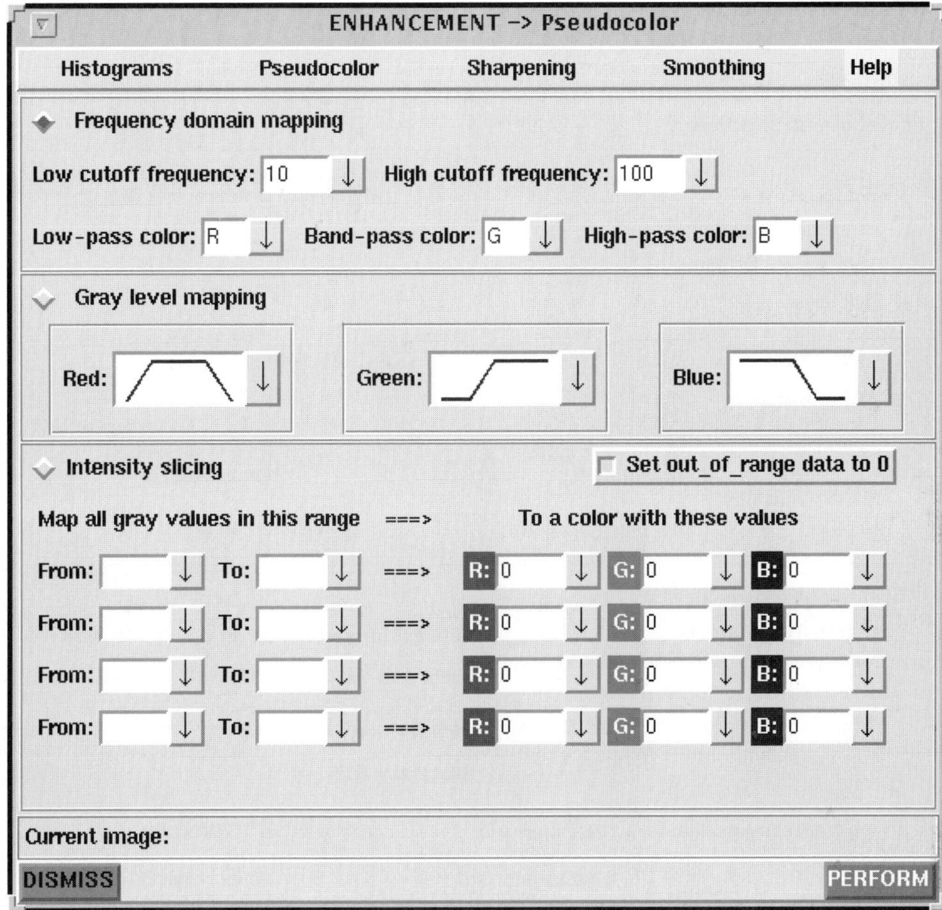

remove gray levels that are not of interest, by making them black (equal to zero), or retain them, but only highlighting the ranges of interest with pseudocolor.

The third selection on the Enhancement selection bar is *Sharpening*; this window is shown in Figure 6.2-20. This window is divided into two main sections; the top section contains spatial domain sharpening methods, and the lower section contains frequency domain methods. The spatial domain methods are based on edge enhancement via spatial convolution operators and histogram modification. For some of these methods, you have the option of adding the result to the original, which will typically enhance the edges. If the result is returned without adding it to the original, then only high-frequency information, edges, will be observed. Additional user parameters include the size of the convolution mask, which determines the width of enhanced edges, and the center value, which determines the amount of original image preserved. For the unsharp masking algorithm you specify the histogram limits and clip percentages, which affect the output in an image-specific manner. In the frequency domain, various highpass and high-frequency emphasis filters are provided. If you desire more control over the filter, use the Analysis→Transforms window.

Figure 6.2-20 GUI Enhancement -> Sharpening Window

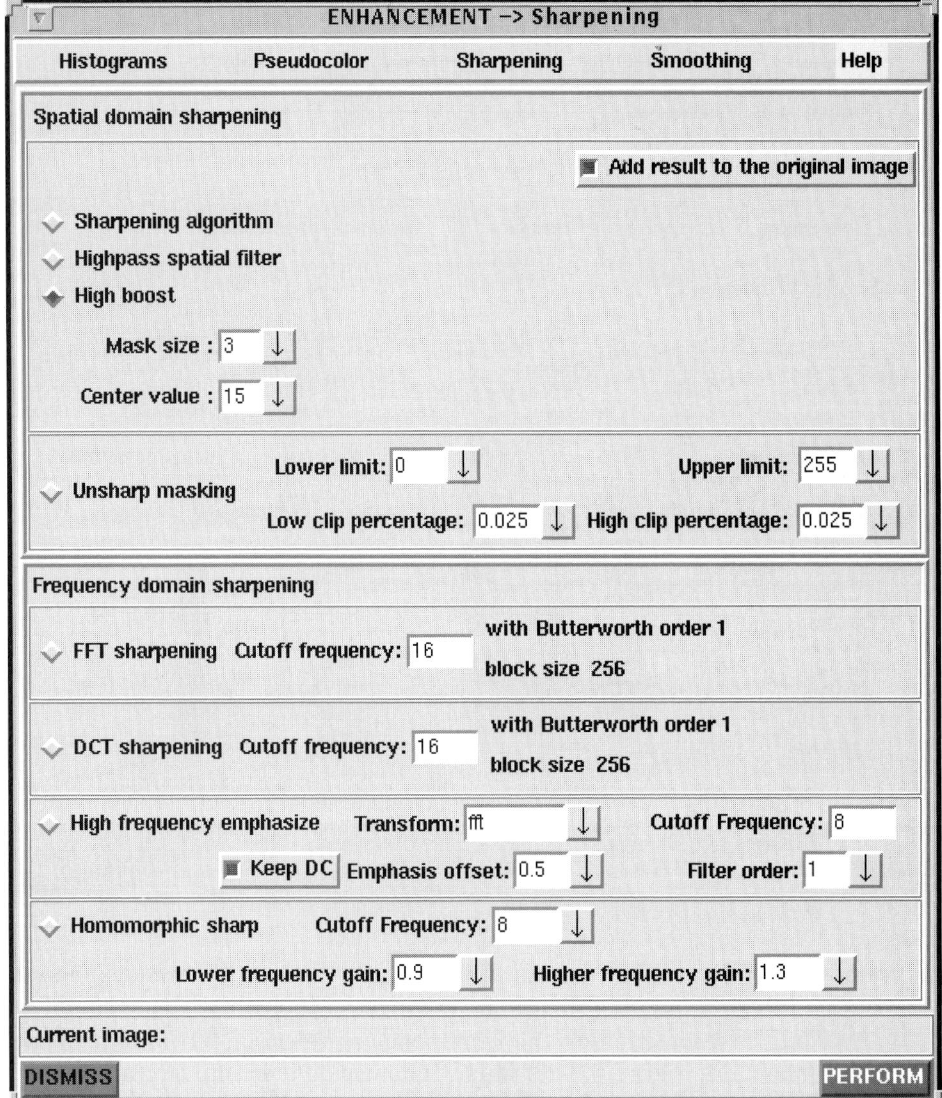

The fourth selection on the Enhancement selection bar is *Smoothing*; this window is shown in Figure 6.2-21. Here, various choices are available in both the spatial and the frequency domains. In the spatial domain are six filters that can be used for image smoothing. The degree of smoothing is specified by the Mask size, a larger mask smooths a larger area, and two of the filters also require a *filter order*, which determines the specific smoothing effect. Smoothing can be achieved in the frequency domain through the use of either the Fourier or discrete cosine transforms in conjunction with lowpass butterworth filters. Select the Cutoff frequency, which determines the degree of smoothing. The lower the cutoff frequency selected, the more smoothing occurs.

Figure 6.2-21 GUI Enhancement -> Smoothing Window

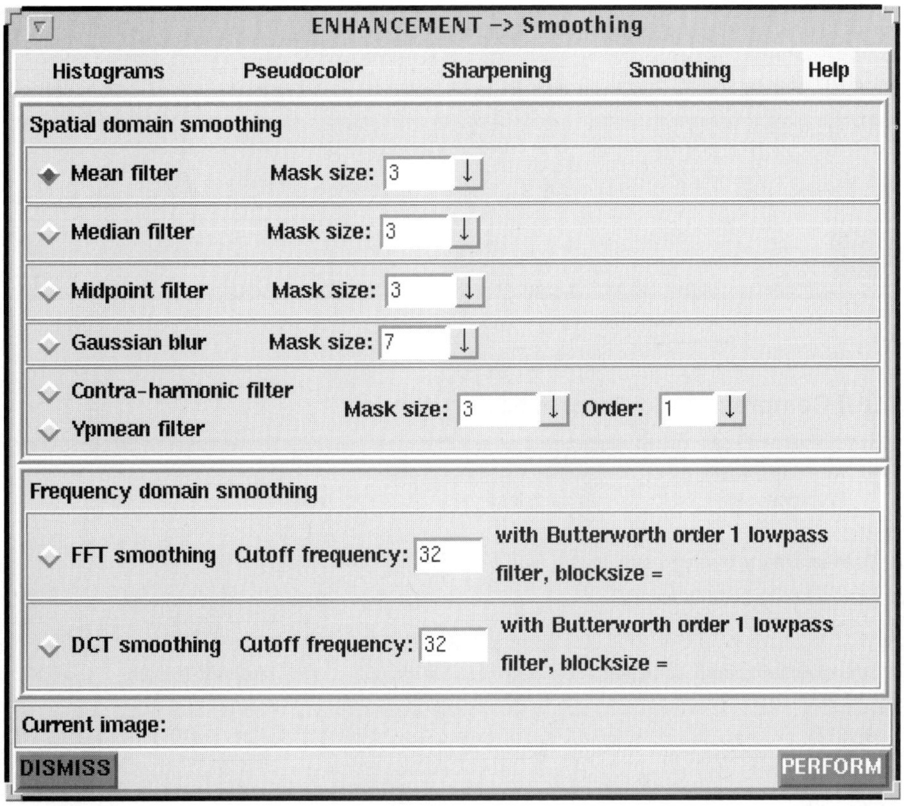

ENHANCEMENT -> Smoothing

Histograms Pseudocolor Sharpening Smoothing Help

Spatial domain smoothing

Mean filter Mask size: 3

Median filter Mask size: 3

Midpoint filter Mask size: 3

Gaussian blur Mask size: 7

Contra-harmonic filter

Ypmean filter

Mask size: 3 Order: 1

Frequency domain smoothing

FFT smoothing Cutoff frequency: 32 with Butterworth order 1 lowpass filter, blocksize =

DCT smoothing Cutoff frequency: 32 with Butterworth order 1 lowpass filter, blocksize =

Current image:

DISMISS PERFORM

6.2.7 Compression Window

The Compression selection bar is shown in Figure 6.2-22. The selection bar has selections across the top that pertain to image compression: *Preprocessing, Lossy,* and *Lossless*. When you select *Preprocessing*, the window expands to include a set of functions that are typically used to reduce the data, or to reorganize the data, before the compression is performed. In Figure 6.2-23, this window is shown. Here we see the natural binary to gray code conversion, which is most useful with run-length coding (and may provide rather random results with any compression methods that are not lossless), as well as quantization techniques. Quantization can be performed in the spatial domain, which will reduce the size of the image, or in the gray-level domain, which will limit the values of the pixels. These types of preprocessing generally require some form of postprocessing during decompression to restore the image. For example, if the data are converted to gray code during compression, they need to be converted back to natural binary code during decompression or, if reduced in size, enlarged during decompression with an interpolative method.

The second selection on the Compression selection bar, *Lossless*, brings up the window shown in Figure 6.2-24. Here we see the lossless compression algorithms that

Figure 6.2-22 GUI Compression Selection Bar

COMPRESSION

Preprocessing Lossless Lossy Help

DISMISS PERFORM

Figure 6.2-23 GUI Compression -> Preprocessing Window

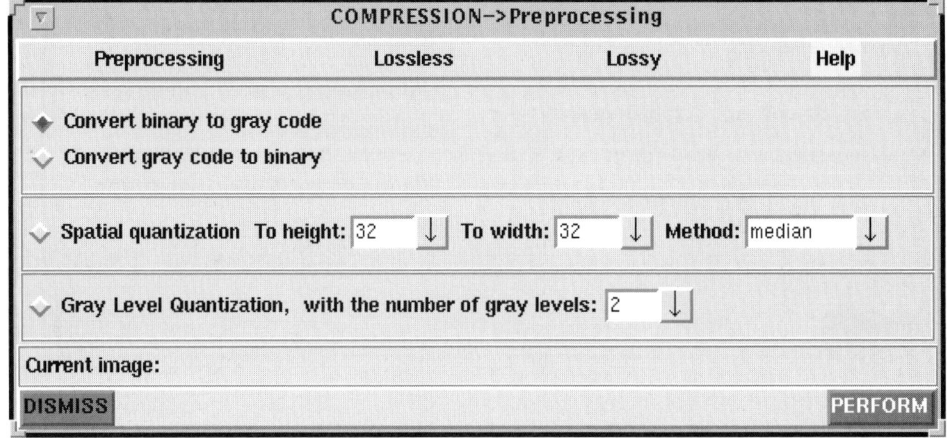

currently exist in CVIPtools. These algorithms allow the original image to be recovered exactly from the original image, but, in some cases, the compressed file may actually be larger than the original file. The third selection on the Compression selection bar, *Lossy*, brings up the window as shown in Figure 6.2-25. Here we see the lossy compression algorithms that currently exist in CVIPtools. These algorithms lose some information in the compression process, so the decompressed image differs from the original uncompressed image. In both the Lossless and Lossy windows, two extra buttons appear at the bottom: *SAVE DECOMPRESSED IMAGE* and *SAVE COMPRESSED DATA*. If the decompressed image is saved, the file is written to disk an image in *pnm* (*P*ortable *A*ny*m*ap) format, and the file will be the same size as the original image. If the compressed data are saved, the file is written in CVIPtools VIP format, which contains header information to allow CVIPtools to decompress it when it is loaded. The size of the compressed VIP file will include the compressed data, plus the size of the header. For objective fidelity metrics, to compare an image before and after compression, use the Utilities→Misc→Compare option, which contains both RMS error and signal-to-noise ratio (SNR) metrics. For more information on the specific algorithms, see Chapter 5, and the help pages.

Figure 6.2-24 GUI Compression -> Lossless Window

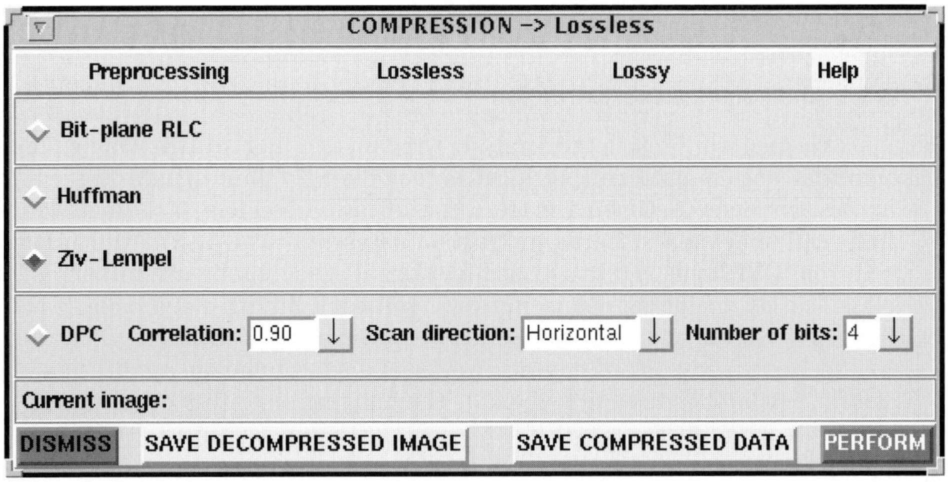

COMPRESSION -> Lossless

Preprocessing Lossless Lossy Help

Bit-plane RLC

Huffman

Ziv-Lempel

DPC Correlation: 0.90 Scan direction: Horizontal Number of bits: 4

Current image:

DISMISS SAVE DECOMPRESSED IMAGE SAVE COMPRESSED DATA PERFORM

Figure 6.2-25 GUI Compression -> Lossy Window

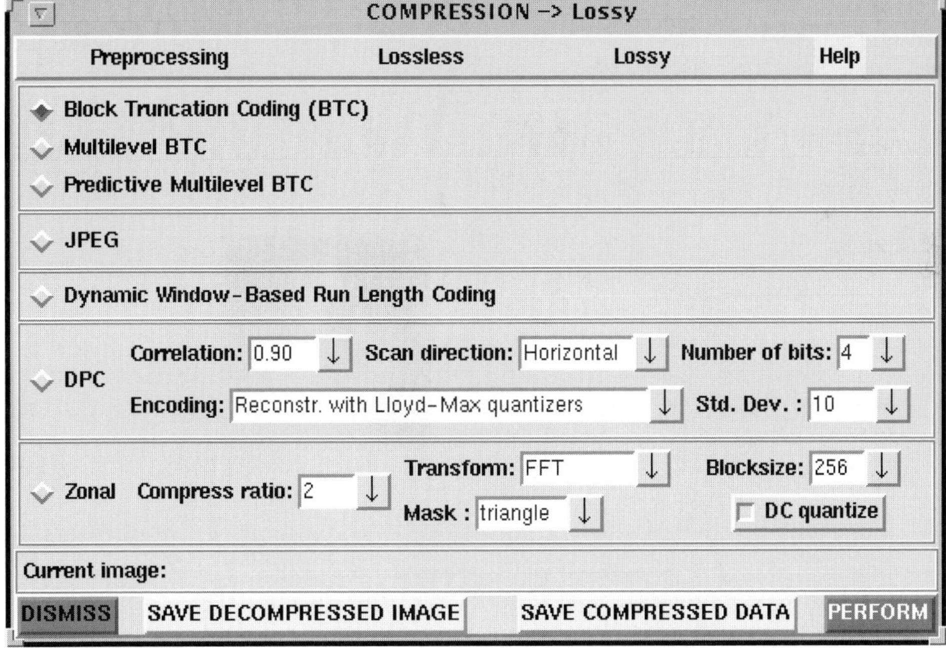

6.3 EXAMPLES

The following examples illustrate how to use CVIPtools by providing a step-by-step procedure for user entry, keyboard and mouse, and the corresponding computer screens. This includes using the main window, the image viewer, the help pages, and the utilities, as well as some of the primary windows. These examples illustrate the various types of user interaction required and should provide a good feel for how

CVIPtools works. By working through these examples, although they use only a small number of the available functions, further exploring all the CVIPtools functions should be easy.

6.3.1 Analysis/Feature Extraction

We may want to use the Analysis window, and its Features window, for an application that entails pattern classification or to simply gain insight regarding a specific application domain. Given a pattern classification problem, CVIPtools can provide the front-end imaging and feature extraction necessary to help solve the problem. The output from CVIPtools is a text-based feature file that can be used in any of the pattern classification, including neural network, software packages available. If pattern classification is not the end goal, the extraction and examination of image features is a useful tool to identify important characteristics of a specific image application domain. This information is then used to help guide the imaging algorithm development process.

Figure 6.3-1 Analysis/Feature Extraction Example, Screen Capture #1

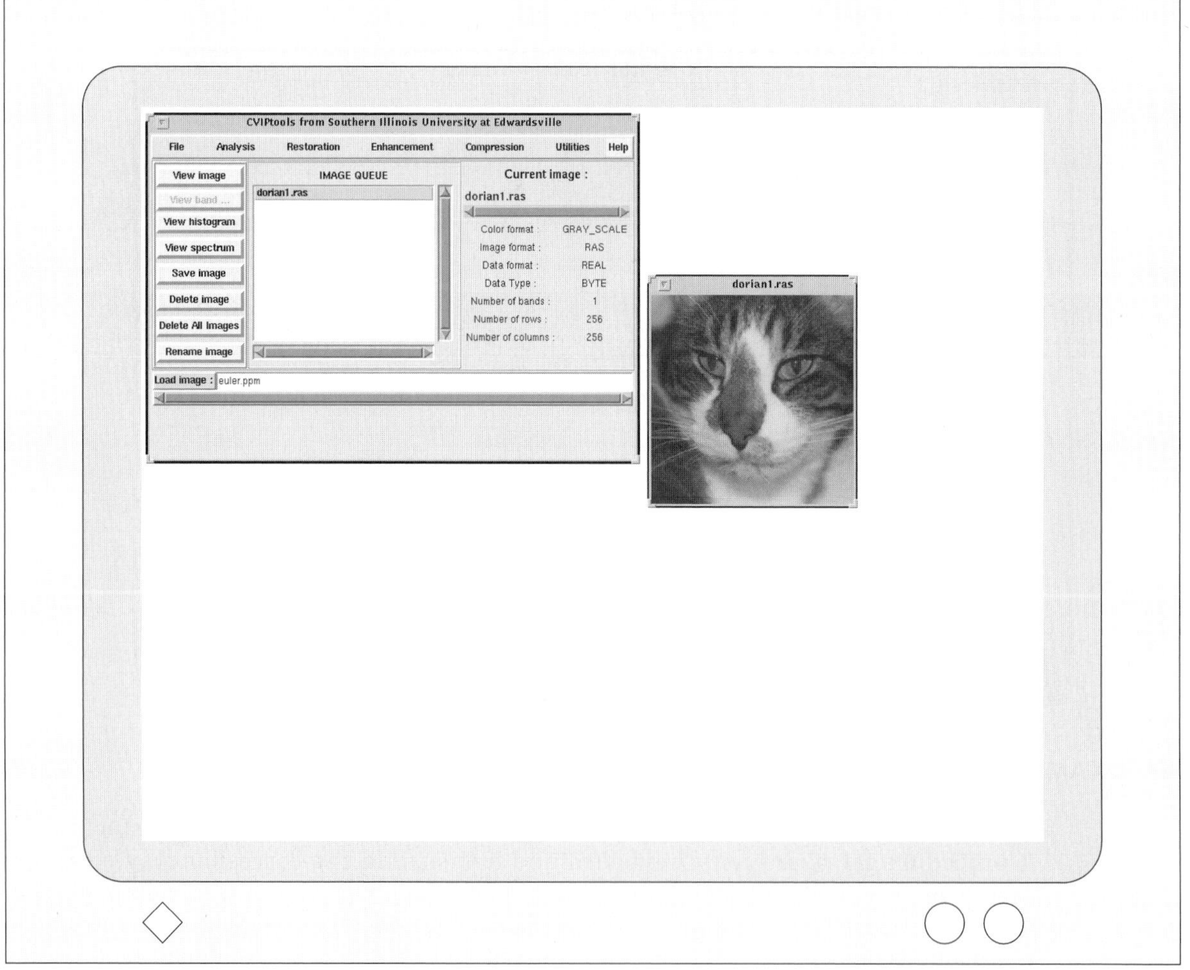

The first step is to run the CVIPtools software by typing *CVIPtools*, and the main window appears. At this point any desired images can be loaded, by typing the names in the *Load image* entry box, or by clicking on the *Load image* button and using the pop-up window to search the directory structure. In Figure 6.3-1, we see an image name typed in the *Load image* entry box. When you press the Enter key, the image will be loaded into the image queue and displayed. Note that one image has already been loaded into the queue and is displayed. If many images are to be loaded from a specific directory, it may be useful to use the *Add a temporary path* option under the *File* button (on left top of main window). If a temporary image path is added, it will be valid throughout the entire CVIPtools session, and you need only type in the image name in the entry box. If you desire a permanent image directory search path name, you must set the environment variable CVIP_IMGPATH accordingly.

Next, click the left mouse button on the *Analysis* button to select it, and the Analysis selection bar appears (Figure 6.3-2). Now select the particular Analysis category of

Figure 6.3-2 Analysis/Feature Extraction Example, Screen Capture #2

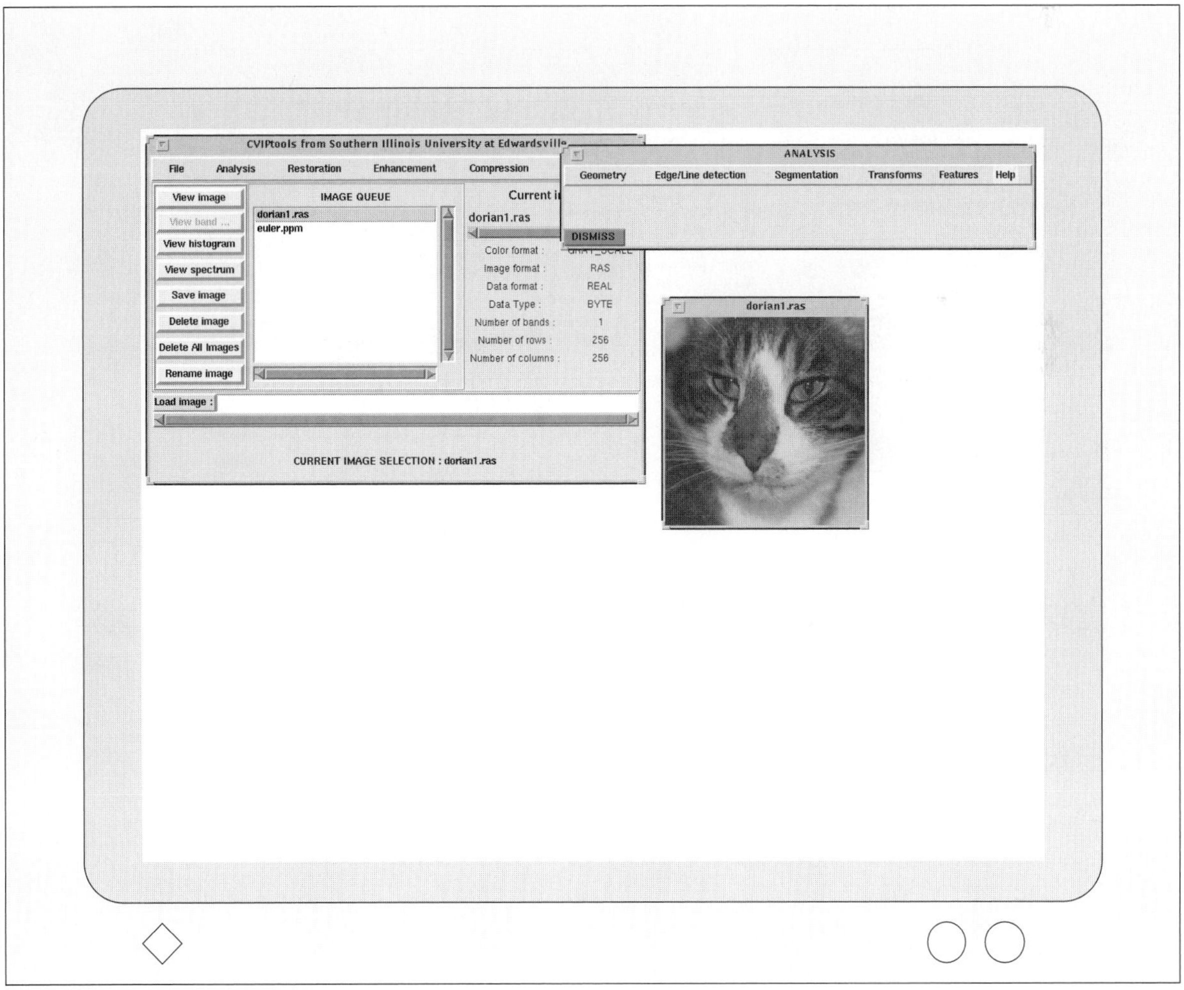

interest. Typically, for feature extraction, some form of segmentation is performed on the image first. Select segmentation by clicking on the *Segmentation* option at the top of the Analysis window. This will cause the Analysis window to expand, and the available segmentation algorithms will appear in it. Next, select the desired segmentation algorithm by clicking on the radiobuttons on the left and set any necessary parameters. When this is done, clicking on the *PERFORM* button in the lower-right corner will perform the function. If you need specific information to use one of the segmentation algorithms, select the help pages by clicking on *Help* at the top of the window. Figure 6.3-3 shows how the screen looks after this sequence of operations. Note that CVIPtools automatically puts the output image in the queue, displays it, and appends a function-related character string to the name.

Figure 6.3-3 Analysis/Feature Extraction Example, Screen Capture #3

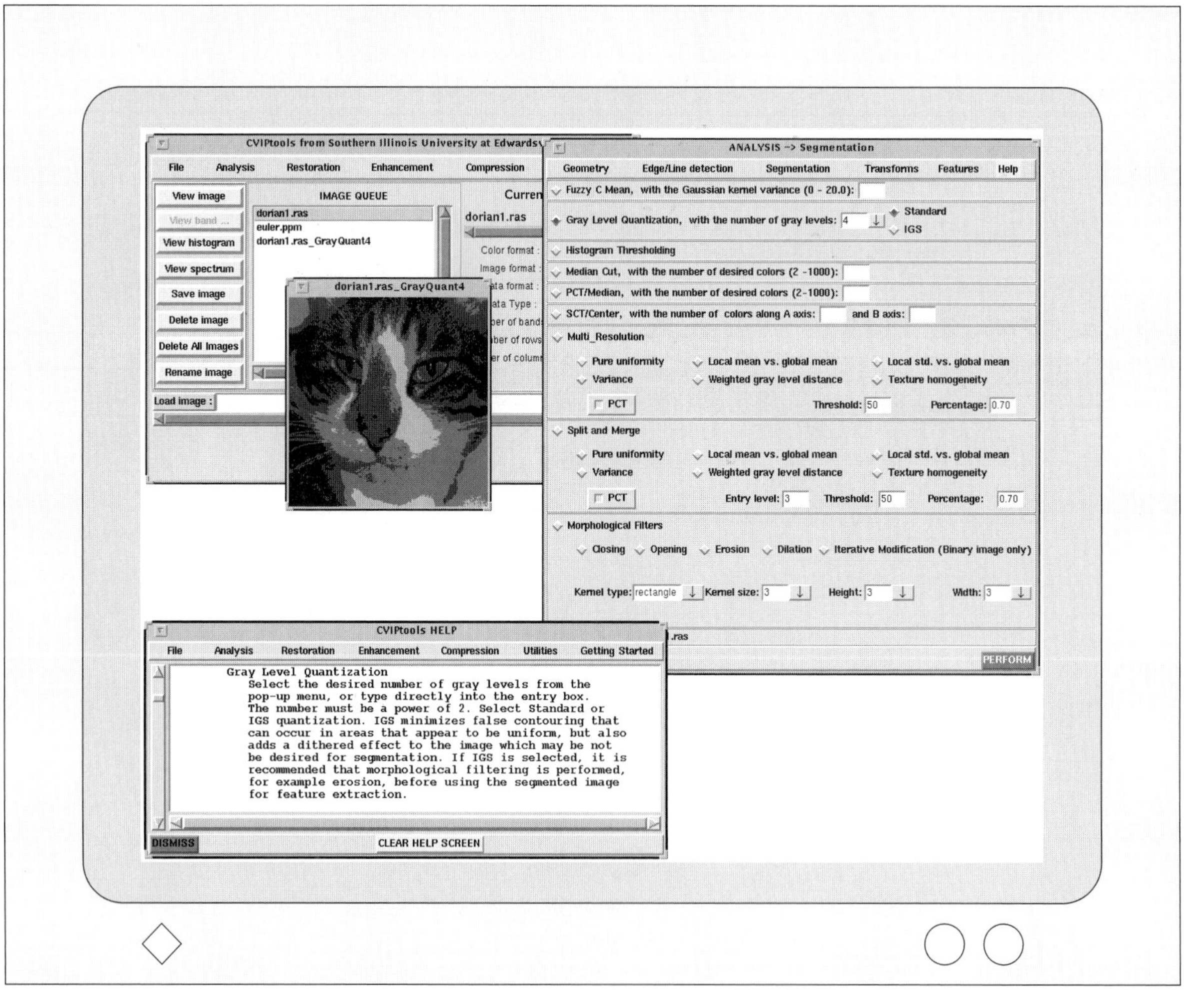

After segmentation we may observe that the resulting image has too many small objects and so want to perform morphological filtering to remove these. First, select the image, by clicking on it in the image queue, and then select *Morphological Filters* at the bottom of the Analysis window. Select the type of filter, set any necessary parameters, and then select *PERFORM*. The computer screen after these operations is shown in Figure 6.3-4.

At this point, to perform feature extraction, select *Features* at the top of the Analysis window. Next, type in the name of the original image, the segmented image, and a feature file name in the entry boxes at the top of the Features window. Next, select the desired features, by clicking on the corresponding checkbuttons. The image object is selected by entering a pair of row and column coordinates within the object in the entry boxes. Alternately, with the CVIPtools image viewer, you can click on the object with the middle mouse button (note that, if the mouse is used, the row and col-

Figure 6.3-4 Analysis/Feature Extraction Example, Screen Capture #4

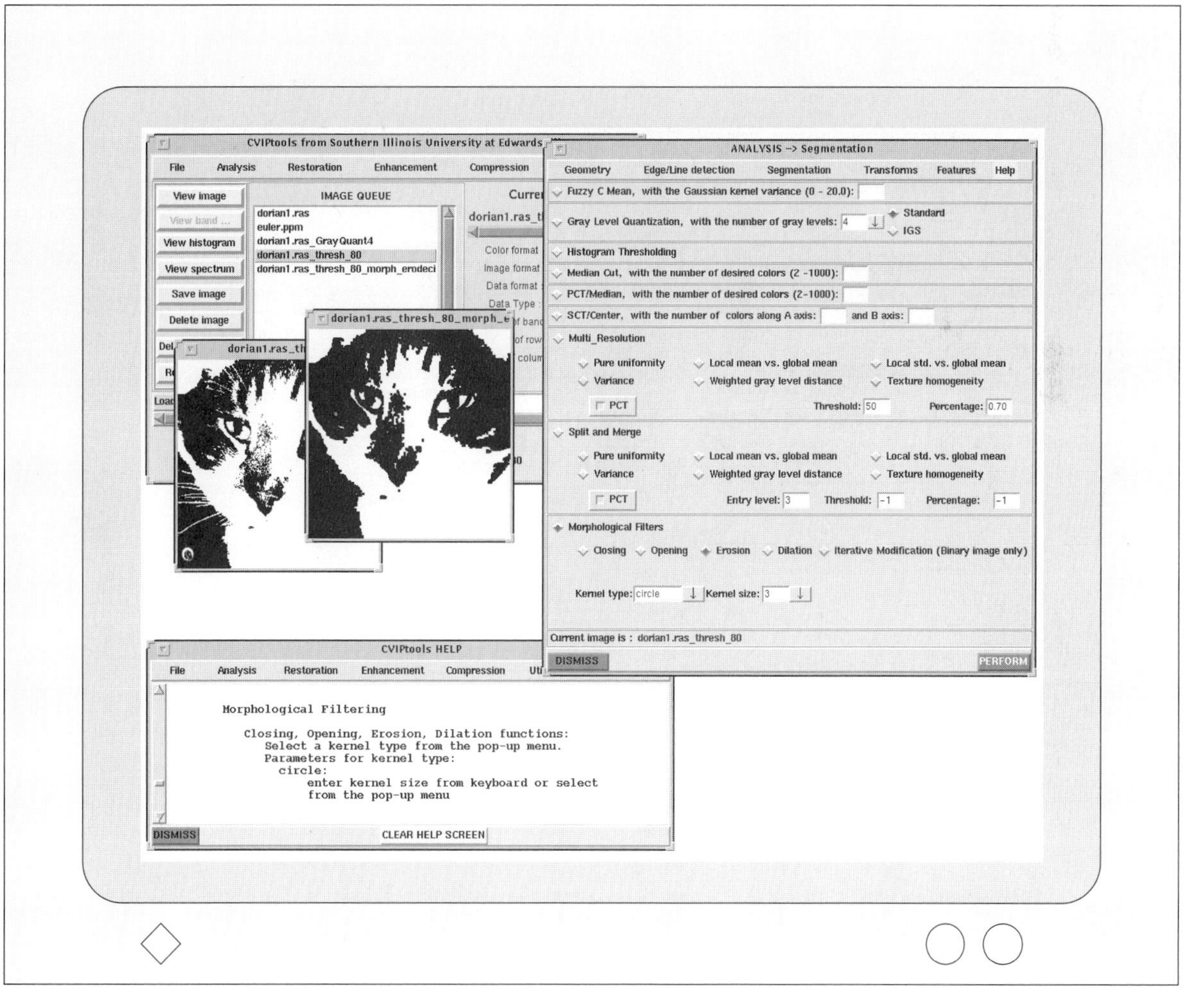

umn coordinates will not appear until you select the *PERFORM* button). A class name for the object can also be entered in the Class entry box. When everything is set up, select the *PERFORM* button, and the features are extracted and written to the specified feature file. This feature file can be viewed with the *View Feature File* button to the right of the feature file name. This has been done in Figure 6.3-5.

Repeat this process for each object from which you want to extract features. You can put object feature information from many different images and classes into the same file, as long as you do not change the features. If you change the features, you must select a new feature file name, or the feature file will be inconsistent. The feature file is written to disk and can then be used as input for pattern classification software, or it can be examined and analyzed by an expert in the particular application domain. When you are done with the Analysis window, you can remove it from the screen by selecting the *DISMISS* button in the lower-left corner.

Figure 6.3-5 Analysis/Feature Extraction Example, Screen Capture #5

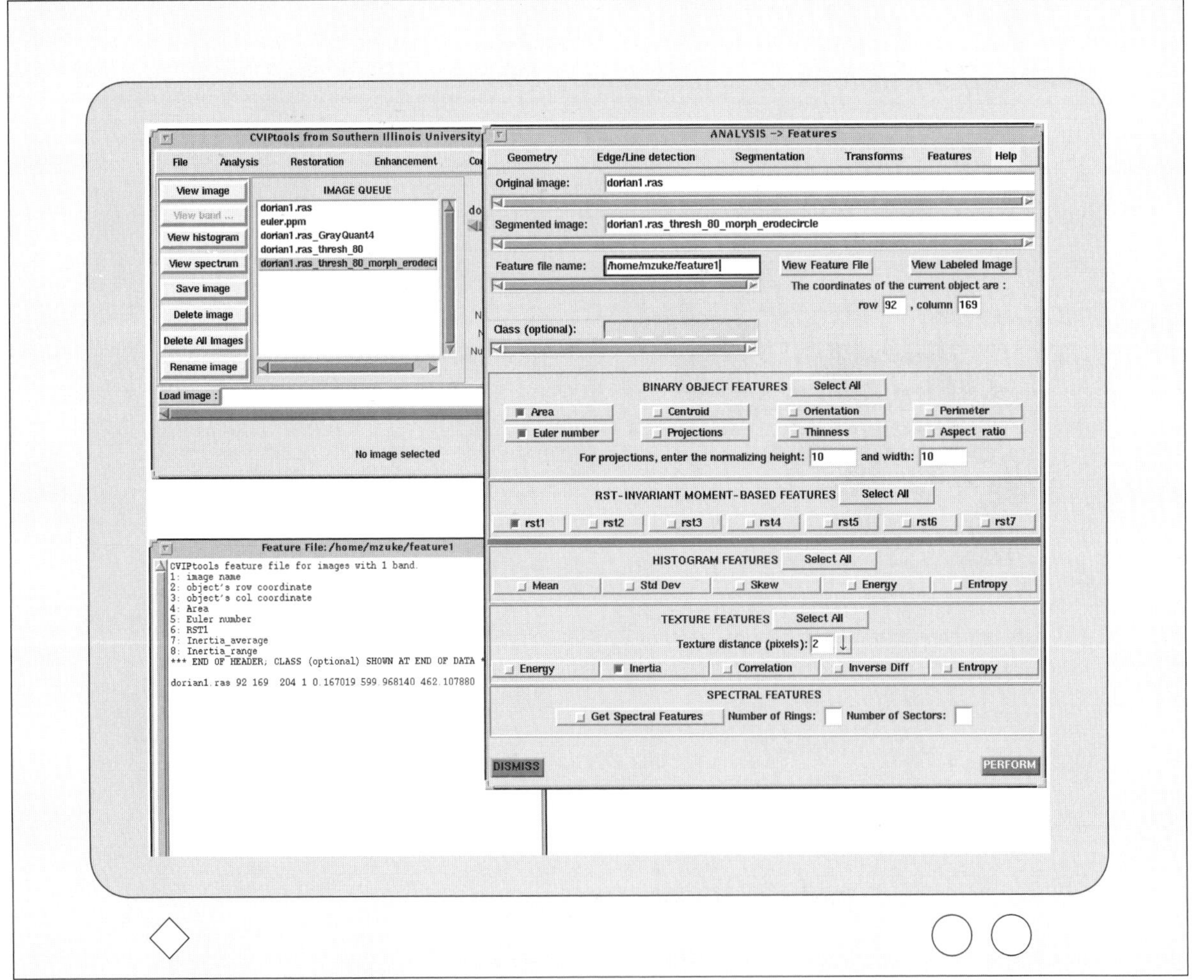

6.3.2 Enhancement/Histograms

Choose the enhancement selection bar by clicking on *Enhancement* at the top of the main window. Next, you will use the pop-up File Select Box to load an image. When you click on the *Load image* button in the main window, the File Select Box appears. To load the image, type a path name in the Pathname entry box, press Enter on the keyboard, and then select the specific image file by double-clicking on the file name. Alternately, traverse the directory structure by double-clicking the mouse on directory names. Figure 6.3-6 shows the screen just before selecting the image. Here the file name is highlighted in the File Select Box, which occurs with a single mouse click; double-click to load it into the image queue.

When you have the image loaded, select *Histograms* at the top of the Enhancement selection bar, and the window expands to show all the histogram operations

Figure 6.3-6 Enhancement/Histograms Example, Screen Capture #1

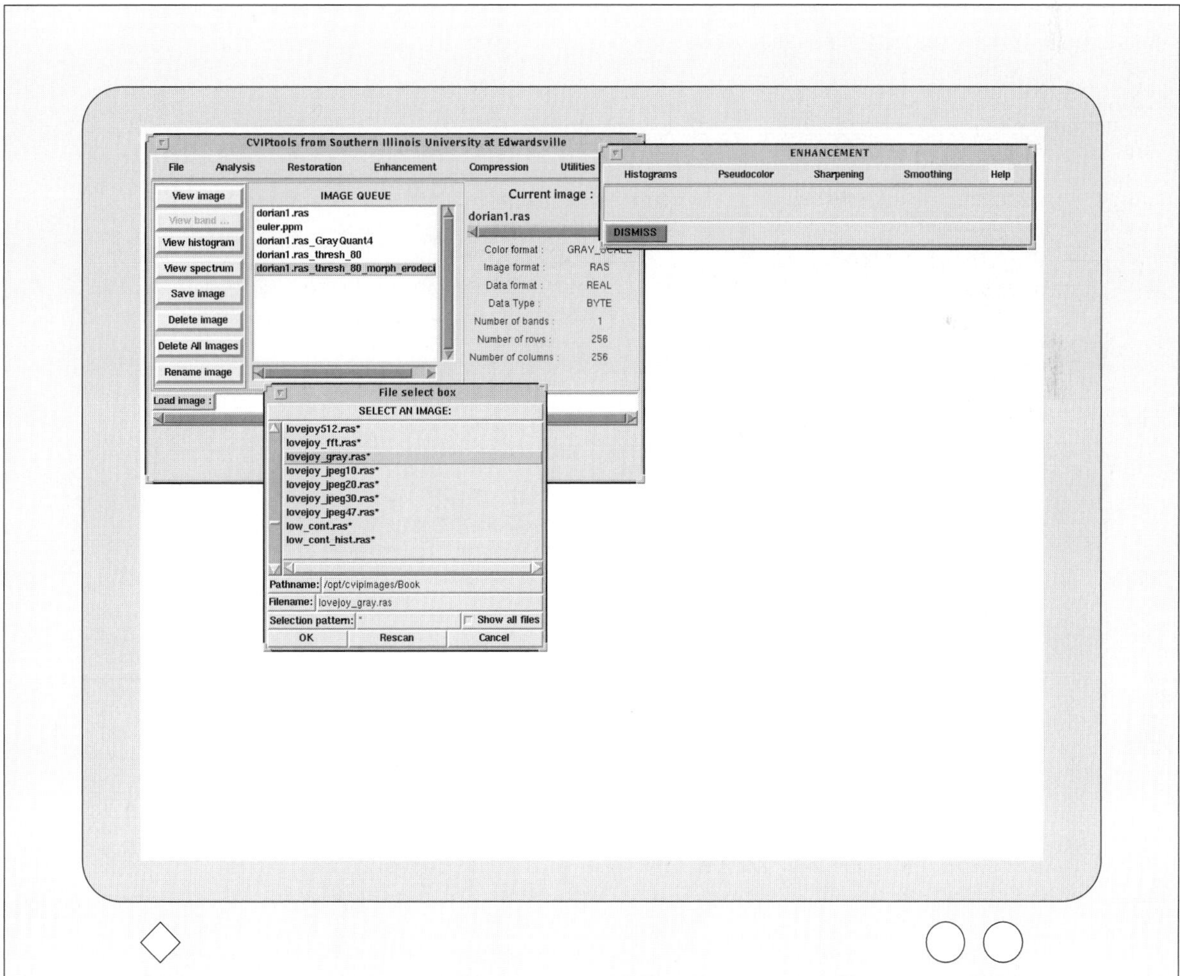

available (see Figure 6.3-7). When you select an image, the name appears in the Current image box at the bottom of the window. Select any desired operations, and enter any necessary parameters, followed by a click on the *PERFORM* button. For example, you can perfrom a histogram slide by selecting the *Histogram slide* radiobutton, choosing the parameters *up/down*, and setting the amount as desired. The histogram of an image can be displayed by selecting it in the image queue, and then clicking on the *View histogram* button at the bottom of this window. Figure 6.3-7 shows how the screen appears after the histogram slide has been increased by 40 and the histograms displayed. Figure 6.3-8 shows how the screen appears after a local histogram equalization with a blocksize of 16. For detailed information on using the functions, see the help pages. For more information on the algorithms, refer to Chapter 4, as well as the references.

Figure 6.3-7 Enhancement/Histograms Example, Screen Capture #2

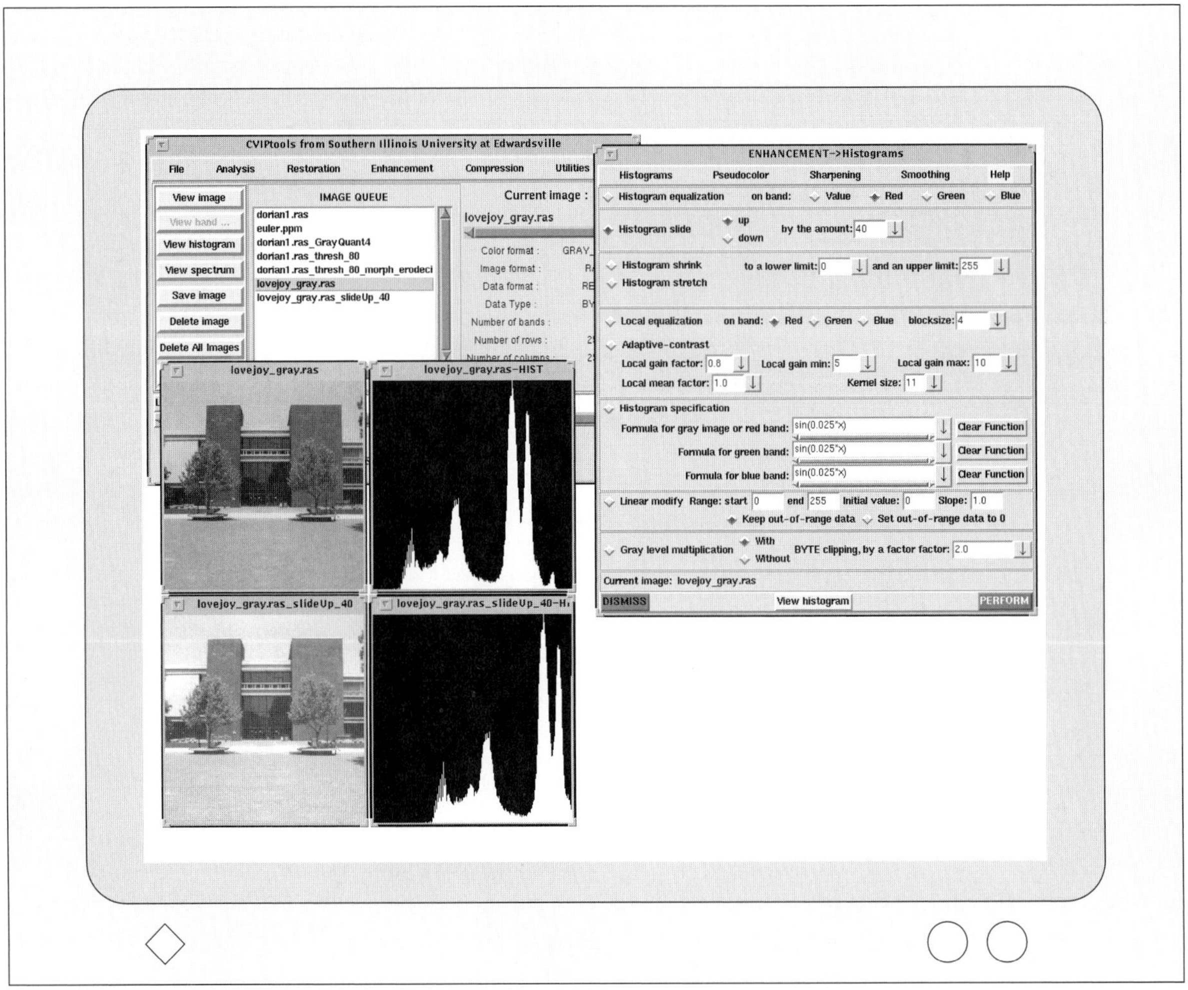

Figure 6.3-8 Enhancement/Histograms Example, Screen Capture #3

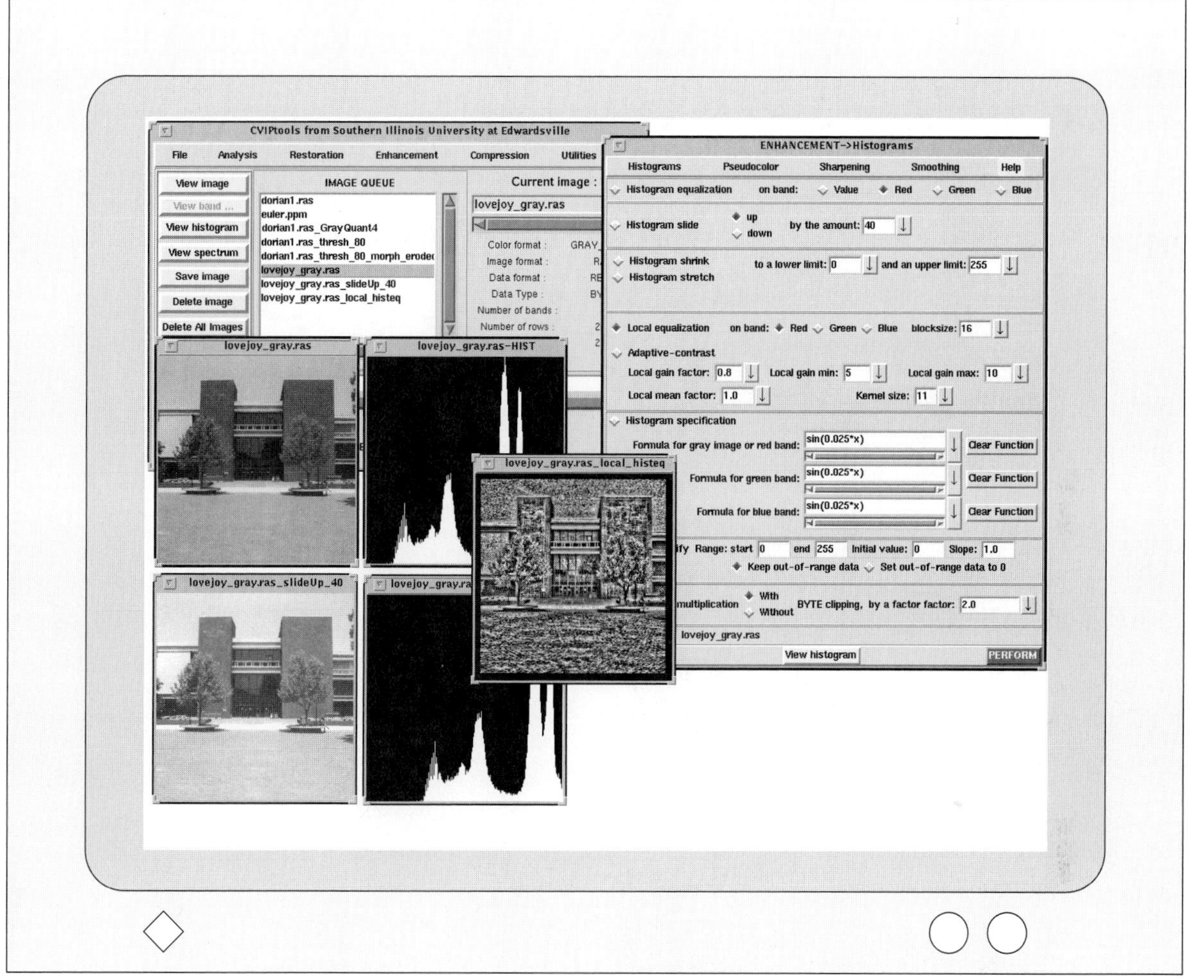

6.3.3 Restoration/Frequency Filters

The restoration selection bar appears when you click on *Restoration* from the top of the main window. Next, make the desired selection from the Restoration selection bar; in this case *Frequency filters* was selected with the left mouse button. Now, to explore the *Parametric Wiener* restoration filter, select this filter with the radiobuttons on the left. When you select a filter, the necessary input parameters appear within this window. Assuming that images have been loaded by one of the previously described methods, select the degradation function next. You have two choices: *Use image* or *Specify the function*. Because you want to do some exploration, select *Specify the function*, which allows you to specify the point spread function (PSF) in terms of window size, shape, and blur method. Figure 6.3-9 shows the computer screen after these selections.

Figure 6.3-9 Restoration/Frequency Filters, Screen Capture #1

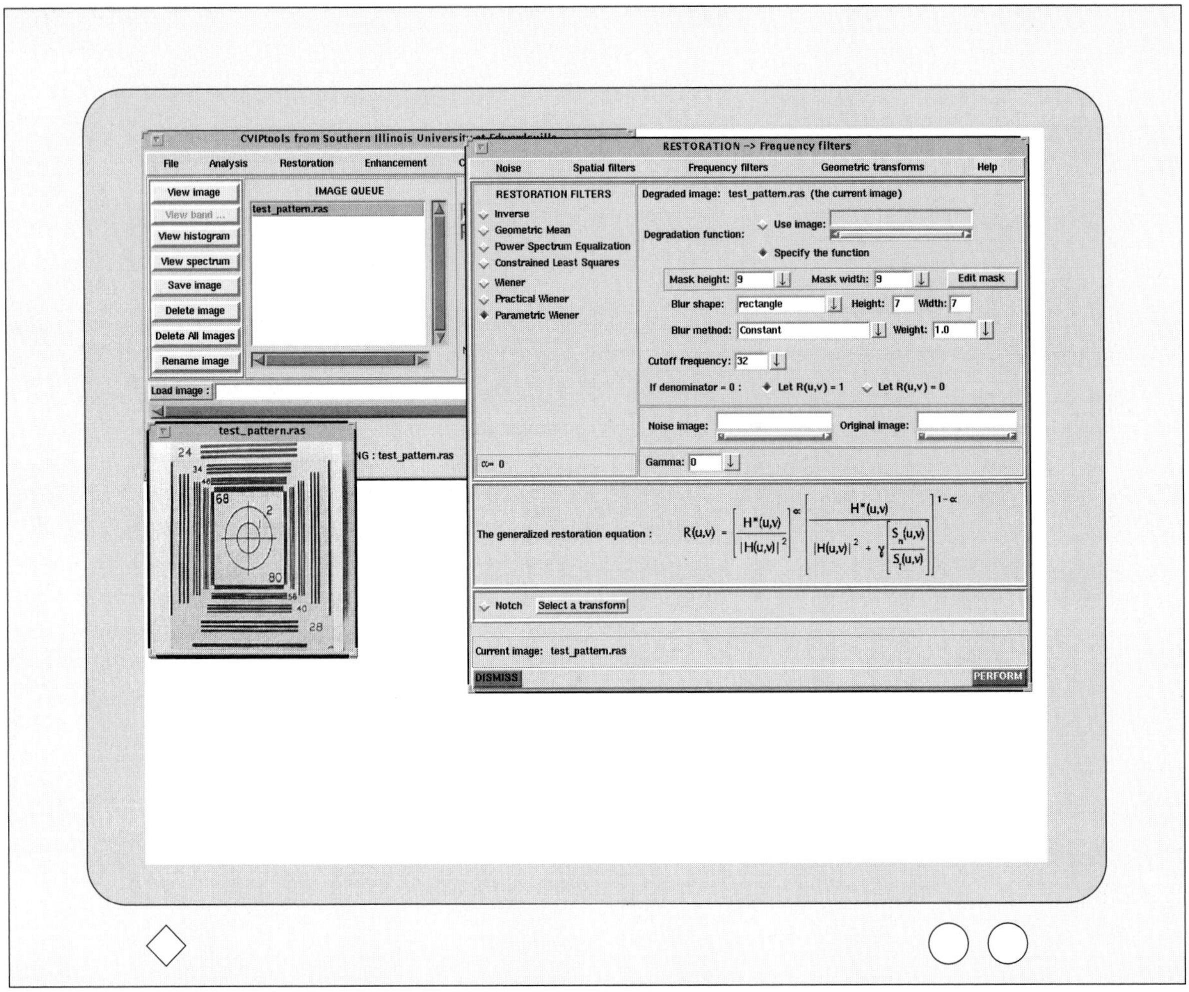

The generalized restoration equation appears in the lower section of the Restoration→Frequency Filters window, and parameters applicable to the specific filter selected appear to the left of the equation. For the *Parametric Wiener* filter, you need the degradation image $H(u, v)$, the power spectrum from the noise image $S_n(u, v)$, and the power spectrum of the original image $S_I(u, v)$. From the entry boxes in the window, you can enter the noise image, from which CVIPtools will find $S_n(u, v)$, and the original image, from which CVIPtools will find $S_I(u, v)$. Our next step is to create or load the required images.

In this case you can create the images because you are performing exploratory imaging. First go back to the main window and select *Utilities*. The next step is to degrade the image; select *Filter→Specify a filter*. Next put the mouse on the down arrow and select a gaussian filter, and then click on the 11 × 11 radiobutton. Figure 6.3-10 shows the computer screen after this has been done. Now select an image from

Figure 6.3-10 Restoration/Frequency Filters, Screen Capture #2

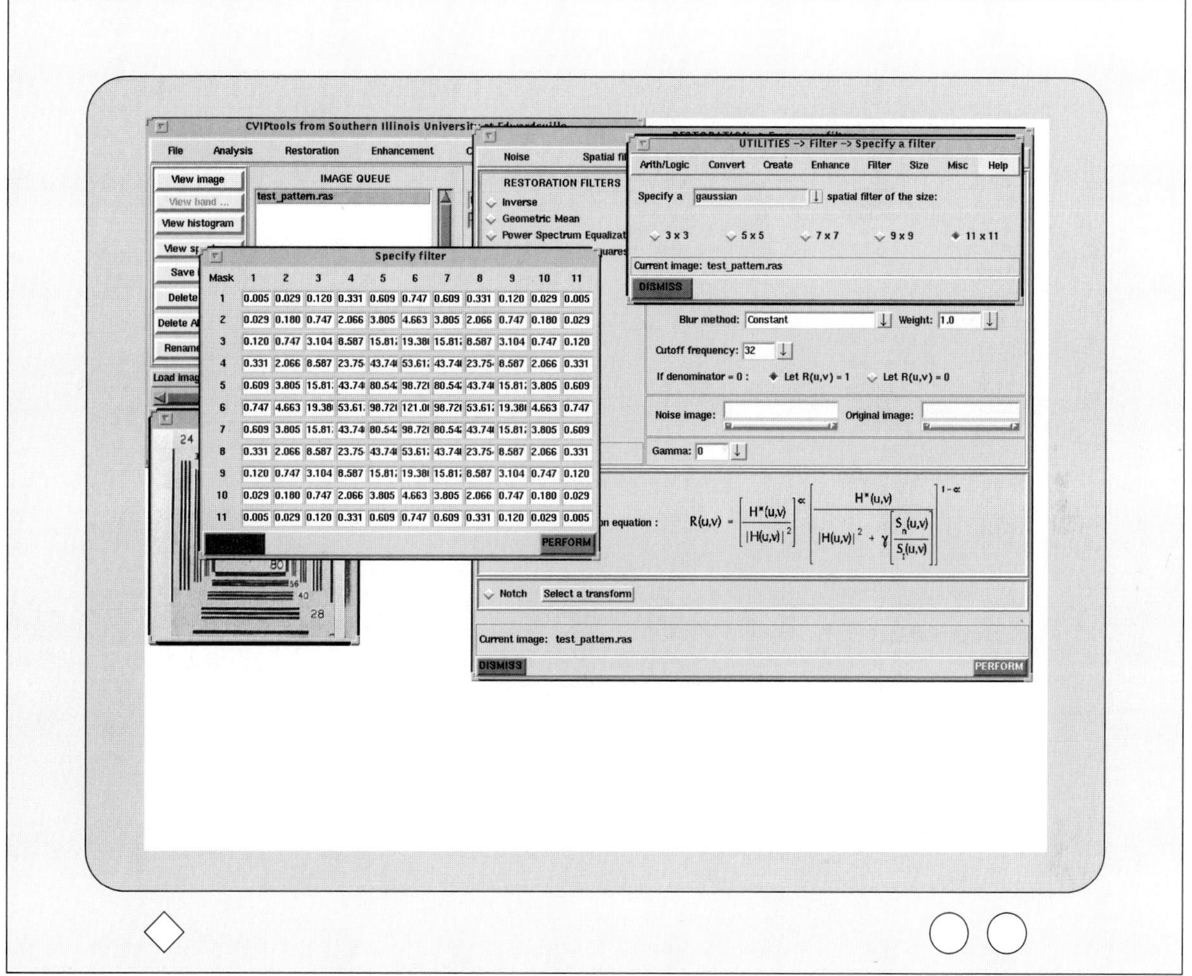

the queue and press the *PERFORM* button on the Specify filter window. This will convolve the 11×11 gaussian filter with the image. Now you can dismiss the filter window, if desired.

To add noise to the image and create the noise image itself, select *Create→Add noise* from Utilities. First, select the desired type of noise and set any necessary parameters. Here, pick gaussian noise, with a mean of 0 and a variance of 100. Next, select the degraded image in the queue, the one you convolved with the gaussian filter, and press *PERFORM* on the Utilities window to add noise to this image. The computer screen after this operation is shown in Figure 6.3-11. Next, select the *Use black image* radiobutton on the Create→Noise window and press *PERFORM* again. This will create the desired noise image. At this point, you can dismiss the Utilities window or leave it on the screen.

Figure 6.3-11 Restoration/Frequency Filters, Screen Capture #3

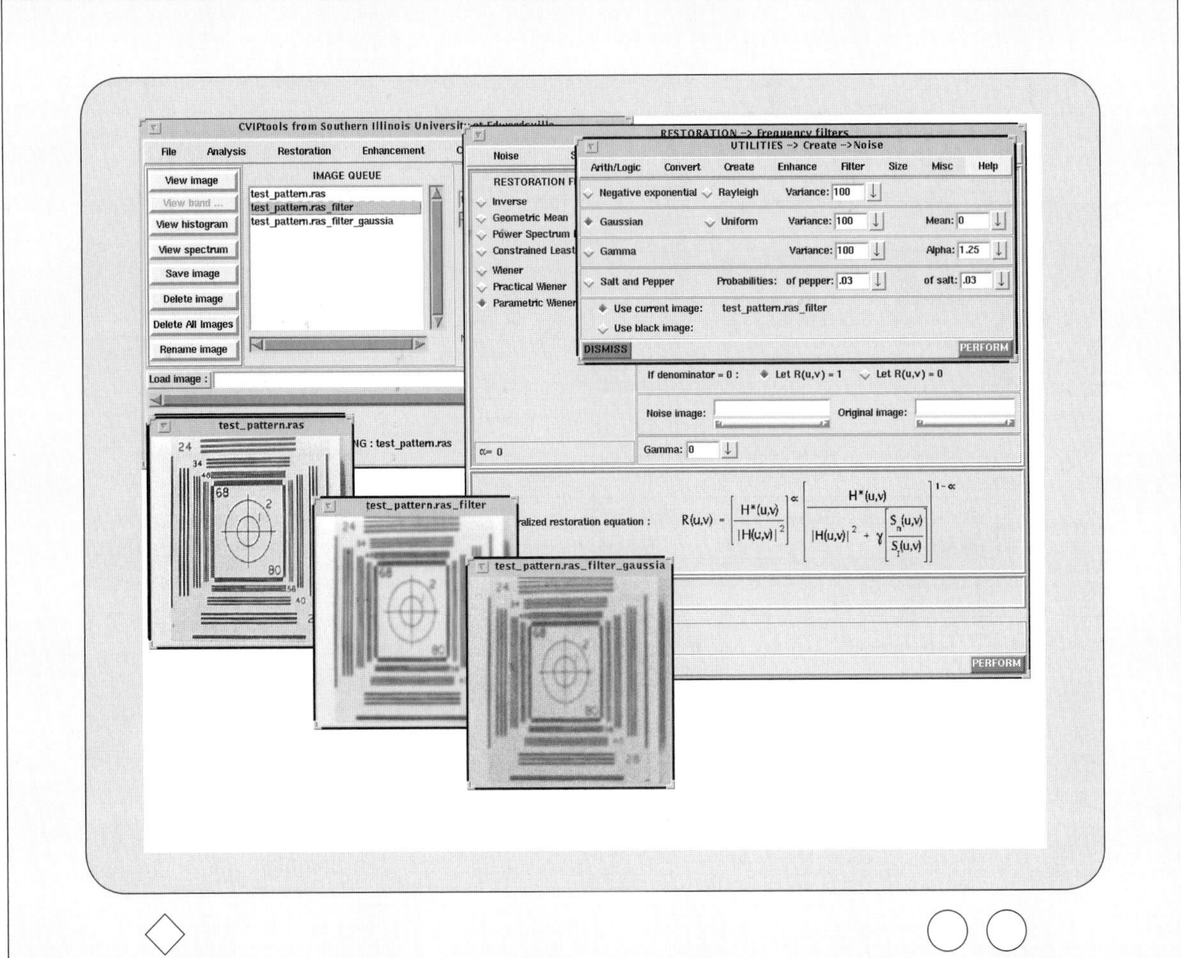

Now that you have created all the necessary images, go back to the Restoration→Frequency Filters window. Here, select from the queue the degraded image with the noise you added to it—this is the image you want to restore. Next, specify the degradation function by selecting an 11×11 mask, set height and width equal to 11, and select a rectangular blur shape and a gaussian blur method. At this point, select *Edit mask* if you want to change any of the mask coefficients. You can also modify the mask coefficients with the Weight parameter, which is used as a multiplier for the mask before $H(u, v)$ is found.

Next, select a cutoff frequency to limit the restoration in the frequency domain—any frequency coefficients above this value are not used in the restoration (set to 0). Additionally, you must decide what to do if the denominator in the restoration filter equation is zero. Here we have two choices: set the filter to either 0 or 1. With the first choice, you are not using zeros; with the second choice, you are retaining the information in the degraded, noisy image.

To specify the noise image and the original image, type in the corresponding image names or select the desired image in the queue and then click on the *Noise image* or *Original image* text button in the window. When this is done, set gamma and press *PERFORM*. CVIPtools then performs all the necessary transforms, filtering and inverse transforms to provide the restored image. Note that you can specify either images or spectrums to these functions, and they automatically perform only the necessary calculations. Figure 6.3-12 shows the computer screen after this has been done.

In practice, you may have a degraded image that you wanted to restore. In this case, you need to estimate the noise function, the degradation function, and the original image. You can do this experimentally with the various types of filters, degradation functions, noise types, and all of their associated parameters available in CVIPtools. This leaves you free to be creative and explore numerous possibilities until you achieve suitable results.

Figure 6.3-12 Restoration/Frequency Filters, Screen Capture #4

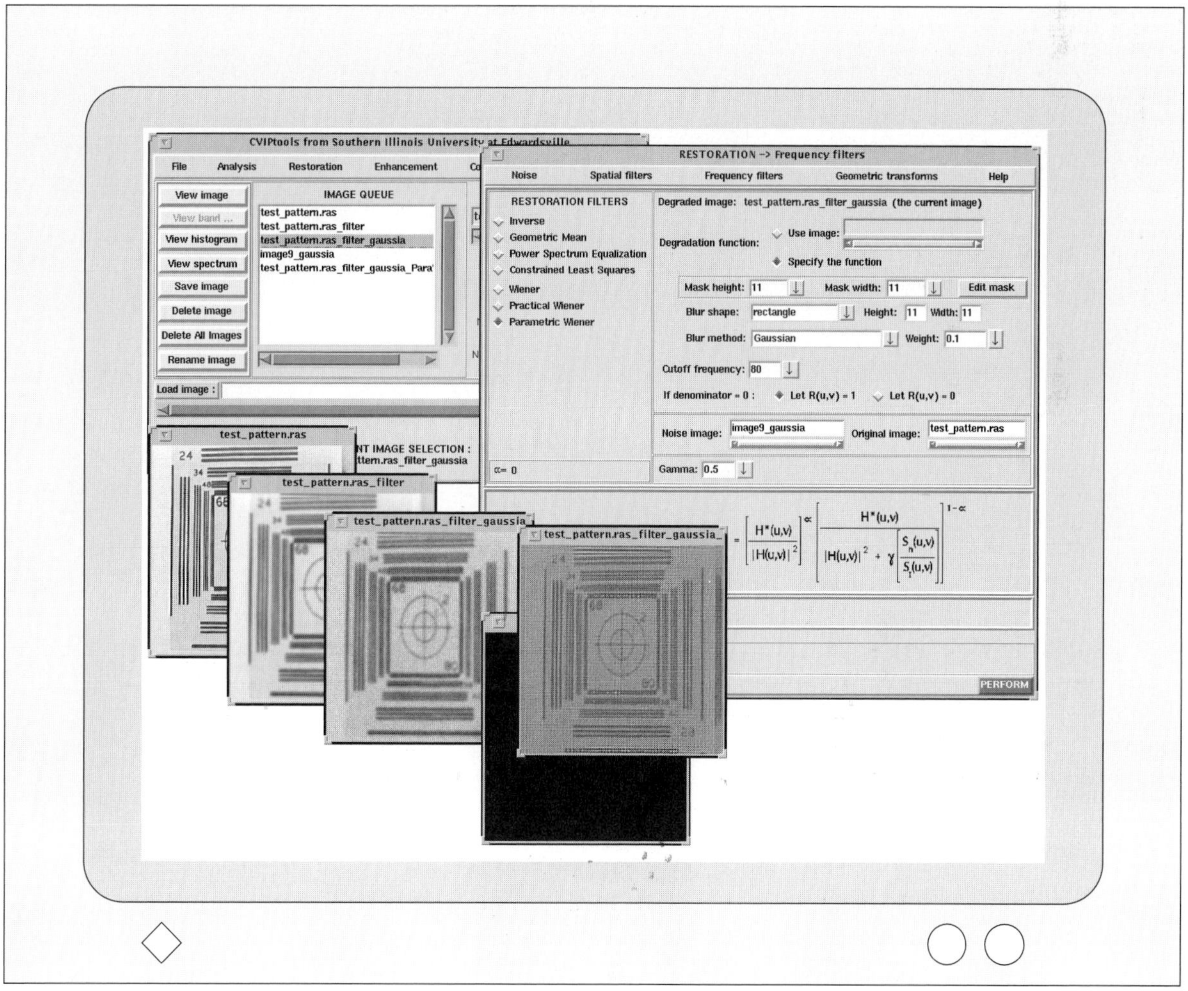

CHAPTER 7

CVIPtools Applications

7.1 INTRODUCTION

The following sections describe some of the applications developed using the CVIPtools software environment—specifically, using the CVIPlab program and the CVIPtools libraries (see Chapter 8). After the application is developed, useful functions can be extracted and put into the proper CVIPtools libraries. This creates an environment where current and future research can build upon previous work.

The application to skin tumor borders was partially funded by the National Institutes of Health (NIH) and was developed concurrently with the first version of CVIPtools; many of the segmentation algorithms came from this research and development. The helicopter enhancement application was the first application developed with a graphical user interface and was funded through the Department of Defense (DoD). The wavelet/vector quantization compression application represents a very active area of research and was partially funded by the NIH. The deformable templates application was initially developed by Stealth Technologies for application to automatic segmentation for computer-aided surgical tools. The final application in this chapter describes a basic model for simulating visual acuity and is also applied to a simple night vision simulation. Although this application is only a preliminary development, it illustrates the ease with which research and development in the CVIPtools environment can lead to a working solution in a very short period of time.

Portions of this section have been published as an article in the *IEEE Engineering in Medicine and Biology Magazine*, January/February 1996. This material is reprinted here with permission of the IEEE.

7.2 Automatic Skin Tumor Border Identification

The detection of boundaries in images is a fundamental problem in computer vision as well as a necessary preliminary step for further image understanding. This research was developed to serve as a front-end for a computer vision system for skin tumor evaluation. The most predictive features for various skin cancers will be targeted by the computer vision system, allowing automatic induction software to classify the tumor. The problem of interest addressed is the identification of skin tumor boundaries; the border is the first and most critical feature to identify. Object boundaries and surface contours are fairly easily detected by the human observer, but automatic border detection can be a more difficult problem. The images may contain reflections, shadows, or extraneous artifacts that make the process of finding the border by computer more difficult.

7.2.1 Border-Finding Algorithm

A skin tumor may be distinguished from surrounding skin by features such as color, brightness or luminance, texture, and shape or any combination thereof. The use of color as a means to identify the tumor border is particularly important because, in some cases, it is difficult to identify the tumor border in a monochrome image. The border-finding algorithm presented here involves a series of preprocessing steps to remove noise from the image followed by color image segmentation, data reduction, object localization, and contour encoding. This process is depicted in Figure 7.2-1.

7.2.2 Noise Removal

The input image may contain noise that will make the segmentation process less accurate. For example, skin tumor images often contain extraneous artifacts such as rulers and hair that make localizing the tumor border more difficult. In addition, the images may contain undesirable color variations such as shadows and reflections that tend to bias the color map when performing color segmentation. In Figure 7.2-2 a ruler and a video frame border are seen at the bottom of the image, and reflections occur directly around the tumor. In order to reduce the effects of such noise, the

Figure 7.2-1 Border-Finding Algorithm

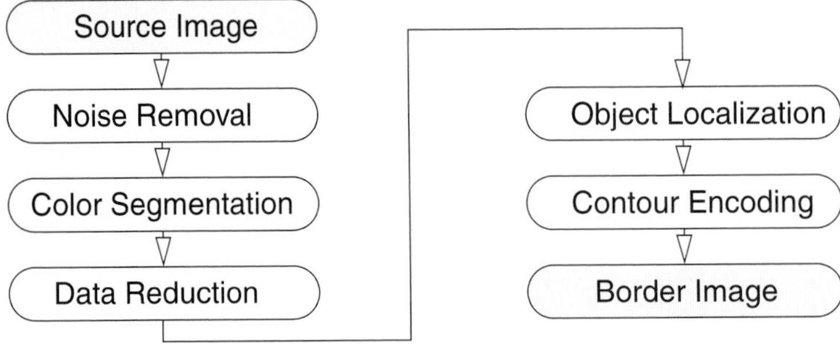

Figure 7.2-2 Border-Finding Example

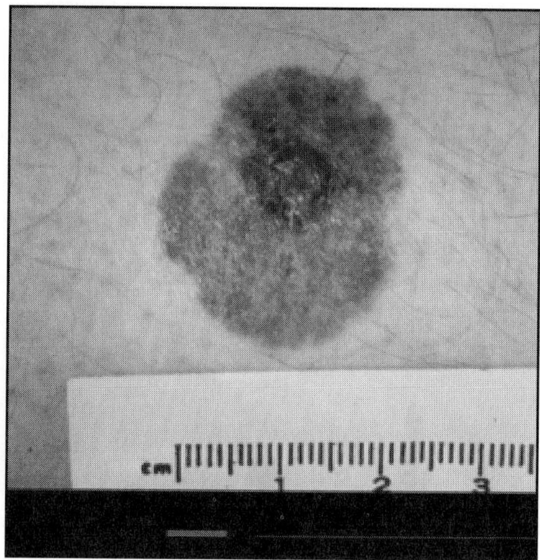

Original image.

images were first processed with a pseudomedian filter and then a nonskin detecting algorithm (based on heuristics, see Figure 7.2-3) was applied to mask out unwanted artifacts such as rulers and reflections. As seen in Figure 7.2-4, the majority of the reflections and ruler artifacts have been masked out by this algorithm; the masked-out pixels are set to black (note also that the video frame border has been cropped prior to this processing).

Median filtering is computationally intensive but can be approximated by a simpler operator called the *pseudomedian filter*. The pseudomedian is defined as follows:

$$\text{PMED}(S_L) = (1/2)\text{MAXIMIN}(S_L) + (1/2)\text{MINIMAX}(S_L)$$

$$\text{where } S_L \text{ denotes a sequence of elements } s_1, s_2, \ldots, s_L$$

$$\text{where for } M = \frac{(L+1)}{2}$$

$$\text{MAXIMIN}(S_L) = \text{MAX}\Big[[\text{MIN}(s_1, \ldots, s_M)], [\text{MIN}(s_2, \ldots, s_{M+1})], \ldots, [\text{MIN}(s_{L-M+1}, \ldots, s_L)]\Big]$$

$$\text{MINIMAX}(S_L) = \text{MIN}\Big[[\text{MAX}(s_1, \ldots, s_M)], [\text{MAX}(s_2, \ldots, s_{M+1})], \ldots, [\text{MAX}(s_{L-M+1}, \ldots, s_L)]\Big]$$

For example, the pseudomedian of length five is defined as

$$\begin{aligned}\text{PMED}(a, b, c, d, e) = {} & (1/2)\text{MAX}[\text{MIN}(a, b, c), \text{MIN}(b, c, d), \text{MIN}(c, d, e)] \\ & + (1/2)\text{MIN}[\text{MAX}(a, b, c), \text{MAX}(b, c, d), \text{MAX}(c, d, e)]\end{aligned}$$

The MIN followed by MAX contributions of the first part of the equation always result in the actual median or a value smaller, whereas the MAX followed by the MIN contributions result in the actual median or a value larger. The average of the two contributions tends to cancel out the biases.

Figure 7.2-3 Nonskin Masking Algorithm

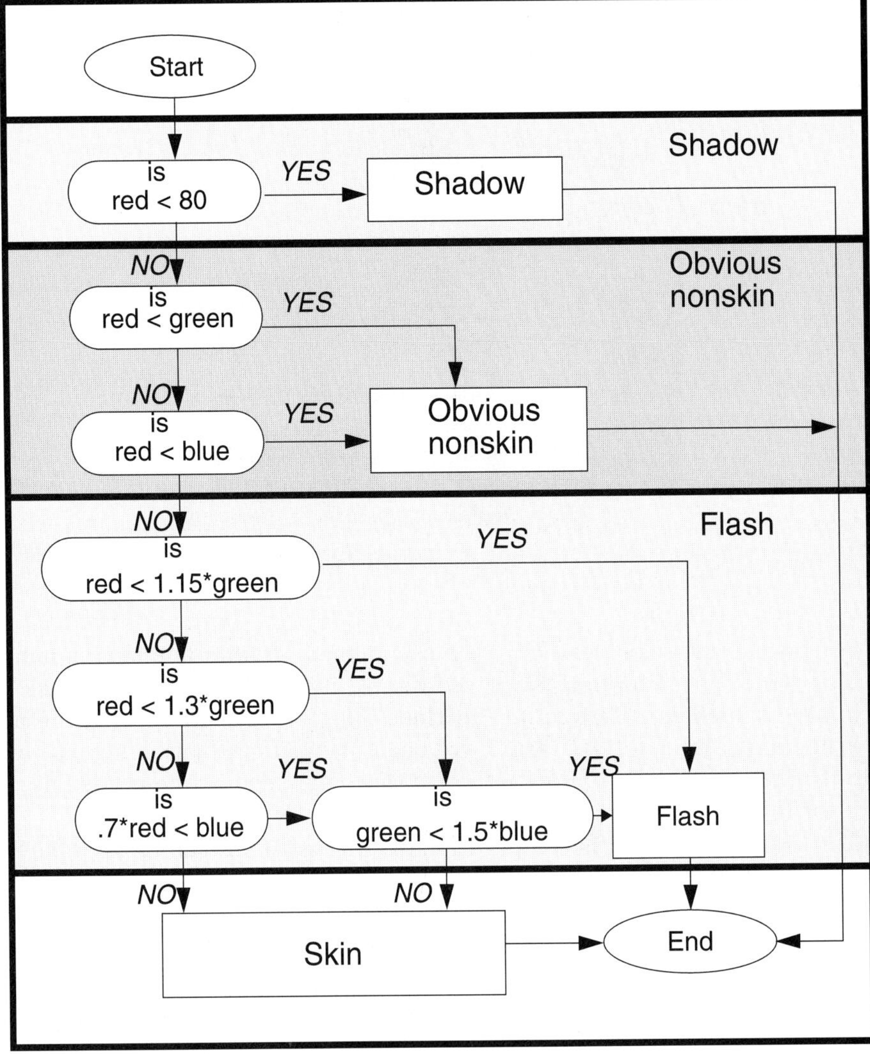

7.2.3 Color Segmentation

The first step in finding the correct border is to segment the image. The border-finding algorithm presented here uses color as the basis for segmenting the tumor images into meaningful regions. Six different color segmentation algorithms were used here: Adaptive Thresholding, Fuzzy c-Means, SCT/Center Split, PCT/Median Cut, Split and Merge, and Multiresolution (these are all described in Chapter 2).

For all methods, the number of colors for segmentation was kept constant at three, an empirically determined optimum based on error criteria for tumor images, with the exception of the SCT/Center Split segmentation method, which, by algorithmic definition, segmented the image into four distinct colors. It should be noted that if the image is segmented into too many colors this may significantly complicate the border-finding task, whereas too few colors may result in border information being lost. Thus, the idea is to find the minimal number of colors while still retaining the maximum amount of border information. Results of the six segmentation methods can be seen in Figure 7.2-5.

Figure 7.2-4 Nonskin Masking Example

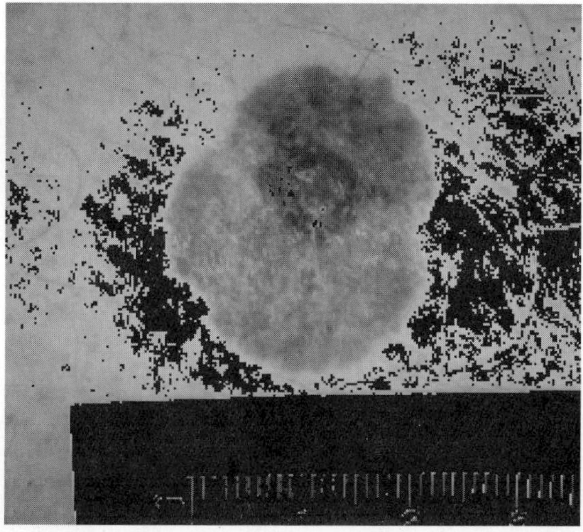

Image after nonskin masking.

Figure 7.2-5 Segmentation Methods

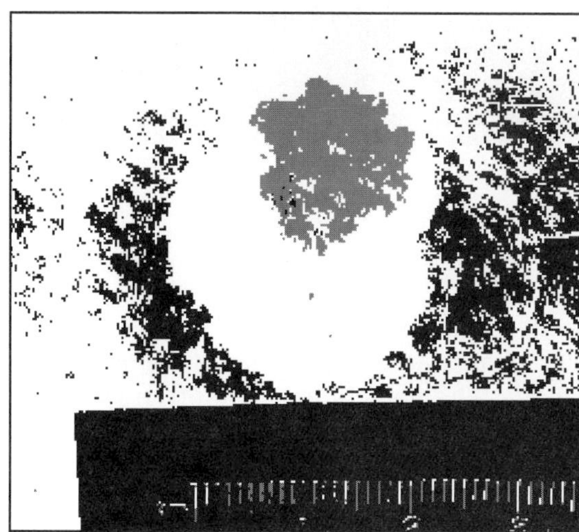

a. Adaptive Thresholding method.

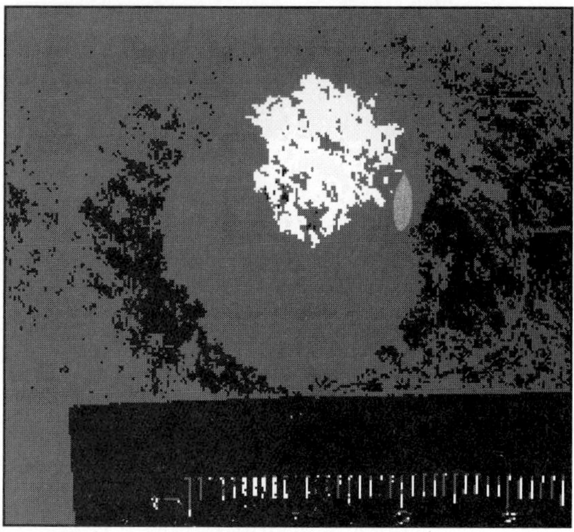

b. Fuzzy c-Means method.

Figure 7.2-5 (Continued)

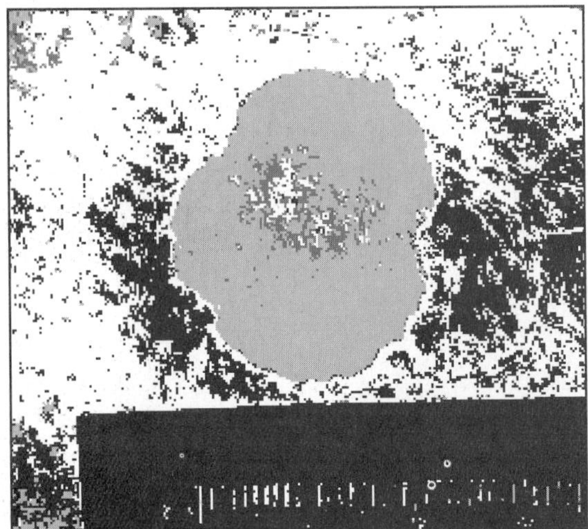

c. SCT/Center Split method.

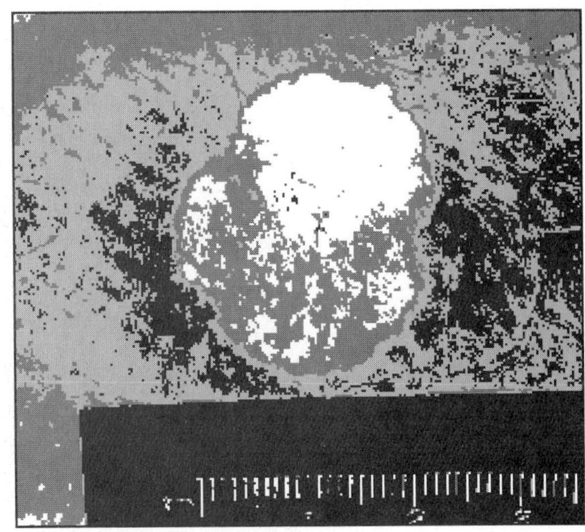

d. PCT/Median Cut method.

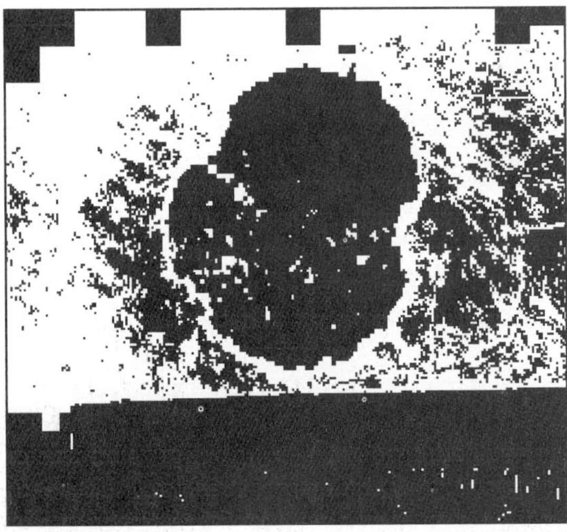

e. Split and Merge method.

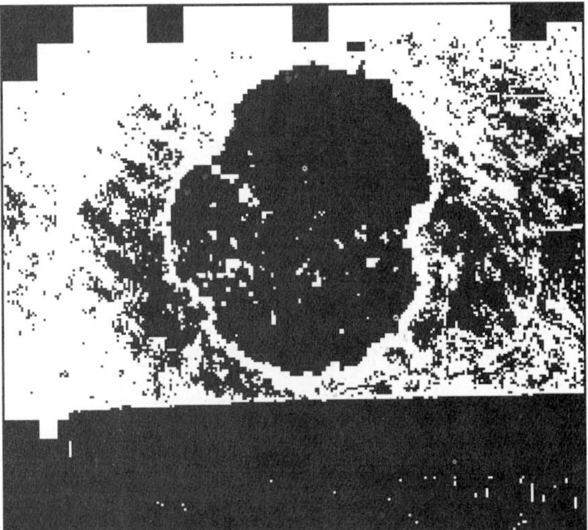

f. Multiresolution method.

7.2.4 Data Reduction

After the image has been segmented, it contains many objects of varying sizes that are represented by a small number of colors. Many of the objects are not of interest and can be eliminated from the image. The objects are also often connected together by narrow isthmuses and contain jagged protrusions. Because the next step is to label and be able to distinguish among all the objects present in the image, which is a computationally intensive process, it is advantageous to reduce the number of objects and to separate them. This is done by representing the segmented image as a gray-scale image, where each color is mapped to a different gray-scale value (this was done for the images shown in Figure 7.2-5), and applying gray-scale morphology to simplify objects.

A gray-scale morphological opening procedure was performed first to smooth the contours of the object, break narrow isthmuses, and eliminate thin protrusions and small objects. Following the opening procedure, gray-scale morphological closing was performed to fill gaps in the contour and eliminate small holes. Programmatically, the dilation and erosion gray-scale morphological operations were accomplished by using extremum operations over $N \times N$ pixel neighborhoods of the input image. According to the extremum method of gray-scale image dilation, if we consider a gray-scale image $I(r, c)$ quantized to an arbitrary number of gray levels, the dilation operation over a 3×3 pixel neighborhood is defined as

$$\text{Dilation}[I(r, c)] = \text{MAX}\ [I(r, c),\ I(r, c+1),\ I(r-1, c+1),\ \ldots,\ I(r+1, c+1)]$$

where MAX[ ] results in the largest amplitude of the nine pixels in the 3×3 neighborhood. A subimage can be used to selectively exclude some of the terms from the extremum operation during each iteration. Based on the size and shape of this subimage, it is possible to synthesize large nonsquare gray-scale structuring elements. If we consider a 3×3 subimage and define a structuring element as $B(j, k)$, then the equation for the dilation operation can be defined as

$$I(r, c) = \text{MAX}[I(r, c) + B(1, 1), I(r, c+1) + B(1, 2), I(r-1, c+1) + B(0, 2), \ldots, I(r+1, c+1) + B(2, 2)]$$

This results in the largest amplitude of the nine pixels in the 3×3 neighborhood of $I(r, c)$ excluding terms not contained within the object identified as the structuring element in the subimage $B(j, k)$.

If we apply this extremum method to gray-scale image erosion, the equation for performing image erosion over a 3×3 pixel neighborhood is defined as

$$I(r, c) = \text{MIN}[I(r, c) - B(1, 1), I(r, c+1) - B(1, 2), I(r-1, c+1) - B(0, 2), \ldots, I(r+1, c+1) - B(2, 2)]$$

This results in the smallest amplitude of the nine pixels in the 3×3 neighborhood of $I(r, c)$ excluding terms not contained within the object identified as the structuring element in the subimage $B(j, k)$.

Figure 7.2-6b shows the result of applying this morphological opening-closing procedure to Figure 7.2-6a, which is the result from application of the SCT/Center segmentation algorithm. We see that the number of objects is drastically reduced, and much of the spatial "noise" is eliminated. This spatial filtering greatly facilitates the search for a likely tumor candidate.

7.2.5 Object Localization

The process of object localization consists of four phases: 1) object labeling, 2) calculating object properties (area and circularity), 3) trimming the object list, and 4) choosing the best tumor object candidate. After the object data have been reduced and smoothed, it is necessary to label the objects and determine which is the most likely tumor object. The sequential labeling algorithm described in Chapter 2 is used to label the objects. If no objects are found, the image is represented as an empty set, and all pixel values are set to 1, which represents an image containing background only and no tumor. After the objects are labeled, the area, centroid, and eigenvalue ratio of each

Figure 7.2-6 Data Reduction

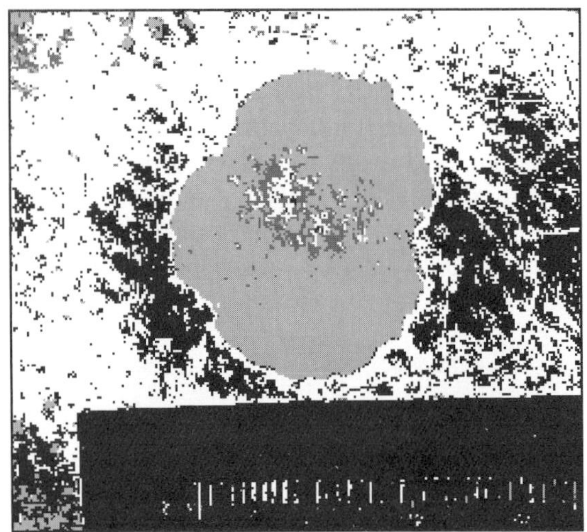

a. Result of SCT/Center Split segmentation.

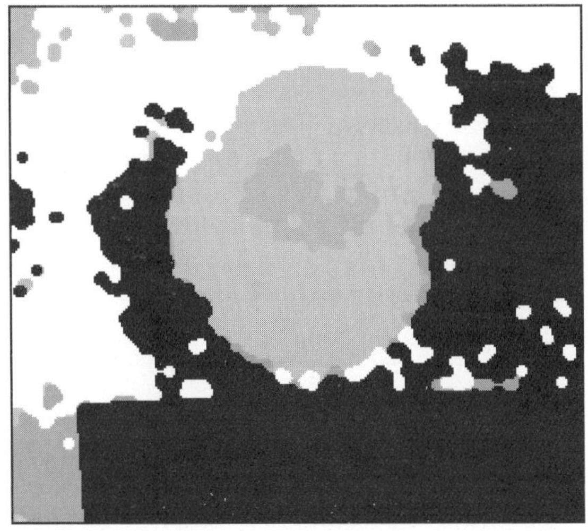

b. Result of applying morphological opening-closing to figure a.

object are found. These binary object features (see Chapter 2 and references) are then used to trim the list of objects and leave only those that are possible tumor candidates. This is based on the assumption that the tumor should exhibit some degree of circularity and also be of significant size relative to the image. The eigenvalue ratio was used to determine circularity and is defined in the references.

It was empirically determined that objects smaller than 0.5% and larger than 90% of the image size could be discarded, and objects with an eigenvalue ratio of less than 0.1 could also be eliminated from consideration. From this list of objects, a border candidate that has its centroid closest to that of a manually determined point, which is approximately the centroid of the original tumor, is selected. In the final application this will not be necessary because the tumor will be in the center of the image.

7.2.6 Contour Encoding

After the tumor object candidate has been selected, it is necessary to encode the object contour. This is accomplished by using Freeman chain coding (see references) to follow and vectorize the contour. After the border contour has been vectorized, the contour is smoothed by subsampling the vector data and using a B-spline to connect the points (shown in Figures 7.2-7a, b). A seed fill algorithm is then used to fill in the border contour, and this image (see Figure 7.2-7c) is used to calculate the border error metric. Figure 7.2-7d shows the image that results from overlaying the border on the original tumor image.

7.2.7 Experimental Results

The following error metric was used to determine the success of the border segmentation relative to the true border. The true border for each tumor was determined manually by a dermatologist.

$$\text{Border Error Metric} = \frac{\text{area}\left(I_A(r,c) \otimes I_B(r,c)\right)}{\text{area}\left(I_A(r,c)\right)}$$

where $\otimes$ is the XOR operator and area( ) is the binary area

$I_A(r, c)$ = the data set representing the actual (manual) border

$I_B(r, c)$ = the data set representing the segmented border

$I_A(r, c)$ and $I_B(r, c)$ are both binary images, 0 = tumor, 1 = background

Figure 7.2-7 Contour Encoding

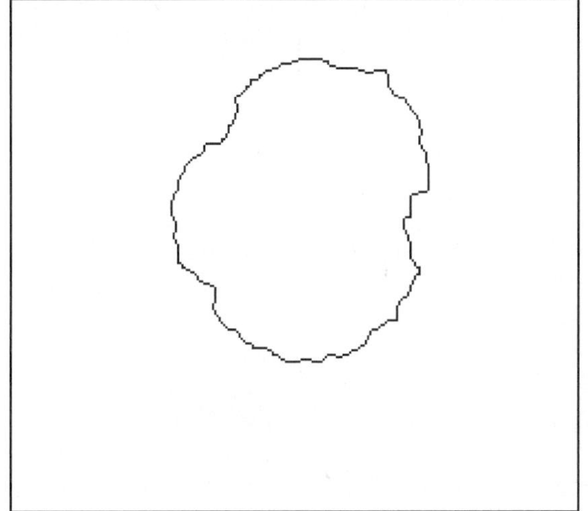

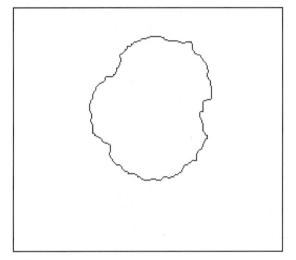

a. Freeman chain code produces a vectorized contour.

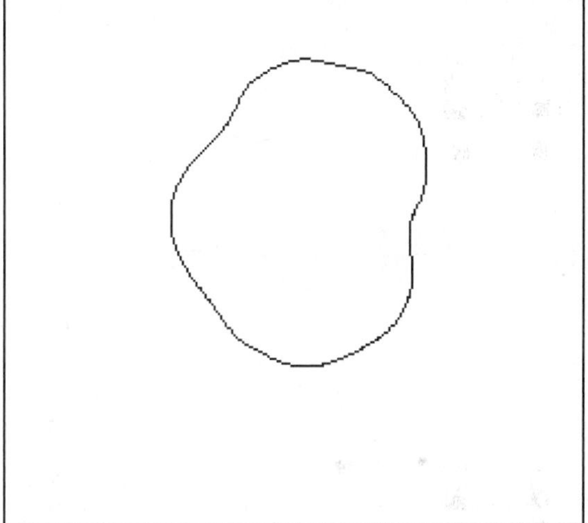

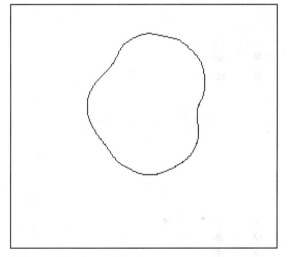

b. The contour is smoothed.

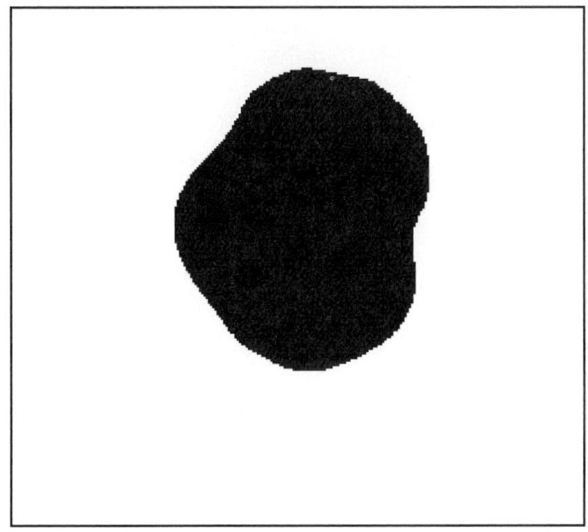

c. Seed fill is applied.

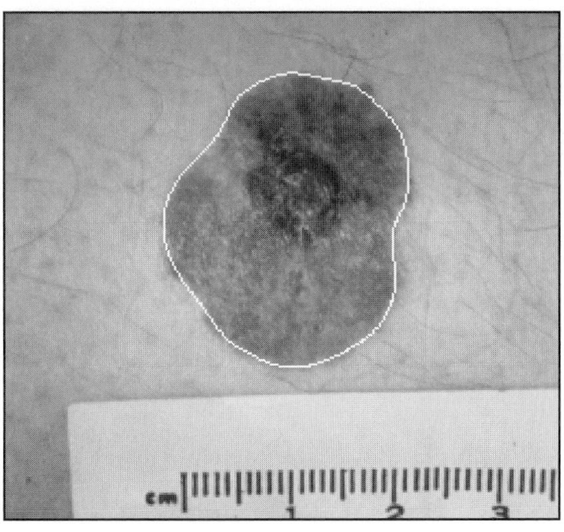

d. Resulting image.

If you use this metric, a value of 0 will result if both the manual and segmented borders are exactly the same, and a value of 1, if the segmented border is the empty set, as would occur with an algorithm that found no information. This indicates that the useful range of values will be in the range 0.0–1.0.

Figure 7.2-8 illustrates a comparison of the result of the border-finding algorithm for each of the six color segmentation methods: (a) Adaptive Thresholding, (b) SCT/Center Split, (c) Fuzzy c-Means, (d) PCT/Median Cut, (e) Multiresolution Segmentation, and (f) Split and Merge. Each image (in Figure 7.2-8) depicts the original skin tumor image with an overlay of the detected border for the respective segmentation method implemented. The PCT/Median Cut segmentation method shown in Fig-

Figure 7.2-8 Experimental Results

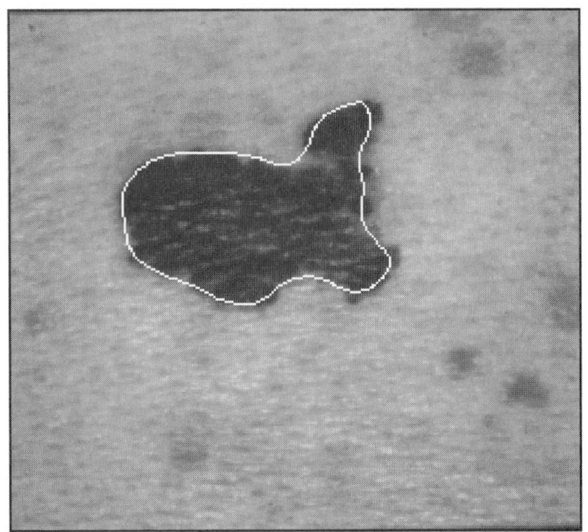

a. Adaptive Thresholding result.

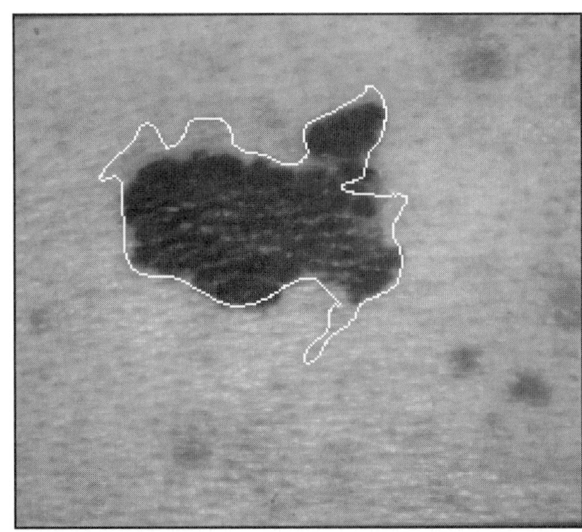

b. SCT/Center Split result.

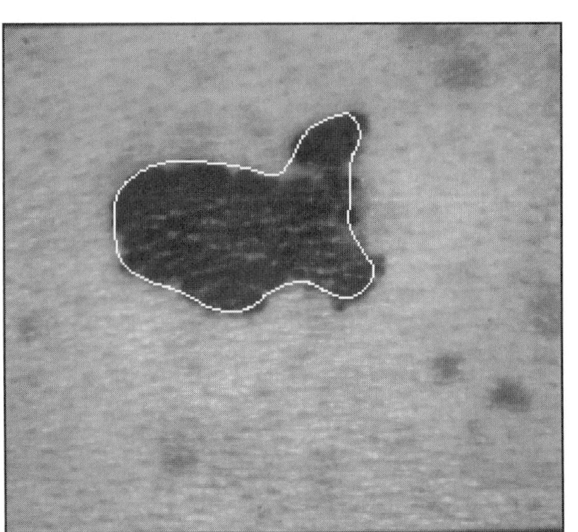

c. Fuzzy c-Means result.

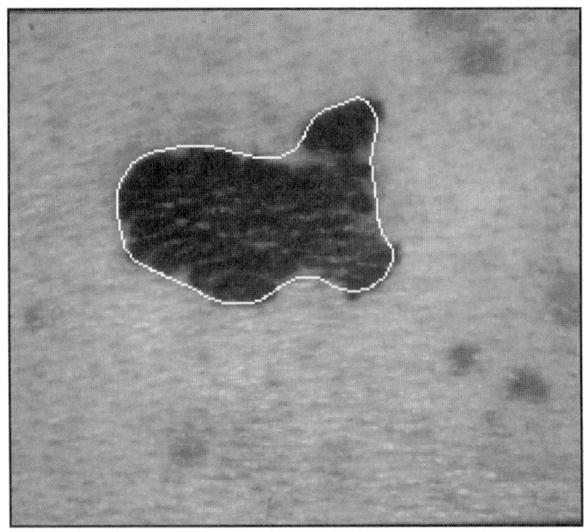

d. PCT/Median Cut result.

Figure 7.2-8 (Continued)

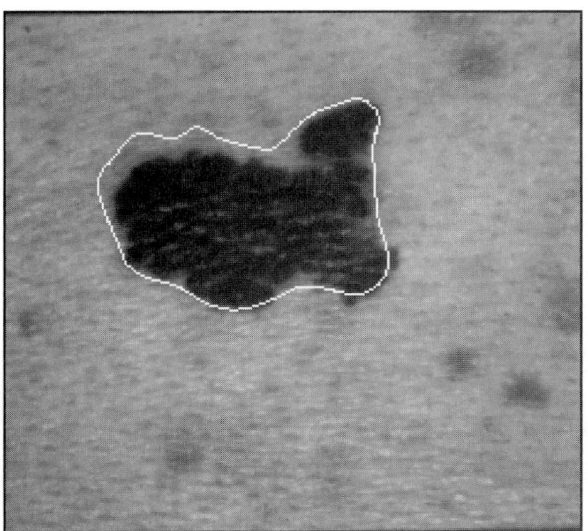

e. Multiresolution result.

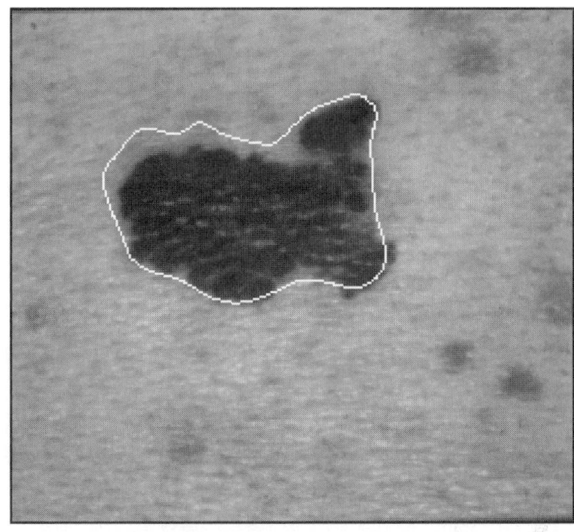

f. Split and Merge result.

ure 7.2-8d resulted in the most accurate border for this tumor with an error metric of 0.104836. The overall results show that the best color segmentation algorithm used in conjunction with the border-finding process is the PCT/Median Cut (more information, including complete statistical results for a set of 66 dermatologist-selected tumor images, can be found in the references).

After analyzing these results, we found that some of the segmentation algorithms were successful with specific tumor types, while others were more successful with other tumor types. As a result of this analysis, a method for combining the segmentation algorithms was explored. This combined method involves simply merging information from each of the six segmentation methods at the object localization stage of the border-finding algorithm. The border objects resulting from each of the segmentation methods are compared, and the object that most accurately conforms to the criteria for an ideal border candidate is chosen. Although this combined segmentation is more computationally expensive, it significantly increases the likelihood of correctly identifying the tumor border. The results indicate that even though the PCT/Median Cut algorithm shows the most promising results, the segmentation algorithms with more evenly distributed errors, Adaptive Thresholding and Fuzzy c-Means, also have potential for future research efforts.

7.2.8 References

Definitions and details for eigenvalue ratio and Freeman chain coding can be found in [Pratt 91] and [Hance 96]. For more detailed experimental results see [Hance 96] and [Hance/Umbaugh/Moss/Stoecker 96]. [Umbaugh/Moss/Stoecker 92] and [McClean 94] provided color analysis and tumor border identification information.

Hance, G. H., Umbaugh, S.E, Moss, R. H., and Stoecker, W. V., "Unsupervised Color Image Segmentation: with Application to Skin Tumor Borders," *IEEE Engineering in Medicine and Biology*, pp. 104–111, January/February 1996.

Hance, G. H., *Development of Computer Vision and Image Processing Tools with Application to Skin Tumor Border Identification*, MSEE Thesis, Electrical Engineering Department, Southern Illinois University at Edwardsville, 1996.

McLean, R., *Tumor Classification Based on Relative Color Analysis of Melanoma and Non-Melanoma Tumor Images*, MS Thesis, Department of Electrical Engineering, University of Missouri-Rolla, 1994.

Pratt, W. K., *Digital Image Processing*, New York: Wiley, 1991.

Umbaugh, S. E, Moss, R. H., and Stoecker, W. V., "An Automatic Color Segmentation Algorithm with Application to Identification of Skin Tumor Borders," *Computerized Medical Imaging and Graphics*, Vol. 16, No. 3, 1992.

7.3 Helicopter Image Enhancement and Analysis

This application was developed to assist in a research project whose goal was the development of a computer simulation of helicopter response to store separation. *Store separation* is the act of releasing an object from the underside of the aircraft. The objects released may include anything from boxes of medical or nutritional supplies to the firing of rockets. When the object ("store") is released, it causes motion in the helicopter rotors, as well as other parts of the aircraft, and it would be very useful to develop a computer simulation to model this process. Without a computer simulation of this process, the only method for testing various stores with different aircraft is to run actual test flights, which is both expensive and dangerous. The computer simulation will allow for much more extensive testing, be more cost effective, and ultimately save lives.

Our part in the project was to develop some image enhancement and analysis tools to be used as aids in the research and to design them in such a way that they could be easily incorporated into the final system. This makes CVIPtools the ideal choice for development because of its minimal requirements of standard ANSI C and its extensive libraries. Our next step was to investigate the specific problems on the project.

Many of the difficulties encountered were caused by the methods used to acquire images. Images were gathered with high-speed film cameras of store separation (object release) during actual flight tests. These flight tests were done at the Army Proving Grounds at Yuma, Arizona. To avoid excessive distortion caused by heat rising from the desert, the tests were performed near sundown; consequently, the images were very dark. Additionally, the images may be blurred as a result of lens distortion, rotor movement, high-velocity store movement, camera resolution, and atmospheric disturbance, among other things. After the high-speed film was collected, it had to be digitized before any computer analysis could be performed. The digitization process also can cause distortion from sampling and quantization; in one specific case it led to severe raster blur.

7.3.1 Ghost Image Removal

One set of images exhibited a severe raster blurring, which was a result of the two image fields being out of synchronization. This resulted in a very poor image set. Because of the high costs of running the test flights, we needed to fix the images instead of performing the flight tests again. We developed a simple method to restore the images by using what we called a raster deblurring filter. The filter was implemented by decimating the image by removing every other pixel in both the horizontal and vertical directions. Next, the image was enlarged by inserting the average of the vertical neighbors between each pixel in the vertical direction and then following the same procedure for the horizontal direction. This is a simple form of enlargement via linear interpolation. It worked quite well for this particular data set. See Figure 7.3-1 for an example.

7.3.2 Image Enhancement

Image enhancement methods were developed primarily for noise removal and to improve image contrast. They consist of spatial convolution filters, histogram methods, unsharp masking, homomorphic filtering, and notch filtering. The notch filter is specifically for sinusoidal noise removal, which often results from mechanical vibration. The spatial convolution filters are the average and median filter for noise removal and the high-frequency emphasis filter for edge enhancement. The histogram methods include equalization, slide, stretch, and a gray-level multiple. The unsharp masking algorithm is used for image deblurring. In the process of putting this application together, we learned much regarding some of the practical realities involved in getting these methods and algorithms to actually do what was desired. Additionally, this was the first application to be implemented with a GUI.

The spatial filters for noise removal worked as expected (see Chapter 3); however, the high-frequency emphasis filter did not provide the desired results. We implemented the filter described in Chapter 4 for high-frequency emphasis:

Figure 7.3-1 Ghost Image Removal

a. Image blurred by scan fields being out of synchronization.

b. Ghost image removed by raster deblur filter.

$$\begin{bmatrix} -1 & -1 & -1 \\ -1 & x & -1 \\ -1 & -1 & -1 \end{bmatrix}$$

Because we could not achieve what we wanted with this filter alone, we developed a simple algorithm that worked for this application. First, we convolved the original image with the high-frequency emphasis filter mask with a value of $x = 9$. Next, because the high-frequency emphasis filter results in an image with primarily small values (and some negative values), we linearly remapped the result to the 0 to 255 range (8 bits/pixel). To further enhance the output image, a histogram equalization was performed. The results can be seen in Figure 7.3-2, where we observe that the original image is retained, but the edges are enhanced. The edges appear enhanced because of the *outlining* effect where edges are outlined with black on one side and white on the other side. The figure also shows that the high-frequency emphasis filter will enhance noise because noise usually appears in high frequencies.

The histogram methods and the unsharp masking algorithm are described in Chapter 4, but in the process of developing this application we did encounter some interesting practical problems. The first was that the histogram equalization and histogram stretch did not always provide us with the desired results. As a result, we added a *view histogram* feature to GUI, which enables you to modify the histogram based on information about the current histogram, which provides you with useful feedback in order to customize the histogram modification. So, if a histogram equalization does not provide the desired results, the slide, stretch, shrink, and multiply options can be used experimentally to develop a sequence suitable for the current image. For example, if the histogram of the image is as shown in Figures 7.3-3a, b, with poor contrast, but a few values near both ends (0 and 255 for 8-bit data), a histo-

Figure 7.3-2 High-Frequency Emphasis

a. Original image.

b. Result of applying high-frequency emphasis to (a). Image has been histogram equalized to make it visible.

gram stretch will not help. However, a histogram slide, which is implemented to clip if slid too far, can eliminate the few gray level values at the ends, and then a stretch can be used to improve the contrast. This is illustrated in Figure 7.3-3e, f. Note that the same result can be achieved by allowing a certain number of pixels at the extremes to be excluded during a histogram stretch operation. The histogram stretch operation in CVIPtools was then expanded to include this option.

Another problem discovered during this development was that the unsharp masking algorithm (see Figure 4.3-5 for a flowchart of this algorithm) did not sharpen

Figure 7.3-3 Histogram Modification-Based Enhancement

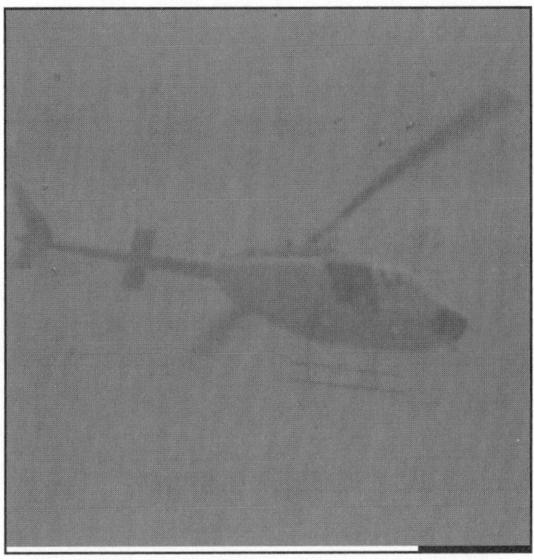

a. Poor contrast image.

b. Histogram of image (a). Note a few values at both ends, so histogram stretch will not help.

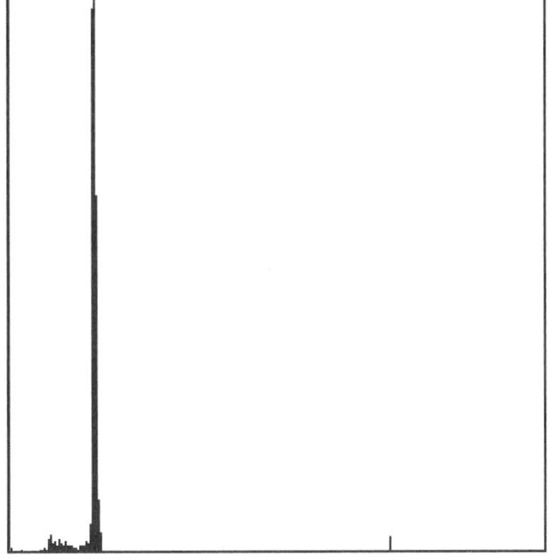

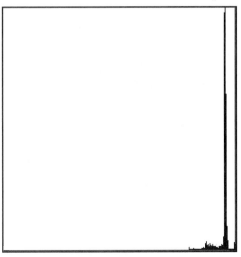

c. Histogram slide down by 75 to remove low gray levels.

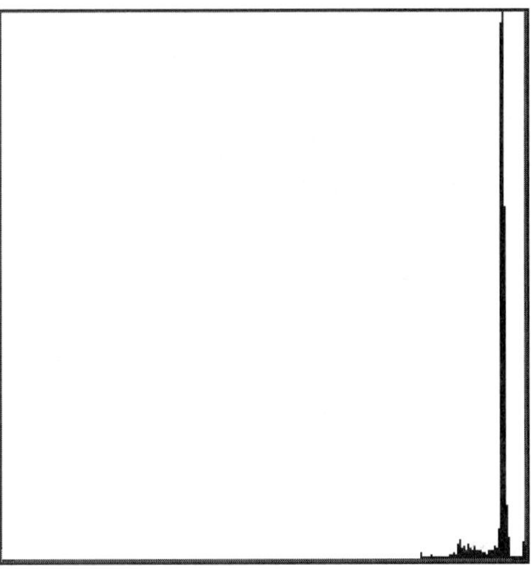

d. Histogram slide up by 205 to remove high gray levels.

Figure 7.3-3 (Continued)

e. Resultant image after histogram stretch.

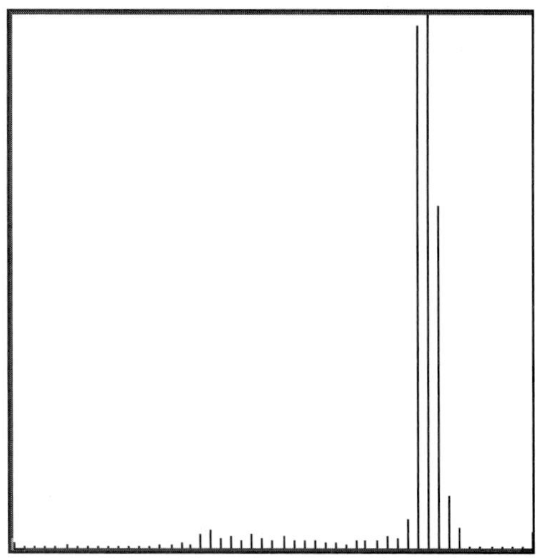

f. Histogram of image (e).

the images adequately. After much experimentation we determined that this problem occurred because the lowpass filter, as implemented with a convolution mask, left the resultant image padded with zeros. These zeros in the image caused the histogram modification operations that followed to have undesirable results. This problem was solved by rewriting the convolution function to assume DCT-type symmetry on the image edges, instead of padding with zeros.

The frequency domain filters specified for this project included the homomorphic filter and the notch filter. The homomorphic filter is described in Chapter 4 and is illustrated in Figure 4.3-4, and the notch filter is described in Chapter 3 and is illustrated in Figure 3.4-6.

7.3.3 Graphical User Interface

The main window for this application is shown in Figure 7.3-4. The title RWSI stands for Rotary Wing Store Integration, which is the official name of the DoD program under which this work was done. Here we see that the previously described tools are accessible from the button bar across the top. The *Filters* button has a drop-down menu that contains the average, median, homomorphic, and high-frequency emphasis filters. The FFT is available so that the spectrum can be obtained and viewed; this may be required for frequency domain analysis. Additionally, the main window contains a lower button bar with utilities for viewing, saving, deleting, and renaming images. Two additional utilities are included and can be accessed via the *Resize* and *Help* buttons located in the upper-right corner. The image queue is shown in the middle of this window, with the *Load image* entry box directly above it.

The resize utility allows you to enter a factor for image resizing; it can be greater than 1 for enlargement or less than 1 for shrinking. The *Help* button pops up the window shown in Figure 7.3-5a and is also accessible from such pop-up windows as the Histogram window shown in Figure 7.3-5b. The top buttons for the other tools all pop-up windows that contain entry boxes for the necessary parameters. The GUI also allows access to the two demonstrations that were developed to illustrate the possibility for future research and developments in this area.

Figure 7.3-4 Helicopter Application GUI

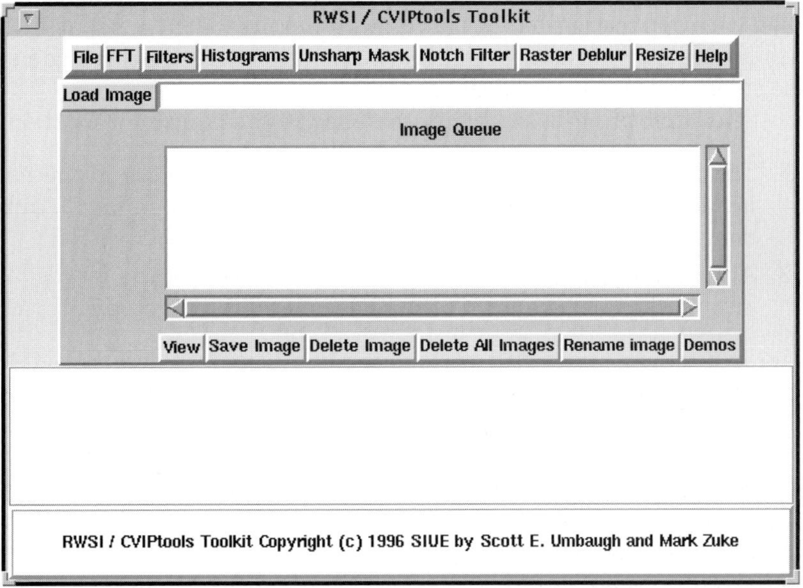

Figure 7.3-5 RWSI Windows

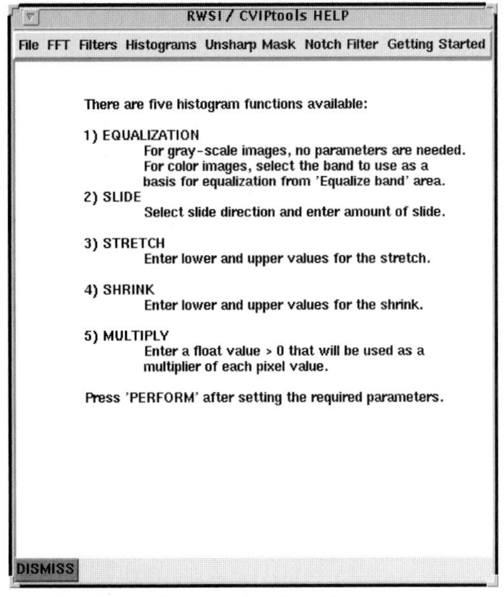

a. RWSI Help window, showing histogram help.

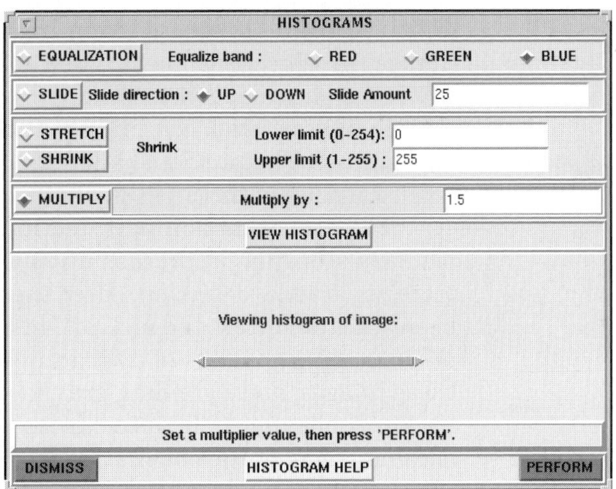

b. RWSI Histogram top-level window.

7.3.4 Demonstrations

These two demonstrations, accessible via the *Demos* button in the lower-right corner of the main window, include one for helicopter locating and one for store separation detection. The motivation for developing these demonstrations is to more fully automate the screening and gathering of flight test image data for use in the simulation development system. The helicopter locating uses a segmentation technique combined with application-based heuristics, whereas the store separation detection works with a sequence of images and identifies the image in which the store is separated (the object is dropped) from the aircraft.

The application-based heuristics used in the helicopter location were as follows:

1. The helicopters are basically dark and difficult to see in the images. This is true because military aircraft are designed to be camouflaged, and the filming was done near dusk.
2. After the image has been segmented into objects, the largest object is the background. This is true for all the images used in this study.

After experimentation with various segmentation, we determined that the following algorithm worked to solve this imaging problem:

1. Find the gray-level average and minimum and maximum gray levels.
2. Slide the histogram up by half the maximum slide allowed that will not cause clipping.
3. Segment via histogram thresholding.
4. Use the mode in the histogram that contains the largest number of pixels; assume that this is the background. Turn all these pixels to white (255).
5. Find the average of the resulting image and use it to reverse threshold the image (values above the threshold are set to black; those below it are set to white).
6. The helicopter object is now the largest white object in the image.

Two examples of this are shown in Figure 7.3-6.

The store separation detection was developed to work with a sequence of images, digitized frame by frame from high-speed film, and to identify the specific frame (image) in which the object is released from the aircraft. It is based on the idea that when the store separates from the aircraft, this location in the image experiences a change in gray level. Specifically, the location will become the same gray level as the background. The previously defined helicopter locator is used to determine the gray-level statistics of the helicopter.

The algorithm is illustrated in the flowchart in Figure 7.3-7. Here we see three primary steps: 1) preprocessing, 2) object identification, and 3) object following and store separation detection. After these three steps, the object is tested to see if it has been separated from the aircraft. If so, the process terminates; if not, it continues. The *preprocessing* consists of 1) reading two sequential images, 2) histogram equalizing these images, and 3) finding gray-level statistics for the image, the helicopter, and the background (non-helicopter areas in the image). Histogram equalization is done because many of the original images are very low contrast, and we want to maximize the amount of "information" available. The gray-level statistics are calculated to be used in the thresholding and in the separation detection.

Figure 7.3-6 Helicopter Locating

a. First original image.

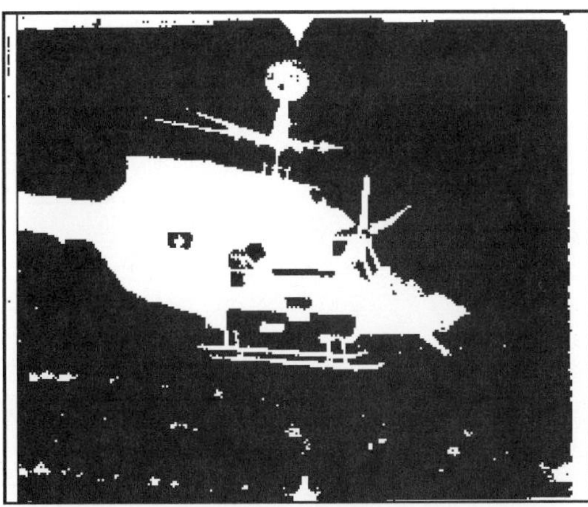

b. Result of helicopter locating algorithm with first image.

c. Second original image.

d. Result of helicopter locating algorithm with second image.

Object identification is accomplished by 1) subtracting the two histogram-equalized images and performing 2) a histogram thresholding segmentation on the result, 3) spatial quantization to reduce size, 4) binary thresholding, 5) object labeling, and 6) object detection. The subtraction is done to remove most of the background information in the image because it is not changing much from frame to frame. The image is segmented to further reduce the amount of information in the image and is then spatially reduced by a factor of 16, by using a 4×4 block average. The resultant image is then thresholded to generate a binary image in which the objects are labeled and the "store" object is identified using application-based heuristics.

The entire process continues for each pair of sequential images. After each iteration the object following and store separation algorithms are performed. Here, the

Figure 7.3-7 Store Separation Detection Flowchart

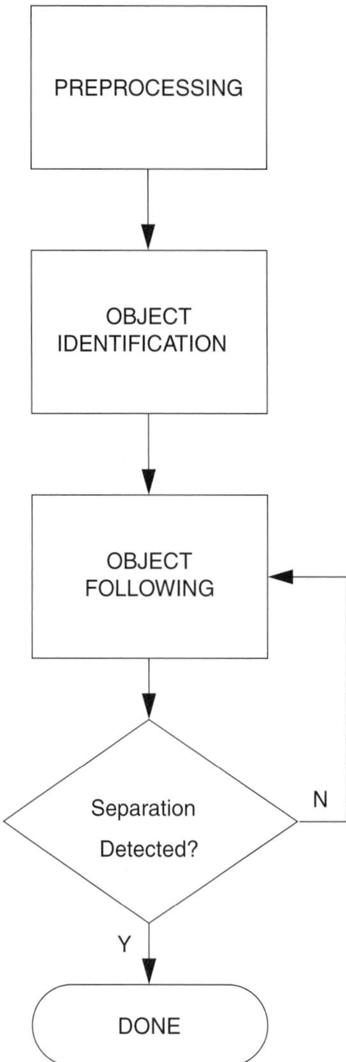

object that has been identified as the store object is used in conjunction with the histogram-equalized image to determine the gray-level statistics for the object. We then get information about the surrounding area. When we observe the background between the helicopter and the store, we determine that separation has occurred. In Figure 7.3-8 this process is illustrated.

7.3.5 References

[Johnson/Umbaugh 93] describes a method to design a spatial filter for image enhancement of these flight images based on sequence domain characteristics of the

Figure 7.3-8 Store Separation Detection

a. First image in a sequence of ten.

Suppressing noise...
Find separation at frame No#4
Reading NO#5 image....
Subtracting two images...
found 3 modes
lower = 1
peak = 3
upper = 99
lower = 100
peak = 118
upper = 125
lower = 126
peak = 130
upper = 255
Suppressing noise...
Find separation at frame No#5
Reading NO#6 image

b. Portion of screen display during application of the algorithm.

c. Enhanced portion of image determined to be last image before separation.

d. Enhanced portion of image determined to be first image after separation. Note the appearance of background between store and helicopter body.

image. The helicopter locating and store separation algorithms are described in more detail in [Luo/Razack/Smith/Taylor/Zuke 95].

Johnson, M. L., Umbaugh, S. E, *Store Separation Flight Test Image Enhancement Tools*, presented at the Joint Aircraft-Stores Compatibility Subgroup of the Joint Ordinance Commanders Group Aircraft-Stores Compatibility Seminar, September 29, 1993, Fort Walton Beach, FL.

Luo, K., Razack, A., Smith, F., Taylor, F., and Zuke, M., *Advanced Image Processing Techniques Applied to Helicopter Images for Helicopter Identification, Store Separation Detection, and Image Deblurring*, EE538: Topics in Image Processing project report, April 27, 1995, Southern Illinois University at Edwardsville.

7.4 Wavelet/Vector Quantization Compression

In recent years, the rapid growth in the field of diagnostic imaging has produced several new classes of digital images, resulting from computerized tomography, magnetic resonance imaging, and other imaging modalities. In addition, well-established imaging modalities such as X-rays and ultrasound will increasingly be processed and stored in a digital format in the future. It has been estimated that a 600-bed hospital would need almost 2 thousand gigabytes, or 2 million megabytes, of storage per year if all images produced at the hospital were to be stored in a digital format. The need for high-performance compression algorithms to reduce storage and transmission costs is evident. The compression techniques described herein are likely to be effective for a far wider range of images than the skin tumor images employed in this research.

A relatively large database consisting of more than 1,000 digital skin tumor images has been established for research and development for a computer vision system to be used for skin tumor evaluation. Each original image has a spatial resolution of 512×512 pixels, 24-bit magnitude (8 bits each for red, green, and blue), and occupies 786K bytes of storage space. Although the original images are stored in uncompressed form, the need for compression of the original images arises when multiple copies are maintained at different sites. In addition, compression reduces the requirements of time, cost, and bandwidth if the images are to be transmitted from a field site or a physician's office to a hospital for evaluation.

7.4.1 Experimental Procedures

The wavelet transform algorithms developed for this research use the Daubechies 4 element basis vectors described in Chapter 2. Bit allocation for the vector quantization was based largely on heuristics and on trial and error, although the optimal bit allocation rule presented in the references was used as a general guideline.

Compressed images were evaluated by three image processing students who were asked to rate their quality in terms of a standard grading scale that is used to evaluate the quality of analog television signals. This scale grades picture impairment on a scale from 1 to 6, where 1 = imperceptible impairment, 2 = just perceptible impairment, 3 = perceptible but not disturbing impairment, 4 = somewhat objectionable impairment, 5 = definitely objectionable impairment, and 6 = extremely objectionable impairment. Wavelet-based vector quantization (WVQ) algorithms were evaluated along with two standard compression algorithms. To eliminate bias, the evaluations were performed blindly; that is, observers were not informed as to which compression algorithm had been applied to an image. The evaluations took place on a 20-inch color monitor. Observers were shown the original images for reference and were allowed to examine the images from any desired distance.

In addition, two dermatologists were consulted for evaluation of the compressed images in terms of their suitability for clinical diagnosis. An image processing professor who has done much research in this area was also consulted. These three comprised our expert evaluators, and two of these experts participated in experiments that used the same grading scale for evaluation as did the image processing students.

7.4.2 Design of Compression Algorithms

Image compression using a combination of the wavelet transform and vector quantization entails an almost infinite number of possible compression algorithms. Numerous variables are involved in the design of a compression algorithm. These variables include the number of subbands in the wavelet, the vector length (corresponding to image blocksize), the size of the codebook (related to bits per pixel), and various preprocessing steps.

The primary preprocessing step used involves application of the principal components transform (PCT) in the color space. A medical color image contains redundancies among its color bands which may be removed by applying the PCT to the image. The PCT decorrelates the color bands of the image and compacts most of the information (variance) into the first color band. As a result, careful coding of the first PCT color band combined with less precise coding of the second and third PCT bands should yield higher fidelity at greater compression rates than separate encoding of each color band without the PCT. It is also possible to perform the wavelet transform first, then apply the PCT, and finally code the result using vector quantization. In both cases, applying the PCT allows for precise coding of the primary band, whereas higher compression ratios may be obtained in the secondary and tertiary bands.

The compression algorithms developed during this research may be divided into three types, based on the various methods of preprocessing with the PCT. The first type consists of compression algorithms that employ no preprocessing. The absence of preprocessing greatly simplifies the design of the rest of the compression algorithm because the same bit-allocation procedure is applied to each color band of the wavelet-transformed image. However, the lack of preprocessing does not result in significant computational savings, and it renders the compression algorithm suboptimal.

The second type of compression algorithm preprocesses the color image by the PCT and then performs spatial quantization (size reduction) in PCT bands two and three. The design of the rest of the compression algorithm remains relatively simple because it is normally possible to use the same compression algorithms in each color band. During decompression, the spatially reduced bands are reconstructed using simple first-order interpolation or a more sophisticated filter. One modification to this type of compression algorithm is to apply the PCT for preprocessing but, instead of spatially quantizing the two least significant bands, to code them at very low bit rates.

In the third type of compression algorithm, the wavelet bands are divided into groups of three bands of equal size, illustrated in Figure 7.4-1. The PCT is then performed on these groups of wavelet bands. Because these bands are highly correlated (they look similar), as is shown in Figure 7.4-2, application of the PCT will help to pack the information into a smaller number of bytes. In addition, the PCT may be applied in the RGB color space as in the previous methods. Preliminary research indicates that compression algorithms developed with these ideas promise the best results at a given compression rate, but they are also the most complicated and computationally intensive. Because we also were interested in relatively fast algorithms for this

Figure 7.4-1 Division of an Image

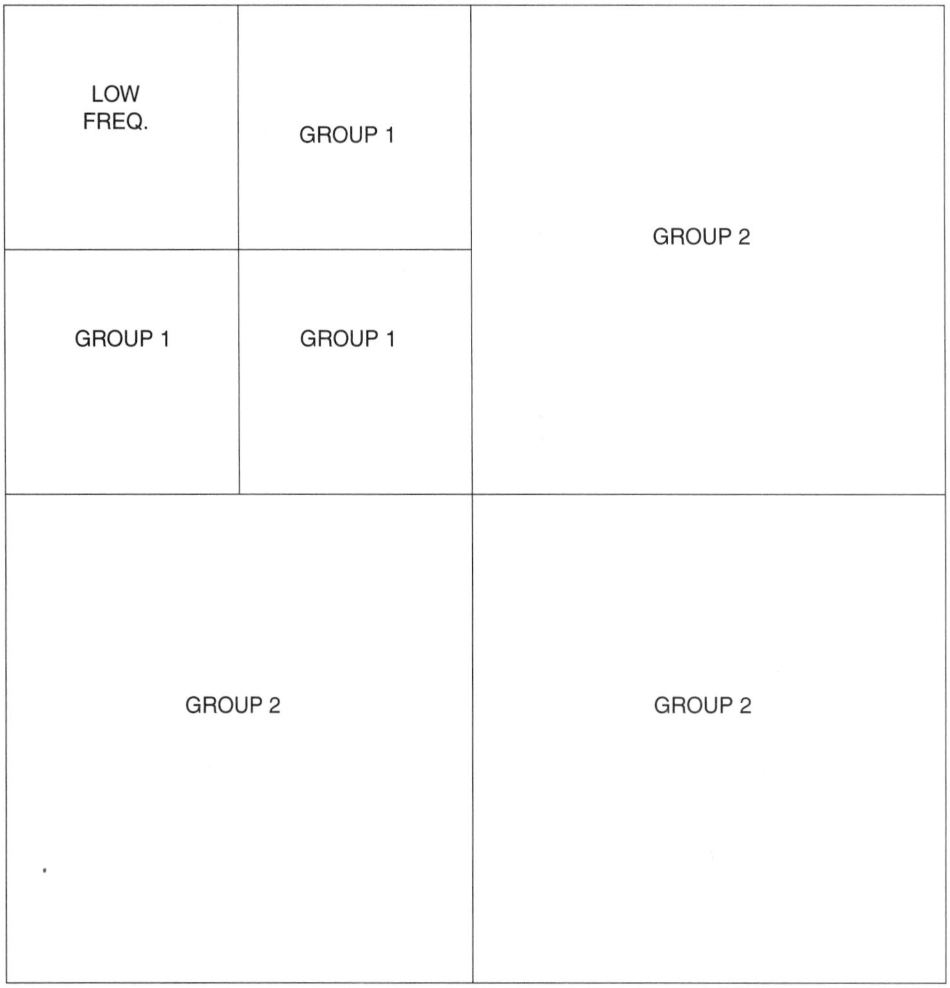

LOW FREQ.	GROUP 1	GROUP 2
GROUP 1	GROUP 1	
GROUP 2		GROUP 2

Division of a wavelet-transformed image into groups for the purpose of applying the Principal Components Transform to each group.

application, the following experimentation was done with the first two types of compression algorithms.

Experiments were performed on 66 images with three of the compression algorithms judged in preliminary evaluations to have the best potential, and their performance was evaluated relative to each other and relative to the CCC (Color Cube Compression, see references) and JPEG compression algorithms.

7.4.3 Vector Quantization

To apply vector quantization, a codebook must be defined to store the coded vectors (see Chapter 5). The standard algorithm for codebook generation is referred to as the Linde, Buzo, Gray (LBG) algorithm, which is defined as follows:

Figure 7.4-2 Correlated Wavelet Bands

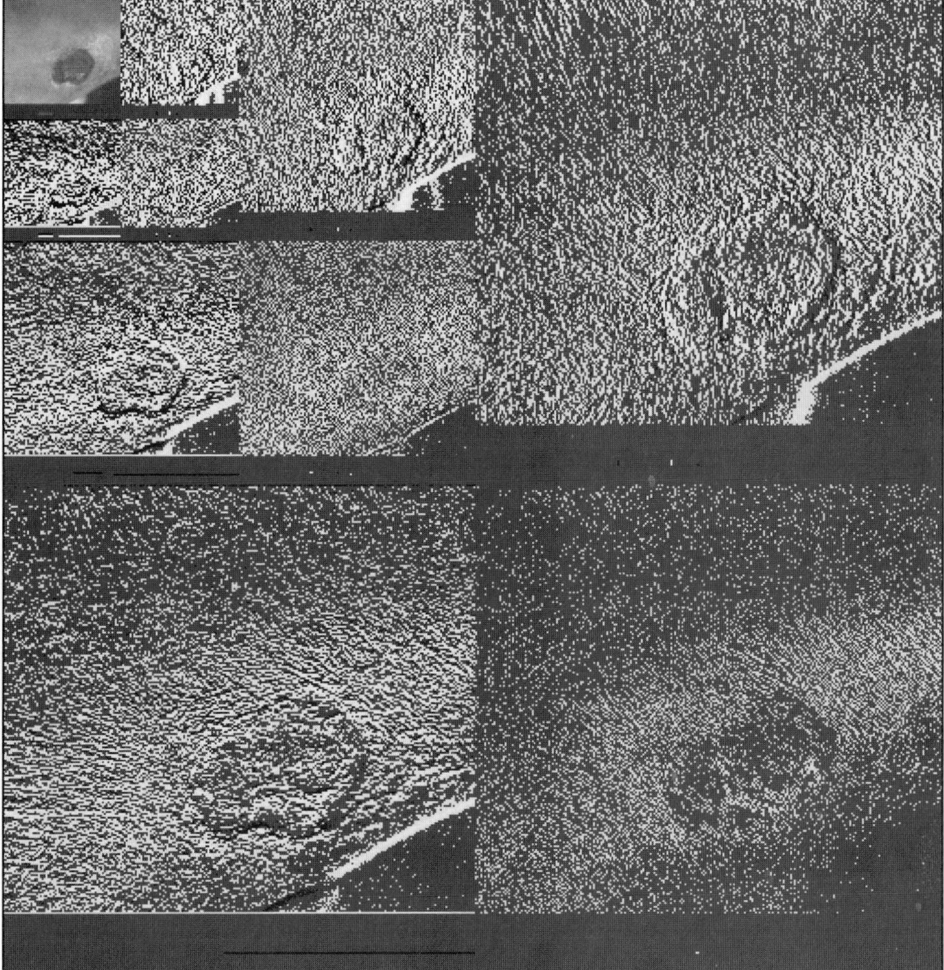

A ten-band wavelet transform of the image 441n.ppm. Note that equal-sized wavelet bands are highly correlated (they look similar). The image has been histogram-equalized to show detail.

1. Given an arbitrary codebook, encode each input vector according to the nearest-neighbor criterion. Use a distance metric to compare all the input vectors to the encoded vectors, and then sum these errors (distances) to provide a distortion measure. If the distortion is small enough (less than a predefined threshold), quit. If not, go to step 2.
2. For each codebook entry, compute the euclidean centroid of all the input vectors encoded into that specific codebook vector.
3. Use the computed centroids as the new codebook, and go to step 1.

The LBG algorithm and other iterative codebook design algorithms do not, in general, yield optimum codes. Subject to certain conditions, the LBG algorithm will yield "good" codes.

The overall compression ratio will be determined by the allocation of bits among the frequency bands of the wavelet-transformed image. To implement a specific bit rate (compression ratio), we must determine a suitable combination of codebook size and blocksize for each frequency band. For example, a 0.75 bit/pixel compression ratio may be implemented by using a 2×2 block size and 8 codebook vectors ($8 = 2^3$, so 3 bits for every 4 pixels), or it may implemented with a 4×4 block size and 4,096 codebook vectors (12 bits for every 16 pixels). The choice between these two alternatives will depend heavily upon whether it is desired to develop a universal codebook to be used by many images or whether a separate codebook will be trained and transmitted with each image.

Two main methods exist for applying vector quantization to a real image compression problem. The classical LBG method consists of developing a universal codebook that remains constant for all images to be compressed. A copy of the codebook is maintained at the transmitting and receiving site, and the size of the codebook is not counted when compression ratios are computed. This method allows for extremely fast compression after the universal codebook has been created. The second method consists of using the LBG algorithm to generate a codebook for each image. This, of course, requires storage and transmission of the codebook indices as well as the codebook, and it becomes imperative that the codebook is kept as small as possible. This method also produces higher-quality images than the universal codebook method because each image is compressed with a codebook developed specifically for that image.

All compression algorithms described here use this second method of vector quantization because it has been observed that significant degradation in image quality results when universal codebooks are used. Additionally, we developed algorithms with codebooks that account for only 10–20% of the total space needed to store a compressed image, so the improvements in image quality were worth the added cost in storage requirements. Designing the compression algorithms by minimizing the size of the codebooks has the added advantage of reducing compression times. Compression times on a low-end Sun workstation (Sparc 2) were reduced from 30–45 minutes to 2–5 minutes when designing for minimum codebook sizes.

7.4.4 Compression Algorithms

Many compression algorithms were developed during preliminary experimentation, but three were selected for more thorough testing. We believe these three are representative of the algorithms developed, as well as meet the speed requirements for this application. These algorithms all utilize a ten-band wavelet decomposition, with the Daubechies four-element basis vectors, in combination with vector quantization. They are called wavelet-based/vector quantization followed by a number (WVQ#); specifically WVQ2, WVQ3, and WVQ4. In addition, one algorithm (WVQ4) employs the PCT for preprocessing, followed by subsampling the second and third PCT bands by a factor of 2:1 in the horizontal and vertical directions. Table 7.4-1 contains the details of the parameters for each of these algorithms; the wavelet frequency band numbers used in this table are shown in Figure 7.4-3.

Table 7.1 lists the wavelet band numbers versus the three WVQ algorithms. For each WVQ algorithm, we have a blocksize, which corresponds to the vector size, and the number of bits, which, for vector quantization, corresponds to the codebook size.

Figure 7.4-3 Ten-Band Wavelet Band Numbering

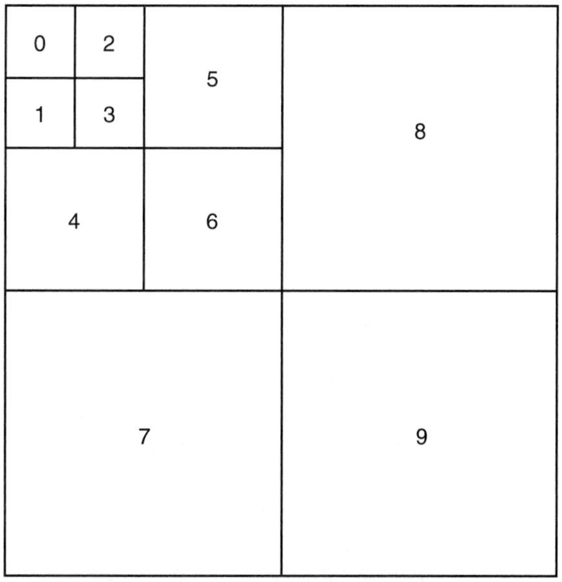

0	2	5	8
1	3		
4		6	
7		9	

These numbers correspond to band numbers used in Table 7.4-1.

Table 7.4-1 WVQ Compression Algorithm Parameters

Band Number	WVQ2		WVQ3		WVQ4 (PCT)	
	blocksize	**# of bits**	**blocksize**	**# of bits**	**blocksize**	**# of bits**
0	(scalar)	8	(scalar)	8	(scalar)	8
1, 2	2x2	8	2x2	8	2x2	8
3	2x2	6	2x2	6	2x2	6
4, 5	2x2	5	2x2	5	2x2	5
6	2x2	5	2x4	4	2x4	5
7, 8	2x4	4	X	0	2x4	5
9	X	0	X	0	X	0

The lowest wavelet band is coded linearly using 8-bit scalar quantization. Vector quantization is used for bands 1–8, where the number of bits/vector defines the size of the codebook. For example, if we use 8 bits/vector for the codebook, this corresponds to a codebook of 256 (2^8) entries. The highest band is completely eliminated (0 bits are used to code them) in WVQ2 and WVQ4, whereas the highest three bands are eliminated in WVQ3. For WVQ2 and WVQ3, each of the red, green, and blue color planes are individually encoded using the parameters in the table. For WVQ4, however, the PCT is used to preprocess the image, and the parameters in Table 7.4-1 are used only for the PCT color band. The other two color bands, which contain much less information, are first subsampled and then encoded with a small number of bits.

Table 7.4-2 contains the speed and compression ratio of these compression algorithms, and the other two compression methods used here for the purpose of evaluating the wavelet-based vector quantization algorithms: JPEG and CCC. Note that the compression ratio of JPEG varies, while the other compression methods have constant ratios. These numbers were generated by running on a Sun Sparc 2 workstation.

Although the WVQ algorithms had significantly longer compression times, the primary factor for this application is the decompression time, and the WVQ methods compared favorably to the other methods with this metric.

The advantages of preprocessing an image using the PCT depend highly upon the cross-correlation among the color bands of the image. Images with high correlation coefficients will yield better results than images with low correlation coefficients when processed with the PCT. The advantage of applying the PCT to an image is thus dependent upon the statistics of the image. We found that, in some cases, the distortion introduced by subsampling of the PCT-transformed image was unacceptable. Because of this, we believe that a future, automated compression tool that uses the PCT to decorrelate its color bands should be artificial intelligence (AI) based so that preprocessing using the PCT takes place only when the image statistics indicate that such preprocessing would be beneficial. Also note that the errors introduced by subsampling or coarsely quantizing bands two and three of a PCT-transformed skin tumor image affect the chrominance components of the image. That is, even though the decompressed image may look much like the original in terms of texture and details, its color will tend to be different. Color is an important feature used by experts to diagnose skin tumors, and changes in the chrominance values of an image may adversely affect attempts at manual and automated diagnosis of the tumor image.

Table 7.4-2 Compression Ratio and Speed

Algorithm	Compression Ratio	Compression Time (seconds)	Decompression Time (seconds)
WVQ2	10.16:1	220	11
WVQ3	15.64:1	150	11
WVQ4	21.76:1	140	15
CCC	11.78:1	14	10
JPEG	17.6-36.5:1	6	3

7.4.5 Quality Assessment of the Compressed Images

The subjective quality of the compressed images as evaluated by three lay observers is shown in Figure 7.4-4. The quality grading ranges from 1 (imperceptible degradation) to 6 (extremely objectionable degradation) for each compression algorithm and represents the average quality grade assigned to all 66 images compressed with the algorithm. The range shown was obtained by plotting the equivalent of one standard deviation above and below the average quality grade of each compression algorithm.

Figure 7.4-4 Subjective Quality Ratings

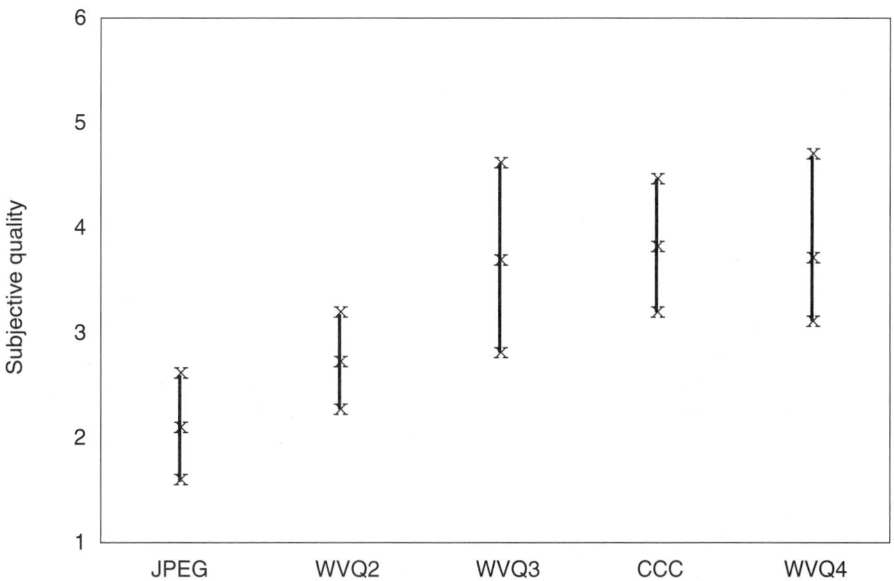

Subjective quality of compressed images, evaluated by lay observers

As Figure 7.4-4 indicates, two of the three WVQ compression algorithms were rated equal to or better than the CCC compression algorithm by the three observers. However, experts expressed concern about some of the compression artifacts resulting from these two algorithms, and in particular from algorithm WVQ3. To the experts, the color quantization and blocking artifacts resulting from the CCC were preferable to the general blurring and loss of detail resulting from the WVQ algorithms, whereas to the layperson, the quantization artifacts resulting from the CCC were more objectionable than the smoothed look of the WVQ algorithms.

JPEG was rated by the lay observers as the best compression method. However, the algorithm WVQ2 was rated equal to or better than JPEG for 19 of the 66 images. For 22 of the remaining images, a majority of the observers stated that the change in hue resulting from the WVQ2 algorithm was the primary reason they considered the JPEG images to be of better quality. The WVQ4 algorithm was consistently rated the worst performer. In some cases, this algorithm resulted in color changes as well as noticeable changes in texture stemming from the subsampling and reconstruction from use of the PCT.

The compressed images were also evaluated by two experts. These were a dermatologist and an image processing professor. The result of their evaluations is depicted in Figure 7.4-5. It is believed that the expert evaluations measure the relative quality of the compression algorithms with a fair degree of accuracy. The experts consistently rated JPEG as the best compression method, followed by the CCC. The WVQ4 algorithm produced several images considered definitely objectionable and extremely objectionable by the experts, whereas the WVQ2 and WVQ3 algorithms produced images of reasonable quality, although of lesser quality than those produced by JPEG and CCC. As Figure 7.4-5 shows, the two experts agreed with one another on the over-

Figure 7.4-5 Subjective Quality Ratings

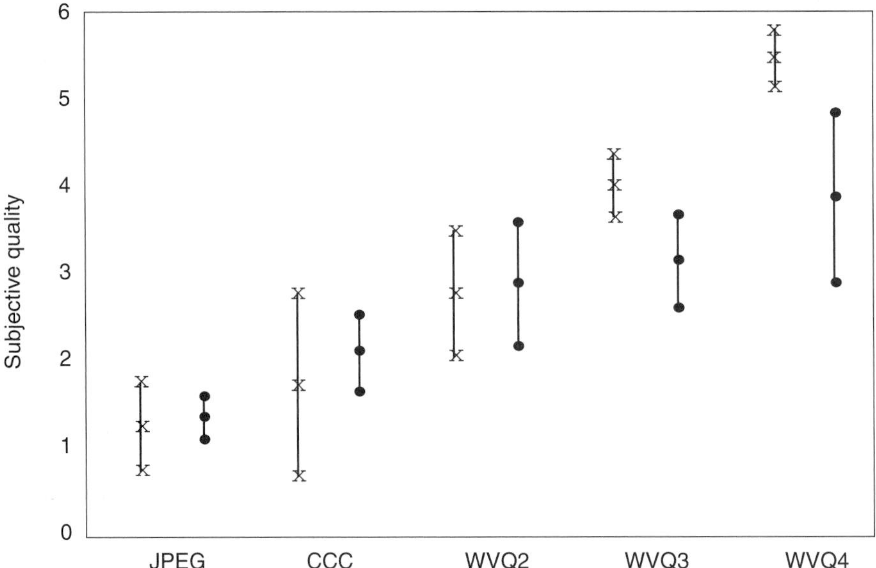

Subjective quality range of compressed images as evaluated by experts. The dermatologist's results are indicated by an x, while the image processing professor's results are denoted by circles.

all assessment of the images, although the dermatologist tended to penalize color changes more heavily than the image processing professor.

The main difference between the experts' assessments of the images and the lay observers' assessments was that the color change resulting from the WVQ algorithms was more objectionable to the experts. This is understandable, considering the importance of color in the diagnosis of skin lesions.

To illustrate the degree of quality loss resulting from the various compression algorithms, monochrome versions of the original and decompressed images of two of the tumors are shown in Figures 7.4-6 and 7.4-7. Although the color changes that sometimes occurred with WVQ4 cannot be observed in these figures, any blurring or textural changes can be observed, and the general quality of the compression algorithms can be compared (see the color images on the CD-ROM).

7.4.6 Conclusions

Two of the WVQ algorithms perform quite well in terms of subjective image quality. The quality ratings of the WVQ algorithms by the human observers indicate that wavelet-based vector quantization methods for color images warrant further investigation. The reservations expressed by an expert about the clinical suitability of the WVQ algorithms point toward possible further improvements in these algorithms: The use of the shorter Haar wavelet filters would serve to decrease the blurring effects that occur when a four-tap filter is used for the wavelet transform. In addition, coding the high-frequency bands with more precision should decrease the averaging effect, although more bits would be needed to code the images.

Coding the high-frequency bands with more precision should also decrease the distortion in hue in some of the WVQ-coded images. It may even be possible to correct the distortion in hue by storing a *correction factor* with the coded image to indicate the original hue of the image. It would then be possible to adjust the hue after decompression. This would work, however, only if the distortion in hue is uniform throughout the entire image.

As seen, one disadvantage with vector quantization methods is the severe computational requirements of the compression stage. Even the fastest WVQ algorithm implemented is ten times slower than JPEG or the CCC. However, the ever-increasing

Figure 7.4-6 Tumor 327 Comparisons

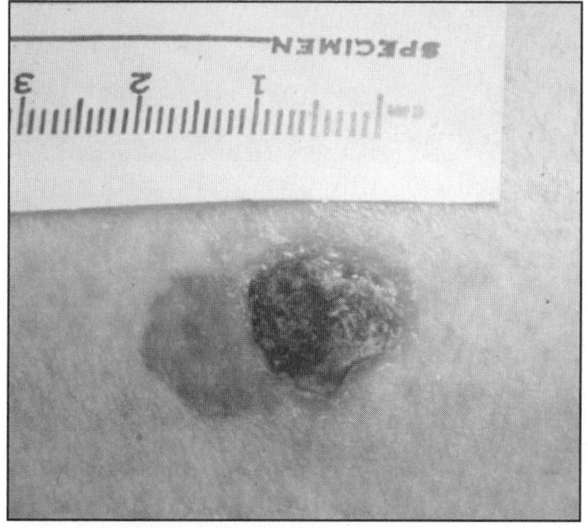

a. Original tumor 327n.ras.

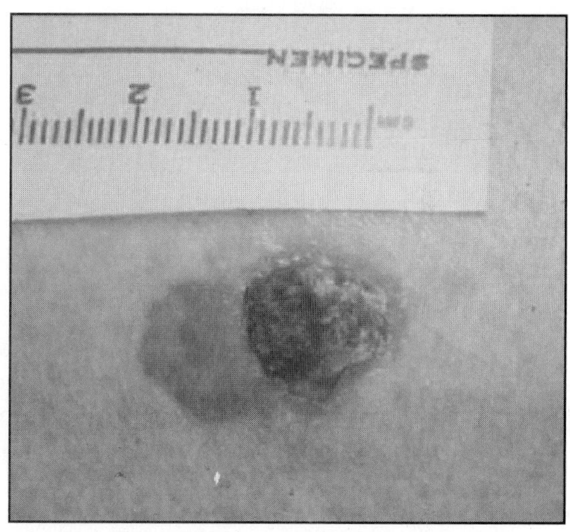

b. 327n.ras compressed using WVQ2.

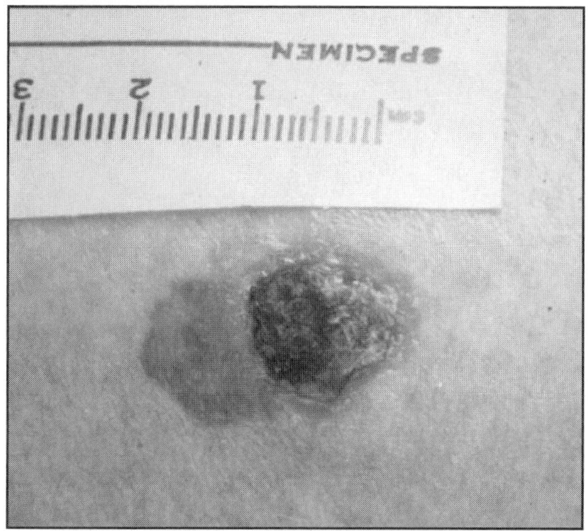

c. 327n.ras compressed using JPEG.

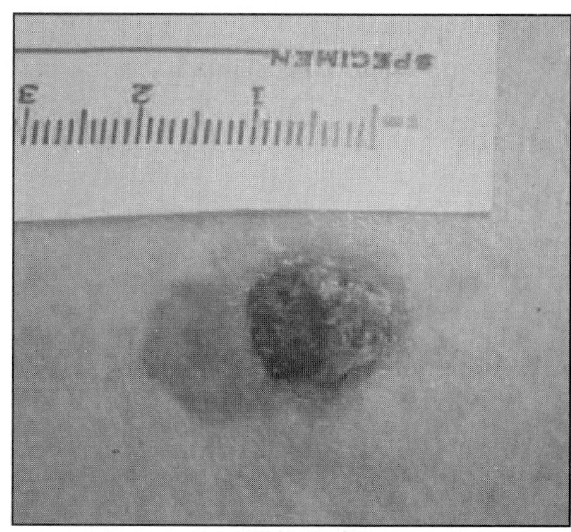

d. 327n.ras compressed using WVQ4.

Figure 7.4-6 (Continued)

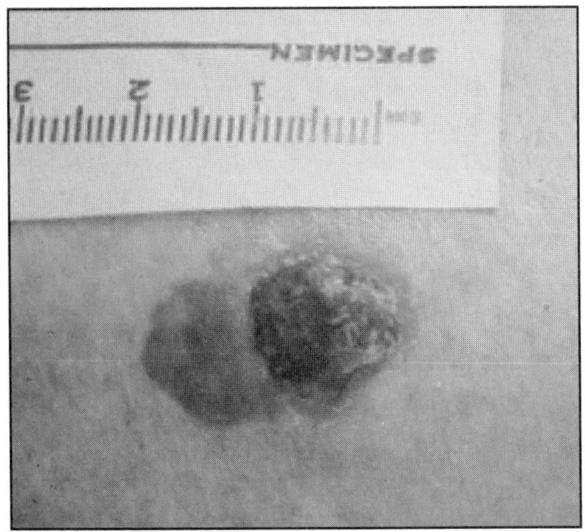

e. 327n.ras compressed using WVQ3.

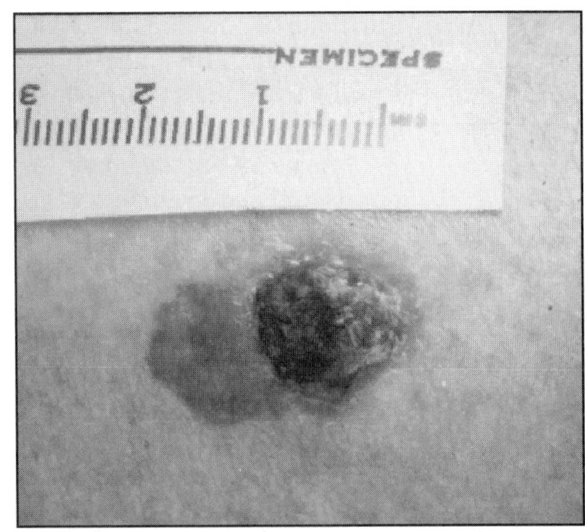

f. 327n.ras compressed using CCC.

speed and compute power available on the desktop may mitigate this factor. For applications where the compression is off-line, compression times of 2–4 minutes are acceptable. The important factor with many compression methods is the decompression time, which is about 10 seconds for these WVQ algorithms.

Figure 7.4-7 Tumor 1277 Comparisons

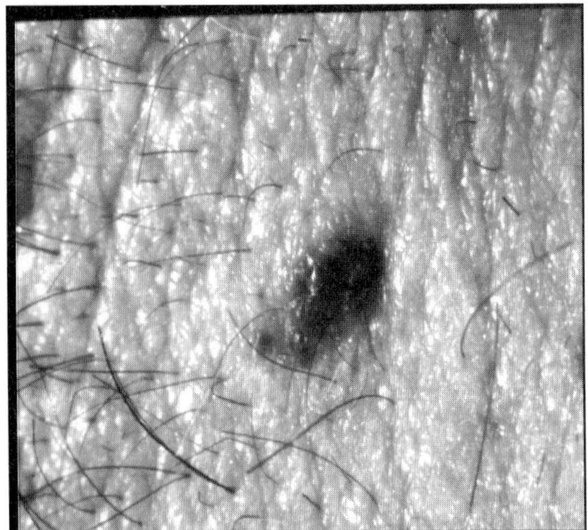

a. Original tumor 1277n.ras.

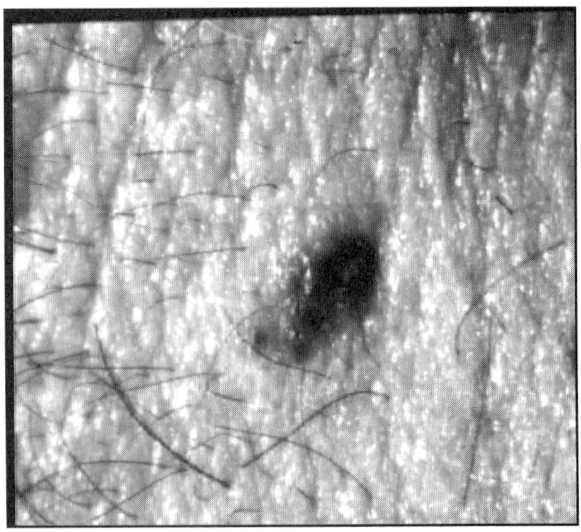

b. 1277n.ras compressed using WVQ2.

Figure 7.4-7 (Continued)

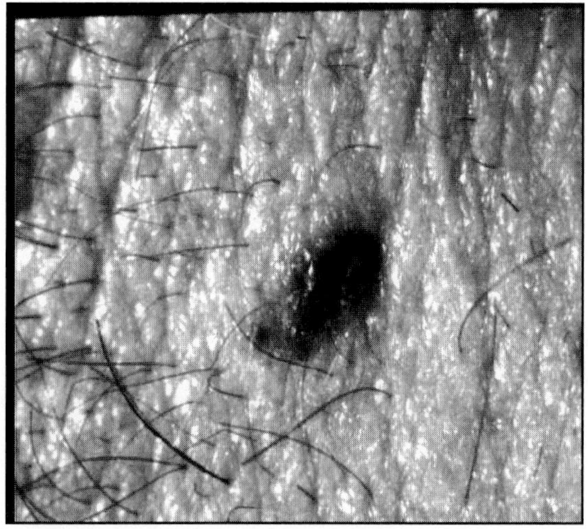

c. 1277n.ras compressed using JPEG.

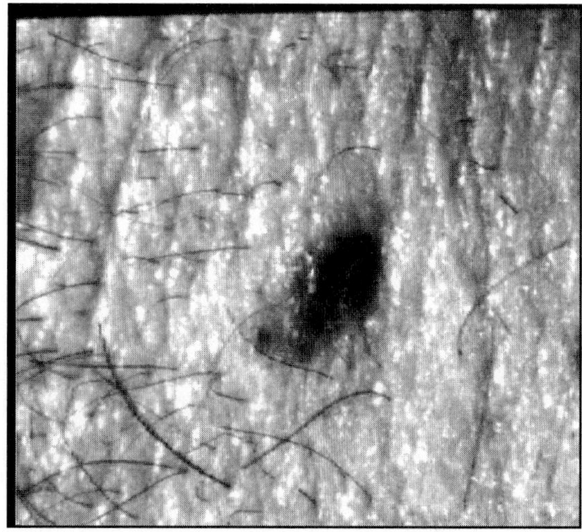

d. 1277n.ras compressed using WVQ4.

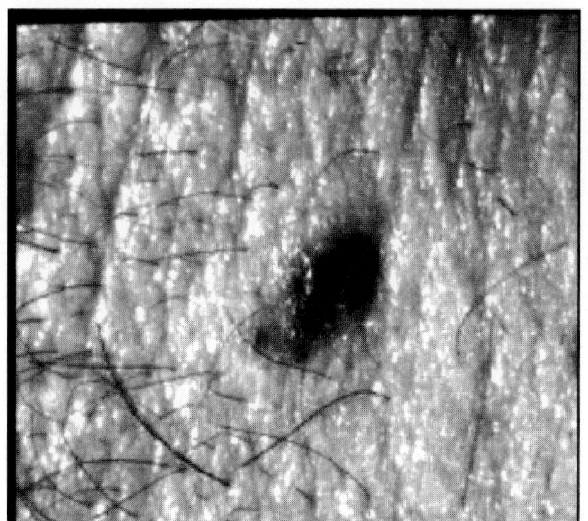

e. 1277n.ras compressed using WVQ3.

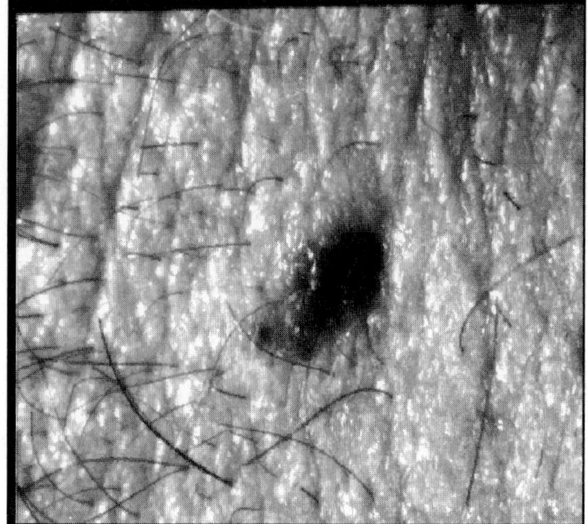

f. 1277n.ras compressed using CCC.

7.4.7 References

[Kjoelen 95] provided the basis for much of this section and more details on this specific application can be found there. [Umbaugh 90], [Golston/Stocker/Moss 92], and [Stoecker/Li/Moss 92] showed the importance of color for skin tumor image analysis. Information regarding numbers of images required by hospitals was found in [Leotta/Kim 93]. Detailed information on the CCC compression algorithm can be found in [Campbell/DeFanti, et al. 86] and [Heckbert 83]. More information on the optimal bit allocation rule is in [Antonini/Barlaud/Mathieu/Daubechies 92]. The PCT is explored more in [Elliot/Rao 82] and [Murakami/Asai/Itoh 86]. More on wavelets can be found in [Daubechies 89], [Chui 92], and [Antonini/Barlaud/Mathieu/Daubechies 92]. [Aleman 93] compares the use of universal codebooks and image specific codebooks. Subjective image analysis measures are described in [Golding 78]. Details on the LBG algorithm can be found in [Linde/Buzo/Gray 80].

Aleman, C., *Still Picture Image Coding Based on Wavelets and Vector Quantization*, M.S. Thesis, Southern Illinois University, Edwardsville, 1993.

Antonini, M., Barlaud, M., Mathieu, P., and Daubechies, I., "Image Coding Using Wavelet Transform," *IEEE Transactions on Image Processing*, Vol. 1, No. 2, 1992.

Campbell, G., DeFanti, T., et al., "Two bit/pixel Full Color Encoding," *Siggraph*, Vol. 20, No. 4, 1986.

Chui, C., *An Introduction to Wavelets*, San Diego: Academic Press, 1992.

Daubechies, I., "Orthonormal Bases of Compactly Supported Wavelets," *Communications on Pure and Applied Mathematics*, Vol. 41, pp. 909–996, 1989.

Elliot, D., and Rao, K., *Fast Transforms—Algorithms, Analyses, Applications*, Orlando, FL: Academic Press, 1982.

Golding, L. S., "Quality Assessment of Digital Television Signals," *SMPTE Journal*, Vol. 87, 153–157, 1978.

Golston, J., Stoecker, W., and Moss, R., "Automatic Detection of Irregular Borders in Melanoma and Other Skin Tumors," *Computerized Medical Imaging and Graphics*, Vol. 16, 199–203, 1992.

Heckbert, P., "Color Image Quantization for Frame Buffer Display," *Computer Graphics*, Vol. 16, No. 3, 297–304, 1983.

Kjoelen, A., *Wavelet Based Compression of Skin Tumor Images*, Master's Thesis in Electrical Engineering, Southern Illinois University at Edwardsville, 1995.

Leotta, D., and Kim, Y., "Requirements for Picture Archiving and Communications," *IEEE Engineering in Medicine and Biology Magazine*, Vol. 12, No. 1, 62–69, 1993.

Linde, Y., Buzo, A., and Gray, R., "An Algorithm for Vector Quantizer Design," *IEEE Transactions on Communications*, Vol. 28, No. 1, 1980.

Murakami, T., Asai, K., and Itoh, A., "Vector Quantization of Color Images," *ICASSP Tokyo Proceedings*, IEEE Conference on Acoustics, Speech and Signal Processing, pp. 133–136, April 1986.

Stoecker, W., Li, W., and Moss, R., "Automatic Detection of Asymmetry in Skin Tumors," *Computerized Medical Imaging and Graphics*, Vol. 16, 191–197, 1992.

Umbaugh, S. E, *Computer Vision in Medicine: Color Metrics and Image Segmentation Methods for Skin Cancer Diagnosis*, Ph.D. Dissertation, Electrical Engineering Department, University of Missouri—Rolla, UMI Dissertation Services, Ann Arbor, MI, 1990.

7.5 Image Segmentation Using a Deformable Template Algorithm

Medical imaging methods, such as magnetic resonance imaging (MRI), are very useful to medical professionals in diagnosis and for surgical procedure planning. MRI works by gathering two-dimensional (2-D) image slices of a three-dimensional (3-D) object. These 2-D images are then combined into a 3-D model for viewing. If this 3-D model could be displayed and manipulated in real time, there is a great potential for use in surgical procedures. For example, providing the surgeon with a probe that can pinpoint an exact location in a patient's MRI scan, and allowing the surgeon to see this on a 3-D model *during surgery*, will greatly facilitate many types of procedures.

In order to ease 3-D manipulation of the model in real time and to aid the medical professional, the images are often segmented into anatomical structures based on models. Current methods of segmenting MRI images require an expert to segment the 2-D images manually, which is very tedious, time-intensive, and error-prone. This application was developed to automate the segmentation process, as part of the preliminary research for a system that has been developed by Surgical Navigation Technologies to be used during neurosurgery.

7.5.1 Deformable Template Algorithm

The deformable template algorithm attempts to match a template image to a data image by stretching and compressing the template. It is as if the template is a rubber sheet and we are trying to fit it into the data image by finding the best match possible. It works by finding a geometric transform to make the template match the data; this is done by deforming the template by mathematically simulating the application of external forces. In order to simplify the process, pixel values are moved around using only regular geometric transforms. In this process we try to 1) minimize the difference between the deformed template and the test data and 2) keep the general shape of objects in the template.

The deformable template algorithm is an application of function optimization using gradient descent (steepest descent trajectory). When using this type of technique, we must define a cost function, which will determine how close we are to our goal. The cost function represents the global difference between the data and the current state of the deformable template, it is represented by a nonlinear equation, and the optimization is done by minimizing this function. The optimization is made computationally efficient by starting in a low dimensional space, corresponding to a shrunken image, and proceeds to higher dimensions only after a local minimum is found. So the process works by matching global features initially, corresponding to the low dimensional space, and then proceeding on to more detail feature matching in a higher dimensional space.

The algorithm incorporates a process called simulated annealing to avoid getting stuck in a local minimum. *Simulated annealing* works by mathematically simulating the physical process referred to as *annealing*, which is performed by heating matter and then cooling it slowly so that it will harden in a minimum energy state—this tends to strengthen the material. Here we also seek a minimum energy state, where the energy is defined by the cost function. If we simply tried to minimize the cost at every step, we could never escape from a local minimum. By using a procedure called *stochastic relaxation*, which randomly disturbs the system, we can simulate the annealing process and allow for occasional "uphill" steps. The particular techniques incorporated here are based on mathematical models for fluid elasticity, which allows for a smooth result.

The deformable template algorithm is typical of nonlinear optimization algorithms in that the result depends on a good first guess; in other words, the data must not be too different from the template. The mathematical details can be explored in the references.

7.5.2 Segmentation Application

The application requires three input images: a template image (original model), a segmented version of the template (segmented model), and a test image. The deformable template algorithm is applied using the template image and the test image. The template is deformed until a good match to the test image is found. After this is done, we have a set of geometric equations that will spatially map the template to the test image. The resulting mapping is then applied to the segmented version of the template, which results in the estimate for the segmented version of the test image.

To illustrate how this algorithm works, Figure 7.5-1 shows application to a

Figure 7.5-1 Deformable Template Example

a. Original image.

b. Segmented original image.

c. Rotated original image.

d. With many parameters fixed, the algorithm results in this image as the closest match to the original.

rotated image. The original image, used as a template, has been segmented by histogram thresholding to provide the model (Figures 7.5-1a, b). The original was then rotated to provide a test image (Figure 7.5-1c). The algorithm works by simulating the application of external imaginary forces to the original image to deform it, until a reasonable match is found to the test image. The result was achieved by fixing many of the parameters to the algorithm, thus limiting the quality of the result to illustrate the application of a single force. In this case, in Figure 7.5-1d, we can readily see that the deformation is caused by application of an imaginary force to the top of the image.

Figure 7.5-2 illustrates application of the deformable template algorithm to a single 2-D MRI image, which is an image of a single slice of a brain. Here we see that the

Figure 7.5-2 MRI Example

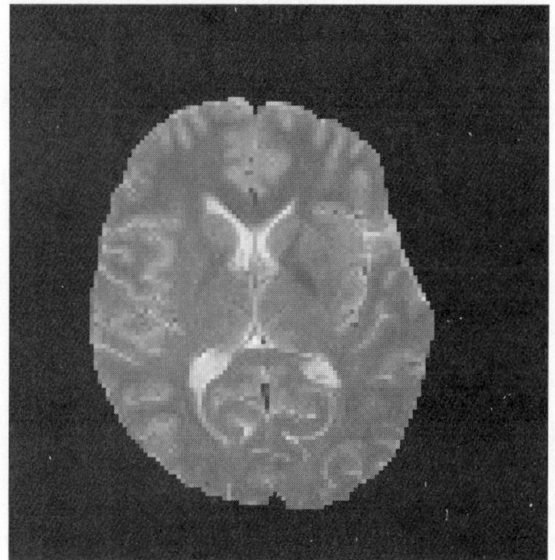

a. Template (original) image.

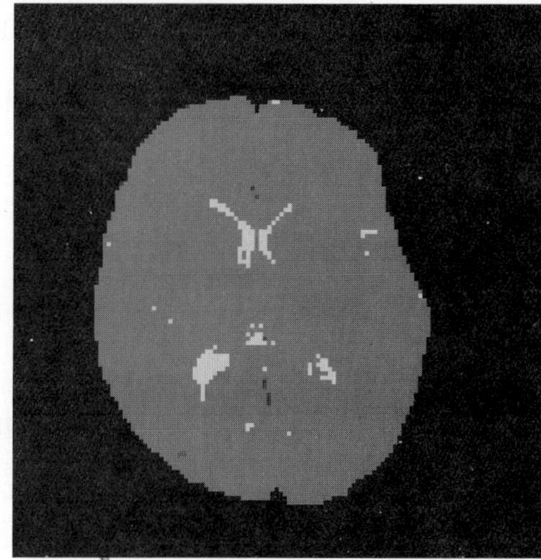

b. Segmented template image.

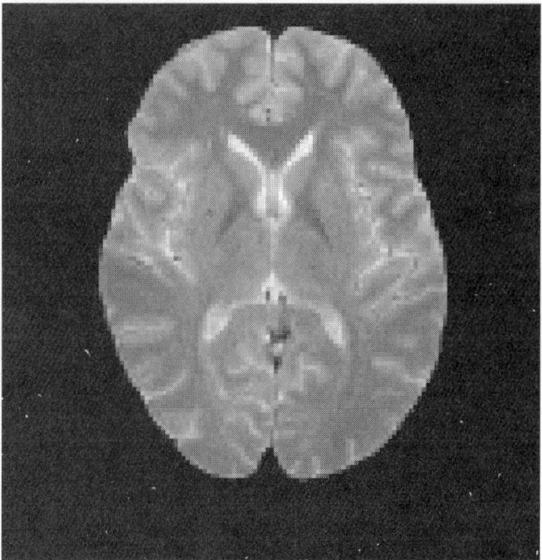

c. Test image.

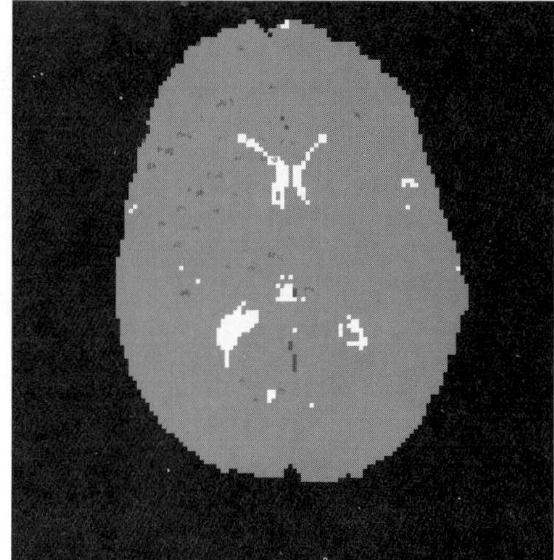

d. Segmentation result from applying the deformable template algorithm.

test data and template are similar, but vary in general shape. By comparing Figures 7.5-2b, d, we can see the segmentation resulting from application of this algorithm.

7.5.3 References

The image-guided surgical system referred to, called the Stealth Station®, is available from Surgical Navigation Technologies [SNT 96]. The software discussed here is detailed in [Kendrick 93]. The mathematical details can be found in [Amit/Grenander/Piccioni 83], [Miller/Christensen/Amit/Grenander 93], and [Kirkpatrick/et al. 83].

Amit, Y., Grenander, U., and Piccioni, M., "Structural Image Restoration Through Deformable Templates," *Journal of the American Statistical Association*, Vol. 86, 376–377, 1991.

Kendrick, L., *Deformable Template Image Models: Application to Segmenting Magnetic Resonance Images*, MSEE Thesis, Electrical Engineering Department, Southern Illinois University at Edwardsville, 1993.

Kirkpatrick, S., et al., "Optimization by Simulated Annealing," *Science*, Vol. 220, No. 4598, May 1983.

Miller, M. I., Christensen, G. E., Amit, Y., and Grenander U., "Mathematical Textbook of Deformable Neuroanatomies," *Proceedings of the National Academy of Science USA*, Vol. 90, No. 11, 944–948, 1993.

Surgical Navigation Technologies (SNT), 530 Compton Street, Broomfield, CO 80020 (Telephone: 303-439-9709, FAX: 303-439-9711), 1996.

7.6 Visual Acuity/Night Vision Simulation

This application was developed by a request from the U.S. Army to do a preliminary investigation of simulating night vision on the computer. The word *preliminary* should be stressed because we had only 2 days to put something together. This being the case, it serves as a good example of what can be easily done in the CVIPtools environment, within a short period of time. The specifications provided were as follows:

1. Simulate 20/50 visual acuity.
2. The images are to be binary (two-valued), with the colors of dark green and black.
3. Use a 30-in. distance from the computer screen to simulate distance in the cockpit of an aircraft of the pilot's eyes from the display.

Using this information we performed some library research to determine our approach to solving this problem. The questions we needed to answer were How can we simulate 20/50 vision? What color of "dark green" should we use? and How do night vision devices work? We decided that the first task was to develop a method to simulate various levels of visual acuity.

7.6.1 Visual Acuity Simulation

Visual acuity is related to the resolving power of the human visual system, but resolving power can actually be measured in two different ways. With the first method we are measuring the smallest spatial detail that can be perceived by considering resolution of high-contrast binary images; this is the standard measure of *visual acuity.* The second measure is a measure of contrast sensitivity and measures resolving power as a function of spatial frequency. This is done by finding the smallest contrast signal required to perceive an alternating pattern of gray and white bars. Although both are measures of the human visual system's response, the first is the measure of interest here.

We determined that in order to simulate 20/50 vision, we would experiment with an image created by digitizing the Snellen chart. The Snellen chart is one of the standard eye charts used to measure visual acuity (see Figure 7.6-1). With the Snellen chart, the terminology 20/50 vision refers to the individual who can just read the 20/50 line on the chart, and the previous lines (20/40, 20/30, 20/20) appear too blurry and cannot be read. We found that although each individual's lens aberration is unique, the typical blurring seen by near-sighted individuals can be modeled by a radial gaussian blur.

We proceeded to experiment with various approximations to a gaussian blur by using 3 × 3, 5 × 5, and 7 × 7 gaussian spatial convolution masks and applying them to a digitized version of the Snellen chart. We had two test subjects, who both wore corrective eyeglasses that corrected their vision to 20/20. The experimental procedure was as follows:

Figure 7.6-1 Snellen Chart

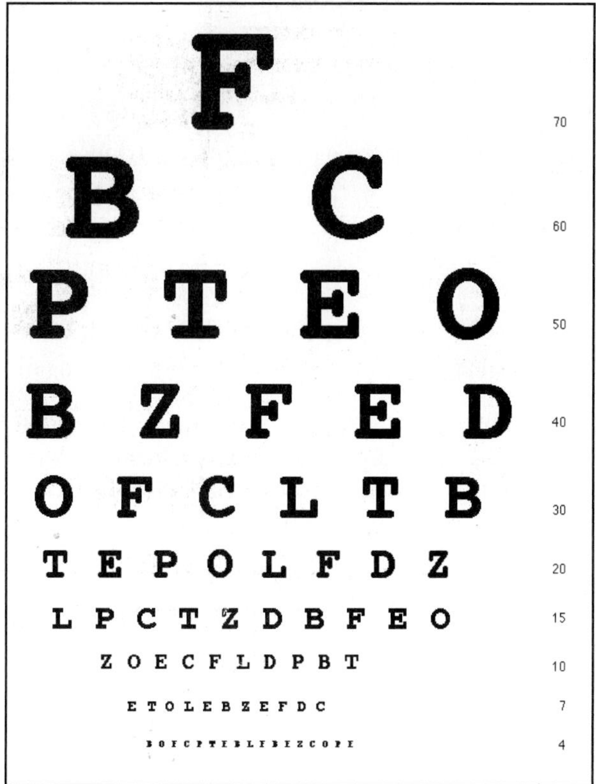

1. Digitize the Snellen eye chart and size it so that the line corresponding to 20/20 vision could be "just read" at a distance of 30 inches from the screen by individuals with corrected 20/20 vision.
2. Convolve the gaussian blur mask with the chart of the image.
3. Continue the blurring (step 2) until the subject cannot read the 20/20 line, but can still read the next line. Call this amount of blurring equivalent to the value of the next line, for example 20/30.
4. Repeat the process (steps 2 and 3) for the next line specified on the Snellen chart.

We collected experimental data for visual acuity levels of 20/20, 20/25, 20/30, 20/40, 20/50, 20/60, 20/80, and 20/120. The entire experiment was an iterative process, with feedback from the individuals, and we performed multiple runs of the experiment until we reached a consensus. Additionally, we attempted to find a linear relationship for the visual acuity as a function of number of times the convolution mask is applied to the image.

After much experimentation we determined that by using a 7×7 convolution mask, we could find a simple relationship that fit our experimental data. This is expressed in the following equation:

$$\begin{aligned} \text{VISUAL ACUITY} &= 20 + 5N \\ \text{where VISUAL ACUITY} &= \text{the denominator in the Snellen metric,} \\ &\quad \text{that is, 20/VISUAL ACUITY} \\ N &= \text{the number of times the } 7 \times 7 \text{ convolution mask} \\ &\quad \text{is convolved with the image} \end{aligned}$$

The 7×7 convolution mask is as follows:

$$\begin{bmatrix} 0.0002 & 0.0015 & 0.0037 & 0.0049 & 0.0037 & 0.0015 & 0.0002 \\ 0.0015 & 0.0088 & 0.0220 & 0.0293 & 0.0220 & 0.0088 & 0.0015 \\ 0.0037 & 0.0220 & 0.0549 & 0.0732 & 0.0549 & 0.0220 & 0.0037 \\ 0.0049 & 0.0293 & 0.0732 & 0.0977 & 0.0732 & 0.0293 & 0.0049 \\ 0.0037 & 0.0220 & 0.0549 & 0.0732 & 0.0549 & 0.0220 & 0.0037 \\ 0.0015 & 0.0088 & 0.0220 & 0.0293 & 0.0220 & 0.0088 & 0.0015 \\ 0.0002 & 0.0015 & 0.0037 & 0.0049 & 0.0037 & 0.0015 & 0.0002 \end{bmatrix}$$

For example, to simulate 20/50 vision, we apply this convolution mask to the image six times. In Figure 7.6-2 are shown images with various levels of blurring and the corresponding visual acuity measure. Note that the size of the image and the distance of the reader from the figure will affect this measure.

7.6.2 Night Vision Simulation

Because of the time constraints previously mentioned, we developed a simple model for night vision simulation based on input from the U.S. Army personnel. We were told that the night vision images were essentially two levels (binary images), with the single color being dark green. We experimented with various green colors and determined that the RGB pixel triple (0,175,0) was suitable. This was based primarily

Figure 7.6-2 Visual Acuity Example

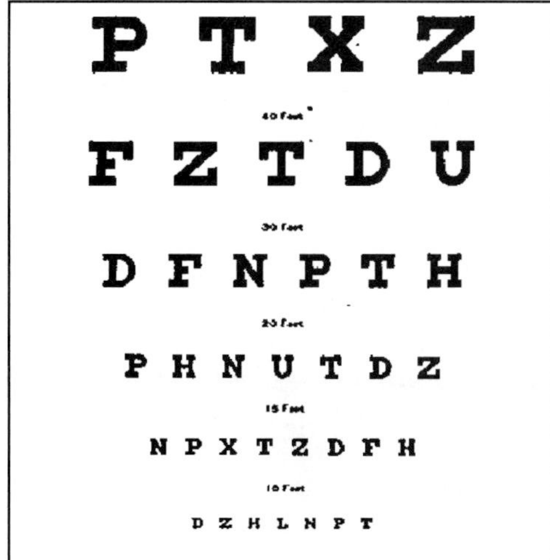

a. Original image representing 20/20 vision.

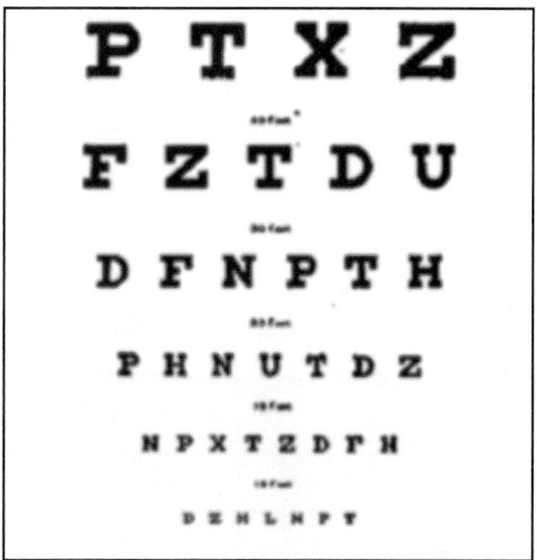

b. Image representing 20/30 vision, $N = 2$.

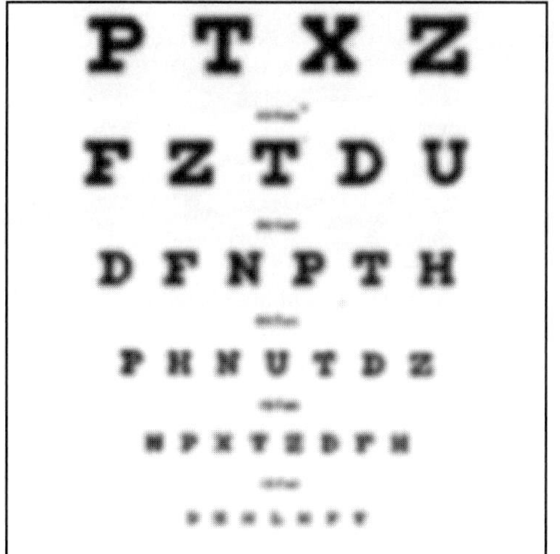

c. Image representing 20/50 vision, $N = 6$.

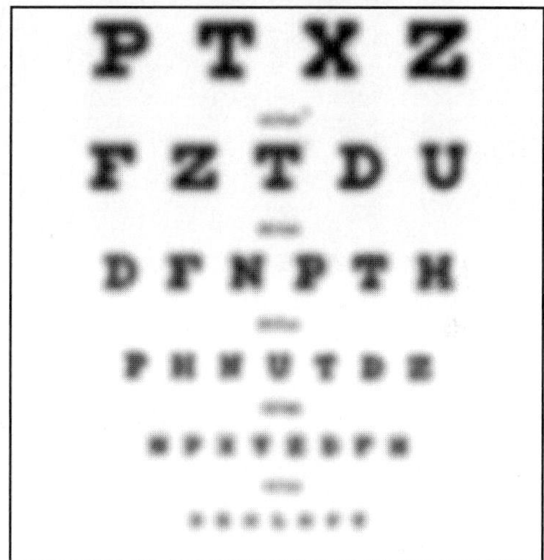

d. Image representing 20/80 vision, $N = 12$.

on input from individuals who were familiar with the night vision systems. We were not supplied with any “real” night vision images, so we developed the application to operate as follows:

1. Ask the user to select a threshold for the image.
2. Threshold the image with the value the user enters. Set the gray levels above the threshold to 255 and those below to 0.
3. Blur the image by using the 7×7 gaussian convolution mask, and applying it to the thresholded image six times (to achieve simulated 20/50 vision).

4. Create a color image and remap the pixels with value 255 to (R,G,B) = (0,175,0), and the 0-valued pixels to (R,G,B) = (0,0,0).
5. Display the simulated night vision image.

In Figure 7.6-3 we see an example of the night vision simulation, with the lighter shade of gray representing the dark green (the color image is on the CD-ROM, called nightvis).

Figure 7.6-3 Night Vision Example

a. Original image.

b. Threshold original image at gray level 50.

c. Result of night vision algorithm, with RGB remapped to (0,175,0).

7.6.3 References

A basic background in human vision is provided in [Davson 62] and [Wade/Swanston 91]. [Marr 82], [Levine 85], and [Arbib/Hanson 87] contain models for the human visual system as well as information relating it to computer vision methods. [Farah 90] relates visual disorders to normal vision and in the process provides insights regarding human visual perception. [Asher 61] and [Zuckerman 64] are useful in describing experimental methods, such as use of the Snellen chart.

Arbib, M. A., and Hanson, A. R. (Eds.), *Vision, Brain, and Cooperative Computation*, Cambridge, MA: MIT Press, 1987.

Asher, H., *Experiments in Seeing*, New York: Basic Books, 1961.

Davson, H. (Ed.), *The Eye, Volume 4, Visual Optics and the Optical Space Sense*, New York: Academic Press, 1962.

Farah, M. J., *Visual Agnosia*, Cambridge, MA: MIT Press, 1990.

Levine, M. D., *Vision in Man and Machine*, New York: McGraw Hill, 1985.

Marr, D., *Vision*, New York: W. H. Freeman and Company, 1982.

Wade, N. J., and Swanston, M., *An Introduction to Visual Perception*, New York: Routledge, Chapman and Hall, 1991.

Zuckerman, J., *Diagnostic Examination of the Eye*, Philadelphia: J. B. Lippincott Company, 1964.

CHAPTER 8

Programming with CVIPtools

8.1 Introduction to CVIPlab

The CVIPlab program was created to allow for experimentation with the CVIPtools functions outside of the CVIPtools environment (see Appendix B for details on setting up the CVIPlab program). It is essentially a prototype program containing a sample CVIP function and a simple menu-driven user interface to ease program use. By following the format of this prototype function, and using library function prototypes (Chapter 9), and the man pages (on CD-ROM), you can implement any algorithms developed in the CVIPtools environment in your own stand-alone program. Additionally, you can incorporate any of your own C functions into this program.

In addition to the CVIPtools libraries, the CVIPlab program requires three files: CVIPlab.c, threshold_lab.c, and CVIPlab.h. The CVIPlab.c file contains the main CVIPlab program, the threshold_lab.c file contains a sample function, and the CVIPlab.h is a header file for function declarations. CVIPlab.c contains a list of header files to include function declarations and three functions: main, input, and threshold_setup. The main function is declared as void, which means it does not return anything and contains code for the menu-driven user interface for CVIPlab. The input function returns an image pointer and illustrates how to read an image file into a CVIPtools image structure and display the resulting image. The threshold_setup function accepts an image pointer as input, gets the threshold value from the user, passes these to the threshold_lab function, and returns an image pointer to the resultant image. The actual processing, in this case performing a threshold operation on an image, is done by the threshold_lab function, which is contained in the file threshold_lab.c. By studying these functions, you can see how to access and process image files using some of the CVIPtools library functions.

The CVIPlab.c program is commented to describe the details more completely and is included here:

```
/************************************************************************
* ======================================================================
*
*   Computer Vision and Image Processing Lab - Dr. Scott Umbaugh SIUE
*
* ======================================================================
*
*             File Name: CVIPlab.c
*           Description: This is the skeleton program for the Computer Vision
*                        and Image Processing Lab Exercises
*   Initial Coding Date: April 23, 1996
*           Portability: Standard (ANSI) C
*             Credit(s): Zhen Li & Kun Luo
*                        Southern Illinois University at Edwardsville
************************************************************************/

/*
** include header files
*/

#include "CVIPtoolkit.h"
#include "CVIPconvert.h"
#include "CVIPdef.h"
#include "CVIPimage.h"
#define CASE_MAX 2

/* Put the command here, as VIDEO_APP, to run your image acquisition
   application program */

#define VIDEO_APP "SunVideo &"

/* Define the image viewer program here as VIEWER. In CVIPtools under UNIX you
** can use the disk based viewer called picture, or the RAM-based viewer
** RamViewer. With Windows NT/95, you can use any disk-based image viewer
** that will display images from the command line. That is, it will work
** when run from a system shell and has the format 'viewer <image_file_name>'.
*/

#define VIEWER "picture"

#include "CVIPlab.h"

/*
** function declaration
*/

Image *threshold_Setup(Image *inputImage);
Image *input();

/*
** start main function
*/

void main(){
    IMAGE_FORMAT  format;           /* the input image format */
    Image         *cvipImage;       /* pointer to the CVIP Image structure */
    char          *outputfile;      /* output file name */
    int           choice;           /* menu choice */
    CVIP_BOOLEAN  done = CVIP_NO;   /* exit from menu loop variable */

/* Set the image viewer
```

```
** 'RamViewer' is the CVIPtools RAM-based viewer for UNIX systems.
** 'picture' is the CVIPtools disk-based viewer for UNIX.
** 'xv' and 'imagetool' are other options on SUN systems.
** Any image viewer will work that can be run from a system shell
** and has the format 'viewer <image_file_name>'.
** VIEWER is defined at the top of the CVIPlab program.
*/
    setDisplay_Image(VIEWER, "Default");

    print_CVIP("\n\n\n\n***************************************");
    print_CVIP("***************************    ");
    print_CVIP("\n*\t\t Computer Vision and Image Processing Lab\t  *");
    print_CVIP("\n*\t\t\t <Insert your name here> \t\t\t  *");
    print_CVIP("\n*****************************************");
    print_CVIP("************************\n\n\n");

    while(!done) {
      print_CVIP("\t\t0.\tExit \n\n");
      print_CVIP("\t\t1.\tGrab and Snap an Image  \n\n");
      print_CVIP("\t\t2.\tThreshold Operation \n\n");
      print_CVIP("\n\nCVIPlab>>");

      /*
      ** obtain an integer between 0 and CASE_MAX from the user
      */
      choice = getInt_CVIP(10, 0, CASE_MAX);

      switch(choice) {

         case 0:
           done=CVIP_YES;
           break;

         case 1:
           /* invokes the Video Application */
           system(VIDEO_APP);
           print_CVIP("\n\tThe Video Application is being invoked\n\n");
           print_CVIP("\t\t\tPlease Wait....\n\n");
           sleep(6);/* keeps the system idle for 6 seconds */
           break;

         case 2:
           /*Get the input image */
           cvipImage = input();
           if(cvipImage == NULL) {
                 error_CVIP("main", "could not read input image");
                 break;
           }

           /* calls the threshold function */
           cvipImage = threshold_Setup(cvipImage);
           if (!cvipImage) {
                 perror_CVIP("main", "threshold fails");
                 break;
           }

           print_CVIP("\n\t\tEnter the Output File Name:  ");
           outputfile = getString_CVIP();

           /*
           ** display the resultant image
```

```
            */
            view_Image(cvipImage,outputfile);

            /*
            ** saves the resulting image data out to disk in <outputfile>
            */
            format = getFileFormat_Image(cvipImage);
            write_Image(cvipImage, outputfile, CVIP_NO, CVIP_NO, format, 1);

            /*
            **IMPORTANT: free the dynamic allocated memory when it is not
            **           needed
            */
           free(outputfile);
               break;

           default:
               print_CVIP("Sorry ! You Entered a wrong choice ");
               break;
      }
     }
}
/*
** end of the function main
*/

/*
** The following function reads in the image file specified by the user,
** stores the data and other image info. in a CVIPtools Image structure,
** and displays the image.
*/

Image* input(){
    char            *inputfile;
    Image           *cvipImage;
    IMAGE_FORMAT    format;

    /*
    ** get the name of the file and store it in the string 'inputfile '
    */
    print_CVIP("\n\t\tEnter the Input File Name:  ");
    inputfile = getString_CVIP();

    /*
    ** create the CVIPtools Image structure from the input file
    */
    cvipImage = read_Image(inputfile,1);
    if(cvipImage == NULL) {
      error_CVIP("init_Image", "could not read image file");
      free(inputfile);
      return NULL;
    }

    /*
    ** display the source image
    */
    view_Image(cvipImage,inputfile);

    /*
    **IMPORTANT: free the dynamic allocated memory when it is not needed
    */
```

```
    free(inputfile);

    return cvipImage;
}

/*
** The following setup function asks the threshold value from the user. After
** it gets the threshold value, it will call the threshold_Image() function.
*/

Image *threshold_Setup(Image *inputImage){
    unsigned int    threshval;      /* Threshold value */

    /*
    ** get a value between between 0 and 255 for threshsold
    */
    print_CVIP("\n\t\tEnter the threshold value:  ");
    threshval = getInt_CVIP(10, 0, 255);

    return threshold_lab(inputImage, threshval);
}
```

The following is the threshold function contained in the threshold_lab.c file:

```
/*************************************************************************
* ======================================================================
* Computer Vision/Image Processing Tool Project - Dr. Scott Umbaugh SIUE
* ======================================================================
*
*             File Name: threshold_lab.c
*           Description: it contains the function to threshold a BYTE image
*   Initial Coding Date: April 23, 1996
*           Portability: Standard (ANSI) C
*             Author(s): Zhen Li & Kun Luo
*                        Southern Illinois University @ Edwardsville
*
** Copyright (C) 1995,1996 SIUE - by Scott Umbaugh, Kun Luo, Yansheng Wei.
*************************************************************************
*/

/*
** include header files
*/

#include "CVIPtoolkit.h"
#include "CVIPconvert.h"
#include "CVIPdef.h"
#include "CVIPimage.h"
#include "CVIPlab.h"

/*
** The following function will compare the actual gray level of the input
** image with the threshold limit. If the gray-level value is greater than the
** threshold limit then the gray level is set to 255 (WHITE_LAB) else to 0
** (BLACK_LAB). Note that the '_LAB' or '_lab' is appended to names used
** in CVIPlab to avoid naming conflicts with existing constant and function
** (e.g. threshold_lab) names.
*/
```

```
#define     WHITE_LAB   255
#define     BLACK_LAB   0

Image *threshold_lab(Image *inputImage, unsigned int threshval){
    byte           **image;         /* 2-d matrix data pointer */
    int            r,               /* row index */
                   c,               /* column index */
                   bands;           /* band index */

    unsigned int   no_of_rows,      /* number of rows in image */
                   no_of_cols,      /* number of columns in image */
                   no_of_bands;     /* number of image bands */

    /*
    ** Gets the number of image bands (planes)
    */
    no_of_bands = getNoOfBands_Image(inputImage);

    /*
    ** Gets the number of rows in the input image
    */
    no_of_rows =  getNoOfRows_Image(inputImage);

    /*
    ** Gets the number of columns in the input image
    */
    no_of_cols =  getNoOfCols_Image(inputImage);

    /*
    ** Compares the pixel value at the location (r,c)
    ** with the threshold value. If it is greater than
    ** the threshold value it writes 255 at the location
    ** else it writes 0. Note that this assumes the input
    ** image is of data type BYTE.
    */
    for(bands=0; bands < no_of_bands; bands++) {
      /*
      ** reference each band of image data in 2-d matrix form;
      ** which is used for reading and writing the pixel values
      */
      image = getData_Image(inputImage, bands);
      for(r=0; r < no_of_rows; r++) {
          for(c=0; c < no_of_cols; c++) {
            if(image[r][c] > (byte) threshval)
                    image[r][c] = WHITE_LAB;
            else
                    image[r][c] = BLACK_LAB;
          }
       }
    }

    return inputImage;
}
/*
** end of function threshold_lab
*/
```

8.1.1 Toolkits, Toolboxes, and Application Libraries

All the functions in the CVIPtools program are accessible to those programming with CVIPlab. The functions are arranged in a hierarchical grouping of libraries, with the Toolkit libraries at the lowest level, the Toolbox libraries at the next level, and the user-generated Application libraries at the highest level, as illustrated in Figure 8.1-1. This hierarchical grouping is devised such that each successive level can use the building blocks (functions) available to it from the previous level(s).

Figure 8.1-1 CVIPtools Libraries

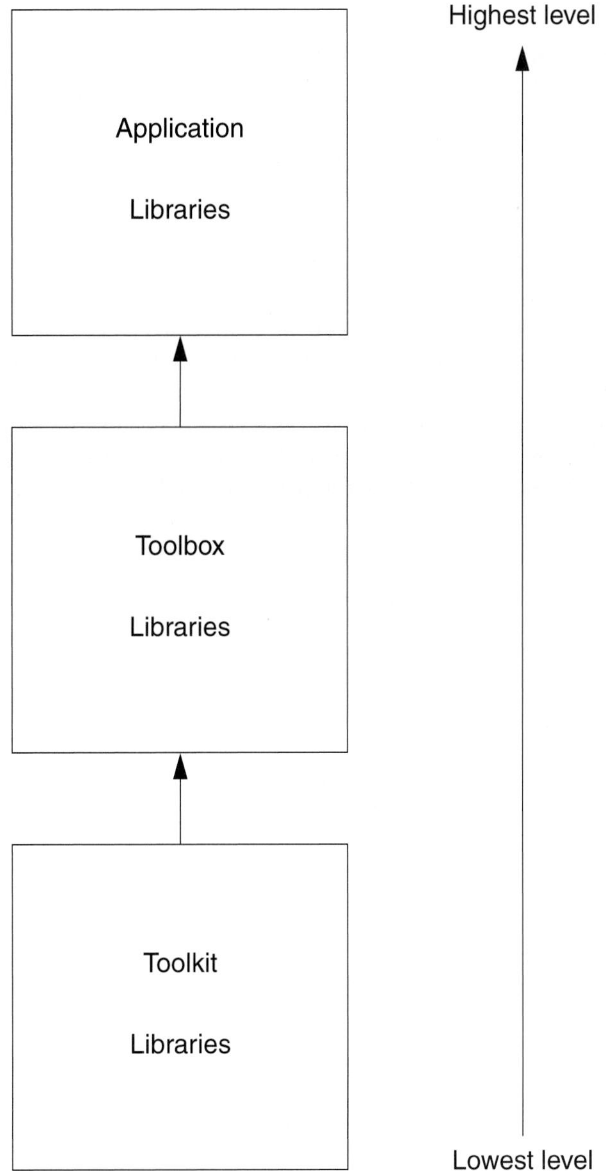

The *Toolkit libraries* contain low-level functions, such as input/output functions, matrix manipulation functions, and memory management functions. The *Toolbox libraries* are the primary libraries for use in application development; they contain the functions that are available from the GUI in CVIPtools, such as the many analysis or enhancement functions. At the highest level, the *Application libraries* are the libraries generated by those using the CVIPtools environment to develop computer imaging applications. In some cases, useful functions are modified and extracted from an application and put into a Toolbox library. Chapter 9 contains function prototypes for all Toolbox library functions and some of the commonly used Toolkit functions. For more details and examples see the man pages on the CD-ROM, and for a quick look at all the available library functions see Appendix C.

8.1.2 Compiling and Linking CVIPlab Under UNIX

CVIPtools and CVIPlab both use the make utility and associated Makefiles to handle the compilation and linking of the program. A Makefile automatically links the proper libraries and allows the programmer to generate the executable (runnable) version of the program simply by typing *make*. We use the standard UNIX/X-windows imake utility to create the Makefile (for details on this process under Windows NT/95, see the README file on the CD). To use the imake utility to create the Makefile requires a template file, CVIPlab.tmpl, which contains the necessary libraries and link paths, and the file Imakefile. Under a UNIX system any required machine-dependent specifications will be automatically inserted into the Makefile through the use of the template file.

To create the Makefile requires two steps: 1) run *xmkmf*, a utility that will generate the Makefile from the Imakefile and CVIPlab.tmpl; 2) run *make depend*, which will handle any dependencies within the compilation and linking process. After the Makefile has been created, you can compile and link the program simply by typing *make*. When adding a new function to the CVIPlab, do the following:

1. Create a file similar to threshold_lab.c for the new_function.
2. Add the new_function_Setup to CVIPlab.c, which is similar to threshold_Setup, and to the CVIPlab menu.
3. Add the function prototype to the CVIPlab.h header file:

   ```
   extern Image *new_function(new_function parameters...)
   ```

4. Add the name to the Imakefile in these two lines:

   ```
   SRCS = CVIPlab.c threshold.c new_function.c
   OBJS = CVIPlab.o threshold.o new_function.o
   ```

5. Type *xmkmf*.
6. Type *make depend*.
7. Type *make*.

8.1.3 Image Data and File Structures

The data and file structures of interest are those that are required to process images. In traditional structured programming, a system can be modeled as a hierarchical set of functional modules, where the modules at one level are built of lower-level modules. Similarly, the information in the system, in this case image data, can use this hierarchical model. In this case we have a five-tiered model with the pixel data at the bottom, the vector data structure at the next level, the matrix data structure at the next level, image data structures next, and finally the image files at the top level. Figure 8.1-2a shows a triangle to illustrate this model because it is naturally larger at the lower levels—it takes many pixels to make up a vector, many vectors to make an image, and so on.

Figure 8.1-2 Image Data and File Structures

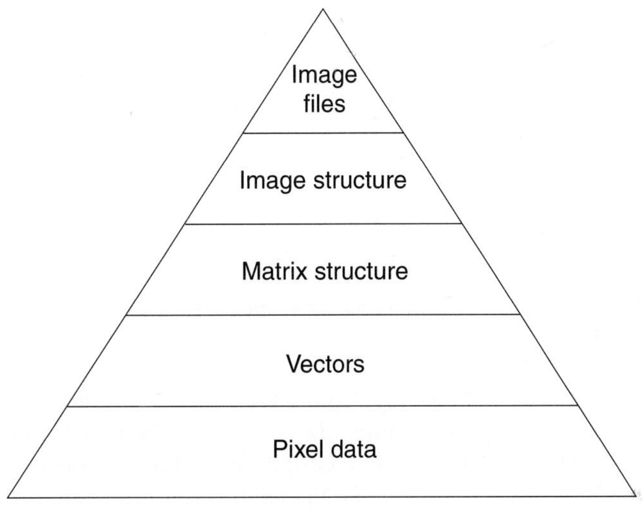

a. Hierarchical model.

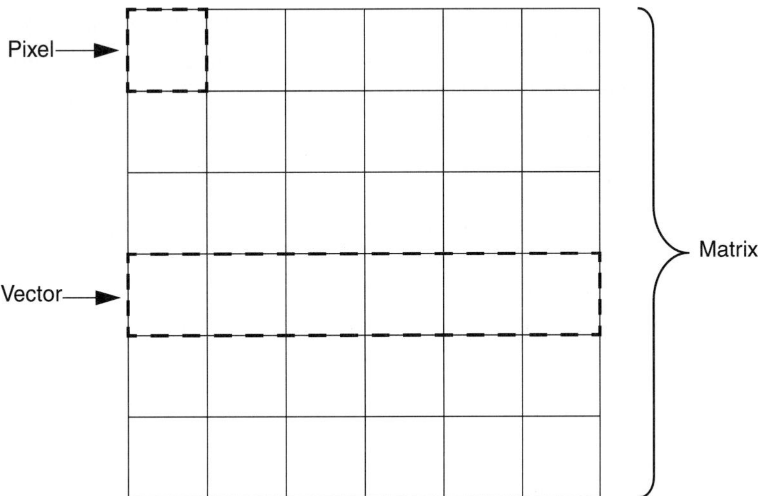

b. Image data representation.

Figure 8.1-2 (Continued)

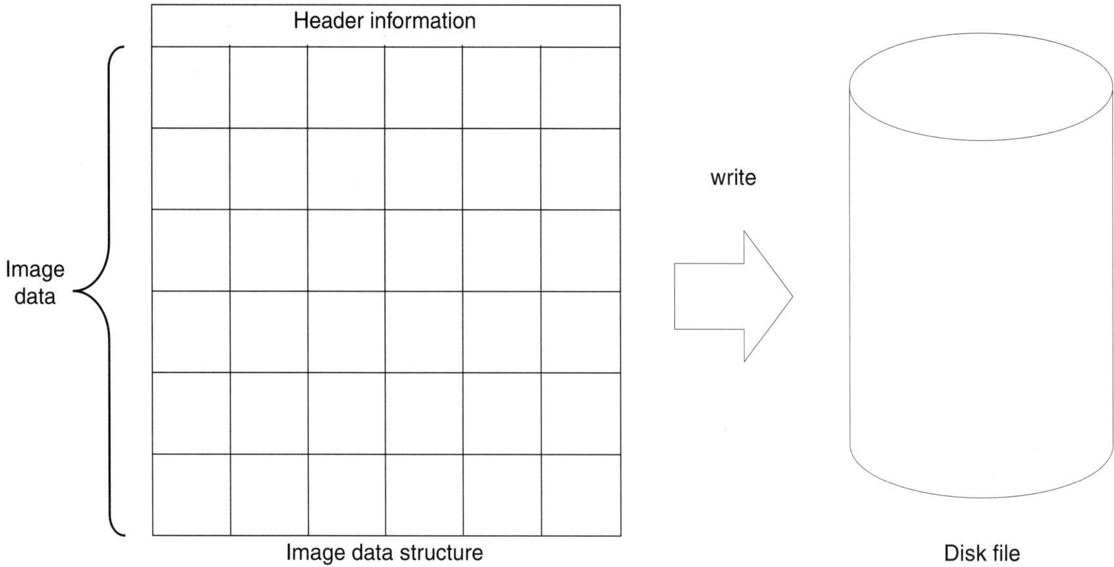

c. Image data structure and disk file.

In Figure 8.1-2b we see that a vector can be used to represent one row or column of an image, and the 2-D image data itself can be modeled by a matrix. The image data structure (Figure 8.1-2c) consists of a header that contains information about the type of image, followed by a matrix for each band of image data values. When the image data structure is written to a disk file, it is translated into the specified file format (for example, PostScript, TIFF, Sun Raster files, and PPM). CVIPtools has its own image file format, the Visualization in Image Processing (VIP) format, and also supports other standard file formats. Because most standard image file formats assume 8-bit data, the VIP format is required for floating point data, complex data, as well as CVIPtools specific information.

The vector data structure can be defined by declaring an array in C of a given type, or by assigning a pointer and allocating a contiguous block of memory for the vector. A *pointer* is simply the address of the memory location where the data reside. In Figure 8.1-3 we see an illustration of a vector; the pointer to the vector is actually the address of the first element in the vector. For images, each element of the vector represents one pixel value, and the entire vector represents one row or column. The vector function library is called libvector.

Figure 8.1-3 Vector Representation

Address	A	A+1	A+2	A+3	…	A+N-2	A+N-1
Datum	255	128	38	234	…	69	10

The matrix structure is at the level above vectors. A matrix can be viewed as a one-dimensional vector, with M multiplied by N elements, that has been mapped into a matrix with M rows and N columns. This is illustrated in Figure 8.1-4, where we see how a one-dimensional array can be mapped to a two-dimensional matrix via a pointer map. The matrix data structure is defined as follows:

```
typedef enum {CVIP_BYTE, CVIP_SHORT, CVIP_INTEGER, CVIP_FLOAT, CVIP_DOUBLE}
     CVIP_TYPE;
typedef enum {REAL, COMPLEX} FORMAT;
typedef struct {
   CVIP_TYPE data_type;
   FORMAT data_format;
   unsigned int rows;
   unsigned int cols;
   void **rptr;   /*real data pointer*/
   void **iptr;   /*imaginary data pointer*/
} Matrix;
```

The data_type field contains the type of data that is stored in the matrix, these are the matrix elements. The data_format field describes whether the matrix elements are real or complex. The next two fields, rows and cols, contain the number of rows and columns in the matrix, and the last two, **rptr and **iptr, are two-dimensional pointers to the matrix elements (if the data_format is REAL, then the imaginary pointer is a null pointer). The matrix function library is called libmatrix, and the associated memory allocation and deallocation functions are called new_Matrix and delete_Matrix, respectively. The data type for the real and imaginary pointers is passed as a parameter to the function that creates and allocates memory for a matrix,

Figure 8.1-4 Matrices and Pointers

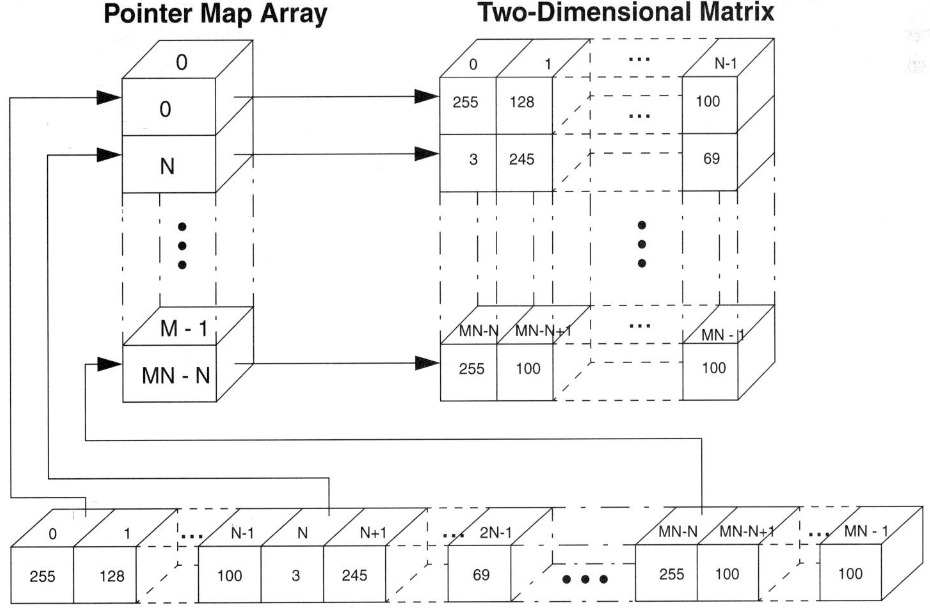

the new_Matrix function. After the matrix has been set up with new_Matrix, the data are accessed as a two-dimensional array by assigning a pointer with the getData_Matrix function; note that care must be taken to cast it to the appropriate data type (for an example see computer vision lab exercise #2).

The image structure is the primary data structure used for processing digital images. It is at the level above the matrix data structure because it consists of a matrix and additional information. The image data structure is defined as follows:

```
typedef enum {PBM, PGM, PPM, EPS, TIF, GIF, RAS, ITX, IRIS, CCC, BIN, VIP,
     GLR, BTC, BRC, HUF, ZVL, ARITH, BTC2, BTC3, DPC, ZON, ZON2, SAFVR, JPG}
     IMAGE_FORMAT;
typedef enum {BINARY, GRAY_SCALE, RGB, HSL, HSV, SCT, CCT, LUV, LAB, XYZ}
     COLOR_FORMAT;
typedef struct {
   IMAGE_FORMAT image_format;
   COLOR_FORMAT color_space;
   int bands;
   Matrix **image_ptr;
   HISTORY story;
} IMAGE;
typedef struct IMAGE Image;
```

The first field, image_format, contains the file type of the original image. When the image is read into CVIPtools, this information is retained for use during a save operation, if you do not specify the desired file type. Note, however, that the image format does not necessarily tell us anything about the actual data in the image queue, especially after it has been processed. The second field, color_space, determines if the image is binary (two-valued), gray scale (typically 8-bit), or color (typically three-plane, 24-bit, RGB). If it is a color image, then this field is updated when a color space conversion is performed. The third field, bands, contains the number of bands in the image; for example a color image has three bands, and a gray-scale image has one band. The next field, **image_ptr, is a pointer to an array of pointers to matrix data structures, where each matrix contains one band of pixel data (see Figure 8.1-5). The last field is for history information and is used by the CVIPtools software to keep track of certain functions such as transforms, which have been performed on an image in the CVIPtools image queue.

The history field in the image structure, story, is a pointer to a history data structure. The history data structure consists of packets of history information, where each packet contains information from a particular function. The history data structure is defined as follows:

```
typedef struct packet PACKET;
   struct packet {
      CVIP_TYPE *dtype;
      unsigned int dsize;
      void **dptr;
   };

typedef struct history *HISTORY;
   struct history {
      PROGRAMS ftag;
      PACKET *packetP;
      HISTORY next;
   };
```

Figure 8.1-5 Image Data

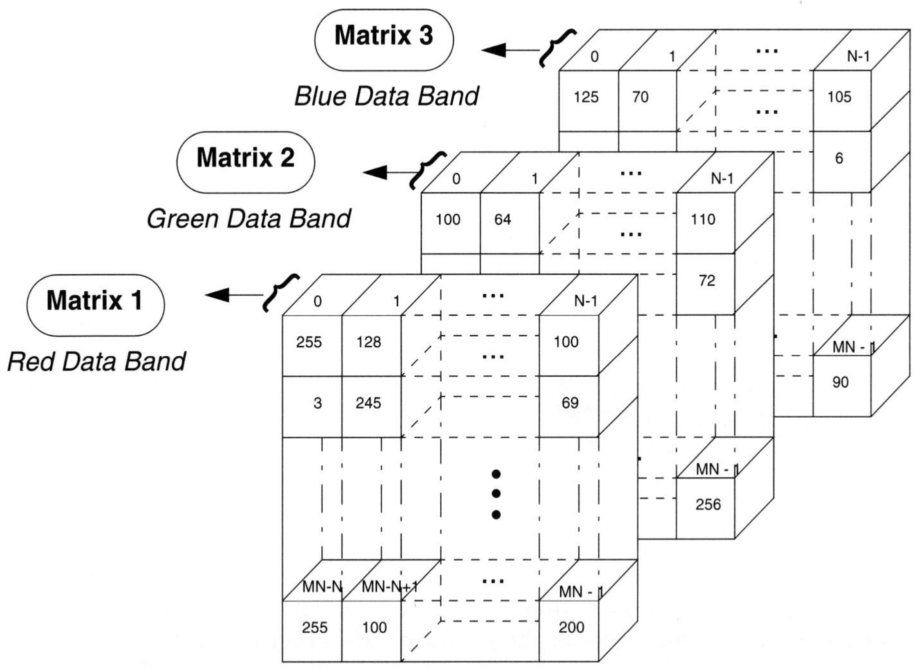

Representation of image data. In addition to the image data shown here, the image structure contains header information.

Functions relating to the history are in libimage, also see CVIPhistory.h.

At the highest level is the image file. The image file can be any of the types previously described as supported by CVIPtools. If the file is an 8-bit/pixel image file, which is typical, then CVIPtools may need to remap the data. For example, if the range of the data is too large for 8-bits, if the data are in floating point format, or the data contain negative numbers, then the data in the image queue must be remapped before it can be written in 8-bit format. This is done automatically, if required. Keep in mind that any image that you see displayed is in the remapped format, so it will automatically be saved as what you see. However, in some cases, this may not be what is desired—we may want to retain the data as it is in the queue. To do this, the image must be saved in the CVIPtools image file format, the VIP format. The VIP file format allows the image data structure to be written to disk and consists of an image header and the image data structure. The VIP structure is as follows:

```
VIP            - 3 bytes (the ASCII letters "ViP")
COMPRESS       - 1 byte, ON or OFF, (depending on whether the data are
                 compressed)
IMAGE_FORMAT   - 1 byte, (e.g. BTC)
COLOR_SPACE    - 1 byte, (e.g. BINARY, GRAYSCALE, RGB, LUV)
DATA_TYPE      - 1 byte, (e.g. CVIP_SHORT, CVIP_BYTE)
NO_OF_BANDS    - 1 byte, (1 for gray-level, 3 for color, other numbers
                 allowed too)
NO_OF_COLS     - 2 bytes,
```

```
NO_OF_ROWS      - 2 bytes,
FORMAT          - 1 byte (REAL or COMPLEX)
SIZEOF HISTORY  - 4 bytes, (Size of history information, in bytes)
HISTORY         - variable size (history information)
RAW DATA
```

Note that if the raw data in a VIP file are examined, the actual number of bytes stored may vary from that listed here. This is due to the fact that we use the standard XDR (External Data Representation) functions to write the VIP files. By using these functions we ensure file portability across computer platforms, but it results in most data types smaller than 32 bits (4 bytes) being written to the file in a standard 32-bit format. To use the VIP image file format, simply use the read_Image and write_Image functions contained in libconverter. (These file read/write functions are in libconverter because a large portion of their functionality is to convert file types to and from the CVIPtools image data structure.) These read/write functions will read/write any of the image file formats supported by CVIPtools and require the programmer to deal with only one data structure, the image data structure.

8.2 CVIP Laboratory Exercises

The CVIP lab exercises are divided into two sections, one for computer vision and one for image processing. Most of the exercises are programming exercises, but a few are exploratory exercises using CVIPtools. The programming exercises allow you to acquire skills in C programming while exploring computer imaging concepts, and the CVIPtools-based exercises let you gain insight into various computer imaging concepts without the need for programming. The CVIPlab program contains the threshold function, which can be used as a prototype for adding the functions in the exercises.

8.2.1 Computer Vision Exercises

Computer Vision Lab Exercise #1: Simple Binary Object Features

1. Write a C function to find the area and the coordinates of the center of the area of a binary image (see Chapter 2). Incorporate this function call into the case statement at the beginning of the CVIPlab program (do this for all the functions written), so that it can be accessed via the menu. Remember that the value that represents 1 for the binary images is actually 255 and 0 is 0.

   The program should display the following:

   ```
   Area = <#> pixels
   Row_center = <row coordinate for center of area>
   Column_center = <column coordinate for center of area>
   ```

2. Test this function using images you create with CVIPtools. Use the Utilities selection bar to create test images with the Create option (Utilities→Create). To create images with multiple objects, use the AND and OR logic functions available from Utilities→Arith/Logic. (Note: You may want to save the images for later use.)

Computer Vision Lab Exercise #2: Sequential Labeling Algorithm

1. Write a C function to implement a sequential labeling algorithm. Follow the flowchart for the labeling algorithm from Chapter 2, Figures 2.6-1 and 2.6-2 (also described in the [Horn 86] reference). You may assume that row and column 0 do not contain objects, so that you can start the scan with row 1 and column 1. Define a two-dimensional array for the labels using a fixed size, for example for a 256×256 image:

```
int label[256][256]; /*declaration*/
```

   NOTE: Be sure to initialize the array elements (for example, via FOR loops), if needed. When the memory is allocated for the array it may contain garbage.

2. Test this function using images you create with CVIPtools. Use the Utilities selection bar to create test images with the Create option (Utilities→Create). To create images with multiple objects, use the AND and OR logic functions available from Utilities→Arith/Logic.
3. When you are certain that the your function implements the algorithm correctly, modify the label function using a Matrix structure for the label array. This will allow the use of any size image, without the need to change the size of the array. This is done as follows:

```
Matrix *label_ptr; /* declaration of pointer to Matrix data structure */
int    **label;    /* declaration of pointer to matrix data */
...
/*allocating the memory for the matrix structure*/
label_ptr = new_Matrix(no_of_rows, no_of_cols, CVIP_INTEGER, REAL);

/* getting the matrix data into the label array*/
label = (int **) getData_Matrix(label_ptr);

...
label[r][c]                  /*accessing the array elements */
...
delete_Matrix(label_ptr);    /*freeing the memory space used by the
                             matrix*/
```

4. Modify the label function so that it will find the Area, Row_center, and Column_center for each object. Note that you can use your previous functions by modifying the variables for Area, Row_center, and Column_center to be arrays and use the label as the index into the array (be sure to initialize the array elements to 0). The information for each object should be printed to the screen, along with the object number.

Additional Work

1. Modify the function so that it will handle objects on the edges of the image.
2. Modify the Update function (see flowchart, Figures 2.6-1 and 2.6-2) so that it does not require multiple image scans, for example keep a linked list of equivalent labels and rescan the image only once (after all the labeling is done).
3. Modify the function to work with gray-level images.
4. Modify the function to work with color images.
5. Modify the function to handle any number of objects.

6. Modify the function so that it will output the labeled image (that is, the label array is written to disk as an image, with appropriate gray levels to make all the objects visible).

Computer Vision Lab Exercise #3: Euler Number Calculation via Local Counting Technique

1. Write a C function to find the number of upstream-facing convexities X, upstream-facing concavities V, and the euler number for a binary image. Use the method discussed in Section 2.6.2, assuming six-connectivity.

   The function should display the following:

```
The number of upstream-facing convexities = <X>
The number of upstream-facing concavities = <V>
The euler number for the image = <X - V>
```

2. Test this function using images you create with CVIPtools. Use the Utilities selection bar to create test images with the Create option (Utilities→Create). To create images with multiple objects, use the AND and OR logic functions available from Utilities→Arith/Logic.

Additional Work

1. Modify the function to find the euler number for each object in a binary image containing multiple objects.
2. Modify your function to handle other connectivity types (four, eight, and four/eight).
3. Modify your function to handle gray-level images.
4. Modify your function to handle color images.

Computer Vision Lab Exercise #4: Iterative Morphological Operators

An iterative image modification method for performing binary image morphology has been described in Section 2.4.6, as follows:

1. A set of surrounds (neighbors) S, where $a = 1$. These surrounds are defined by number in Figure 2.4-16.
2. The logic function $L(a, b)$, where b is the current pixel value, specifies the output of the morphological function.
3. The number of iterations n.

   The function $L(\ )$ and the values of a and b are all functions of the row and column, (r, c), but for concise notation this is implied. The set of surrounds S is based on six connectivity [SE and NW diagonals], where N = neighbor and X = nonneighbor diagonals, as follows:

$$\begin{bmatrix} N & N & X \\ N & B_{ij} & N \\ X & N & N \end{bmatrix}$$

For example, for the image:

$$\begin{bmatrix} 0 & 0 & 0 & 0 & 0 & 0 & 1 & 1 \\ 0 & 1 & 1 & 1 & 1 & 1 & 1 & 0 \\ 0 & 1 & 0 & 1 & 1 & 1 & 1 & 0 \\ 0 & 0 & 0 & 0 & 1 & 1 & 1 & 1 \\ 0 & 1 & 1 & 1 & 1 & 1 & 1 & 1 \\ 0 & 0 & 0 & 0 & 0 & 0 & 0 & 0 \end{bmatrix}$$

$$\text{Let } S = \begin{bmatrix} 1 & 1 & x \\ 1 & x & 1 \\ x & 0 & 0 \end{bmatrix}$$

where x means this value is not considered, since we are assuming six-connectivity.

EXAMPLE 8-1

Let $L(a, b) = ab$ (logical AND operation).

The window S (3×3 window) is scanned across the image. If a match is found, then $a = 1$ and the output is computed by performing the specified $L(a, b)$ function, in this case by ANDing a with b (b is the center pixel of the subimage under the window). This gives the value of our new image, which will equal $(1)b = b$. If the window S does not match the underlying subimage, then $a = 0$ (false) and $L(a, b) = ab = (0)b = 0$. In either case, the resulting value is written to the new image at the location corresponding to the center of the window.

The window S is scanned across the entire image in this manner and the resultant image is as follows:

$$\begin{bmatrix} 0 & 0 & 0 & 0 & 0 & 0 & 0 & 0 \\ 0 & 0 & 0 & 0 & 0 & 0 & 0 & 0 \\ 0 & 0 & 0 & 0 & 0 & 0 & 0 & 0 \\ 0 & 0 & 0 & 0 & 0 & 0 & 0 & 0 \\ 0 & 0 & 0 & 0 & 0 & 0 & 0 & 0 \\ 0 & 0 & 0 & 0 & 0 & 1 & 1 & 0 \\ 0 & 0 & 0 & 0 & 0 & 0 & 0 & 0 \end{bmatrix}$$

The set S can contain more than one surround. If it does, then $a = 1$ when the underlying neighborhood matches *any* of the surrounds in the set S. Another parameter which can be considered is the rotation of the surround S. In the preceding example, the number of possible rotated combinations of the given surround S is 5. If the surround S is rotated once counterclockwise it looks like

$$\begin{bmatrix} 1 & 1 & x \\ 1 & x & 0 \\ x & 1 & 0 \end{bmatrix}$$

EXAMPLE 8-2

$S = \{\,\}, L(a, b) = 0, n = 1, f = 1$

The set of surrounds (neighbors) is a null set. This implies that $a = 0$ because a surround is not specified. The boolean function $L(a, b) = 0$. For this combination, all the cells of the image are set to zero, i.e., we have a black image as output.

EXAMPLE 8-3

$S = \{\,\}, L(a, b) = (!b), n = 1, f = 1$

In this case $a = 0$, but this is irrelevant because $L(a, b) = !b$, which implies that the center pixel is negated (complimented).

If $b = 1, L(a, b)(!1) = 0;$

Elseif $b = 0, L(a, b) = (!0) = 1.$

EXAMPLE 8-4

$S = \{7\}, L(a, b) = ab, n = 1, f = 1$

Consider the following image with the surround S as follows:

$$\begin{bmatrix} 0 & 0 & 0 & 0 & 0 & 0 & 0 & 0 \\ 0 & 1 & 1 & 1 & 1 & 1 & 1 & 0 \\ 0 & 1 & 1 & 1 & 1 & 1 & 1 & 0 \\ 0 & 1 & 1 & 1 & 1 & 1 & 1 & 0 \\ 0 & 1 & 1 & 1 & 1 & 1 & 1 & 0 \\ 0 & 1 & 1 & 0 & 0 & 0 & 0 & 0 \\ 0 & 1 & 1 & 0 & 0 & 0 & 0 & 0 \end{bmatrix}$$

$$\text{Let } S = \begin{bmatrix} 1 & 1 & x \\ 1 & x & 1 \\ x & 1 & 1 \end{bmatrix}$$

In this case, $a = 1$ for the surround shown here. If the surround does not match, then $L(a, b) = 0(b) = 0$. If there is a match, then $L(a, b) = 1(b) = b$. The resultant image is as follows:

$$\begin{bmatrix} 0 & 0 & 0 & 0 & 0 & 0 & 0 & 0 \\ 0 & 0 & 0 & 0 & 0 & 0 & 0 & 0 \\ 0 & 0 & 1 & 1 & 1 & 1 & 0 & 0 \\ 0 & 0 & 1 & 1 & 1 & 1 & 0 & 0 \\ 0 & 0 & 0 & 0 & 0 & 0 & 0 & 0 \\ 0 & 0 & 0 & 0 & 0 & 0 & 0 & 0 \\ 0 & 0 & 0 & 0 & 0 & 0 & 0 & 0 \end{bmatrix}$$

Because the logic function is a logical AND operation, if the edge pixels are not 1's, the edges are removed. This operation retains a cluster of 1's with the edge pixels removed. So, the appendages (thin lines) are removed from the original image—this is an erosion operation.

EXAMPLE 8-5

$S = \{1,7\}, L(a, b) = (!a)b, n = 1, f = 1$

Consider the following image with the surrounds {S} as follows:

$$\begin{bmatrix} 0 & 0 & 0 & 0 & 0 & 0 & 0 & 0 \\ 0 & 1 & 1 & 1 & 1 & 1 & 1 & 0 \\ 0 & 1 & 1 & 1 & 1 & 1 & 1 & 0 \\ 0 & 1 & 1 & 1 & 1 & 1 & 1 & 0 \\ 0 & 1 & 1 & 1 & 0 & 0 & 0 & 0 \\ 0 & 1 & 1 & 1 & 0 & 0 & 0 & 0 \\ 0 & 0 & 0 & 0 & 0 & 0 & 0 & 0 \end{bmatrix}$$

Let $S1 = \begin{bmatrix} 0 & 0 & d \\ 0 & d & 0 \\ d & 0 & 0 \end{bmatrix}$ and $S7 = \begin{bmatrix} 1 & 1 & d \\ 1 & d & 1 \\ d & 1 & 1 \end{bmatrix}$ (Note: These are defined in Figure 2.4-16.)

If $b = 1, L(a, b) = (!a)1 = !a$;

Elseif $b = 0, L(a, b) = (!a)0 = 0$.

The new image after the preceding operation is

$$\begin{bmatrix} 0 & 0 & 0 & 0 & 0 & 0 & 0 & 0 \\ 0 & 1 & 1 & 1 & 1 & 1 & 1 & 0 \\ 0 & 1 & 0 & 0 & 0 & 0 & 1 & 0 \\ 0 & 1 & 0 & 1 & 1 & 1 & 1 & 0 \\ 0 & 1 & 0 & 1 & 0 & 0 & 0 & 0 \\ 0 & 1 & 1 & 1 & 0 & 0 & 0 & 0 \\ 0 & 0 & 0 & 0 & 0 & 0 & 0 & 0 \end{bmatrix}$$

We can see that this operation removes interior blobs and keeps the edges only. Hence, this is an edge detection operation.

Do the following:

1. Run CVIPtools and load binary images of your choice. Select Analysis→Segmentation and perform morphological filtering, using the Morphological Filters→Iterative Modification selection, with the following parameters: $S = \{1, 7\}, L(a, b) =$

$(!a)b$, $n = 1, f = 1$. Note that the two parameters, n and f, are defined as follows in CVIPtools:

- ☞ The parameter n is the number of iterations. For $n = 1$, the window is scanned once across the entire image; for $n = 2$, twice and so on. For $n = \infty$, the window is scanned till there can be *no* changes in the resultant image.
- ☞ The parameter f is the number of subfields into which the image tessellation is divided. For our purposes we will use the default, $f = 1$ (multiple subfields applies only to parallel operation, which is not covered here).

See Figure 8.2-1 for the expected results. Here we see that this operation works as an edge detector.

2. Apply to the images: $S = \{7\}, L(a, b) = ab, n = 1, f = 1$.
3. Apply to the images: $S = \{4,8\}, L(a, b) = ab, n = 1, f = 1$.
4. Apply to the images: $S = \{7\}, L(a, b) = (!a)b, n = 1, f = 1$.
5. Apply to the images: $S = \{1,2,3\}, L(a, b) = (!a)b, n = \infty, f = 1$.
6. Experiment with other images and parameters.

Computer Vision Lab Exercise #5: Edge Detection—Roberts and Sobel

1. Write a function to implement the Roberts edge detector from Chapter 2. Let the user select either the square root, or absolute value form. The C functions for absolute value and square root are abs() and sqrt(). Note that you will need to deal with potential overflow problems because the results may be greater than 255. This may be dealt with by using a floating point image structure as an intermediate image and then remapping the image when completed. This is done as follows:

```
Image  *outputimage;    /*declaration of image structure pointer*/
float **image_data;     /* declaration of image data pointer*/
...
outputimage = new_Image(PGM, GRAY_SCALE, no_of_bands, no_of_rows,
     no_of_cols, CVIP_FLOAT, REAL);    /*creating a new image structure*/
image_data = getData_Image(outputimage, bands);    /*getting the data into an
     array that can be accessed as: image_data[r][c] */
...
outputimage = remap_Image(outputimage, CVIP_BYTE, 0, 255); /*remapping a
     float image to byte size, this is done before writing the image to disk
     with the write_Image function*/
```

   Test the function on gray-level images. Compare the results from the two methods by using the Utilities→Misc→Compare two images selection in CVIPtools.

2. Write a function to implement the Sobel edge detector from Chapter 2. The function should output an image that contains the Sobel magnitude and is remapped as with the Roberts. Test the function on gray-level images of your choice.

Additional Work

1. Modify the functions to handle color images.
2. Implement other edge detectors described in Chapter 2.

Figure 8.2-1 Iterative Morphological Filtering

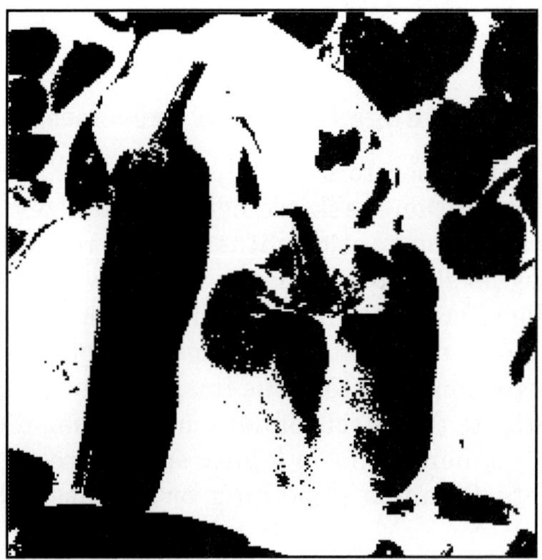

a. Image pepper.pbm.

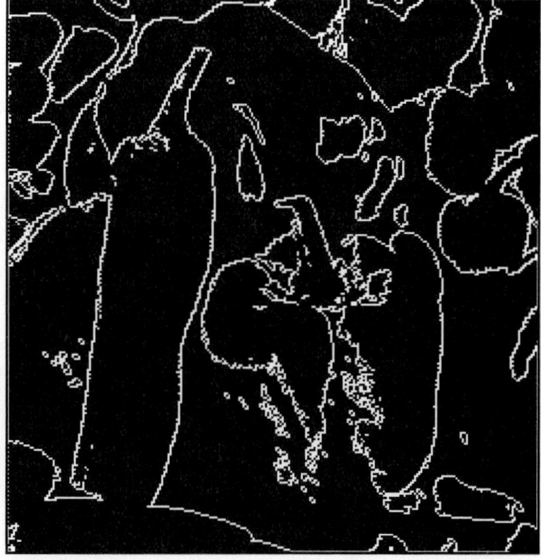

b. Resultant image from (a)—$S = \{1, 7\}$; $L(a, b) = (!a)b$; $n = f = 1$.

c. Image morph_rect2.pgm.

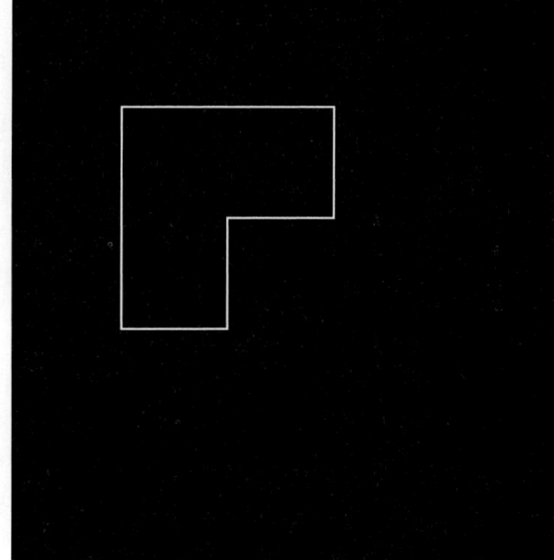

d. Resultant image from (c)—$S = \{1, 7\}$; $L(a, b) = (!a)b$; $n = f = 1$.

3. Use the Analysis→Edge/Line detection→Edge link selection in CVIPtools to connect the lines in the output images. Note that this requires a binary image, so be sure to apply a threshold operation to the images first. Thresholding can be performed directly in this window by typing the threshold value in the entry box and clicking on the *Threshold current image* button at the bottom of the window.

4. Experiment with the Hough transform via the Analysis→Edge/Line detection→Hough selection in CVIPtools. Note that this performs automatic edge linking as a postprocessing step.

Computer Vision Lab Exercise #6: CVIPtools Edge Detection Exercises

As described in Chapter 2, many edge detection operators are implemented with convolution masks. The summations of the convolution mask coefficients equal zero for many edge detectors. This results in regions of constant brightness being reduced to zero, and regions of abrupt changes in gray level returning large values, which indicate prominent edges. Because this has the effect of eliminating the zero spatial frequency component, it makes the operator independent of average image brightness.

Prior to the application of the edge detection operators, the image can be preprocessed by using a lowpass filter to help mitigate the effects of noise in the image. This may be effective because noise in an image generally has a higher spatial frequency content than the normal image components. Common smoothing, or lowpass, filters used are

$$\frac{1}{9}\begin{bmatrix} 1 & 1 & 1 \\ 1 & 1 & 1 \\ 1 & 1 & 1 \end{bmatrix} \qquad \frac{1}{16}\begin{bmatrix} 1 & 2 & 1 \\ 2 & 4 & 2 \\ 1 & 2 & 1 \end{bmatrix} \qquad \frac{1}{10}\begin{bmatrix} 1 & 1 & 1 \\ 1 & 2 & 1 \\ 1 & 1 & 1 \end{bmatrix}$$

average *gaussian* *generic lowpass*

The amount of blurring that a lowpass filter causes is determined by the size of the mask and the mask coefficients. Notice that the sum of the kernel values for all the lowpass filters is 1. This fact is important for understanding how lowpass filters operate; they retain the average value component.

Consider a portion of an image without high-frequency content. This means that the pixel values are of constant value or that they are changing slowly. As a lowpass mask is passed over this portion of the image, the new value for the pixel of interest is calculated as the sum of the coefficients multiplied by the neighborhood pixel values. If all the neighborhood pixel values are the same (constant), the new pixel value is the same as the old value. This is the reason the sum of the coefficients is chosen to be 1. Low-frequency content has been preserved. As the kernel is moved over a portion of the image with high-frequency content, any rapid changes in intensity get averaged out with the remaining pixels in the neighborhood, thereby lowering the high-frequency content. The visual result of lowpass filtering is a slight blur of the image. This blur results because any sharp pixel transitions are averaged with their surroundings as the high-frequency content is attenuated.

Depending upon the degree of noise corruption, the appropriate filter can be used for noise smoothing. For low noise content images any one of these filters can be used for noise smoothing. Because the choice is optional in CVIPtools, you should use the filters if necessary or bypass the option.

The DC bias is the average brightness of the image. For edge detection we want the edges of objects in the image to be visible while setting nonedge regions to zero. This can be achieved by removing the DC bias by using convolution masks whose coefficients sum to zero. Using edge detection and keeping the DC component is an enhancement technique in which regions of constant brightness levels are preserved rather than reducing the regions to black (zero gray level). In this case, the detected edges overlap the original image, which tends to enhance the edges visually.

Do the following:

1. Run CVIPtools, and load test images of your choice. As one of the test images, create a binary image of a circle with Utilities→Create→Circle.
2. Select Utilities→Create→Add Noise. Add noise to your test images. Use both gaussian and salt-and-pepper noise.
3. Select Analysis→Edge/Line Detection. Compare thresholding the output at different levels with the Kirsch, Pyramid, and Robinson edge detection operators. In CVIPtools, select the desired edge detector, select None for prefiltering, select the desired threshold with postthreshold option. Alternately, select None for postthreshold and use the *Threshold* button at the bottom of the window (this allows for the testing of different threshold levels without the need to rerun the edge detection operation).
4. Compare using different size kernels with the Sobel and Prewitt operators. In CVIPtools, select the desired edge detection operator; then select the kernel size.
5. Compare keeping the DC bias versus not keeping it, using the Roberts and laplacian.
6. Using the Frei-Chen, compare the line subspace versus the edge subspace, with the projection option in CVIPtools.
7. Select the edge detection operators that work the best for the test images. For the images containing noise, compare the resultant images with and without applying a lowpass filter as a preprocessing step (use the prefilter option in CVIPtools).

Computer Vision Lab Exercise #7: Template Matching

Template matching is a simple method for finding a specific object within an image, where the template contains the object for which we are searching (see Section 2.6.6). The template is moved across the image in order to find subimages that match the template. Due to the many possible sources of noise and distortion in images, we use a similarity or distance (error) measure (see Section 2.6.1) to determine how well the template matches the image data.

1. Write a C function to perform template matching. The function should take two input images: the image of interest $I(r, c)$ and the template image $T(r, c)$. The function will move the template across the image of interest, searching for pattern matches by calculating the error at each point. The distance measure to be used for this exercise is the Euclidean distance measure defined by

$$D(\bar{r}, \bar{c}) = \sqrt{\sum_{r}\sum_{c}[I(r', c') - T(r, c)]^2}$$

If we overlay the template on the image, then $\bar{r}, \bar{c}$ are the row and column coordinates of $I(r, c)$ corresponding to the center of the template where a match occurs. The r', c' designation is used to illustrate that as we slide the template across the image, the limits on the row and column coordinates of $I(r, c)$ will vary depending on: 1) where we are in the image and 2) the size of the template. You need only consider parts of the image that fully contain the template image. Your function should handle any size image and template, but you may assume that the template is smaller than the image. A match will occur when the error measure is less than a specified threshold. In your function, the threshold should be specified by user input. Where a match occurs, the program should display the following:

```
Error = <D(r̄, c̄)>
Row = <row coordinate for center of object>
Column = <column coordinate for center of object>
```

2. Test this function with images you create using CVIPtools. For example, create a small image for the template with a single object, and then create a larger image with multiple objects for the test image.

Additional Work

1. Create more images with CVIPtools, and test your function more extensively.
2. Expand the function by allowing for the rotation of the template. Consider the error to be the minimum error from all rotations.
3. Modify the function for efficiency by comparing the template only to image objects, not every subimage.
4. Make your function more useful by adding size invariance to the template matching. This is done by growing, or shrinking, the object to the size of the template before calculating the error.
5. Experiment with using different error and similarity measures described in Section 2.6.1.

Computer Vision Lab Exercise #8: Computer Vision Project

The following process can be followed to streamline project development:

1. Use CVIPtools to explore algorithm development, using primarily the Analysis window for these types of projects.
2. Perform segmentation by using the Segmentation window and/or edge detection.
3. Perform morphological filtering (in the lower portion of the Analysis→Segmentation window) to reduce and solidify spatial objects.
4. Use the Analysis→Features window to do feature extraction.
5. Examine the feature file to come up with a classification algorithm. Or, if available, use the feature files as input to pattern classification or neural network software.
6. Code the algorithm developed into your CVIPlab program, as follows:
   a. Find the function name that corresponds to the CVIPtools function (see Chapter 9 and the appendices).

b. If desired, check the manual page on the CD-ROM to see an example of how the function is used in a C program.

c. Do this (a and b) for all functions needed.

d. Code the feature extraction algorithm with these functions in your CVIPlab program.

e. Code your classification algorithm into your CVIPlab program, or get the C function from your pattern classification software.

7. Test your program on real images input from an imaging device.

Project Topics

1. Implement a program for the recognition of geometric shapes such as circles, squares, rectangles, and triangles. Make it robust so that it can determine if a shape is unknown.
2. Experiment with the classification of tools such as screwdrivers, wrenches, and hammers. Find features that will differentiate the classes of tools. Design a method for system calibration and then identify various sizes of nuts and bolts, as well as number of threads per inch.
3. Implement a program to identify different coins and bills. Make it robust so that it cannot be fooled by counterfeits.
4. Design a program to read bar codes. Experiment with different types of codes. Experiment with different methods to bring images into the system. In general, scanners will be easier to work with than cameras—verify this.
5. Implement a program to perform character recognition. Start with a subset, such as the numbers 0–9, and then expand using what you have learned. Experiment with different fonts, printers, and lighting conditions.

8.2.2 Image Processing Exercises

Digital Image Processing Lab Exercise #1: Arithmetic/Logic Operations

1. The arithmetic and logic operations are useful utilities in image processing. For example, the process of complementing an image can be useful as an enhancement technique. Because the eye responds logarithmically to brightness changes, details characterized by small brightness changes in the dark regions may not be visible. Complementing the image converts these small deviations in the dark regions to the bright regions, where they may be easier to detect. Partial complementing of an image also produces potentially useful results. An example would be to leave the upper half of the gray scale untouched while complementing the lower half. Bright regions in the original image are unaffected while dark regions are complemented.

   Write a C function that performs a partial complement on an image. Have the user input the gray-level range over which the complement is performed. NOTE: The C symbol for bitwise NOT is ~.

2. The AND and OR image combinations are normally used to mask and put together images. We use the AND function to mask off portions of an image. Given an image of which we wish to retain only a small section, a second image is generated to be a mask. The mask image is composed of black or white pixels (where each pixel contains 8 bits). A black pixel has all 8 bits set to 0. A white pixel has all bits set to 1. Black pixels are used where the original image should be masked, and white pixels are used where the original image should appear in the output image. The AND combination of the mask and the original image produces the final masked image. Image OR combinations are used to put together two images into a composite output image. Given that the objects in the two images do not spatially overlap and the nonobject areas are masked, the images may be ORed together combining both into a single output image. The eXclusive-OR combination may sometimes be useful as a simple image comparison tool. Pixels are compared bit by bit, producing an output image displaying black where the two pixels are exactly identical. Where the pixels are not perfectly identical, the output pixel will be something other than black, depending on the actual bit-by-bit comparison.

   Write C functions that perform the following logical operations on two images: AND, OR, XOR. NOTE: C symbols for these bitwise operators are &, |, and ^, respectively.

3. Image subtraction is used in any application where we want to remove unchanging background information from images. This requires sequential images that contain the same background information. Examples include using two sequential frames of video for motion detection, or using medical images taken at different times to detect any physical changes that have occurred. When the subtraction is performed, identical portions of the images will be black in the output image. Differences in the two images, such as objects that have changed or moved, will show up clearly in the output image. With motion detection this enables the measurement of both the motion and the direction. If the time between image acquisition is known, then speed of the moved object may also be calculated. For medical images, this may aid the physician in diagnosis and treatment.

   Write a C function to subtract two images, and put this function in a separate file from the logic functions. Initially, use BYTE data types, which will result in clipping at zero for negative results. Next, modify the function to use FLOAT data types (use the cast_Image function in libimage) and then remap when the process is completed (use the remap_Image function in libmap). Note that these two methods will result in different output images.

Additional Work

1. Extend the logic operations to work with data types other than BYTE.
2. Extend the logic operations to include NAND, NOR, and more complex boolean expressions.
3. Extend the subtraction function to perform addition, multiplication, and division.

4. Experiment with different methods of handling overflow and underflow with the arithmetic operations.

Digital Image Processing Lab Exercise #2: Imaging Geometry

1. Image enlargement is useful in a variety of applications. For example, some image processes require that two input images be in precise geometrical alignment prior to their combination. Enlarging a portion of an image can aid in this process. Additionally, enlargement may allow visible recognition of image degradation, helping in the selection of a restoration model. In graphic arts, images may be enlarged for a variety of reasons.

   Incorporate the CVIPtools function zoom (libgeometry) into your CVIPlab program. Experiment with enlarging an image by different factors. The minimum and maximum factors allowed are 1 and 10, respectively. You have the option of choosing the whole of the image, or any particular quadrant, or you can specify the starting row and column, and the width and height for the enlargement of the particular region of the image.
2. Image rotation allows spatial rotation of an input image. Using the equations from Chapter 2, write a function to rotate an image. Experiment with various degrees of rotation. Incorporate the CVIPtools rotate (libgeometry) function into your CVIPlab program. Does this differ from how your rotate function works?

Additional Work

1. Incorporate the CVIPtools functions crop and bilinear_interp (libgeometry) into your CVIPlab program. These two will provide similar functionality to the zoom function. Compare the results of using bilinear_interp to zoom. The zoom function performs a zero-order hold, while bilinear_interp performs a bilinear interpolation, providing a smoother appearance in the resulting image.
2. Enhance your rotate function to select the center portion of rotated image and enlarge it to the original image size.
3. Put the CVIPtools function spatial_quant into your CVIPlab program. Compare using the three different reduction methods available: average, median, and decimation.

Digital Image Processing Lab Exercise #3: Fourier Transform Exercise

For digital images, the gray level of each pixel represents the brightness of the image at that point. We can apply signal transform techniques, such as the Fourier transform, to convert digital images from the spatial domain to the spatial frequency (spectral) domain. This may be done to analyze the spatial frequency content of the digital image and to analyze the performance of a system like a lowpass filter or a highpass filter. The Fourier transform equations are given in Chapter 2.

Do the following:

1. Run CVIPtools and create simple geometric images with Utilities→Create. Perform a Fourier transform with the blocksize equal to the image size using Analysis→Transforms. In most cases it is observed that the frequency components are

perpendicular to the direction of the flat edges of the original image. Can you explain this? Use the *View spectrum* button in the main window to display the magnitude of an FFT output. Where is most of the energy concentrated? View the phase for several images. Can you see any correlation between the phase and the image?

2. Repeat step 1 using smaller blocksizes for the FFT; for example, use an 8×8 blocksize.
3. Repeat steps 1 and 2 using complex images.

Properties of 2-D Fourier Transform

A Fourier transform pair refers to an equation in a one domain, either spatial or spectral, and its corresponding equation in the other domain. This implies that if we know what is done in one domain, we know what will occur in the other domain.

Convolution

Convolution in one domain is the equivalent of multiplication in the other domain; this is what allows us to perform filtering in the spatial domain with convolution masks. These transform pairs (the double arrow indicates a Fourier transform pair) define this property:

$$I_1(r, c) * I_2(r, c) \Leftrightarrow F_1(u, v)F_2(u, v)$$
$$I_1(r, c)I_2(r, c) \Leftrightarrow F_1(u, v) * F_2(u, v)$$

where $*$ denotes the convolution operation

Note that it may be computationally less intensive to apply filters in the spatial domain of the image rather than the frequency domain of the image, especially if parallel hardware is available.

Do the following:

1. Apply a lowpass spatial filter mask, using the mean filter under Utilities→Filter, to an image. Apply lowpass filtering in the frequency domain with Analysis/Transforms (first perform the FFT, followed by the filter on the FFT output image). Compare the resultant images. Experiment with the mask size of the spatial filter and with the cutoff frequency and type of the frequency domain lowpass filter. Adjust these parameters until the resultant images look similar. Perform a Fourier transform on the resultant images and compare the spectra.
2. Apply a highpass spatial filter mask, using Utilities/Filter→Specify a Filter, to an image (hold the mouse on the pop-up menu to select a filter type). Apply highpass filtering in the frequency domain. Follow a process similar to what was done previously in step 1, and then compare the resultant images and spectra.

Translation

The translation properties of the Fourier transform are given by the following pairs:

$$I(r, c)\, e^{j2\pi\frac{(u_0 r + v_0 c)}{N}} \Leftrightarrow F(u - u_0, v - v_0)$$

and

$$I(r - r_0, c - c_0) \Leftrightarrow F(u, v)\, e^{-j2\pi\frac{(u r_0 + v c_0)}{N}}$$

where the double arrow indicates a Fourier transform pair. The first equation tells us that if the image is multiplied by a complex exponential (remember this is really a form of a sinusoidal wave), its corresponding spectrum is shifted. The second equation tells us that if the image is moved, the resulting Fourier spectrum undergoes a phase shift, but the magnitude of the spectrum remains the same.

Do the following:

1. Translate an image in the spatial domain, using Analysis/Geometry→Translate an Image with the default Wrap-around option. Now perform an FFT on the original image and the translated image. Compare the spectra of these images. In addition to the log remapped magnitude (the default spectral display), use the *View spectrum* button in the main window to compare the phase images.
2. Repeat step 1, but select the Fill option instead of Wrap-around. Experiment with different Fill values.

Rotation

The rotation property can be easily illustrated by using polar coordinates:

$$r = x\cos(\theta),\ c = x\sin(\theta)$$
$$u = w\cos(\phi),\ v = w\sin(\phi)$$

The Fourier transform pair $I(r, c)$ and $F(u, v)$ become $I(x, \theta)$ and $F(w, \phi)$, respectively, and we can write a Fourier transform pair to illustrate the rotation property as follows:

$$I(x, \theta + \theta_0) \Leftrightarrow F(w, \phi + \theta_0)$$

This property tells us that if an image is rotated by an angle θ_0, then $F(u, v)$ is rotated by the same angle, and vice versa.

Do the following:

1. Create an image of a vertical line with Utilities→Create. Perform the FFT on this image. In which direction do you see the frequency components? Now create an image of a horizontal line and perform the FFT. Which direction do you see the frequency components?
2. Create a 256 × 256 image of a rectangle from row 100 to 150, and column 100 to 150. Perform an FFT on this image. Now rotate the image, using Analysis→Geometry, by 45 degrees. Next, crop a 256 × 256 image from the center of this rotated image with Utilities→Size, and perform the FFT. Compare the resulting spectra. Did the rotation cause any artifacts that may be affecting the spectrum?

Digital Image Processing Lab Exercise #4: Basic Image Enhancement Techniques

1. Write a program to implement spatial convolution masks. Let the user select from one of the following masks:

   Lowpass filter masks (smoothing):

$$\frac{1}{9}\begin{bmatrix}1 & 1 & 1\\ 1 & 1 & 1\\ 1 & 1 & 1\end{bmatrix} \quad \frac{1}{10}\begin{bmatrix}1 & 1 & 1\\ 1 & 2 & 1\\ 1 & 1 & 1\end{bmatrix} \quad \frac{1}{16}\begin{bmatrix}1 & 2 & 1\\ 2 & 4 & 2\\ 1 & 2 & 1\end{bmatrix}$$

   Highpass filter masks (sharpening):

$$\begin{bmatrix}-1 & -1 & -1\\ -1 & 9 & -1\\ -1 & -1 & -1\end{bmatrix} \quad \begin{bmatrix}1 & -2 & 1\\ -2 & 5 & -2\\ 1 & -2 & 1\end{bmatrix} \quad \begin{bmatrix}0 & -1 & 0\\ -1 & 5 & -1\\ 0 & -1 & 0\end{bmatrix}$$

2. Modify the program to allow the user to input the coefficients for a 3×3 mask.
3. Experiment with using the smoothing and sharpening masks. Try images with and without added noise.

Additional Work

1. Incorporate the CVIPtools function median_filter into your CVIPlab program. Is it faster or slower than your median filtering function?
2. Compare the results of using your spatial masks to frequency domain filtering using CVIPtools.
3. Modify the program to handle larger masks. Expand the preceding masks as described in Chapters 2 and 4.
4. Write a median filtering function. Compare the median filter to lowpass filter masks for image smoothing.

Digital Image Processing Lab Exercise #5: Transform Filtering

The CVIPtools transform filtering routines use the filter definitions as defined in Chapter 2. The origin for the FFT, corresponding to the zero-frequency term (DC component), is in the center, while for all other transforms the origin is in the upper-left corner. The user chooses the blocksize before performing the transform. Because all the transforms implemented are fast transforms based on powers of 2, the blocksize must be a power of 2. If the selected blocksize will not evenly cover the image, then the image is zero-padded as required. For DC components located in the upper-left corner (such as the cosine and Walsh transform), CVIPtools lets you specify cutoff frequencies ranging from 1 to blocksize. For the FFT, with the DC term in the center, the range is from 1 to blocksize/2. This is a result of the nature of the spectrum as illustrated in Figures 2.5-13, 2.5-15, and 2.5-18.

Do the following:

1. This exercise illustrates some basic frequency-domain concepts on an artificially constructed image.

   a. Create a 256 × 256 image of a horizontal sinusoidal wave using Utilities→Create in CVIPtools. Perform the FFT on the image with a blocksize of 256 (the entire image). View the magnitude of the FFT output with the *View spectrum* button on the main window. Notice the frequency component corresponding to the horizontal frequency of the sinusoid. What is the approximate location of this frequency component? Notice also the location of the DC component (center). As is the case in most images, the DC component is the largest component in the transformed image.

   b. Now filter the transformed image with a bandreject filter. Leave the *Keep DC* button on. Choose an ideal bandreject filter. Choose cutoff frequencies to remove the horizontal sinusoidal frequency. Perform the inverse FFT on the image. Compare the resulting image with the original image. Did you succeed in removing most of the sinusoid?

2. This exercise illustrates the effects of lowpass filtering on a real image and compares the effects of ideal filters to those of butterworth filters.

   a. Load a 256 × 256 complex gray-scale image and perform the FFT on it. Use a blocksize of 16. Filter the transformed image using an ideal lowpass filter with a cutoff of 4. Keep the DC value. Note that CVIP tools will automatically perform the inverse transform. Notice the absence of high-frequency information and the ringing effect, which appears as waves emanating from edges in the image, caused by the sharp frequency cutoff.

   b. Now apply a butterworth lowpass filter of order 1 to the Fourier-transformed image. Use the same cutoff as in part 2a. Keep the DC during filtering; then note that CVIP tools will automatically perform the inverse FFT on the resulting image. Compare the result with that of part a. Is the ringing effect as noticeable now? Why/why not?

   c. Repeat part b using a sixth-order butterworth filter instead of a first-order filter. Compare the result to the result in parts a and b. Because the frequency response of a sixth-order butterworth filter is close to that of an ideal filter, the image should be similar to that of part a.

   d. Repeat a, b, and c using a blocksize of 256 × 256 and cutoff of 64.

3. This exercise illustrates the relationship between transform-domain filtering in the DCT, FFT, and WHT domain. It also gives you an opportunity to experiment with the CVIPtools transforms and filtering routines.

   a. Load any image of your choice. Remember that the time it takes for the transform to be performed depends on the image size; larger images will take longer, as will color images.

   b. Choose the ideal lowpass filter type, any blocksize, and any cutoff frequency (CF) which is evenly divisible by two. Apply this filter to the image in the Walsh domain. Repeat this procedure using the DCT and the FFT. Remember to use a cutoff frequency equal to CF/2 for the FFT.

   c. Compare the images resulting from filtering with different transforms.

d. Which transform resulted in the best quality of the filtered image? Which transform resulted in the poorest quality filtered image? Compare your answers with what you know about the properties of the DCT, FFT, and Walsh transforms from Chapter 2. Do your answers agree with what you would expect?

Digital Image Processing Lab Exercise #6: Histogram Modification I

1. Write a C function to implement a histogram stretch/shrink. Clip if the numbers go out of BYTE range.
2. Write a C function to perform a histogram slide on an image. Have your program find the maximum and minimum gray-level values in the image and calculate the largest value of a left or right slide that is possible before gray-level saturation occurs. Warn the users of this, but let them clip if desired.

Additional Work

1. Enhance your histogram stretch to allow for a specified percentage of pixels to be clipped at either the low end (set to zero) or high end (set to 255), or both.
2. Modify the histogram stretch to allow for out-of-range results, followed by a remap (see libmap).
3. Incorporate the CVIPtools histogram functions (libhisto) into your program. Compare them to the functions you wrote. Are the results the same? Are they faster or slower?

Digital Image Processing Lab Exercise #7: Histogram Modification II

1. Write a function to perform unsharp masking enhancement. Use the flowchart given in Section 4.3. Note that all the functions needed have already been written—lowpass filtering (via spatial convolution masks), subtraction, and histogram shrink and stretch. Experiment with various spatial lowpass filter masks and various values for the histogram shrink limits.
2. Incorporate the CVIPtools function histeq (libhisto) into your CVIPlab program.

Additional Work

1. Write your own histogram equalization function.
2. Experiment with histogram specification in CVIPtools.
3. Incorporate the CVIPtools function local_histeq into your CVIPlab. Compare the results of using this to the histeq function.

Digital Image Processing Lab Exercise #8: Project

The following process can be followed to streamline project development:

1. Use CVIPtools to explore algorithm development. Explore various options, sequences of operations, various parameter values, and so on, until desired results are achieved. Be creative; the beauty of CVIPtools is that anything can be tried and tested in a matter of seconds, and the results from different parameter values can easily be compared side by side.
2. Code the algorithm developed into the CVIPlab program, as follows:
   - **a.** Find the function name that corresponds to the CVIPtools function (see Chapter 9 and the appendices).
   - **b.** If desired, check the manual page on the CD-ROM to see an example of how the function is used in a C program.
   - **c.** Do a and b for all functions needed.
3. Put the functions into the CVIPlab program.
4. Write the necessary drivers to use the functions to implement the algorithm developed.
5. Test your program on images suitable for the application.

Project Topics

1. Find images that you want to process and improve. These may be images of poor contrast, blurred images, and so on, for example, personal photos taken by "Uncle Bob," medical images such as X-ray images, and "UFO" images. Use CVIPtools to explore image enhancement.
2. Incorporate the CVIPtools function gray_linear into your CVIPlab. Use this function to implement gray-level mapping pseudocolor. Apply it to X-ray images.
3. Implement a program for frequency domain pseudocolor enhancement. Apply it to ultrasound images.
4. Find images that have been degraded and for which a model is available, or can be developed, for the degradation. Examples include satellite images, medical images, and images from news stories (e.g., the JFK assassination). Use CVIPtools to explore image restoration.
5. Define a specific image domain of interest. Collect a number of these types of images. Explore image compression with CVIPtools. Design and perform subjective tests and compare the results to objective measures available in CVIPtools such as signal-to-noise ratio and RMS error.

8.3 The CVIPtcl and CVIPwish Shells

The CVIPtcl and CVIPwish shells are extensions to the Tcl and Tk application package. Tcl stands for Tool Command Language and is a scripting language as well as an interpreter for the language. Tk is an X-windows-based toolkit associated with Tcl, which can be used to integrate graphical interfaces into Tcl-based programs. Tcl/Tk was first developed by John Ousterhout at the University of California, Berkeley, in 1988 and is continually being enhanced and upgraded.

CVIPtcl is a Tcl shell into which all the CVIPtools functions have been incorporated. CVIPwish is basically CVIPtcl with the Tk libraries linked into it; this shell is the underlying basis for the current version of CVIPtools. These shells were developed using Tcl 7.6 and Tk 4.2. After an algorithm has been developed with CVIPtools, Tcl/Tk scripts are easily created and can be run under CVIPtcl and CVIPwish. This is a particularly attractive method for rapid prototyping because the scripts are very straightforward to create (for more information on creating Tcl/Tk scripts, see the references).

8.3.1 Using the CVIPtcl Shell

The CVIPtcl shell can be run by changing directories to the $CVIPHOME/ bin directory (/Src/CVIPTCL directory or the CD) and typing in the program name *cviptcl*. When this is done, the signon message and CVIPtcl prompt % will appear:

```
*****************************************************
*                  CVIPTCL/CVIPWISH                 *
*                   Version: 1.7d                   *
*           Scott Umbaugh, Kun Luo, Zhen Li         *
*               Arve Kjoelen, Mark Zuke             *
*               Wenxing Li, Yansheng Wei            *
*          Electrical Engineering Department        *
*   Southern Illinois University at Edwardsville    *
*****************************************************
%
```

Now, any of the Tcl, and specifically CVIPtcl, functions can be used. To see a list of all the CVIPtcl functions available, simply type *cvip* at the % prompt:

```
% cvip
   Available for cvip:

LIBARITHLOGIC:
add                 and                 divide              multiply
not                 or                  subtract            xor

LIBAND:
assemble_bands      extract_band

LIBCOLOR:
cxform              pseudocol_freq      luminance           lum_average
pct

LIBCOMPRESS:
bitplane_rlc        btc                 btc2                btc3
dpc                 fractal             gray_rlc            huffman
jpeg                vector_quanity      wvq                 zonal
zonal2              zvl                 rms_error           snr

LIBCONVERTER:
convert             gray_binary         halftone            read
write

LIBFEATURE:
area                aspect              centroid            euler
hist_feature        irregular           label               orientation
perimeter           projection          rst_invariant       spectral_feature
texture             thinness
```

```
LIBGEOMETRY:
bilinear_interp     black               circle              cosine_wave
create_checkboard   create_mesh         crop                copy_paste
cut_paste           display_mesh        enlarge             keyboard_mesh
line                rectangle           rotate              shrink
sine_wave           solve_bilinear      spatial_quant       square_wave
translate           warp                zoom

LIBHISTOGRAM:
gray_linear         gray_multiply       hist_shrink         hist_slide
hist_stretch        histeq              histo_spec          local_histeq

LIBMORPH:
morph_close         morph_dilate        morph_erode         morph_open
morpho

LIBNOISE:
gamma_noise         gaussian_noise      negative_exp_noise  rayleigh_noise
speckle_noise       uniform_noise

LIBSEGMENT:
fuzzyc              gray_quant          hist_thresh         igs
median_cut          multi_resolution    pct_median          sct_split
split_merge         threshold

LIBSPATIALFILTER:
ace_filter          adapt_median_filter adaptive_contrast   alpha_filter
gauss_blur          contra_filter       diff_spatial        edge_link
frei_chen           geometric_filter    harmonic_filter     highpass_spatial
hough               kirsch              laplacian           laplacian_spatial
lowpass_spatial     maximum_filter      mean_filter         median_filter
midpoint_filter     minimum_filter      mmse_filter         nightvision
prewitt             pyramid             raster_deblur       roberts
robinson            sharp               single_filter       smooth
sobel               specify_spatial     unsharp             ypmean_filter

LIBTRANSFORM:
dct                 fft                 fft_phase           haar
hadamard            walsh               wavelet

LIBXFORMFILTER:
bpf_filter          brf_filter          data_h_image        h_image
hfe_filter          homomorphic         hpf_filter        inverse_xformfilter
simple_wiener       wiener              parametric_wiener   least_squares
power_spect_eq      geometric_mean      lpf_filter          notch

UTILITY:
add_path            assemble_bands      datarange           data_get
data_remap          display             dq                  extract_band
fft_phase           gray_binary         history             info
list                load                read                rename
rms_error           save                snr                 view
viewer              write

%
```

These functions are divided into libraries, similar to the CVIPtools libraries. These functions are executed by typing *cvip*, the command name, followed by any necessary input parameters. To see the required input parameters for any given function simply type *cvip <function_name>*, as in the following example for the fft function:

```
% cvip fft
Usage: cvip fft <image> <forward/inverse(1/0)> <block_size(2^n)>
For argument(s) you don't need, pass -1.
%
```

This will provide the usage note, which describes how to use the function. In this case, the fft function requires three inputs: the name of the image to be processed, the forward or inverse transform, and the desired blocksize. Note that in CVIPtcl, as in CVIPtools, an image must be loaded into the image queue with the load (or read) command before it can be processed. The output image is automatically named and the name is printed to the shell; this image can then be viewed with the view (or display) command. A complete list of the usage notes and a cross-reference listing for the CVIPtools and CVIPtcl functions are included in the appendices.

8.3.2 References

Because CVIPtcl is an extension to the Tcl shell, it can be used for programming in the Tcl scripting language. The CVIPwish shell can be used for Tcl/Tk programming. Books on Tcl/Tk programming include:

Ousterhout, J. K., *Tcl and the Tk Toolkit*, Reading, MA: Addison-Wesley, 1994.

Welch, B. B., *Practical Programming in Tcl and Tk*, Upper Saddle River, NJ: Prentice Hall PTR, 1995.

Applicable Web sites include:

http://sunscript.sun.com

http://www.cimetrix.com/sven/xf.html

http://www.sco.com/Techology/tcl/Tcl.html

http://cuiwww.unige.ch/eao/www/TclTk.html

CHAPTER 9

CVIPtools Library Functions

9.1 Introduction

This chapter contains a brief description of each of the CVIPtools Toolbox libraries, and prototypes for all the functions. Some of the commonly used Toolkit functions from libband, libimage, and libmap are also included. The libraries are in alphabetical order, as are all the functions contained in each library. Additionally, pointers to related functions are included to ease the function search process. This information will facilitate the use of these functions in the CVIPlab program, or any other C program.

In general, many functions return pointers to CVIPtools Image structures (IMAGE or Image are both valid designations). If the return value is NULL, an error has occurred. The general philosophy regarding memory management is that *whoever has control is responsible.* This means that the memory used for any parameters passed to a function will be either used for return data or freed by that function. It also means that if programmers want to retain a data structure, they should pass a *copy* of it to any CVIPtools function. By clearly following this simple rule, memory leaks can be avoided. For details of the operation of a specific function, see the man pages on the CD-ROM. For a quick look at the function list see Appendix C.

9.2 Arithmetic and Logic Library—libarithlogic

Functions for the application of arithmetic and logic operations to images are contained in this library. These functions require one or two Image pointers as input and all return an Image pointer. Related functions, specifically multiplication functions that perform gray-level mapping, are in the library libhisto.

LIBARITHLOGIC FUNCTION PROTOTYPES

Image ***add_Image**(Image *inputImage1, Image *inputImage2)
<inputImage1> - pointer to an image
<inputImage2> - pointer to an image

Image ***and_Image**(Image *inputImage1, Image *inputImage2)
<inputImage1> - pointer to an image
<inputImage2> - pointer to an image

Image ***divide_Image**(Image *inputImage1, Image *inputImage2, CVIP_BOOLEAN zero2num)
<inputImage1> - pointer to an image
<inputImage2> - pointer to an image
<zero2num> - method of handling zeros in denominator

Image ***multiply_Image**(Image *inputImage1, Image *inputImage2)
<inputImage1> - pointer to an image
<inputImage2> - pointer to an image

Image ***not_Image**(Image *inputImage)
<inputImage> - pointer to an image

Image ***or_Image**(Image *inputImage1, Image *inputImage2)
<inputImage1> - pointer to an image
<inputImage2> - pointer to an image

Image ***subtract_Image**(Image *inputImage1, Image *inputImage2)
<inputImage1> - pointer to an image
<inputImage2> - pointer to an image

Image ***xor_Image**(Image *inputImage1, Image *inputImage2)
<inputImage1> - pointer to an image
<inputImage2> - pointer to an image

9.3 BAND IMAGE LIBRARY—LIBBAND

Although libband is a Toolkit library consisting of lower-level functions, these two are particularly useful so they are listed here. These functions allow for processing of individual bands of multiband images.

Image ***assemble_bands**(Image **inImgs, int noimgs);
<inImgs> - pointer to array of image pointers
<noimgs> - number of image pointers contained in the array.

Image ***extract_band**(Image *inImg, int bandno);
<inImgs> - pointer to image
<bandno> - band number to be extracted from the image.

9.4 COLOR IMAGE LIBRARY—LIBCOLOR

The color library, libcolor, primarily contains functions that modify color image information by a color transform. This includes principal components, luminance, and various color space transforms. In addition, the frequency domain pseudocolor is contained here, but the gray-level mapping pseudocolor that appears in CVIPtools uses the gray-level linear transform contained in libhisto.

LIBCOLOR FUNCTION PROTOTYPES

Image ***colorxform**(const Image *rgbImage, COLOR_FORMAT newcspace, float *norm, float *refwhite, int dir)

<rgbImage> - pointer to an image (data type equal to or less precise than type CVIP_FLOAT)

<newcspace> - desired color space, one of: RGB, HSL, HSV, SCT, CCT, LUV, LAB, XYZ

<norm> - pointer to a normalization vector

<refwhite> - pointer to reference white values (for LUV and LAB only)

<dir> - direction of transform (1 => (RGB→newcspace) else (newcspace→RGB)

Image ***ipct**(Image *imgP, CVIP_BOOLEAN is_mask, float *maskP)

<imgP> - pointer to an image

<is_mask> - whether to ignore a background color (CVIP_YES or CVIP_NO)

<maskP> - background color to ignore

Image ***luminance_Image**(Image *inIm)

<inIm> - pointer to an image

Image ***lum_average**(Image *input_Image)

<input_Image> - pointer to an image

Image ***pct**(Image *imgP, CVIP_BOOLEAN is_mask, float *maskP)

<imgP> - pointer to an image

<is_mask> - whether to ignore a background color (CVIP_YES or CVIP_NO)

<maskP> - background color to ignore

Image ***pct_color**(Image *imgP, CVIP_BOOLEAN is_mask, float *maskP, int choice)

<imgP> - pointer to Image structure

<is_mask> - whether to ignore a background color (CVIP_YES or CVIP_NO)

<maskP> - background color to ignore

<choice> - 1 = perform PCT, 2 = perform IPCT

Image ***pseudocol_freq**(Image *grayImage, int inner, int outer, int blow, int bband, int bhigh)

<grayImage> - input gray image

<inner> - low cutoff frequency
<outer> - high cutoff frequency
<blow> - map lowpass results to band # (R = 0, G = 1, B = 2)
<bband> - map bandpass results to band # (R = 0, G = 1, B = 2)
<bhigh> - map highpass results to band # (R = 0, G = 1, B = 2)
(note: blow != bband != bhigh)

9.5 COMPRESSION LIBRARY—LIBCOMPRESS

The image compression library contains functions that compress and decompress images, as well as associated functions. The compression functions write the compressed data file to disk; they return a 0 upon succesful completion and a –1 if an error occurs. The compressed data file is either in CVIPtools VIP format or a standard compression file format such as JPEG. The decompression functions take file names as input and output Image pointers to the decompressed image. Two utility functions, rms_error and srn, which return the root-mean-square error and signal-to-noise ratio are included.

LIBCOMPRESS FUNCTION PROTOTYPES

int **bit_compress**(Image *inputImage, char *filename, byte sect)
<inputImage> - pointer to the image
<filename> - pointer to a character array containing output file name
<sect> - bitmask of planes to retain

Image ***bit_decompress**(char *filename)
<filename> - pointer to a character array

Image ***bit_planeadd**(char *filename)
<filename> - pointer to a character array

int **btc_compress**(Image *inputImage, char *filename, int blocksize)
<inputImage> - pointer to the image
<filename> - pointer to character array containing output file name
<blocksize> - size of the block on which to operate

Image ***btc_decompress**(char *filename)
<filename> - pointer to a character array

int **btc2_compress**(Image *inputImage, char *filename, int blocksize)
<inputImage> - pointer to an image
<filename> - pointer to character string containing output file name
<blocksize> - blocksize

Image ***btc2_decompress**(char *filename)
<filename> - pointer to character string containing file name

int **btc3_compress**(Image *inputImage, char *filename, int blocksize)
<inputImage> - pointer to an image
<filename> - pointer to character string containing output file name
<blocksize> - blocksize

Image ***btc3_decompress**(char *filename)
<filename> - pointer to character array

int **dpc_compress**(Image *inputImage, char *filename, float ratio, int bit_length, int clipping, int direction, int origin)
<inputImage> - pointer to an image
<filename> - pointer to character string containing output file name
<ratio> - the correlation factor
<bit_length> - number of bits for compression (1 to 8)
<clipping> - clip to maximum value (1), otherwise 0
<direction> - scan image horizontally (0) or vertically (1)
<origin> - use original (1) or reconstructed (0) values

Image ***dpc_decompress**(char *filename)
<filename> - compressed file

int **glr_compress**(Image *inputImage, char *filename, int win)
<inputImage> - pointer to an image
<filename> - pointer to character string containing output file name
<win> - size of window (1–128)

Image ***glr_decompress**(char *filename)
<filename> - name of the compressed file

int **huf_compress**(Image *inputImage, char *filename)
<inputImage> - pointer to the image
<filename> - pointer to character string containing output file name

Image ***huf_decompress**(char *filename)
<filename> - pointer to character string containing file name

int **jpg_compress**(Image *cvipImage, char *filename, int quality, CVIP_BOOLEAN grayscale, CVIP_BOOLEAN optimize, int smooth, CVIP_BOOLEAN verbose, char *qtablesFile)
<cvipImage> - pointer to the image
<filename> - pointer to character string containing output file name
<quality> - quality factor, determines amount of compression
<grayscale> - output image gray scale only (CVIP_YES or CVIP_NO)?
<optimize> - fast or slower (better results) (CVIP_YES or CVIP_NO)?
<smooth> - smooth out artifacts (CVIP_YES or CVIP_NO)?
<verbose> - text messages during compression (CVIP_YES or CVIP_NO)?

<qtablesFile> - pointer to file containing user specified quantization tables (NULL pointer will use default tables)

Image ***jpg_decompress**(char *filename, int colors, CVIP_BOOLEAN blocksmooth, CVIP_BOOLEAN grayscale, CVIP_BOOLEAN nodither, CVIP_BOOLEAN verbose)

<filename> - pointer to character string containing file name

<colors> - number of colors to use

<blocksmooth> - postprocess to improve visual results for block artifacts (CVIP_YES or CVIP_NO)?

<grayscale> - output image gray scale (CVIP_YES or CVIP_NO)?

<nodither> - use no dithering on the output mage (CVIP_YES or CVIP_NO)?

<verbose> - text messages during compression (CVIP_YES or CVIP_NO)?

float ***rms_error**(Image *im1, Image *im2)

<im1> - pointer to the image

<im2> - pointer to the image

float ***snr**(Image *im1, Image *im2)

<im1> - pointer to the image

<im2> - pointer to the image

int **zon_compress**(Image *inputImage, char *filename, int block_size, int choice, int mask_type, float compress_ratio)

<inputImage> - pointer to an image

<filename> - pointer to character string containing output file name

<block_size> - a power of 2; kernel size is <block_size>^2

<choice> - transform to use: 1 = FFT 2 = DCT 3 = Walsh 4 = Hadamard

<mask_type> - type of kernel to use: 1 = triangle 2 = square 3 = circle

<compress_ratio> - compression ratio, from 1.0 (min) to (block_size*block_size/4) (max) for all kinds of transforms

Image ***zon_decompress**(char *filename)

<filename> - path name of the file that stores the compressed image

int **zon2_compress**(Image *inputImage, char *filename, int block_size, int choice, int mask_type, float compress_ratio)

<inputImage> - pointer to an image

<filename> - pointer to character string containing file name

<block_size> - a power of 2; kernel size is <block_size>^2

<choice> - transform to use: 1 = FFT, 2 = DCT, 3 = Walsh, 4 = Hadamard

<mask_type> - type of kernel to use: 1 = triangle, 2 = square, 3 = circle

<compress_ratio> - compression ratio, from 1.0 (min) to (block_size*block_size/4) (max) for all kinds of transforms

Image ***zon2_decompress**(char *filename)

<filename> - path name of the file that stores the compressed image

int **zvl_compress**(Image *inputImage, char *filename)
<inputImage> - pointer to an image
<filename> - pointer to character string containing output file name

Image ***zvl_decompress**(char *filename)
<filename> - pointer to character string containing file name

9.6 CONVERSION LIBRARY—LIBCONVERTER

The conversion library contains all the functions that convert the various image file types to the CVIPtools Image structure, and back from the Image structure to the file type. However, the programmer does not need to use these functions directly because the higher level read and write image functions (read_Image and write_Image) take care of any required overhead. The function that converts between gray code and natural binary code and a halftoning function are also in this library.

LIBCONVERTER FUNCTION PROTOTYPES

Image ***bintocvip**(char *raw_image, FILE *inputfile, int data_bands, COLOR_ORDER color_order, INTERLEAVE_SCHEME interleaved, int height, int width, CVIP_BOOLEAN verbose)
Image ***ccctocvip**(char *prog_name, FILE *cccfile, int verbose)

Image ***CVIPhalftone**(Image *cvip_Image, int halftone, int maxval, float fthreshval, CVIP_BOOLEAN retain_image, CVIP_BOOLEAN verbose)
<cvip_Image> - pointer to input image
<halftone> - indicates method used to convert from gray scale to binary (one of QT_FS, QT_THRESH, QT_DITHER8, QT_CLUSTER3, QT_CLUSTER4, QT_CLUSTER8)
<maxval> - specifies maximum range of input image (usually 255)
<fthreshval> - threshold value (for QT_THRESH) between [0.0 . 1.0].
<retain_image> - retain image after writing
<verbose> - shall I be verbose (CVIP_YES or CVIP_NO)?

void **cviptobin**(Image *raw_Image, char *raw_image, FILE *outputfile, COLOR_ORDER color_order, INTERLEAVE_SCHEME interleaved, CVIP_BOOLEAN verbose)

void **cviptoccc**(Image *cvip_Image, char *ccc_name, FILE *cccfile, int maxcolor, int dermvis, int verbose)

void **cviptoeps**(Image *cvip_Image, char *eps_name, FILE * outputfile, float scale_x, float scale_y, int band, CVIP_BOOLEAN verbose)

void **cviptogif**(Image *gif_Image, char *gif_name, FILE *outfp, int interlace, int verbose) void **cviptoitex**(Image *cvip_Image, char *cvip_name, FILE *outputfile, char *image_comment, int verbose)

void **cviptoiris**(Image *cvipImage, char *f_name,FILE *fp, int prt_type, int verb)

int **cviptojpg**(Image *cvipImage, char *filename, int quality, CVIP_BOOLEAN grayscale, CVIP_BOOLEAN optimize, int smooth, CVIP_BOOLEAN verbose, char *qtablesFile)

void **cviptopnm**(Image *cvip_Image, char *pnm_name, FILE *outfp, int verbose)

void **cviptoras**(Image *ras_Image, char * ras_name,FILE *outfp, int pr_type, int verbose)

void **cviptotiff**(Image *cvip_Image, char *tiff_name, unsigned short compression, unsigned short fillorder, long g3options, unsigned short predictor, long rowsperstrip, int verbose)

CVIP_BOOLEAN **cviptovip**(Image *cvipImage, char *filename, FILE *file, CVIP_BOOLEAN save_history, CVIP_BOOLEAN is_compressed, CVIP_BOOLEAN verbose)

Image ***epstocvip**(char *eps_image, FILE *inputfile, CVIP_BOOLEAN verbose)

Image ***giftocvip**(char *name, FILE *in, int imageNumber, int showmessage)

Image ***gray_binary**(Image *inputImage, int direction)
<inputImage> - pointer to an Image
<direction> - direction (0 = gray→binary 1 = binary→gray)

Image ***iristocvip**(char *f_name,FILE *fp, int format, int verb)

Image ***itextocvip**(char *itex_image, FILE *inputfile, CVIP_BOOLEAN verbose)

Image ***jpgtocvip**(char *filename, int colors, CVIP_BOOLEAN blocksmooth, CVIP_BOOLEAN grayscale, CVIP_BOOLEAN nodither, CVIP_BOOLEAN verbose)

Image ***pnmtocvip**(char *pnm_file, FILE *ifp, int format, int verbose)

Image ***rastocvip**(char *rasterfile, FILE * ifp, int verbose)

Image ***read_Image**(char *filename, int showmessages)
<filename> - pointer to a character string containing the file name
<showmessages> - shall I be verbose?

Image ***tifftocvip**(char *tiff_file, int verbose)

Image ***viptocvip**(char *filename, FILE *file, CVIP_BOOLEAN verbose) int **write_Image**(Image *cvip_Image, char *filename, CVIP_BOOLEAN retain_image, CVIP_BOOLEAN set_up, IMAGE_FORMAT new_format, CVIP_BOOLEAN showmessages)

<cvip_Image> - pointer to valid CVIP Image structure

<filename> - pointer to a character string containing the file name

<retain_image> - retain image after writing (CVIP_YES or CVIP_NO)?

<set_up> - run setup (CVIP_YES or CVIP_NO)?

<new_format> - enumeration constant specifying the format of the file to be read

<showmessages> - shall I be verbose (CVIP_YES or CVIP_NO)?

9.7 DISPLAY LIBRARY—LIBDISPLAY

The display library contains functions relating to image display and viewing. The view_Image function provides the interface for image viewing that is most accessible and flexible for the programmer. The view_Image function will use the current image viewer as set by the function setDisplay_Image. The CVIPtools RAM-based image viewer, RamViewer (which calls the function displayRAMImage), or any available disk-based viewer can be selected.

LIBDISPLAY FUNCTION PROTOTYPES

void **display_Image**(const char *image_name, IMAGE_FORMAT format);

<image_name> - name of the image file

<format> - the image format

void **display_RAMImage**(Image *inputImage, char *name)

<inputImage> - pointer to input image structure

<name> - character string used as image name in display window

char ***getDisplay_Image**(void);

void **setDisplay_Image**(char *viewer, char *format);

<viewer> - viewer name

<format> - intermediate image format for display

void **view_Image**(Image *inputImage, char *imagename)

<inputImage> - pointer to the input image structure

<imagename> - character string as the image name in display window

9.8 FEATURE EXTRACTION LIBRARY—LIBFEATURE

The feature extraction library contains the functions that extract binary (object), histogram, texture, and spectral features from images. All the feature functions require a labeled image, generated by the label function, and any spatial coordinate in the object of interest as input parameters. In addition, the spectral, texture, and histogram features require the original image as input. These three functions, called spectral_feature, texture, and hist_feature, use the object selected in the labeled image as a mask on the original image so that only the selected object is included in the calculations. Features can be extracted from any of the bands of a multiband image by first using the function extract_band in libband (in CVIPtools, this is done automatically for a multiband image).

LIBFEATURE FUNCTION PROTOTYPES

long **area**(Image *labeledImage, int r, int c)
<labeledImage> - pointer to the labeled image
<r> - row coordinate of a point on the labeled image
<c> - column coordinate of a point on the labeled image

double **aspect**(Image *labeledImage, int r, int c)
<labeledImage> - pointer to the labeled image
<r> - row coordinate of the point on the labeled image
<c> - column coordinate of the point on the labeled image

int ***centroid**(Image *labeledImage, int r, int c)
<labeledImage> - pointer to a labeled image
<r> - row coordinate of a point on the labeled image
<c> - column coordinate of a point on the labeled image

int **euler**(Image * labeledImage, int r, int c)
<labeledImage> - pointer to a labeled image
<r> - row coordinate of a point on the labeled image
<c> - column coordinate of a point on the labeled image

double ***hist_feature**(Image *originalImage, Image *labeledImage, int r, int c)
<originalImage> - pointer to the original image
<labeledImage> - pointer to the labeled image
<r> - row coordinate of a point on the labeled image
<c> - column coordinate of a point on the labeled image
Note: Returns five histogram features—mean, standard deviation, skew, energy, and entropy via a pointer to double. Its value is equal to the initial address of a one-dimensional array, which contains the five histogram features for each band. If the original image is a color image, the first five values are for band 0, the next five data are for band 1, and so on.

double **irregular**(Image *labeledImage, int r, int c)
<labeledImage> - pointer to a labeled image
<r> - row coordinate of a point on a labeled image
<c> - column coordinate of a point on a labeled image

Image ***label**(const Image *imageP)
<imageP> - pointer to an image

double **orientation**(Image * labeledImage, int r, int c)
<labeledImage> - pointer to a labeled image
<r> - row coordinate of a point on a labeled image
<c> - column coordinate of a point on a labeled image

int **perimeter**(Image *labeledImage, int r, int c)
<labeledImage> - pointer to a labeled image
<r> - row coordinate of a point on the labeled image
<c> - column coordinate of a point on the labeled image

int ***projection**(Image * labeledImage, int r, int c, int height, int width)
<labeledImage> - pointer to a labeled image
<r> - row coordinate of a point on the labeled image
<c> - column coordinate of a point on the labeled image
<height> - image height after the object of interest is normalized
<width> - image width after the object of interest is normalized

double ***rst_invariant**(Image *label_image, int row, int col)
<label_image> - pointer to a labeled image structure
<row> - a row coordinate within the object of interest
<column> - a column coordinate within the object of interest

POWER ***spectral_feature**(Image *originalImage, Image *labeledImage, int no_of_rings, int no_of_sectors, int r, int c)
<originalImage> - pointer to the original image
<labeledImage> - pointer to the labeled image
<no_of_rings> - number of rings
<no_of_sectors> - number of sectors
<r> - row coordinate of a point on the labeled image
<c> - column coordinate of a point on the labeled image

POWER data structure:

```
typedef struct
{
int         no_of_sectors;
int         no_of_bands;
int         imagebands;
double      *dc;
double*     sector;
double*     band;
} POWER;
```

TEXTURE ***texture**(const Image *ImgP, const Image *segP, int band, int r, int c, long int hex_equiv, int distance)
<ImgP> - pointer to source image structure

<segP> - pointer to labeled image structure

<band> - the band of the source image to be worked on

<r> - the row coordinate of the object

<c> - the column coordinate of the object

<hex_equiv> - the hex equivalent of the texture feature map

<distance> - the pixel distance to calculate the co-occurence matrix

TEXTURE data structure:

```
typedef struct {
/* [0] → 0 degree, [1] → 45 degree, [2] → 90 degree, [3] → 135 degree,
     [4] → average, [5] → range (max - min) */
float ASM[6];              /* (1) Angular Second Moment */
float contrast[6];         /* (2) Contrast */
float correlation[6];      /* (3) Correlation */
float variance[6];         /* (4) Variance */
float IDM[6];              /* (5) Inverse Difference Moment */
float sum_avg[6];          /* (6) Sum Average */
float sum_var[6];          /* (7) Sum Variance */
float sum_entropy[6];      /* (8) Sum Entropy */
float entropy[6];          /* (9) Entropy */
float diff_var[6];         /* (10) Difference Variance */
float diff_entropy[6];     /* (11) Difference Entropy */
float meas_corr1[6];       /* (12) Measure of Correlation 1 */
float meas_corr2[6];       /* (13) Measure of Correlation 2 */
float max_corr_coef[6];    /* (14) Maximal Correlation Coefficient */
} TEXTURE;
```

double **thinness**(Image *labeledImage, int r, int c)

<labeledImage> - pointer to a labeled image

<r> - row coordinate of a point on a labeled image

<c> - column coordinate of a point on a labeled image

9.9 Geometry Library—libgeometry

Libgeometry contains all functions relating to changing image size and orientation, as well as functions that create images of geometric shapes and sinusiodal waves. These functions all return Image structures, except mesh_to_file and display_mesh which are used as utility functions by the image warping and restoration function, mesh_warping.

LIBGEOMETRY FUNCTION PROTOTYPES

Image ***bilinear_interp**(Image *inImg, float factor)

<inImg> - pointer to an image

<factor> - factor > 1 to enlarge, factor < 1 to shrink

Image ***copy_paste**(Image *srcImg, Image *destImg, unsigned start_r, unsigned start_c, unsigned height, unsigned width, unsigned dest_r, unsigned dest_c, CVIP_BOOLEAN transparent);

<srcImg> - source image to copy the subimage
<destImg> - destination image for pasting
<start_r> - row value of the upper-left corner of the subimage on srcImg
<start_c> - column value of the upper-left corner of the subimage on srcImg
<height> - height of desired subimage
<width> - width of desired subimage
<dest_r> - row value of the upper-left corner of the destImg area to paste the subimage
<dest_c> - column value of the upper-left corner of the destImg area to paste the subimage
<transparent> - whether the paste is transparent or not

Image ***create_black**(int width, int height)
<width> - desired image width
<height> - desired image height

Image ***create_checkboard**(int im_width, int im_height, int first_c, int first_r, int block_c, int block_r)
<im_width> - image width, number of columns
<im_height> - image height, number of rows
<first_c> - first column of checkerboard
<first_r> - first row of checkerboard
<block_c> - width of checkerboard blocks
<block_r> - height of checkerboard blocks

Image ***create_circle**(int im_width, int im_height, int center_c, int center_r, int radius)
<im_width> - image width, number of columns
<im_height> - image height, number of rows
<center_c> - circle center column coordinate
<center_r> - circle center row coordinate
<radius> - radius of circle

Image ***create_cosine**(int img_size, int frequency, int choice)
<img_size> - number of rows (and columns) in new image
<frequency> - sine wave frequency
<choice> - enter 1 for horizontal, 2 for vertical cosine wave

Image ***create_line**(int im_width, int im_height, int start_c, int start_r, int end_c, int end_r)
<im_width> - image width, number of columns
<im_height> - image height, number of rows
<start_c> - first column coordinate of line
<start_r> - first row coordinate of line
<end_c> - last column of line

<end_r> - last row of line

Image ***create_rectangle**(int im_width, int im_height, int start_c, int start_r, int rect_width, int rect_height)
<im_width> - image width, number of columns
<im_height> - image height, number of rows
<start_c> - first column for rectangle
<start_r> - first row for rectangle
<rect_width> - width of rectangle
<rect_height> - rectangle height

Image ***create_sine**(int img_size, int frequency, int choice)
<img_size> - number of rows (and columns) in new image
<frequency> - sine wave frequency
<choice> - enter 1 for horizontal, 2 for vertical sine wave

Image ***create_squarewave**(int img_size, int frequency, int choice)
<img_size> - number of rows (and columns) in new image
<frequency> - sine wave frequency
<choice> - enter 1 for horizontal, 2 for vertical square wave

Image ***crop**(Image *imgP, unsigned row_offset, unsigned col_offset, unsigned rows, unsigned cols)
<imgP> - pointer to an image
<row_offset> - row coordinate of upper-left corner
<col_offset> - column coordinate of upper-left corner
<rows> - height of desired subimage
<cols> - width of desired subimage

int **display_mesh**(Image* inputImage, struct mesh *inmesh)
<inputImage> - pointer to an image structure
<inmesh> - pointer to input mesh structure

struct mesh ***keyboard_to_mesh**()
(mesh structure defined under mesh_warping function)

Image ***enlarge**(Image *cvipImage, int row, int col)
<cvipImage> - pointer to an image
<row> - number of rows for enlarged image
<column> - number of columns for enlarged image

void **mesh_to_file**(struct mesh *mesh_matrix, char* mesh_file);
<mesh_matrix> - mesh structure
<mesh_file> - file name for mesh file created

Image ***mesh_warping**(Image *inputImage, struct mesh *inmesh, int method);

<inputImage> - pointer to image structure

<inmesh> - mesh structure

<method> - method used for gray-level interpolation: 1 = nearest neighbor, 2 = bilinear interpolation, 3 = neighborhood average

Mesh data structure:

```
struct mesh_node {
        int x;
        int y;
    };
    struct mesh {
        int width;
        int height;
        struct mesh_node **nodes;
    };
```

Image ***rotate**(Image *input_Image, float degree)

<input_Image> - pointer to an image

<degree> - amount to rotate image (1–360)

Image ***shrink**(Image *input_Image, float factor)

<input_Image> - pointer to an image

<factor> - scaling factor (0.1–1.0)

int **solve_c**(struct mesh_node intie[4], struct mesh_node outtie[4], float *c)

<intie> - input tiepoints

<outtie> - output tiepoints

<c> - pointer to result array

Image ***spatial_quant**(Image *cvipImage, int row, int col, int method)

<cvipImage> - pointer to an image

<row> - number of rows for reduced image

<column> - number of columns for reduced image

<method> - reduction method to use where 1 = average, 2 = median, 3 = decimation

Image ***translate**(Image *cvipImage, CVIP_BOOLEAN do_wrap, int r_off, int c_off, int r_mount, int c_mount, int r_slide, int c_slide, float fill_out)

<cvipImage> - pointer to an image

<do_wrap> - wrap image during translation if CVIP_YES

<r_off> - row number of upper-left pixel in area to move

<c_off> - column number of upper-left pixel in area to move

<r_mount> - height of area to move

<c_mount> - width of area to move

<r_slide> - distance to slide vertically

<c_slide> - distance to slide horizontally

<fill_out> - value to fill vacated area in cut-and-paste

Image ***zoom**(Image *input_Image, int quadrant, int r, int c, int width, int height, float temp_factor)

<input_Image> - pointer to an image

<quadrant> - 1 = UL, 2 = UR, 3 = LL, 4 = LR, 5 = ALL, 6 = Specify (x, y), dx, dy

<r> - column coordinate of area's upper-left corner

<c> - row coordinate of area's upper-left corner

<width> - width of area to enlarge

<height> - height of area to enlarge

<temp_factor> - degree of enlargement

9.10 HISTOGRAM LIBRARY—LIBHISTO

The histogram library, libhisto, contains functions relating to modifying the image by histogram manipulation or gray-level mapping. The histogram shrink operation in CVIPtools is performed with the remap_Image function (see libmap).

LIBHISTO FUNCTION PROTOTYPES

float ****define_histogram**(int bands, int mode, char **eq)

<bands> - number of bands in the image

<mode> - prompt the user for input (mode = 1), or use <eq> (mode = 0)

<eq> - string for mapping equation

float ****get_histogram**(Image *inputP)

<input_image> - pointer to an image structure from which a histogram is obtained

Image ***get_histogram_Image**(Image *inputP)

<input_image> - pointer to an image structure from which a histogram is obtained

Image ***gray_linear**(Image * inputImage, double start, double end, double s_gray, double slope, int change, int band)

<inputImage> - pointer to an image

<start> - initial gray level to modify

<end> - final gray level to modify

<s_gray> - new initial gray level

<slope> - slope of modifying line

<change> - 0 = change out-of-range pixels to black, 1 = don't modify out-of-range pixel values

<band> - the band on which to operate

Image ***gray_multiply**(Image *input, float ratio)

<input> - pointer to an image

<ratio> - multiplier

Image ***gray_multiply2**(Image *input, float ratio)
<input> - pointer to an image
<ratio> - multiplier

Image ***histeq**(Image *in, int band)
<in> - pointer to an image
<band> - which band (0, 1, or 2) to operate on; use 0 for gray

void **histogram_show**(float **histogram)
<histogram> - a 2-D array containing a histogram for each image band

Image ***hist_spec**(Image *imageP, int mode, char **input)
<imageP> - pointer to an image
<mode> - prompt the user for input (mode = 1), or use <eq> (mode = 0)
<input> - a 2-D string array for mapping equation for each image band

Image ***histogram_spec**(Image* imageP, float **histogram)
<imageP> - pointer to the input image structure
<histogram> - the specified histogram

Image ***hist_slide**(Image *input, int slide)
<input> - pointer to an image
<slide> - amount of histogram slide

Image ***hist_stretch**(Image * inputImage, int low_limit, int high_limit, float low_clip, float high_clip)
<inputImage> - pointer to an image
<low_limit> - lower limit for stretch
<high_limit> - high limit for stretch
<low_clip> - percentage of low values to clip before stretching
<high_clip> - percentage of high values to clip before stretching

Image ***local_histeq**(Image *in, int size, int mb)
<in> - pointer to an image structure
<size> - desired blocksize
<mb> - RGB band on which to calculate histogram (0,1,2)

Image ***make_histogram**(float **histogram, IMAGE_FORMAT image_format, COLOR_FORMAT color_format)
<histogram> - a 2-D float array of the histogram data
<image_format> - the image format of the resulting image
<color_format> - the color format of the resulting image

void **showMax_histogram**(float **histogram, char *title)
<histogram> - pointer to a histogram pointer
<title> - name given to histogram image

9.11 Image Library—Libimage

Although libimage is a Toolkit library, consisting of lower level functions, some of the commonly used functions are included here for reference. For a complete listing of function prototypes and data structure details, see the man pages on the CD-ROM.

int **cast_Image**(Image *src, CVIP_TYPE dtype)
<src> - pointer to image structure
<type> - new data type

void **delete_Image**(Image *A)
<A> - pointer to image structure

Image ***duplicate_Image**(const Image *a)
<a> - pointer to image structure

unsigned **getNoOfBands_Image**(Image *image)
<image> - pointer to an image

unsigned **getNoOfCols_Image**(Image *image)
<image> - pointer to an image

unsigned **getNoOfRows_Image**(Image *image)
<image> - pointer to an image

CVIP_TYPE **getDataType_Image**(Image *image)
<image> - pointer to an image

Image ***new_Image**(IMAGE_FORMAT image_format, COLOR_FORMAT color_space, int bands, int height, int width, CVIP_TYPE data_type, FORMAT data_format)
<image_format> - original file format of image
<color_space> - current color space of image
<bands> - number of spectral bands
<height> - height of image (number of rows)
<width> - width of image (number of cols)
<data_type> - current data type of image
<data_format> - specifies real or complex data

9.12 Data Mapping Library—Libmap

Although libmap is a Toolkit library, consisting of lower level functions, some of the commonly used functions are included here for reference. For a complete listing of function prototypes, see the man pages on the CD-ROM.

Image ***condRemap_Image**(const Image *imageP, CVIP_TYPE dtype, unsigned dmin, unsigned dmax)

<imageP> - pointer to an image

<dtype> - datatype of data to be mapped

<dmin> - minimum value for range

<dmax> - maximum value for range

Image *__logMap_Image__(Image *image, int band)

<image> - pointer to image structure

<band> - the band to do log mapping: -1 = all bands, 0 = 1st band, 1 = 2nd band, etc.

Image *__remap_Image__(const Image *imageP, CVIP_TYPE dtype, unsigned dmin, unsigned dmax)

<imageP> - pointer to an Image

<dtype> - datatype of data to be mapped

<dmin> - minimum value for range

<dmax> - maximum value for range

9.13 MORPHOLOGICAL LIBRARY—LIBMORPH

The morphological library, libmorph, contains all functions relating to image morphology. All the functions return Image structures and will operate on binary images. All the functions but morphIterMod_Image and morpho, which implement the iterative method described in Chapter 2, will accept gray-level images as input. Note that these functions can be used on multiband images with the use of the extract_band and assemble_bands functions in libband.

These functions are available in two forms: those that allow the user to set up the matrix structure for the morphological kernel, and those that only require an integer to specify one of the predefined kernels. The first type, with an **_Image** extension appended to the function name, are more flexible but more difficult to use. The second type, without the **_Image** extension, are easier to use because they require parameters like the morphological functions in CVIPtools.

LIBMORPH FUNCTION PROTOTYPES

Image *__MorphClose_Image__(Image *inputImage, Matrix *kernelP, CVIP_BOOLEAN user_org, int row, int col)

<inputImage> - pointer to an image

<kernelP> - a pointer to a matrix structure

<user_org> - define center of kernel

<row> - user-defined row of kernel center

<col> - user-defined column of kernel center

Image *__MorphClose__(Image *inputImage, int k_type, int ksize, int height, int width)

<inputImage> - pointer to an input image structure

<k_type> - kernel type (1 = disk 2 = square 3 = rectangle 4 = cross)
<ksize> - size of the kernel (height and width of mask)
<height> - height for square and rectangle, thickness of lines for cross
<width> - width for rectangle, size of cross

Image ***MorphDilate_Image**(Image *inputImage, Matrix *kernelP, CVIP_BOOLEAN user_org, int row, int col)
<inputImage> - pointer to an image
<kernelP> - a pointer to a matrix structure
<user_org> - define center of kernel
<row> - user-defined row of kernel center
<col> - user-defined column of kernel center

Image ***MorphDilate**(Image *inputImage, int k_type, int ksize, int height, int width)
<inputImage> - pointer to an input image structure
<k_type> - kernel type (1 = disk 2 = square 3 = rectangle 4 = cross)
<ksize> - size of the kernel (height and width of mask)
<height> - height for square and rectangle, thickness of lines for cross
<width> - width for rectangle, size of cross

Image ***MorphErode_Image**(Image *inputImage, Matrix *kernelP, CVIP_BOOLEAN user_org, int row, int col)
<inputImage> - pointer to an image
<kernelP> - a pointer to a matrix structure
<user_org> - define center of kernel
<row> - user-defined row of kernel center
<col> - user-defined column of kernel center

Image ***MorphErode**(Image *inputImage, int k_type, int ksize, int height, int width)
<inputImage> - pointer to an input image structure
<k_type> - kernel type (1 = disk 2 = square 3 = rectangle 4 = cross)
<ksize> - size of the kernel (height and width of mask)
<height> - height for square and rectangle, thickness of lines for cross
<width> - width for rectangle, size of cross

Image ***morphIterMod_Image**(Image *binImage, const Matrix **surMATS, CVIP_BOOLEAN(*const boolFUNC)(CVIP_BOOLEAN a, CVIP_BOOLEAN b), int no_of_sur, int connectedness, int no_of_iter, int f)
<binImage> - pointer to an image (binary image)
<surMATS> - pointer to set S (surrounds) for which a_ij = 1 (See Figure 2.4-16 for definitions.)
<boolFUNC> - pointer to Boolean function of form L(a,b) (c_ij = L(a_ij,b_ij))
<no_of_sur> - number of surrounds

<connectedness> - the connectivity scheme being used; one of the constants: FOUR, EIGHT, SIX_NWSE, SIX_NESW

<no_of_iter> - number of iterations to perform

<f> - number of subfields into which the image tesselation will be divided

Image ***morpho**(const Image *binImage, const char *surround_str, CVIP_BOOLEAN rotate, int boolFUNC, int connectedness, unsigned no_of_iter, int fields)

<binImage> - pointer to image structure (binary image)

<surround_str> - pointer to a string holding the set of surrounds, such as "1, 7, 8" (See Figure 2.4-16 for definitions.)

<rotate> - rotate or not (CVIP_YES, CVIP_NO)

<boolFUNC> - integer number for the Boolean function (1-6): 1: 0, 2: !a, 3: ab, 4: a+b, 5: a^b, 6: (!a)b

<connectedness> - the connectivity scheme being used (FOUR, EIGHT, SIX_NWSE, or SIX_NESW)

<no_of_iter> - number of iterations to perform

<fields> - number of subfields into which the image tesselation will be divided

Image ***MorphOpen_Image**(Image *inputImage, Matrix *kernelP, CVIP_BOOLEAN user_org, int row, int col)

<inputImage> - pointer to an image

<kernelP> - a pointer to a matrix structure

<user_org> - define center of kernel

<row> - user-defined row of kernel center

<col> - user-defined column of kernel center

Image ***MorphOpen**(Image *inputImage, int k_type, int ksize, int height, int width)

<inputImage> - pointer to an input image structure

<k_type> - kernel type (1 = disk 2 = square 3 = rectangle 4 = cross)

<ksize> - size of the kernel (height and width of mask)

<height> - height for square and rectangle, thickness of lines for cross

<width> - width for rectangle, size of cross

9.14 NOISE LIBRARY—LIBNOISE

The noise library, libnoise, contains all functions that add noise to an image. The amount of noise added to the image, which will determine the signal-to-noise ratio, can be controlled through the variance parameter. The larger the variance, the more noise will be added. Note that a noise-alone image can be created by adding noise to an all black image (an all black image can be created using the function create_black in libgeometry).

LIBNOISE FUNCTION PROTOTYPES

Image *gamma_noise(Image *imageP, float *var, int *alpha)
<imageP> - pointer to an image structure
<var> - variance of the noise distribution
<alpha> - alpha parameter for gamma distribution

Image *gaussian_noise(Image *imageP, float *var, float *mean)
<imageP> - pointer to an image structure
<var> - variance of the noise distribution
<mean> - mean or average value for distribution

Image *neg_exp_noise(Image *imageP, float *var)
<imageP> - pointer to an image structure
<var> - variance of the noise distribution

Image *rayleigh_noise(Image *imageP, float *var)
<imageP> - pointer to an image structure
<var> - variance of the noise distribution

Image *speckle_noise(Image *imageP, float *psalt, float *ppepper)
<imageP> - pointer to an image structure
<psalt> - probability of salt noise (high gray level = 255)
<ppepper> - probability of pepper noise (low gray level = 0)

Image *uniform_noise(Image *imageP, float *var, float *mean)
<imageP> - pointer to an image structure
<var> - variance of the noise distribution
<mean> - mean or average value of distribution

9.15 SEGMENTATION LIBRARY—LIBSEGMENT

The segmentation library, libsegment, contains all functions that perform image segmentation. These functions all require an input Image structure and any parameters for the specific algorithm; they return the segmented images as Image structures.

LIBSEGMENT FUNCTION PROTOTYPES

Image *fuzzyc_segment(Image *srcImage, float variance)
<srcImage> - pointer to an image
<variance> - value for gaussian kernel variance

Image *gray_quant_segment(Image *cvipImage, int num_bits)
<cvipImage> - pointer to an image
<num_bits> - number of gray levels desired (2, 4, 8, ..., 128)

Image ***hist_thresh_segment**(Image *imgP)

<imgP> - pointer to an image structure

Image ***igs_segment**(Image *inputImage, int gray_level)

<inputImage> - input image pointer

<gray_level> - the number of gray levels desired (2, 4, 8, ..., 256)

Image ***median_cut_segment**(Image *imgP, int newcolors, CVIP_BOOLEAN is_bg, Color bg)

<impP> - pointer to an image

<newcolors> - desired number of colors

<is_bg> - is background color?

<bg> - background color

Color data structure:

```
struct ColorType {
  byte r, g, b;
};
typedef struct ColorType Color;
```

Image ***multi_resolution_segment**(Image *imgP, unsigned int choice, void *parameters, CVIP_BOOLEAN Run_PCT)

<imgP> - pointer to source image structure

<choice> - predicate test chosen: (1) pure uniformity; (2) local mean vs. global; (3) local standard deviation vs. global mean; (4) Number of pixels within two times standard deviation; (5) weighted gray-level distance test; (6) texture homogeneity test

<parameters> - cutoff value usage determined by predicate test

<Run_PCT> - choice to run PCT on color images

Image ***pct_median_segment**(Image *imgP, unsigned colors)

<impP> - pointer to an image

<colors> - desired number of colors

Image ***sct_split_segment**(Image *imgP, int A_split, B_split)

<imgP> - a pointer to an image structure

<A_split> - number of colors to divide along angle A

<B_split> - number of colors to divide along angle B

Image ***split_merge_segment**(Image *imgP, unsigned int level, unsigned int choice, void *parameters, CVIP_BOOLEAN Run_PCT)

<imgP> - pointer to source image structure

<level> - the level to begin procedure

<choice> - predicate test chosen: (1) pure uniformity; (2) local mean vs. global; (3) local standard deviation vs. global mean; (4) Number of pixels within two times standard deviation; (5) weighted gray-level distance test; (6) texture homogeneity test

<parameters> - cutoff value usage determined by predicate test

<Run_PCT> - choice to run PCT on color images

Image ***threshold_segment**(Image *inputImage, unsigned int threshval, CVIP_BOOLEAN thresh_inbyte)

<inputImage> - pointer to image structure

<threshval> - threshold value

<thresh_inbyte> - CVIP_NO (0) apply threshval directly to image data; CVIP_YES (1) threshval is CVIP_BYTE range; remap to image data range before thresholding

9.16 Spatial Filter Library—libspatialfilter

The spatial filter library, libspatialfilter, contains all functions relating to spatial filtering. The edge detection functions—Kirsch, Robinson, pyramid, laplacian, Sobel, Roberts, Prewitt, and Frei-Chen—can all be accessed via the function, edge_detect_filter, which has preprocessing and postprocessing functions built in. This library also contains the Hough transform, the unsharp masking algorithm, and the visual acuity/night vision simulation function described in Chapter 7. Note that some of the spatial filtering functions in CVIPtools (and in the CVIPtcl shell) are implemented using convolve_filter and get_default_filter.

LIBSPATIALFILTER FUNCTION PROTOTYPES

Image ***acuity_nightvision_filter**(Image *cvipImage, char reason, int threshold, int choice)

<cvipImage> - pointer to an image

<reason> - y = night vision, n = acuity simulation

<threshold> - binary threshold for night vision simulation (pass –1 if acuity selected)

<choice> - visual acuity value, 20, 30, 40, ... (pass –1 if night vision selected)

Image ***adaptive_contrast_filter**(Image *inputImage, float k1, float k2, unsigned int kernel_size, float min_gain, float max_gain)

<inputImage> - pointer to image structure

<k1> - local gain factor multiplier

<k2> - local mean multiplier

<kernel_size> - size of local window (must be odd)

<min_gain> - local gain factor minimum

<max_gain> - local gain factor maximum

Image ***alpha_filter**(Image *imageP, int mask_size, int p)

<imageP> - pointer to an image

<mask_size> - size of the filtering window ($3 \rightarrow 3 \times 3$)

<p> - number of maximum and minimum pixels to be excluded from the mean calculation

Image ***contra_filter**(Image *imageP, int mask_size, int p)

<imageP> - pointer to an image

<mask_size> - size of the filtering window ($3 \to 3 \times 3$)

<p> - filter order

Image ***convolve_filter**(Image *imageP, Matrix *filP)

<imageP> - pointer to an image

<filP> - pointer to a matrix containing the kernel to be convolved with <imageP>

Image ***edge_detect_filter**(Image *imageP, int program, int mask_choice, int mask_size, int keep_dc, int threshold, int threshold1, int thresh, int thr)

<imageP> - pointer to an image

<program> - desired edge detector: EDGE_KIRSCH, EDGE_ROBINSON, EDGE_PYRAMID, EDGE_LAPLACIAN, EDGE_SOBEL, EDGE_ROBERTS, EDGE_PREWITT, EDGE_FREI

<mask_choice> - type of smoothing filter: 1 = gaussian blur, 2 = generic lowpass 1, 3 = generic lowpass 2, 4 = neighborhood average

<mask_size> - laplacian/Roberts (1,2); Sobel/Prewitt (3, 5, 7)

<keep_dc> - 0 (no) or 1 (yes)

<threshold> - value for postprocessing binary threshold

<threshold1> - Frei-Chen projection method: 1 = project onto edge subspace, 2 = project onto line subspace, 3 = show complete projection

<thresh> - Frei-Chen projection threshold: 1 = set threshold on edge projection, 2 = set threshold on line projection, 3 = smallest angle between the above

<thr>—if <thresh> = 1 or 2, set threshold for angle (in radians) for Frei-Chen

Image ***edge_link_filter**(Image *cvipImage, int connection)

<cvipImage> - pointer to an image

<connection> - maximum connect distance

Image ***geometric_filter**(Image *imageP, int mask_size)

<imageP> - pointer to an image

<mask_size> - size of the filtering window ($3 \to 3 \times 3$)

Matrix ***get_default_filter**(PROGRAMS type, int dimension, int direction)

<type> - type of filter needed: BLUR_SPATIAL, DIFFERENCE_SPATIAL, LOWPASS_SPATIAL, LAPLACIAN_SPATIAL, HIGHPASS_SPATIAL

<dimension> - size of blur filter needed

<direction> - direction for difference filter; 0 = horizontal, 1 = vertical

Image ***harmonic_filter**(Image *imageP, int mask_size)

<imageP> - pointer to an image

<mask_size> - size of the filtering window ($3 \to 3 \times 3$)

Image ***hough_filter**(Image *cvipImage, char *name, char *degree_string, int threshold, int connection, int interactive)

<cvipImage> - pointer to a binary image structure
<name> - name of the input image
<degree_string> - a string indicating angles of interest
<threshold> - minimum number of pixels to define a line
<connection> - maximum distance to link on a line
<interactive> - 0 = use above parameters; 1 = read degree_string, threshold, and connection from standard input

Image ***maximum_filter**(Image *imageP, int mask_size)
<imageP> - pointer to an image
<mask_size> - size of the filtering window (3 → 3 x 3)

Image ***mean_filter**(Image *imageP, int mask_size)
<imageP> - pointer to an image
<mask_size> - size of the filtering window (3 → 3 × 3)

Image ***median_filter**(Image *inputImage, int size)
<inputImage> - pointer to an image
<size> - mask size (3, 5, 7, 9, ...)

Image ***midpoint_filter**(Image *imageP, int mask_size)
<imageP> - pointer to an image
<mask_size> - size of filtering window (3 → 3 × 3)

Image ***minimum_filter**(Image *imageP, int mask_size)
<imageP> - pointer to an image
<mask_size> - size of filtering window (3 → 3 × 3)

Image ***mmse_filter**(Image * inputImage, float noise_var, unsigned int kernel_size)
<inputImage> - pointer to an image
<noise_var> - noise variance of input image
<kernel_size> - kernel size (an odd number)

Image ***raster_deblur_filter**(Image *cvip_image)
<cvip_image> - pointer to an image structure

Image ***single_filter**(Image *orig_image, float s_c, float s_r, int r_cen, int c_cen, float rot, float beta, int N, float *h, int choice)
<orig_image> - pointer to an image
<s_c> - horizontal sizing factor, 1 for no change
<s_r> - vertical sizing factor, 1 for no change
<r_cen> - row coordinate for new center, 0 for no change
<c_cen> - column coordinate for new center, 0 for no change
<rot> - angle of rotation, 0 for no change
<beta> - value for beta, typically 0.3–0.8

<N> - kernel size (3, 5, 7, ...)

<h> - kernel array (of size N*N)

<choice> - operation of filter: 1 = (– –); 2 = (+ +); 3 = (+ –); 4 = (– +)

Image ***smooth_filter**(Image *inputImage, int kernel)

<inputImage> - pointer to an image

<kernel> - kernel size, from 2 to 10

Matrix ***specify_filter**(int row, int col, float **temp)

<row> - number of rows of mask

<col> - number of columns of mask

<temp> - mask value array

Image ***unsharp_filter**(Image *inputImage, int lower, int upper, float low_clip, float high_clip)

<inputImage> - pointer to an image structure

<lower> - lower limit for histogram shrink (0–254)

<upper> - upper limit for histogram shrink (1–255)

<low_clip> - percentage of low values to clip during hist_stretch

<high_clip> - percentage of high values to clip during hist_stretch

Image ***Ypmean_filter**(Image *imageP, int mask_size, int p)

<imageP> - pointer to an image structure

<mask_size> - size of the filtering window ($3 \rightarrow 3 \times 3$)

<p> - filter order

9.17 TRANSFORM LIBRARY—LIBTRANSFORM

The transform library, libtransform, contains all functions relating to frequency or sequency domain transforms. The transforms have all been implemented with fast algorithms, so they require the input images to have dimensions that are powers of two. If the input images have non-power-of-two dimensions, the images will be automatically padded with zeros to conform to this criterion. If zero-padding is required, using a small block size will minimize it. These functions all return image structures.

LIBTRANSFORM FUNCTION PROTOTYPES

Image ***fft_transform**(Image *input_Image, int block_size)

<input_Image> - pointer to an image

<block_size> - size of the subimages on which to perform the transform (e.g., 8 for 8×8 blocks)

Image ***fft_phase**(Image *fftImage, int remap_norm, float k)

<fftImage> - pointer to a complex image structure

<remap_norm> - 0 = remaps the phase data and returns a CVIP_BYTE image; 1 = normalizes the magnitude, using value of k, returns a complex image

<k> - constant to normalize the magnitude

Image ***dct_transform**(Image* inputImage, int blocksize)

<inputImage> - pointer to an image

<blocksize> - size of the subimages on which to perform the transform (e.g., 8 for 8×8 blocks)

Image ***haar_transform**(Image *in_Image, int ibit, int block_size)

<in_Image> - pointer to an image

<ibit> - 1 (forward transform) or 0 (inverse transform)

<block_size> - blocksize (4, 8, 16, ..., largest_dimension/2)

Image ***idct_transform**(Image* inputImage, int blocksize)

<inputImage> - pointer to an image

<blocksize> - blocksize used for forward transform

Image ***ifft_transform**(Image *in_Image, int block_size)

<in_Image> - pointer to an image

<block_size> - blocksize used for forward transform

Image ***wavdaub4_transform**(Image *image, int isign, int lowband)

<image> - pointer to an image

<isign> - 1 (forward transform) or 2 (inverse transform)

<lowband> - number of rows/(2^([(number of bands desired –1)/3]–1))

Image ***wavhaar_transform**(Image *image, int isign, int lowband)

<image> - pointer to an image

<isign> - 1 (forward transform) or 2 (inverse transform)

<lowband> - number of rows/(2^([(number of bands desired –1)/3]–1))

Image ***walhad_transform**(Image *in_Image, int ibit, int block_size)

<in_Image> - pointer to an image

<ibit> - 0 = inverse Walsh transform, 1 = Walsh transform, 2 = inverse Hadamard transform, 3 = Hadamard transform

<block_size> - blocksize (4, 8, 16, ..., largest_dimension/2)

9.18 Transform Filter Library—LIBXFORMFILTER

The transform filter library, libxformfilter, contains all functions relating to transform domain filtering. Lowpass, highpass, bandpass, and bandreject filters are available in both ideal and butterworth filter types. A high-frequency emphasis filter is also included, which uses a butterworth highpass filter. These filters take an image structure as input, which is assumed to be the output from a transform function, and they output the filtered transform data as an image structure. In order to get the filtered image back, the corresponding inverse transform must be applied to the output from these functions. These standard filters assume FFT symmetry; consequently, use with a non-FFT requires the use of the function nonfft_xformfilter. These filters all use a circular filter shape.

Frequency domain restoration filters, such as wiener filters and inverse filters are contained in this library. The restoration filters will accept either the original images or the transformed images as inputs. To obtain the restored image, the inverse FFT (ifft_transform) must be applied to the restoration filter output. A utility function to create various point spread function (PSF) images for these filters is called h_image. The homomorphic filter, typically used in image enhancement to equalize uneven contrast, is also contained in this library.

LIBXFORMFILTER FUNCTION PROTOTYPES

Image *__Butterworth_Band_Pass__(Image *in_Image, int block_size, int dc, int inner, int outer, int order)

<in_Image> - pointer to an image

<block_size> - desired blocksize

<dc> - drop (0) or retain (1) dc component

<inner> - inner cutoff frequency

<outer> - outer cutoff frequency

<order> - filter order

Image *__Butterworth_Band_Reject__(Image *in_Image, int block_size, int dc, int inner, int outer, int order)

<in_Image> - pointer to an image

<block_size> - desired blocksize

<dc> - drop (0) or retain (1) dc component

<inner> - inner cutoff frequency

<outer> - outer cutoff frequency

<order> - filter order

Image *__Butterworth_High__(Image *in_Image, int block_size, int dc, int cutoff, int order)

<in_Image> - pointer to an image

<block_size> - desired blocksize

<dc> - drop (0) or retain (1) dc component

<cutoff> - cutoff frequency

<order> - filter order

Image *__Butterworth_Low__(Image *in_Image, int block_size, int dc, int cutoff, int order)

<in_Image> - pointer to an image

<block_size> - desired blocksize

<dc> - drop (0) or retain (1) dc component

<cutoff> - cutoff frequency
<order> - filter order

Image ***geometric_mean**(Image *degr, Image *degr_fn, Image *p_noise, Image *p_orig, float gamma, float alpha, int choice, int cutoff)
<degr> - pointer to the degraded image
<degr_fn> - pointer to the degradation function
<p_noise> - pointer to the noise power spectral density
<p_orig> - pointer to the original image power spectral density
<gamma> - 'gamma' in the generalized restoration equation
<alpha> - 'alpha' in the generalized restoration equation
<choice> - sets the maximum gain using the DC value as a baseline
<cutoff> - cutoff frequency for filtering

Image ***High_Freq_Emphasis**(Image *in_Image, int block_size, int dc, int Cutoff, float alfa, int order)
<in_Image> - pointer to an image
<block_size> - desired blocksize
<dc> - drop (0) or retain (1) dc component
<Cutoff> - cutoff frequency
<alfa> - a constant (typically 1.0 to 2.0)
<order> - filter order

Image ***h_image**(int type, unsigned int height, unsigned int width)
<type> - mask type: 1 = Constant, 2 = Center weighted, 3 = Gaussian
<height> - height of the mask image
<width> - width of the mask image

Image ***homomorphic**(Image *cvipImage, float upper, float lower, int cutoff)
<cvipImage> - pointer to an image
<upper> - upper limit, > 1
<lower> - lower limit,< 1
<cutoff> - cutoff frequency

Image ***Ideal_Band_Pass**(Image *in_Image, int block_size, int dc, int inner, int outer)
<in_Image> - pointer to an image
<block_size> - desired blocksize
<dc> - drop (0) or retain (1) dc component
<inner> - inner cutoff frequency
<outer> - outer cutoff frequency

Image ***Ideal_Band_Reject**(Image *in_Image, int block_size, int dc, int inner, int outer)

<in_Image> - pointer to an image
<block_size> - desired blocksize
<dc> - drop (0) or retain (1) dc component
<inner> - inner cutoff frequency
<outer> - outer cutoff frequency

Image ***Ideal_High**(Image *in_Image, int block_size, int dc, int cutoff)
<in_Image> - pointer to an image
<block_size> - desired blocksize
<dc> - drop (0) or retain (1) dc component
<cutoff> - cutoff frequency

Image ***Ideal_Low**(Image *in_Image, int block_size, int dc, int cutoff)
<in_Image> - pointer to an image
<block_size> - desired blocksize
<dc> - drop (0) or retain (1) dc component
<cutoff> - cutoff frequency

Image ***inverse_xformfilter**(Image *numP, Image *denP, int choice, float cutoff)
<numP> - pointer to the numerator, the degraded image
<denP> - pointer to the denominator, the inverse filter (PSF)
<choice> - sets the maximum gain using the DC value as a baseline
<cutoff> - cutoff frequency

Image ***least_squares**(Image *degr, Image *degr_fn, Image *snr_approx, float gamma, int choice, int cutoff)
<degr> - pointer to the degraded image
<degr_fn> - pointer to the degradation function
<snr_approx> - pointer to smoothness function image
<gamma> - gamma in least-squares equation
<choice> - sets the maximum gain using the DC value as a baseline
<cutoff> - cutoff frequency

Image ***nonfft_xformfilter**(Image *imgP, int block_size, int dc, int filtertype, int p1, float p2, int order)
<imgP> - pointer to an image structure
<block_size> - size of blocks used in the transform
<dc> - retain the dc term (1) or not (0)
<filtertype> - one of IDEAL_LOW, BUTTER_LOW, IDEAL_HIGH, BUTTER_HIGH, IDEAL_BAND, BUTTER_BAND, IDEAL_REJECT, BUTTER_REJECT, HIGH_FREQ_EMPHASIS
<p1> - cutoff frequency for lowpass and highpass, lower cutoff for bandpass
<p2> - upper cutoff for bandpass filters, offset for high-frequency emphasis
<order> - filter order, if butterworth filter selected

Image *__notch__(Image *cvipImage, char *name, NOTCH_ZONE *zone, int number, CVIP_BOOLEAN interactive)

<cvipImage> - input image data

<name> - the name of the image

<zone> - a data structure containing information about which part of the image to remove

<number> - number of notches to perform (ignored if interactive = CVIP_YES)

<interactive> - ask for input from keyboard (CVIP_YES or CVIP_NO)

NOTCH_ZONE data structure:

```
typedef struct {
        int x;
        int y;
        int radius;
    } NOTCH_ZONE
```

Image *__parametric_wiener__(Image *degr, Image *degr_fn, Image *p_noise, Image *p_orig, float gamma, int choice, int cutoff)

<degr> - pointer to the degraded image

<degr_fn> - pointer to the degradation function

<p_noise> - pointer to the noise power spectral density

<p_orig> - pointer to the original image power spectral density

<gamma> - 'gamma' in the parametric wiener filter equation

<choice> - sets the maximum gain using the DC value as a baseline

<cutoff> - cutoff frequency for filtering

Image *__power_spect_eq__(Image *degr, Image *degr_fn, Image *p_noise, Image *p_orig, int choice, int cutoff)

<degr> - pointer to the degraded image

<degr_fn> - pointer to the degradation function

<p_noise> - pointer to the noise power spectral density

<p_orig> - pointer to the original image power spectral density

<choice> - sets the maximum gain using the DC value as a baseline

<cutoff> - cutoff frequency for filtering

Image *__simple_wiener__(Image *degr, Image *degr_fn, float k)

<degr> - pointer to the degraded image

<degr_fn> - pointer to the degradation function

<k> - a constant

Image *__wiener__(Image *degr, Image *degr_fn, Image *p_noise, Image *p_orig, int choice, int cutoff)

<degr> - pointer to the degraded image

<degr_fn> - pointer to the degradation function

<p_noise> - pointer to the noise power spectral density

<p_orig> - pointer to the original image power spectral density
<choice> - sets the maximum gain using the DC value as a baseline
<cutoff> - cutoff frequency for filtering

III

Appendices

APPENDIX A

The CVIPtools CD-ROM

The CVIPtools CD-ROM contains all the necessary files and information to set up and maintain a CVIPtools environment. This includes:

- ☞ A README file, which contains detailed information regarding the installation of CVIPtools and other useful information
- ☞ CVIPtools source code
- ☞ CVIPtools libraries for supported operating systems
- ☞ CVIPtools executables for supported operating systems
- ☞ CVIPtools environment Install program
- ☞ CVIPtcl interface source code
- ☞ CVIPwish interface source code
- ☞ CVIPtools GUI Tcl/Tk scripts
- ☞ Tcl/Tk source code
- ☞ Tcl/Tk libraries for supported operating systems
- ☞ Makefiles for Win32 and Imakefiles for UNIX, for CVIPtools compilation
- ☞ CVIPlab source code
- ☞ CVIPlab executables for supported operating systems
- ☞ Makefiles and Imakefiles for CVIPlab compilation
- ☞ UNIX man pages
- ☞ HTML versions of man pages
- ☞ Scripts for setting up the CVIPlab, adding libraries, adding CVIPtools headers, creating man pages, and other useful things
- ☞ Images

To get CVIPtools updates and other useful information see the CVIPtools home page on the World Wide Web: *http://www.ee.siue.edu/CVIPtools.*

APPENDIX B

Setting Up and Updating Your CVIPtools Environment

The entire CVIPtools environment can be set up on any computer running UNIX or Windows NT/95 operating systems; executables for supported operating systems are included on the CD-ROM. The CVIPlab and the CVIPtools libraries can be set up on any computer with an ANSI-C compiler. On the CD-ROM are README files which contain the specific information for the various operating systems and configurations. There are separate README files for UNIX and Windows NT/95 operating systems, named README.unix and README.win32.

Installation files are included that will guide you in setting up your CVIPtools environment. For UNIX operating systems, use the shell script called install.csh. For Windows NT/95, use the installation batch file called install.bat. These installation files will prompt you for any system-specific or configuration information required that it cannot find automatically. Note that under UNIX the install process can automatically initialize system environment variables, whereas under Windows NT/95 this must be done manually after the installation process.

B.1 GETTING CVIPTOOLS SOFTWARE UPDATES

CVIPtools software updates can be obtained from the Internet via the World Wide Web, or directly over Internet with ftp.

B.1.1 To get via the WWW

☞ Access the CVIPtools Homepage at *http://www.ee.siue.edu/CVIPtools.*

☞ Follow the directions for getting CVIPtools.

To get from the CVIPtools ftp site:

- ☞ In a command shell, type in *ftp ftp.ee.siue.edu.*
- ☞ Login as *ftp* and use your email address as the password.
- ☞ Go to the /dist/CVIPtools directory and get the README file.
- ☞ Read the README file, which contains all the necessary information for getting and installing CVIPtools.

APPENDIX C

CVIPtools Functions

This document contains the most recent listing of all functions available to CVIPtools developers. The functions are grouped by class and library. There are two classes of libraries within CVIPtools—Toolkit and Toolbox.

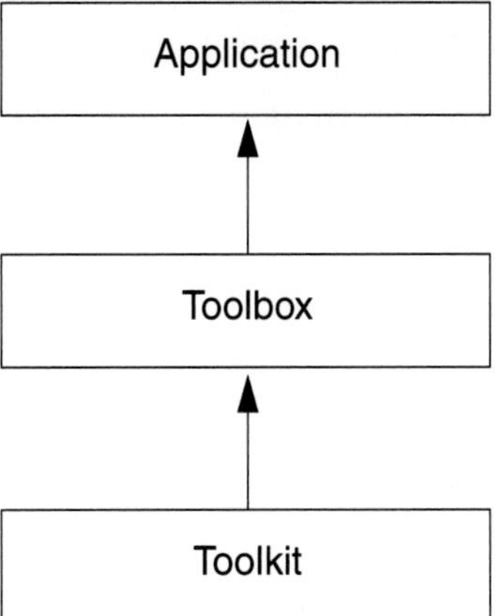

The Toolkit libraries contain low-level functions, such as data handling and memory management. The Toolbox libraries contain the functions that are typically used by CVIPlab programmers, such as transforms or segmentation routines. This organization is a hierarchical grouping of libraries in which each class successively builds upon the previous class by using the lower-level functions to create higher-level functions. For more detailed information on a particular function see the associated UNIX manual page (on CD-ROM). These pages are in standard man page format for UNIX users and in HTML format for Windows NT/95 users.

C.1 TOOLKIT LIBRARIES

LIBBAND—data handling of multispectral imagery

assemble_bands	- assembles multiband image from single-band images
bandcast	- casts image data to greater precision
bandcopy	- copies band data
band_minmax	- finds the min and max values of each band
extract_band	- extracts one band from a multiband image
matalloc	- allocates an array of matrices
matfree	- frees memory allocated by matalloc
vecalloc	- allocates an array of vectors
vecfree	- frees memory allocated by vecalloc

LIBIMAGE—basic image class methods for type conversion, memory management, etc. (*Note: See $CVIPHOME/include/CVIPimage.h for the* ***get*** *and* ***set*** *macros, such as getData_Image.*)

cast_Image	- casts an image
delete_Image	- Image class destructor
dump_Image	- prints image information
duplicate_Image	- creates a new instance of an existing image
getBand_Image	- references a band of matrix data
getBandVector_Image	- unloads image bands into a vector
getColorSpace_Image	- gets color space of image (e.g., RGB, GRAY, etc.)
getDataFormat_Image	- gets data format (i.e., REAL or COMPLEX)
getDataType_Image	- gets data type of image (e.g., CVIP_BYTE, CVIP_FLOAT, etc.)
getData_Image	- returns pointer to data (macro in CVIPimage.h)
getFileFormat_Image	- gets file format of image (e.g., PPM, PGM, etc.)
getImagPixel_Image	- reads an imaginary pixel sample from the image
getImagRow_Image	- references an imaginary row of the image
getNoOfBands_Image	- gets number of data bands of image
getNoOfCols_Image	- gets width of image
getNoOfRows_Image	- gets height of image
getPixel_Image	- same as "getRealPixel_Image"
getRealPixel_Image	- reads a real pixel sample from the image
getRealRow_Image	- references a real row of the image
getRow_Image	- same as "getRealRow_Image"

makeComplex_Image	- makes real image complex
makeReal_Image	- makes complex image real
new_Image	- Image class constructor
setBand_Image	- adds a new reference to a band of matrix data
setImagPixel_Image	- writes an imaginary pixel sample to the image
setPixel_Image	- same as "setRealPixel_Image"
setRealPixel_Image	- writes a real pixel sample to the image data
history_add	- adds info to image history structure
history_check	- checks if an operation has been done on an image
history_copy	- copies information from old_story
history_get	- gets info about an operation done on an image
history_show	- sets up routine for history print
history_print	- performs output of history structure to h_story

LIBIO—general-purpose I/O, memory management routines

allocMatrix3D_CVIP	- allocates memory for volume matrix
allocMatrix_CVIP	- allocates memory for regular matrix
close_CVIP	- closes a file for reading or writing
error_CVIP	- prints error message to terminal
freeMatrix3D_CVIP	- frees memory associated with volume matrix
freeMatrix_CVIP	- frees memory associated with regular matrix
getFloat_CVIP	- gets floating point value from the user
getInt_CVIP	- gets integer value from user
getString_CVIP	- gets character string
getUInt_CVIP	- gets unsigned integer value from user
init_CVIP	- parses standard info from command line
msg_CVIP	- prints regular message to terminal
openRead_CVIP	- opens a file for reading (handles "stdin")
openWrite_CVIP	- opens a file for writing (handles "stdout")
perror_CVIP	- prints system error message to terminal
print_CVIP	- same as "msg_CVIP" minus extra argument
quiet_CVIP	- turns off messaging
usage_CVIP	- prints usage message
verbose_CVIP	- turns on messaging

LIBMANAGER—object managers/handlers

addhead_DLL	- adds link to head of list

addnext_DLL	- adds link following the current link
addtail_DLL	- adds link to tail of list
delete_DLL	- double linked list class destructor
find_DLL	- finds a particular object in the list
head_DLL	- sets current link to head of list
isempty_DLL	- is the list empty?
ishead_DLL	- is current link pointing to head?
istail_DLL	- is current link pointing to tail?
new_DLL	- double linked list class constructor
next_DLL	- points to next link
previous_DLL	- points to previous link
print_DLL	- prints list
print_reverse_DLL	- prints list in reverse order
promote_DLL	- promotes current link to head of list
removecurr_DLL	- removes current link
removehead_DLL	- removes link from head of list
removetail_DLL	- removes link from tail of list
replace_DLL	- replaces object pointed to by current link
retrieve_DLL	- retrieves object pointed to by current link
size_DLL	- gets size of list (number of links)
tail_DLL	- sets current link to tail of list
addhead_LL	- adds link to head of list
addnext_LL	- adds link following the current link
delete_LL	- linked list class destructor
find_LL	- finds a particular object in the list
head_LL	- sets current link to head of list
isempty_LL	- is the list empty?
ishead_LL	- is current link pointing to head?
istail_LL	- is current link pointing to tail?
new_LL	- linked list class constructor
next_LL	- points to next link
previous_LL	- points to previous link
print_LL	- prints list
promote_LL	- promotes current link to head of list
removehead_LL	- removes head link
removenext_LL	- removes next link

replace_LL	- replaces object pointed to by current link
retrieve_LL	- retrieves object pointed to by current link
size_LL	- returns size of list
tail_LL	- sets current link to tail of list
addobject_HT	- adds object using separate chaining technique
delete_HT	- Hash Table class destructor
findobject_HT	- finds object
new_HT	- Hash Table class constructor
setkey_HT	- sets the hash table key
isempty_Stack	- determines whether a stack is empty
new_Stack	- creates a new instance of an object stack
pop_Stack	- pops an object off the stack
push_Stack	- pushes an object onto the stack

LIBMAP—image data mapping functions

condRemap_Image	- if the range is 0–255, no remap is done; if it exceeds this range, it is linearly remapped to 0–255
linearTrans_Image	- performs linear mapping of an image through a transformation matrix
logMap_Image	- maps image data logarithmically for better display of FFT spectral images
remap_Image	- maps image data into a specified range
trun_Image	- remaps image data, maintains relative size of each data band

LIBMATRIX—matrix algebra, manipulation and numerical analysis routines (*Note: See $CVIPHOME/include/CVIPmatrix.h for the* ***get*** *and* ***set*** *macros, such as getData_Matrix.*)

add_Matrix	- adds two matrices
and_Matrix	- performs a bitwise AND on two matrices
cbrt_Matrix	- finds cube root of a matrix (real/complex)
clone_Matrix	- returns a new matrix
conj_Matrix	- finds complex conjugate of matrix
copy_Matrix	- copies matrix a to matrix b
covariance_Matrix	- finds the covariance estimate of N data bands
crop_Matrix	- creates a new matrix from region of original
delete_Matrix	- Matrix class destructor
det_Matrix	- finds the determinant of a matrix

duplicate_Matrix	- creates new instance of an existing matrix
eigenSystem_Matrix	- finds the eigenvectors of a matrix
fastCopy_Matrix	- faster copy if data types are the same
getDataFormat_Matrix	- gets data format (i.e., REAL or COMPLEX)
getData_Matrix	- same as "getRealData_Matrix"
getDataType_Matrix	- gets data type of matrix (e.g., CVIP_BYTE, CVIP_FLOAT, etc.)
getImagData_Matrix	- references imaginary data (mapped into rows)
getImagRow_Matrix	- gets row of imaginary row
getImagVal_Matrix	- gets an "imaginary" matrix element
getNoOfCols_Matrix	- gets number of columns in matrix
getNoOfRows_Matrix	- gets number of rows in matrix
getRealData_Matrix	- references real data (mapped into rows)
getRealRow_Matrix	- gets row of real data
getRealVal_Matrix	- gets a "real" matrix element
getRow_Matrix	- same as "getRealRow_Matrix"
getVal_Matrix	- same as "getRealVal_Matrix"
invert_Matrix	- inverts a matrix
mag_Matrix	- finds magnitude of a matrix (real/complex)
makeComplex_Matrix	- makes real matrix complex
makeReal_Matrix	- makes complex matrix real
mult_Matrix	- performs vector multiplication of two matrices
multPWise_Matrix	- performs piecewise multiplication
new_Matrix	- Matrix class constructor
print_Matrix	- prints contents of matrix in row major form
read_Matrix	- reads a matrix structure from disk
rect2pol_Matrix	- converts from rectangular to polar coordinates
scale_Matrix	- scales a matrix by some factor
setImagVal_Matrix	- sets an "imaginary" matrix element
setRealVal_Matrix	- sets a "real" matrix element
setVal_Matrix	- same as "setRealVal_Matrix"
sqrt_Matrix	- finds square root of matrix (real/complex)
square_mag_Matrix	- finds magnitude squared of a matrix (real/complex)
sub_Matrix	- subtracts two matrices
transpose_Matrix	- finds the transpose of a matrix
write_Matrix	- writes a matrix structure to disk

LIBOBJECT—object analysis and identification routines

build_ChainCode	- finds the chain-code "contour" of an object
delete_ChainCode	- deletes an instance of a chain code object
delete_Object	- object class destructor
drawBB_Objects	- draws a bounding box around all objects
draw_ChainCode	- draws the contour of an object onto an image using the object's chain code
getProp_Object	- finds object moment properties
getProp_Objects	- finds moment properties of multiple objects
getXY_ChainCode	- turns a chain code into a list of X-Y coordinates
label_Objects	- sequentially labels objects (used by label function in libfeature)
listToVector_Objects	- creates an object vector from an object list
match_Object	- matches an object
new_ChainCode	- creates a new instance of a chain code object
new_Object	- Object class constructor
print_ChainCode	- prints the chain code results to a file
printLabel_Objects	- prints an object list to a file
print_Object	- prints object statistics to file
printProp_Objects	- prints a list of object properties to a file
read_ChainCode	- reads a chain code from a file
readLabel_Objects	- reads an object list from a file
read_Object	- reads object statistics from disk
readProp_Objects	- reads a list of object properties from a file
report_ChainCode	- prints out the chain code values
trimList_Objects	- trims an object list based on properties

LIBROI—region/area of interest designation, manipulation of an image

asgnFullImage_ROI	- assigns ROI as full image dimension
asgnImage_ROI	- assigns a ROI to an image
delete_ROI	- ROI class destructor
getDataFormat_ROI	- gets data format of ROI
getDataType_ROI	- gets data type of ROI
getHorOffset_ROI	- gets horizontal offset from pixel (0, 0)
getHorSize_ROI	- gets height/horizontal size of region
getImagPixel_ROI	- gets/reads imaginary pixel sample from ROI
getImagRow_ROI	- references imaginary row from the ROI

getNoOfBands_ROI	- gets number of data bands in ROI
getNoOfCols_ROI	- same as "getHorSize_ROI"
getNoOfRows_ROI	- same as "getVerSize_ROI"
getPixel_ROI	- same as "getRealPixel_ROI"
getRealPixel_ROI	- gets/reads real pixel sample from ROI
getRealRow_ROI	- references real row from the ROI
getRow_ROI	- same as "getRealRow_ROI"
getVerOffset_ROI	- gets vertical offset from pixel (0,0)
getVerSize_ROI	- gets width/vertical size of region
loadRow_ROI	- loads data from a buffer into ROI
new_ROI	- ROI class constructor
setImagPixel_RO	- sets/writes imaginary pixel sample to ROI
setPixel_ROI	- same as "setRealPixel_ROI"
setRealPixel_ROI	- sets/writes real pixel sample to ROI
unloadRow_ROI	- unloads row of data from ROI into buffer

LIBVECTOR—vector algebra and manipulation routines

band2pixel_Vector	- converts a band vector to a pixel vector
convolve_Vector	- convolves two vectors
copy_Vector	- copies vector *a* to vector *b*
findHisto_Vector	- finds the histogram of a vector
findMaxVal_Vector	- returns maximum value in vector
findMinVal_Vector	- returns minimum value in vector
normalize_Vector	- normalizes a vector between 0 and 1
pixel2band_Vector	- converts a pixel vector to a band vector
printHisto_Vector	- prints histogram values out to a file
subSample_Vector	- subsamples a list of vector points

C.2 Toolbox Libraries

LIBARITHLOGIC—arithmetic and logical operations on images

add_Image	- adds two images
and_Image	- performs a logical AND on two images
divide_Image	- divides one image by another
multiply_Image	- multiplies two images
not_Image	- performs a logical NOT on an image

or_Image	- performs a logical OR on two images
subtract_Image	- subtracts one image from another
xor_Image	- performs a logical XOR on two images

LIBCOLOR—color map utilities and color transforms

colorxform	- performs seven color transforms, and inverse transforms
ipct	- performs the inverse principal components transform
luminance_Image	- performs color to luminance transform
lum_average	- performs color to monochrome using average of all bands
pct	- performs the principal components transform in RGB-space
pct_color	- handles both forward and inverse PCT
pseudocol_freq	- pseudocolor transform using FFT spectrum and filters

LIBCOMPRESS—image compression/data reduction routines

bit_compress	- decomposes gray-level image into eight bit planes; each bitplane is then run-length coded and stored in a binary file
bit_decompress	- decompresses each binary file (corresponding to a particular bitplane) into corresponding binary images
bit_planeadd	- decompresses bitplane files and adds any combinations of bitplanes to produce the resultant gray-level image
btc_compress	- compresses the image with two-level BTC
btc_decompress	- decompresses the image from two-level BTC file
btc2_compress	- multilevel block truncation coding (BTC) image compression
btc2_decompress	- decompress multilevel BTC encoded image
btc3_compress	- predictive BTC compression
btc3_decompress	- decompresses predictive BTC encoded image
dpc_compress	- differential predictive coding compression
dpc_decompress	- differential predictive coding decompression
glr_compress	- performs gray-level run-length coding for any window length specified by the user (window range 1–125)
glr_decompress	- performs gray-level run-length decoding from the encoded binary file

huf_compress	- performs Huffman coding and stores the probability table and encoded data into a binary file
huf_decompress	- performs Huffman decoding from the encoded binary file
jpg_compress	- JPEG compression
jpg_decompress	- JPEG decompression
rms_error	- calculates root-mean-square between two images
snr	- calculates peak signal-to-noise ratio in decibels
zon_compress	- zonal coding based compression, DC quantize
zon_decompress	- zonal coding based decompression, DC quantize
zon2_compress	- zonal coding based compression, no DC quantize
zon2_decompress	- zonal coding based decompression, no DC quantize
zvl_compress	- Ziv-Lempel compression
zvl_decompress	- Ziv-Lempel decompression

LIBCONVERTER—image conversion, I/O utilities

bintocvip	- converts binary (raw) file format to CVIPtools data structure
ccctocvip	- converts CCC file format to CVIPtools data structure
CVIPhalftone	- quantizes gray image to binary, dithering options
cviptobin	- converts CVIPtools data structure to binary (raw) file format
cviptoccc	- converts CVIPtools data structure to CCC file format
cviptoeps	- converts CVIPtools data structure to EPS file format; handles only monochrome images
cviptogif	- converts CVIPtools data structure to GIF file format
cviptoitex	- converts CVIPtools data structure to ITEX file format
cviptojpg	- converts CVIPtools data structure to JPEG file format
cviptoiris	- converts CVIPtools data structure to SGI IRIS file format
cviptopnm	- converts CVIPtools data structure to PNM file format
cviptoras	- converts CVIPtools data structure to Sun RAS file format
cviptotiff	- converts CVIPtools data structure to TIFF file format
cviptovip	- converts CVIPtools data structure to VIP file format
epstocvip	- converts EPS file format to CVIPtools data structure; handles only monochrome images
giftocvip	- converts GIF file format to CVIPtools data structure

gray_binary	- converts natural binary code to gray code and gray to binary
itextocvip	- converts ITEX file format to CVIPtools data structure
iristocvip	- converts IRIS file format to CVIPtools data structure
jpgtocvip	- converts JPEG file format to CVIPtools data structure
pnmtocvip	- converts PNM file format to CVIPtools data structure
rastocvip	- converts Sun RAS file format to CVIPtools data structure
read_Image	- reads image from disk
tifftocvip	- converts TIFF file format to CVIPtools data structure
viptocvip	- converts VIP file format to CVIPtools data structure
write_Image	- writes image to disk

LIBDISPLAY—display and view functions

display_Image	- displays an image using external software
display_RAMImage	- displays an image in X-windows from the CVIPtools image queue directly
getDisplay_Image	- gets the program (viewer) used to display images
setDisplay_Image	- sets the program (viewer) used to display images
view_Image	- general-purpose image view function

LIBFEATURE—feature extraction functions

area	- finds area of binary object (number of pixels)
aspect	- finds aspect ratio (based on bounding box) of binary object
centroid	- finds row and column coordinates of a binary object
euler	- finds euler number of a binary object
hist_feature	- finds histogram features: mean, standard deviation, skew, energy, entropy
irregular	- finds irregularity (1/thinness ratio) of binary object
label	- labels connected objects in an image
orientation	- finds orientation of a binary object via axis of least second moment
perimeter	- finds the perimeter length of a binary object
projection	- finds row and column projections of size-normalized object
rst_invariant	- finds seven rotation/scale/translation-invariant moment based on features of binary object

spectral_feature	- finds spectral features based on FFT power in rings and sectors
texture	- finds 14 texture features for four orientations
thinness	- finds thinness ratio (4π(area/perimeter^2)) of binary object

LIBGEOMETRY—geometry manipulation routines

bilinear_interp	- shrinks or enlarges an image using bilinear interpolation to calculate the gray-level value of new pixels
copy_paste	- copies a subimage from one image and pastes it into another
create_black	- creates an all-black image
create_checkboard	- creates a checkerboard image
create_circle	- creates a circular image
create_cosine	- creates a cosine wave image of any size and desired frequency
create_line	- creates a line image
create_rectangle	- creates a rectangular image
create_sine	- creates a sine wave image of any size and desired frequency
create_squarewave	- creates a square wave image of any size and desired frequency
crop	- crops a subimage from an image
display_mesh	- displays a mesh file as an image; used in mesh_warping
enlarge	- enlarges image to a user-specified size
keyboard_to_mesh	- creates a mesh structure from keyboard entry for mesh_warping
mesh_to_file	- saves a mesh structure to a file; used in mesh_warping
mesh_warping	- geometrically warps and restores an image
rotate	- rotates the given image by an angle specified by the user (range 1–360 degrees)
shrink	- shrinks the given image by a factor specified by the user (range 0.1–1)
solve_c	- solves bilinear equation; used with warp
spatial_quant	- quantizes an image by one of the following methods: average, median, or decimation
translate	- moves the entire image horizontally and/or vertically; also used for cut-and-paste of subimage
zoom	- zooms the given image by a factor specified by the user (range 1–10)

LIBHISTO—image histogram modification/contrast manipulation routines

define_histogram	- allows user to specify an equation for histogram modification
get_histogram	- generates a histogram array from an image
get_histogram_Image	- generates a histogram image from an image
gray_linear	- gray-level linear modification
gray_multiply	- remaps (if necessary) to byte and multiply, clip at 255
gray_multiply2	- casts image to float, multiplies by constant, outputs float image
histeq	- histogram equalization
histogram_show	- prints ASCII representation of a histogram
hist_spec	- performs histogram manipulation using formula specified by character string(s) for the equation(s)
histogram_spec	- performs histogram manipulation using formula specified by define_histogram
hist_slide	- histogram slide
hist_stretch	- histogram stretch, specify range and percent, clip both ends
local_histeq	- local histogram equalization
make_histogram	- generates an image of a histogram
showMax_histogram	- creates an image of a histogram of an image

LIBMORPH—morphological image processing routines

MorphClose_Image	- performs gray-scale morphological closing
MorphClose	- performs gray-scale morphological closing
MorphDilate_Image	- performs gray-scale morphological dilation
MorphDilate	- performs gray-scale morphological dilation
MorphErode_Image	- performs gray-scale morphological erosion
MorphErode	- performs gray-scale morphological erosion
morphIterMod_Image	- performs iterative morphological modification of an image
morpho	- performs iterative modification of an image
MorphOpen_Image	- performs gray-scale morphological opening
MorphOpen	- perform grayscale morphological opening

LIBNOISE—noise-generating routines

gamma_noise	- adds gamma noise to an image
gaussian_noise	- adds gaussian noise

neg_exp_noise	- adds negative-exponential noise
rayleigh_noise	- adds rayleigh noise
speckle_noise	- adds speckle (salt-and-pepper) noise
uniform_noise	- adds uniform noise

LIBSEGMENT—image segmentation routines

fuzzyc_segment	- performs Fuzzy c-Means color segmentation
gray_quant_segment	- performs gray-level quantization
hist_thresh_segment	- performs adaptive thresholding segmentation
igs_segment	- performs improved gray-scale (IGS) quantization
median_cut_segment	- performs median cut segmentation
multi_resolution_segment	- performs multiresolution segmentation
pct_median_segment	- performs PCT/Median Cut segmentation
sct_split_segment	- performs SCT/Center Split segmentation
split_merge_segment	- performs split and merge, and multiresolution segmentation
threshold_segment	- performs binary threshold on an image

LIBSPATIALFILTER—spatial filtering routines

acuity_nightvision_filter	- visual acuity and night vision application (various blur levels)
adaptive_contrast_filter	- adaptive contrast filter; adapts to local gray-level statistics
alpha_filter	- performs an alpha-trimmed mean filter
contra_filter	- performs a contra-harmonic mean filter
convolve_filter	- convolves an image with a matrix
edge_detect_filter	- performs edge detection on an image (Frei-Chen, Kirsch, laplacian, Prewitt, Pyramid, Roberts, Robinson, or Sobel)
edge_link_filter	- links edge points into lines
get_default_filter	- gets matrix for predefined spatial masks; used with convolve_filter
geometric_filter	- performs a geometric mean filter
harmonic_filter	- performs a harmonic mean filter
hough_filter	- performs an Hough transform; links specified lines
maximum_filter	- performs a maximum filter
mean_filter	- performs a mean filter
median_filter	- performs a fast histogram-method median filter

midpoint_filter	- performs a midpoint filter
minimum_filter	- performs a minimum filter on an image
mmse_filter	- minimum mean-squared error restoration filter
raster_deblur_filter	- raster deblurring filter
single_filter	- performs geometric manipulation and enhancement with a single spatial filter
smooth_filter	- smooths the given image (kernel size in the range 2–10)
specify_filter	- creates a convolution mask matrix from float array
unsharp_filter	- performs unsharp masking algorithm
Ypmean_filter	- performs a Yp mean filter

LIBTRANSFORM—two-dimensional unitary transforms

fft_transform	- performs blockwise Fast Fourier Transform
fft_phase	- extracts phase only information from a complex image
dct_transform	- performs blockwise Discrete Cosine Transform
haar_transform	- performs forward or inverse Haar transform
idct_transform	- performs inverse Discrete Cosine Transform
ifft_transform	- performs inverse Fast Fourier Transform
wavdaub4_transform	- performs wavelet xform based on Daubechies wavelet
wavhaar_transform	- performs wavelet transform based on Haar wavelet
walhad_transform	- performs Walsh/Hadamard transform (forward or inverse)

LIBXFORMFILTER—transform filtering routines

Butterworth_Band_Pass	- applies Butterworth bandpass filter in transform domain
Butterworth_Band_Reject	- applies Butterworth bandreject filter in transform domain
Butterworth_High	- applies Butterworth highpass filter in transform domain
Butterworth_Low	- applies Butterworth lowpass filter in transform domain
geometric_mean	- geometric mean restoration filter
High_Freq_Emphasis	- performs a high-frequency emphasis (HP Butterworth + offset)
h_image	- creates an image for the degradation function, $h(r, c)$
homomorphic	- performs homomorphic filtering

Ideal_Band_Pass	- applies ideal bandpass filter in transform domain
Ideal_Band_Reject	- applies an ideal bandreject filter in transform domain
Ideal_High	- applies ideal highpass filter in transform domain
Ideal_Low	- applies ideal lowpass filter in transform domain
inverse_xformfilter	- performs inverse restoration filter
least_squares	- performs least-squares restoration filter
nonfft_xformfilter	- performs standard filters (lowpass, highpass, etc.) on non-FFT symmetry transforms
notch	- performs a notch filter
parametric_wiener	- parametric wiener restoration filter, variable gamma
power_spect_eq	- power spectrum equalization restoration filter
simple_wiener	- simple wiener restoration filter (K parameter)
wiener	- wiener restoration filter

APPENDIX D

CVIPtcl Command List and Corresponding CVIPtools Functions

This list is arranged by library with the CVIPtcl commands on the left (in **bold**), and the corresponding CVIPtools C library function(s) on the right (in *italics*), along with a brief description of the function. Some of the utility functions are specific to CVIPtcl, so they do not have corresponding CVIPtools functions.

LIBARITHLOGIC

add — *add_Image*
adds two images

and — *and_Image*
logical AND two images

divide — *divide_Image*
divides one image by another

multiply — *multiply_Image*
multiplies two images

not — *not_Image*
logical NOT (invert) image

or — *or_Image*
logical OR two images

subtract — *subtract_Image*
subtracts one image from another

xor — *xor_Image*
logical EXCLUSIVE OR two images

LIBBAND

assemble_bands — *assemble_bands*
creates a multiband image from single-band images

extract_band — *extract_band*
extracts a single-band image from a multiband image

LIBCOLOR

cxform — *colorxform*
color space transforms

lum_average — *lum_average*
converts color image into gray-level image via averaging all bands

luminance — *luminance_Image*
converts color image into gray level via standard luminance transform

pct — *pct, ipct, pct_color*
principal components transform

pseudocol_freq — *pseudocol_freq*
creates a pseudocolor image using FFT spectrum and filters

LIBCOMPRESS

bitplane_rlc — *bit_compress, bit_decompress, bit_planeadd*
bitplane run-length coding

btc — *btc_compress, btc_decompress*
block truncation coding (BTC)

btc2 — *btc2_compress, btc2_decompress*
multilevel BTC

btc3 — *btc3_compress, btc3_decompress*
predictive multilevel BTC

dpc — *dpc_compress, dpc_decompress*
differential predictive coding

fractal — *frac_compress, frac_decompression*
Fractal compression

gray_rlc — *glr_compress, glr_decompress*
gray-level run-length coding

huffman — *huf_compress, huf_decompress*
Huffman compression

jpeg — *jpg_compress, jpg_decompress*
JPEG compression

rms_error — *rms_error*
calculates root-mean-square error between two images

snr — *snr*
peak signal-to-noise ratio, using two images

vector_quant — *vq_compress, vq_decompress*
Vector quantization compression

zvl — *zvl_compress, zvl_decompress*
Ziv-Lempel compression

zonal — *zon_compress, zon_decompress*
zonal coding-based compression, DC quantization

zonal2 — *zon2_compress, zon2_decompress*
zonal coding-based compression, no DC quantization

LIBCONVERTER

convert — *bintocvip*
converts binary (raw) file format to CVIPtools data structure

cviptobin
converts CVIPtools data structure to binary (raw) file format

ccctocvip
converts CCC file format to CVIPtools data structure

cviptoccc
converts CVIPtools data structure to CCC file format

epstocvip
converts EPS file format to CVIPtools data structure (only monochrome)

cviptoeps
converts CVIPtools data structure to EPS file format (only monochrome)

giftocvip
converts GIF file format to CVIPtools data structure

cviptogif
converts CVIPtools data structure to GIF file format

itextocvip
converts ITEX file format to CVIPtools data structure

cviptoitex
converts CVIPtools data structure to ITEX file format

jpgtocvip
converts JPEG file to CVIPtools data structure

	cviptojpg converts CVIPtools data structure to JPEG file format
	pnmtocvip converts PNM file format to CVIPtools data structure
	cviptopnm converts CVIPtools data structure to PNM file format
	rastocvip converts Sun RAS file format to CVIPtools data structure
	cviptoras converts CVIPtools data structure to Sun RAS file format
	tifftocvip converts TIFF file format to CVIPtools data structure
	cviptotiff converts CVIPtools data structure to TIFF file format
	viptocvip converts VIP file format to CVIPtools data structure
	cviptovip converts CVIPtools data structure to VIP file format
	cvipandiris converts IRIS file format to/from CVIPtools data structure
halftone	*CVIPhalftone* gray scale to binary via six halftone methods
gray_binary	*gray_binary* converts natural binary code to gray code and gray to binary
read, load	*read_Image* reads an image into queue
write, save	*write_Image* writes an image to disk

LIBFEATURE

area	*area* finds area in pixels of binary object
aspect	*aspect* finds aspect ratio, based on bounding box, of binary object
centroid	*centroid* finds center of area of binary object, row and column

euler — *euler*
calculates the euler number of a binary object

hist_feature — *hist_feature*
calculates histogram features: mean, standard deviation, skew, energy, entropy

irregular — *irregular*
calculates the irregularity of a binary object

label — *label*
labels objects in an image

orientation — *orientation*
determines orientation of a binary via axis of least second moment

perimeter — *perimeter*
calculates the opf a binary object perimeter

projection — *projection*
finds horizontal and vertical projections of size-normalized binary object

rst_invariant — *rst_invariant*
finds seven rotation/scale/translation-invariant moment-based features

spectral_feature — *spectral_feature*
calculates power in different rings and sectors of the Fourier spectrum

texture — *texture*
finds 14 texture features for four orientations based on second-order histogram (gray-level dependence matrix—GLDM)

thinness — *thinness*
finds the thinness of a binary object

LIBGEOMETRY

bilinear_interp — *bilinear_interp*
enlarges or shrinks an image using bilinear interpolation for gray levels

black — *create_black*
creates an all-black image filled with zeros, byte data type

circle — *create_circle*
creates circle image

copy_paste — *copy_paste*
copies a subimage from one image and pastes it in another image

cosine_wave — *create_cosine*
creates a cosine wave image

create_checkboard — *create_checkboard*
creates a checkboard image

create_mesh — *display_mesh, mesh_to_file*
creates sine wave mesh

crop — *crop*
crops an image by specifying starting coordinates and size

display_mesh — *display_mesh*
dislays an original mesh (mesh file) and the regular one

enlarge — *enlarge*
enlarges an image (one with fixed ratio)

keyboard_mesh — *mesh_to_file, display_mesh, keyboard_to_mesh*
keyboard input of a mesh

line — *create_line*
creates a line binary image

rectangle — *create_rectangle*
creates a binary image with a rectangle

rotate — *rotate*
rotates image

shrink — *shrink*
scales image by factor of .1 to 1.0

sine_wave — *create_sine*
creates a sine wave image

solve_bilinear — *solve_c*
solves bilinear equation

spatial_quant — *spatial_quant*
spatial quantization (mean, median, or decimate)

square_wave — *create_squarewave*
creates a square wave image

translate — *translate*
translates an image (wrap or fill with constant)

warp — *mesh_warping*
image warping and geometric restoration

zoom — *zoom*
image zoom

LIBHISTOGRAM

gray_linear — *gray_linear*
gray-level linear modification

gray_multiply — *gray_multiply, gray_multiply2*
multiplies image by constant, optional byte clipping

histeq — *histeq*
histogram equalization

hist_slide — *hist_slide*
histogram slide

hist_shrink — *remap_Image* (in Toolkit library libmap)
histogram shrink

hist_spec — *hist_spec*
histogram specification

hist_stretch — *hist_stretch*
histogram stretch

local_histeq — *local_histeq*
histogram equalization based on local neighborhoods

LIBMORPH

morpho — *morpho, morphIterMod_Image*
Iterative morphological filters for binary images

morph_close — *MorphClose, MorphClose_Image*
morphological close for gray-level images

morph_dilate — *MorphDilate, MorphDilate_Image*
morphological dilate for gray-level images

morph_erode — *MorphErode, MorphErode_Image*
morphological erode for gray-level images

morph_open — *MorphOpen, MorphOpen_Image*
morphological open for gray-level images

LIBNOISE

gamma_noise — *gamma_noise*
adds gamma noise to an image

gaussian_noise — *gaussian_noise*
adds gaussian noise to an image

negative_exp_noise — *neg_exp_noise*
adds negative exponential noise

rayleigh_noise — *rayleigh_noise*
adds rayleigh noise to an image

speckle_noise — *speckle_noise*
adds speckle noise to an image

uniform_noise — *uniform_noise*
adds uniform noise to an image

LIBSEGMENT

fuzzyc	*fuzzyc_segment* fuzzy C-mean segmentation
gray_quant	*gray_quant_segment* gray-level quantization
hist_thresh	*hist_thresh_segment* adaptive histogram thresholding segmentation
igs	*igs_segment* gray-level quantization using improved gray scale (IGS)
median_cut	*median_cut_segment* median cut a color image
multi_resolution	*multi_resolution_segment* multiresolution segmentation
pct_median	*pct_median_segment* PCT/Median Cut segmentation
sct_split	*sct_split_segment* SCT/Center Split
split_merge	*split_merge_segment* split and merge segmentation
threshold	*threshold_segment* thresholds an image

LIBSPATIALFILTER

(includes all edge detection functions)

ace_filter	*exp_ace_filter, log_ace_filter, ace2_filter* three types of adaptive contrast filters
adapt_median_filter	*adapt_median_filter* adaptive median filter, adapts to local statistics
adaptive_contrast	*adaptive_contrast_filter* adaptive contrast filter; adapts to local gray-level statistics
alpha_filter	*alpha_filter* alpha-trimmed mean/order filter
contra_filter	*contra_filter* contra-harmonic mean filter
diff_spatial	*get_default_filter, convolve_filter* difference spatial filter
edge_link	*edge_link_filter* links the edges in an image

frei_chen	*edge_detect_filter* Frei-Chen edge detector
gauss_blur	*get_default_filter, convolve_filter* gaussian blur spatial filter
geometric_filter	*geometric_filter* geometric mean filter
harmonic_filter	*harmonic_filter* harmonic mean filter
highpass_spatial	*get_default_filter, convolve_filter* highpass spatial filter
hough	*hough* Hough transform
kirsch	*edge_detect_filter, kirsch_edge* Krisch edge detector
laplacian	*edge_detect_filter* laplacian edge detector
laplacian_spatial	*get_default_filter, convolve_filter* laplacian spatial filter
lowpass_spatial	*get_default_filter, convolve_filter* lowpass spatial filter
maximum_filter	*maximum_filter* maximum order filter
mean_filter	*mean_filter* mean (spatial averaging) filter
median_filter	*median_filter* median order filter
midpoint_filter	*midpoint_filter* midpoint order filter
minimum_filter	*minimum_filter* minimum order filter
mmse_filter	*mmse_filter* minimum mean-squared error restoration filter; uses noise model
nightvision	*acuity_nightvision_filter* night vision and visual acuity simulation
prewitt	*edge_detect_filter* Prewitt edge detector
pyramid	*edge_detect_filter, pyramid_Image* pyramid edge detector
raster_deblur	*raster_deblur_filter* raster deblurring filter

roberts — *edge_detect_filter*
Roberts edge detector

robinson — *edge_detect_filter, robinson_edge*
Robinson edge detector

sharp — *image_sharp*
spatial sharpening filter

single_filter — *single_filter*
a single filter that allows for resizing, translating, rotating, filtering

smooth — *smooth_filter*
spatial smoothing filter

sobel — *edge_detect_filter, sobel_Image*
Sobel edge detector

specify_spatial — *specify_filter, convolve_filter*
user-specified convolution mask

unsharp — *unsharp_filter*
unsharp masking image sharpening

ypmean_filter — *Ypmean_filter*
Yp mean filter

LIBTRANSFORM

dct — *dct_transform, idct_transform*
discrete cosine and its inverse transform

fft — *fft_transform, ifft_transform*
fast Fourier and its inverse transform

fft_phase — *fft_phase*
extracts FFT phase information from an FFT spectrum, two options: normalize magnitude and leave complex, remap to byte for display

haar — *haar_transform*
Haar and its inverse transform

hadamard — *walhad_transform*
Hadamard and its inverse transform

walsh — *walhad_transform*
Walsh and its inverse transform

wavelet — *wavdaub4_transform, wavhaar_transform*
wavelet transform, with Daubechies or Haar basis vectors

LIBXFORMFILTER

bpf_filter *Ideal_Band_Pass, Butterworth_Band_Pass, nonfft_xformfilter*
bandpass filter

brf_filter *Ideal_Band_Reject, Butterworth_Band_Reject, nonfft_xformfilter*
bandreject filter

data_h_image *new_Image, getData_Image*
creates a degradation image from user input

hfe_filter *High_Freq_Emphasis, nonfft_xformfilter*
high-frequency emphasis filter

h_image *h_image*
creates an image of one of three types for degradation image $h(r, c)$

geometric_mean *geometric_mean*
geometric mean restoration filter

homomorphic *homomorphic*
homomorphic filter

hpf_filter *Ideal_High, Butterworth_High, nonfft_xformfilter*
highpass filter

inverse_xformfilter *inverse_xformfilter*
inverse restoration filter

least_squares *least_squares*
least-squares restoration filter

lpf_filter *Ideal_Low, Butterworth_Low, nonfft_xformfilter*
lowpass filter

notch *notch*
notch filter

parametric_wiener *parametric_wiener*
parametric wiener restoration filter, variable gamma

power_spect_eq *power_spect_eq*
power spectrum equalization restoration filter

simple_wiener *simple_wiener*
practical wiener (K parameter) restoration filter

wiener *wiener*
wiener restoration filter

UTILITY

add_path allows the user to add an automatic search path for image files

assemble_bands	*assemble_bands* (libband) assembles a multiband image from single-band images (libband)
datarange	*findMinVal_Vector, findMaxVal_Vector* (libvector) finds the minimum and maximum in an image of any data type
data_get	returns image data as a string, limited to 900 pixels
data_remap	*cast_Image* (libimage), *remap_Image* (libmap) maps an image to a different data type
dq	removes an image from the image queue
extract_band	*extract_band* (libband) extracts a single band from a multiband image
fft_phase	*fft_phase* (libtransform) extracts FFT spectrum phase information
gray_binary	*gray_binary* (libconverter) converts natural binary code to gray code and gray to binary
history	*history_check* (libimage) retrieves processing history information for an image
info	retrieves file type, color space, bands, size, and data type of an image (see libimage)
list	lists images in the queue
rename	renames an image in the queue
read, load	*read_Image* (libconverter) reads an image into queue
viewer	*setDisplay_Image* (libdisplay) allows the user to select the image viewer to be used by CVIPtcl
view, display	*view_Image, display_Image, display_RAMImage* (libdisplay) new options rgbhl
write, save	*write_Image* (libconverter) writes an image to disk

CVIPtcl Function Usage Notes

The function names are in ***bold italics***, followed by the necessary input parameters enclosed in <>. In general, for unneeded or unknown parameters, use –1. If a single input parameter actually requires more than one entry, it is typically "enclosed in curly brackets," which refers to { }. For example in bitplane_rlc, the <*retained_bitplane(s) (enclosed in curly brackets)*> parameter might be entered as follows:

{ 1 3 5 }

if we wished to retain the bitplanes 1, 3, and 5.

LIBARITHLOGIC

Usage: cvip ***add*** <image1> <image2>

Usage: cvip ***and*** <image1> <image2>

Usage: cvip ***divide*** <image1> <image2>

Usage: cvip ***multiply*** <image1> <image2>

Usage: cvip ***not*** <image>

Usage: cvip ***or*** <image1> <image2>

Usage: cvip ***subtract*** <image1> <image2>

Usage: cvip ***xor*** <image1> <image2>

LIBBAND

Usage: cvip ***assemble_bands*** <image1> <image2>...<imagen>

Usage: cvip ***extract_band*** <image> <bandno>

LIBCOLOR

Usage: cvip ***cxform*** <image> <color_space> <forward/inverse (1/0)> <normalize (1/0)> <do_remap (1/0)>

Usage: cvip ***lum_average*** <gray or color image>

Usage: cvip ***luminance*** <color_image>

Usage: cvip ***pseudocol_freq*** <gray image> <lowpass cutoff> <bandpass upper cutoff> <lp color(0, 1, 2 = RGB)> <bp color> <hp color> <dc>

Usage: cvip ***pct*** <color_image> <forward/inverse (1/2)>

LIBCOMPRESS

Usage: cvip ***bitplane_rlc*** <image> <retained_bitplane(s) (enclosed in curly brackets)>

Usage: cvip ***btc*** <image>

Usage: cvip ***btc2*** <image> <block_size>

Usage: cvip ***btc3*** <image> <block_size>

Usage: cvip ***dpc*** <image> <ratio (0–1)> <do_clipping (1/0)> <number_bits (1–8)> <vertical_scan (1/0)> <original/reconstructed/llyod_max (1/2/3)> <deviation>

Usage: cvip ***fractal*** <image> <tolerance (2–20)> <domain minimum size(2 means that the image is partitioned recursively at least twice)> <domain maximum size(5 means that the image is partitioned recursively to a maximum of 5 times)> <Domain pool type (0-2)> <Domain step size (1-15)> <boolean (y/n) search 3 domain classes> <boolean (y/n) search 24 domain subclasses> <scaling bits(default 5)> <offset bits(default 7)> For arguments you don't need pass -1

Usage: cvip ***gray_rlc*** <image> <window_length>

Usage: cvip ***huffman*** <image>

Usage: cvip ***jpeg*** <image> <quality (integer)> <smooth (1/0)> <grayscale (1/0)> <quantfile>

Usage: cvip ***rms_error*** <image1> <image2>

Usage: cvip ***snr*** <image1> <image2>

Usage: cvip ***vector_quant*** <image, read_codebook, cdbook_infile, error_thres, no_of_entries, rows_per_vector, cols_per_vector, CodeFileName, Scheme number> <image> color for gray <read_codebook 0/1> 0 generate new codebook; 1 using existing codebook; <cdbook_infile 0/1>0 save codebook in seperate file; 1 save codebook in compressed file<error_thres> typical value 5~10<no_of _entries> number of entries in the codebook, typical 8-128<rows_per_vector>vertical vector size\n\t<cols_per_vector> horizontal vector size<CodeFileName> file name for the codebook, note this parameter is only used if you specify to 'use existing codebook' or to 'save codebook in seperate file' t<Scheme number0/1/2> 0 REGULAR_FORMAT; 1 WVQ3; 2 WVQ2; Note function will ignore CodeFileName if not needed.

Usage: cvip ***zonal*** <gray or color image> <block size (power of 2)> <transform type (1–4)> <kernel type (1-3)> <compression ratio>
where transform: 1 = FFT, 2 = DCT, 3 = Walsh, 4 = Hadamard
kernel type: 1 = triangle, 2 = square, 3 = circle

compression ratio is from 1.0 (min) to (block_size*block_size/4) (max) for all transforms

Usage: cvip ***zonal2*** <gray or color image> <blocksize (power of 2)> <transform type (1–4)> <kernel type (1–3)> <compression ratio>
where transform: 1 = FFT, 2 = DCT, 3 = Walsh, 4 = Hadamard
kernel type: 1 = triangle, 2 = square, 3 = circle
compression ratio is from 1.0 (min) to (block_size*block_size/4) (max) for all transforms

Usage: cvip ***zvl*** <image>

LIBCONVERTER

Usage: cvip ***convert*** <image> <new_format>

Usage: cvip ***gray_binary*** <gray or color image> <direction(1/0)>

Usage: cvip ***halftone*** <gray or color image> <halftone - indicates method used to convert from grayscale to binary (1:QT_FS, 2:QT_THRESH, 3: QT_DITHER8, 4: QT_CLUSTER3, 5: QT_CLUSTER4, 6: QT_CLUSTER8)> <maxval - specifies maximum range of input image (usually 255)> <fthreshval - threshold value (for QT_THRESH) between [0.0... 1.0]> <retain_image - retain image after writing (1: CVIP_YES or 0: CVIP_NO)?> <verbose - (1: CVIP_YES or 0: CVIP_NO)?>

Usage: cvip ***read*** <image> /* same as ***load*** */

Usage: cvip ***write*** <image> [format] /* same as ***save*** */

LIBFEATURE

Usage: cvip ***area*** <labeled image> <row coordi> <col coordi>

Usage: cvip ***aspect*** <labeled image> <row coordi> <col coordi>

Usage: cvip ***centroid*** <labeled image> <row coordi> <col coordi>

Usage: cvip ***euler*** <labeled image> <row coordi> <col coordi>

Usage: cvip ***hist_feature*** <original image, color, or gray> <labeled image> <row coordi> <col coordi>

Usage: cvip ***irregular*** <labeled image> <row coordi> <col coordi>

Usage: cvip ***label*** <image>

Usage: cvip ***orientation*** <labeled image> <row coordi> <col coordi>

Usage: cvip ***perimeter*** <labeled image> <row coordi> <col coordi>

Usage: cvip ***projection*** <labeled image> <row coordi> <col coordi> <height> <width>

Usage: cvip ***rst_invariant*** <labeled image> <row coordi> <col coordi>

Usage: cvip ***spectral_feature*** <original image, color, or gray> <labeled image> <no_of_rings> <no_of_sectors> <row coordi> <col coordi>

Usage: cvip ***texture*** <original image, color, or gray> <labeled image> <ith band of original image> <row coordi> <col coordi> <hex number> <distance>

Usage: cvip ***thinness*** <labeled image> <row coordi> <col coordi>

LIBGEOMETRY

Usage: cvip ***bilinear_interp*** <image> <factor (> 1 enlarges, < 1 shrinks)>

Usage: cvip ***black*** <width> [<height>, if omitted, square image]

Usage: cvip ***circle*** <width> <height> <centerx> <centery> <radius>

Usage: cvip ***copy_paste*** <src_image> <dest_image> <r_start> <c_start> <height> <width> <r_dest> <c_dest> <transparent (1/0)>

Usage: cvip ***cosine_wave*** <image size> <1 for horizontal, 2 for vertical> <desired frequency>

Usage: cvip ***create_checkboard*** <width> <height> <starting column> <starting row> <cell width> <cell height>

Usage: cvip ***create_mesh*** <matrix row> <matrix col> <mesh_file_name>

Usage: cvip ***crop*** <image> <row coordi> <col coordi> <row size> <col size>

Usage: cvip ***display_mesh*** <mesh_file>

Usage: cvip ***enlarge*** <image> <rows> <cols>

Usage: cvip ***keyboard_mesh*** <mesh_file_name> <col> <row> <data for the col mesh matrix COL by row enclosed in curly brackets> <data for the ROW mesh matrix col by row enclosed in curly brackets>

Usage: cvip ***line*** <width> <height> <starting column> <starting row> <ending column> <ending row>

Usage: cvip ***rectangle*** <width> <height> <starting column> <starting row> <rectangle width> <rectangle height>

Usage: cvip ***rotate*** <image> <degree (1–360)>

Usage: cvip ***shrink*** <image> <factor (0.1–1.0)>

Usage: cvip ***sine_wave*** <image size> <1 for horizontal, 2 for vertical> <desired frequency>

Usage: cvip ***solve_bilinear*** {X_in[0] X_in[1] X_in[2] X_in[3]} {Y_in[0] Y_in[1] Y_in[2] Y_in[3]} {X_out[0] X_out[1] X_out[2] X_out[3]} {Y_out[0] Y_out[1] Y_out[2] Y_out[3]}

Usage: cvip ***spatial_quant*** <image> <method (1–3) (1. Average, 2. Median, 3. Decimate) <result_height> <result_width>

Usage: cvip ***square_wave*** <image size> <1 for horizontal, 2 for vertical> <desired frequency>

Usage: cvip ***translate*** <image> <wrap(y/n)> <y_slide> <x_slide> <fill_out_value (0–255)>

Usage: cvip ***warp*** <image> <mesh_file_name> <direction (either (1) from irregular mesh to a regular mesh or (2) from regular mesh to an irregular mesh)> <Gray-value interpolation method (1. Nearest neighbor 2. Bilinear interpolation 3. Neighbor average)>

Usage: cvip ***zoom*** <image> <quadrant(1-6)> <zoom_factor> <col offset> <row offset> <width> <height>

LIBHISTOGRAM

Usage: cvip ***gray_linear*** <image, any data type> <begin range> <end range> <start grayvalue> <slope> <set out-of-range to 0/do not change out-of-range(0/1)> <band>

Usage: cvip ***gray_multiply*** <image> <multiplier> <BYTE clipping/Not (1/0)>

Usage: cvip ***histeq*** <image> [0 = r,1 = g, 2 = b]

Usage: cvip ***hist_shrink*** <image> <lower limit (0–244)> <upper limit (1–255)>

Usage: cvip ***hist_slide*** <image> < 1 for slide up/ –1 for slide down> <slide value>

Usage: cvip ***hist_stretch*** <image> <lower limit (0–244)> <upper limit (1–255)> <low clip (percent)> <high clip (percent)>

Usage: cvip ***hist_spec*** <image> <spec eq...>

Usage: cvip ***local_histeq*** <image> [0 = r,1 = g, 2 = b] <blocksize>

LIBMORPH

Usage: cvip ***morph_close*** <image> <kernel type (1-4)> <kernel dimension> <height> <width>
For argument(s) you don't need, pass -1
where for kernel type, 1 means disk; 2 means square; 3 means rectangle; 4 means cross
disk: doesn't need height and width
square: doesn't need width
cross: height is used for thickness of lines and width is used for cross size

Usage: cvip ***morph_dilate*** <image><kernel type (1-4)> <kernel dimension> <height> <width>
For argument(s) you don't need, pass –1
where for kernel type, 1 means disk; 2 means square; 3 means rectangle; 4 means cross
disk: doesn't need height and width
square: doesn't need width
cross: height is used for thickness of lines and width is used for cross size

Usage: cvip ***morph_erode*** <image><kernel type (1–4)> <kernel dimension> <height> <width>
For argument(s) you don't need, pass –1
where for kernel type, 1 means disk; 2 means square; 3 means rectangle; 4 means cross
disk: doesn't need height and width
square: doesn't need width
cross: height is used for thickness of lines and width is used for cross size

Usage: cvip ***morph_open*** <image> <kernel type (1–4)> <kernel dimension> <height> <width>
For argument(s) you don't need, pass –1
where for kernel type, 1 means disk; 2 means square; 3 means rectangle; 4 means cross

disk: doesn't need height and width
square: doesn't need width
cross: height is used for thickness of lines and width is used for cross size

Usage: cvip ***morpho*** <image> {<set>} <rotation> <boolean> <iteration> <subfields>

where {<set>} is a set of numbers from 1 to 14 for surrounds
(numbers separated by spaces, commas or semicolons, e.g., {1 5 10});
<rotation> is y/n;
<boolean> is a logic function of (a,b) from 1 to 7;
<iteration> is a number between 1 and infinity (use 0 for infinity);
<subfields> is a number of subfields (1–3);

LIBNOISE

Usage: cvip ***gamma_noise*** <image> <variance (100–400)> <alpha (1–6)>

Usage: cvip ***gaussian_noise*** <image> <variance (100–400)> <mean (negative will darken image, positive will brighten image)>

Usage: cvip ***negative_exp_noise*** <image> <variance (100–400)>

Usage: cvip ***rayleigh_noise*** <image> <variance (100–400)>

Usage: cvip ***speckle_noise*** <image> <probability of high gray levels (0.01–0.05)> <probability of low gray levels (0.01–0.05)>

Usage: cvip ***uniform_noise*** <image> <variance (100-400)> <mean (negative will darken image, positive will brighten image)>

LIBSEGMENT

Usage: cvip ***fuzzyc*** <color_image> <gaussian kernel variance (0.0–20.0)

Usage: cvip ***gray_quant*** <image> <gray_level (choose one from 2, 4, 8, 16, 32, 64, 128)>

Usage: cvip ***hist_thresh*** <image>

Usage: cvip ***igs*** <gray or color image> <gray level (2, 4, 8, 16, 32, 64, 128, 256)>

Usage: cvip ***median_cut*** <image> <number of the color (2–1000)>

Usage: cvip ***multi_resolution*** <image, color or gray> <choice (1–6)> <PCT (1/0 = y/n)> <threshold/similarity> <percentage/pix distance>
For choice 1,2 & 3: <threshold> & <percentage> are –1
For choice 5: <percentage> is –1
For choice 4 & 6: no –1

Usage: cvip ***pct_median*** <color_image> <number of the colors>

Usage: cvip ***sct_split*** <color_image> <num_color on A axis> <num_color on B axis>

Usage: cvip ***split_merge*** <image, color or gray> <choice (1–6)> <level> <PCT (1/0 = y/n)> <threshold/similarity> <percentage/pix distance>
For choice 1,2 & 3: <threshold> & <percentage> are –1

For choice 5: <percentage> is –1
For choice 4 & 6: no –1

Usage: cvip ***threshold*** <image> <threshval>

LIBSPATIALFILTER

Usage: cvip ***ace_filter*** <image> <mask_size (>=3)> <type (1~3)> <mean> <gain>

Usage: cvip ***adapt_median_filter*** <image> <an odd number for mask size (>=3)>

Usage: cvip ***adaptive_contrast*** <gray or color image> <k1, local gain factor multiplier> <k2, local mean multiplier> <kernel size (odd number)> <local gain factor Min> <local gain factor Max>

Usage: cvip ***alpha_filter*** <image> <mask_size (3, 5, 7, 9 or 11)> <trim_size (a number between 0 and mask_size)>

Usage: cvip ***gauss_blur*** <image> <kernel (3, 5, 7 only)>

Usage: cvip ***contra_filter*** <image> <mask_size (3, 5, 7, 9 or 11)> <filter order [–5 to +5]>

Usage: cvip ***diff_spatial*** <image> <filter ('1' for horizontal filter, '2' for vertical filter)>

Usage: cvip ***edge_link*** <image> <maximum link distance>

Usage: cvip ***frei_chen*** <image> <smooth_filter (1–4)> <keep_DC (0/1)> <projection (1–3)> <threshold_projection (1–3)> <threshold> <radian_threshold>

Usage: cvip ***geometric_filter*** <image> <mask_size (3, 5, 7, 9 or 11)>

Usage: cvip ***harmonic_filter*** <image> <mask_size (3, 5, 7, 9 or 11)>

Usage: cvip ***highpass_spatial*** <image>

Usage: cvip ***hough*** <binary image> <degrees (enclosed in curly brackets)> <line threshold> <connection threshold>

Usage: cvip ***kirsch*** <image> <smooth_filter (1–4)> <threshold>

Usage: cvip ***laplacian*** <image> <smooth_filter (1–4)> <keep_DC (0/1)> <laplacian mask (1, 2)> <threshold>

Usage: cvip ***laplacian_spatial*** <image>

Usage: cvip ***lowpass_spatial*** <image>

Usage: cvip ***maximum_filter*** <image> <mask_size (3, 5, 7, 9 or 11)>

Usage: cvip ***mean_filter*** <image> <mask_size (3, 5, 7, 9 or 11)>

Usage: cvip ***median_filter*** <image> <an odd number for mask size (>=3)>

Usage: cvip ***midpoint_filter*** <image> <mask_size (3, 5, 7, 9 or 11)>

Usage: cvip ***minimum_filter*** <image> <mask_size (3, 5, 7, 9 or 11)>

Usage: cvip ***mmse_filter*** <gray or color image> <noise variance> <kernel size (odd number)>

Usage: cvip ***nightvision*** <image> <nightvision/acuity (y/n)> <threshold> <acuity>

Usage: cvip ***prewitt*** <image> <smooth_filter (1–4)> <keep_DC (0/1)> <kernel> <threshold>

Usage: cvip ***pyramid*** <image> <smooth_filter (1–4)> <threshold>

Usage: cvip ***raster_deblur*** <image>

Usage: cvip ***roberts*** <image> <smooth_filter (1–4)> <keep_DC (0/1)> <detector type (1, 2)> <threshold>

Usage: cvip ***robinson*** <image> <smooth_filter (1–4)> <threshold>

Usage: cvip ***sharp*** <image>

Usage: cvip ***single_filter*** <image> <horizontal sizing factor> <vertical sizing factor> <x coord for new center> < y coord for new center> <angle of rotation> <beta> <kernel size (3, 5, 7)> <kernel array (enclosed by curly brackets)> <filter operation (1–4)>

Usage: cvip ***smooth*** <image> <kernel>
where <kernel> is an integer among 3, 5, and 7

Usage: cvip ***sobel*** <image> <smooth_filter (1–4)> <keep_DC (0/1)> <kernel (3, 5, 7)> <threshold>

Usage: cvip ***specify_spatial*** <image> <rows of mask> <cols of mask> <mask array (enclosed by curly brackets)>

Usage: cvip ***unsharp*** <image> <lower limit> <upper limit> <stretch low clip (percent)> <high clip (percent)>

Usage: cvip ***ypmean_filter*** <image> <mask_size (3, 5, 7, 9 or 11)> <filter order (–5, 0) or (0, +5)>

LIBTRANSFORM

Usage: cvip ***dct*** <image> <forward/inverse (1/0)> <block_size (2^n)>

Usage: cvip ***fft*** <image> <forward/inverse (1/0)> <block_size (2^n)>

Usage: cvip ***fft_phase*** <complex image> <remap_norm (1/0)> <k>
where when remap_norm is 0, remaps the phase data; when 1, normalizes the magnitude

Usage: cvip ***haar*** <image> <forward/inverse (1/0)> <block_size (2^n)>

Usage: cvip ***hadamard*** <image> <forward/inverse (1/0)> <block_size(2^n)>

Usage: cvip ***walsh*** <image> <forward/inverse (1/0)> <block_size (2^n)>

Usage: cvip ***wavelet*** <image> <forward/inverse (1/0)> <basis wavelet> <decomposition level>

LIBXFORMFILTER

Usage: cvip ***bpf_filter*** <image> <keep_dc (y/n)> <Ideal bandpass/Butterworth bandpass (0/1)> <lower cutoff_frequency(1–block_size)> <higher cutoff_frequency (1–block_size)> <Butterworth filter order (1–8)>

Usage: cvip ***brf_filter*** <image> <keep_dc (y/n)> <Ideal bandreject/Butterworth bandreject (0/1)> <lower cutoff_frequency(1–block_size)> <higher cutoff_frequency (1–block_size)> <Butterworth filter order (1–8)>

Usage: cvip ***data_h_image*** <height> <width> <height × width data items>

Usage: cvip ***geometric_mean*** <D image> <H image> <Noise Power Image> <Original Power Image> <gamma> <alpha> <Maximum gain(times DC gain)> <cutoff frequency>

Usage: cvip ***hfe_filter*** <image> <keep_dc (y/n)> <cutoff_frequency (1–block_size)> <Emphasis offset Value> <Butterworth filter order (1–8)>

Usage: cvip ***h_image*** <type (1 = constant, 2 = fixed, 3 = gaussian)> <height> <width>

Usage: cvip ***homomorphic*** <image> <upper frequency gain> <lower frequency gain> <cutoff frequency>

Usage: cvip ***hpf_filter*** <image> <keep_dc (y/n)> <Ideal highpass/Butterworth highpass (0/1)> <cutoff_frequency (1–block_size)> <Butterworth filter order (1–8)>

Usage: cvip ***inverse_xformfilter*** <D image> <H image> < Maximum gain(times DC gain)> <cut-off frequency>

Usage: cvip ***least_squares*** <D image> <H image> <SNR Image> <gamma> <Maximum gain(times DC gain)> <filter cutoff>

Usage: cvip ***lpf_filter*** <image> <keep_dc (y/n)> <Ideal lowpass/Butterworth lowpass (0/1)> <cutoff_frequency (1–block_size)> <Butterworth filter order (1–8)>

Usage: cvip ***notch*** <image> {<row> <col> <size>}...

Usage: cvip ***parametric_wiener*** <D image> <H image> <Noise Power Image> <Original Power Image> <gamma> <Maximum gain(times DC gain)> <cutoff frequency>

Usage: cvip ***power_spect_eq*** <D image> <H image> <Noise Power Image> <Original Power Image> <Maximum gain(times DC gain)> <cutoff frequency>

Usage: cvip ***simple_wiener*** <D image> <H image> <gamma>

Usage: cvip ***wiener*** <D image> <H image> <Noise Power Image> <Original Power Image> <Maximum gain(times DC gain)> <cutoff frequency>

UTILITY

Usage: cvip ***add_path*** <path>

Usage: cvip ***assemble_bands*** <image1> <image2>...<imagen>

Usage: cvip ***data_get*** <single band real image>

Usage: cvip ***datarange*** <image>

Usage: cvip ***data_remap*** <image> <choice for data_type (0, 1, 2, or 3)>
where 0 for byte, 1 for short, 2 for integer, 3 for float

Usage: cvip ***display*** <image> [rgblh] /* same as ***view*** */

Usage: cvip ***dq*** <image-name>

Usage: cvip ***extract_band*** <image> <bandno>

Usage: cvip ***fft_phase*** <complex image> <remap_norm (1/0)> <k>
where when remap_norm is 0, remaps the phase data; when 1, normalizes the magnitude

Usage: cvip ***gray_binary*** <gray or color image> <direction (1/0)>

Usage: cvip ***history*** <image> [PROGRAM]

Usage: cvip ***info*** <image>

Usage: cvip ***list***

Usage: cvip ***read*** <image> /* same as ***load*** */

Usage: cvip ***rename*** <old-name> <new-name>

Usage: cvip ***snr*** <image1> <image2>

Usage: cvip ***rms_error*** <image1> <image2>

Usage: cvip ***viewer*** [picture/RamViewer]

Usage: cvip ***write*** <image> [format] /* same as ***save*** */

APPENDIX F

CVIP Resources

USEFUL CVIP SOFTWARE (free or shareware)

CVIPtools *http://www.ee.siue.edu/CVIPtools*

ImageMagick UNIX-based image processing software from Visioneering Research Laboratory, Inc.
http://sunsite.unc.edu/pub/X11/contrib/applications/ImageMagick/
http://www.vrl.com/Imaging/index.html
http://ftp.wizards.dupont.com/pub/ImageMagick/

IMDISP PC/DOS image processing from NASA
http://nssdc.gsfc.nasa.gov/cd-rom/software/imdisp.html

Image Viewers Freeware and shareware viewers for Windows NT/95
http://www.shadow.net/tucows/softgrap.html

JPEG JPEG C source code
http://www.ijg.org
http://www.disc.org.uk/jpeg/

Khoros image processing, user interface, and graphics tools using X-windows
http://www.khoros.unm.edu/khoros
ftp from *ftp.khoral.com* (*pub/khoros*)

LIBTIFF Portable library of routines for TIFF files
ftp from *sgi.com* (*graphics/tiff*)

NetPBM Internet supported version of pbmplus
ftp from *ftp: wuarchive.wustl.edu* (*/graphics/graphics/packages/NetPBM/*)

NIH Image Image processing and analysis for Macintosh platform from National Institutes of Health
http://ipt.lpl.arizona.edu/IPT/NIHImage/

pbmplus image conversion routines (used in CVIPtools)
http://www.arc.umn.edu/GVL/

tcl/tk scripting language for graphical user interface (used in CVIPtools)
http://sunscript.sun.com/
http://www.arc.umn.edu/GVL

UCFImage Image processing package for PCs from University of Central Florida
http://macmyler.engr.ucf.edu/UCFImage.html
ftp from *ipg.engr.ucf.edu* (*/pub/UCFImage*)

XF a GUI-based tool for creating GUI's in tcl/tk (used in creating CVIPtools)
http://www.cimetrix.com/sven/xf.html

xv image viewer for X-windows
http://www.sun.com/software/catlinkxv/xv.html
ftp from *ftp://ftp.cis.upenn.edu/pub/xv*

Vista computer vision software
http://www.ius.cs.cmu.edu/help/ImageViewers/vista-help.html

PROFESSIONAL SOCIETIES

Institute of Electrical and Electronic Engineers (IEEE)
445 Hoes Lane
PO Box 1331
Piscataway, NJ 08855
World Wide Web: *http://www.ieee.org*

The Pattern Recognition Society (PRS)
National Biomedical Research Foundation
Georgetown University Medical Center
39090 Reservoir Road, NW
Washington, DC 20007

Society of Motion Picture and Television Engineers (SMPTE)
595 West Hartsdale Avenue
White Plains, NY 10607
World Wide Web: *http://www.smpte.org*

Society of Photo-Optical Instrumentation Engineers (SPIE)
PO Box 10
Bellingham, WA 98227-0010
World Wide Web: *http://www.spie.org*

Association for Computing Machinery (ACM)
One Astor Plaza
1515 Broadway
New York, NY 10036
World Wide Web: *http://www.acm.org*

CVIP-RELATED STANDARDS

National Television Standards Committee (NTSC)

Sequential Coleur Avec Memoire (Sequential Color with Memory—SECAM)

Phase Alternation Line (PAL)

International Telecommunications Union (ITU, formerly CCITT): ITU Group 3 and 4, ITU H.261

International Telecommunications Union—Radio (ITU-R, formerly CCIR): ITU-R 601

International Standards Organization (ISO): JPEG/MPEG

American Standards Institute (ANSI), 1430 Broadway, New York, NY 10018

Commission Internationale de l'Eclairage (CIE)

Video Electronics Standards Association (VESA)

Federal Communication Commission (FCC)

Most Standards are available in the United States from American National Standards Institute:
ANSI
11 West 42nd Street
New York, NY 10036
Tel: (212) 642-4900
http://www.ansi.org

ISO Standards can be obtained directly from:
ISO Central Office
Case Postale
1211 Geneva 20
Switzerland
Tel: (22) 749-0111
http://www.iso.ch/welcome.html

ITU (formerly CCITT) Standards can be obtained directly from:
ITU Sales Section
Place Des Nations
CH-1211 Geneva 20
Switzerland
Tel: (22) 730-5111
http://www.itu.int/publications

More information on the JPEG standard and the JFIF file format can be obtained from:

C-Cube Miscrosystems
Attn: Scott St. Clair
Corporate Communications
1778 McCarthy Boulevard
Milpitas, CA 95035
Tel: (408) 944-6300

WORLD WIDE WEB

http://peipa.essex.ac.uk
Pilot European Image Processing Archive

http://www.cs.cmu.edu/afs/cs/project/cil/ftp/html/vision.html
Computer Vision Resource

http://www-sci.lib.uci.edu/HSG/MedicalImage.html
Medical Imaging Center

http://www.rz.go.dlr.de:8081/softarch.html
DLR (German Aerospace Research Establishment)
Freeware Contributed Software Information

http://www.eecs.wsu.edu/IPdb/title.html
Digital Image Processing Instructional Database at Washington State University

http://www-isis.ecs.soton.ac.uk/research/visinfo/rgroup.html
Image Processing Research Groups

http://george.lbl.gov/computer_vision.html
Lawrence Berkeley National Laboratory, ITG-Computer Vision and Robotics

http://george.lbl.gov/ITG.html
Imaging and Distributed Computing Group, Berkeley Lab

http://www-video.eecs.berkeley.edu/
Video and Image Processing Lab, Berkeley

JOURNALS

IEEE Transactions on Image Processing

IEEE Spectrum

IEEE Engineering in Medicine and Biology Magazine

IEEE Signal Processing Magazine

IEEE Expert Magazine

Computer Vision, Graphics and Image Processing

Graphical Models and Image Processing

Image Understanding

IEEE Transactions on Medical Imaging

Computerized Medical Imaging and Graphics

IEEE Transactions on Pattern Analysis and Machine Intelligence

IEEE Transactions on Computers

Pattern Recognition

IEEE Transactions on Signal Processing

IEEE Transactions on Neural Networks

IEEE Transactions on Geoscience and Remote Sensing

Photogrammetric Engineering and Remote Sensing

International Journal of Remote Sensing

Journal of Visual Communication and Image Representation

Numerous conference proceedings and other journals from IEEE, SPIE, SMPTE, PRS

TRADE MAGAZINES

Advanced Imaging, PTN Publishing, 445 Broad Hollow Road, Melville, NY 11747

AV Video & multimedia producer, Knowledge Industry Publications, Inc., 701 Westchester Avenue, White Plains, NY 10604, email: *AVVMMO@kipi.com*, WWW: *http://www.KIPINET.com*

Biophotonics International, Laurin Publishing, Berkshire Common, 2 South Street, P.O. Box 4949, Pittsfield, MA 01202-4949, email: *Photonics@MCIMail.com*

Computer Graphics World (CGW), PennWell Publishing Co., 10 Tara Boulevard, 5th Floor, Nashua, NH 03062-2801, WWW: *http://www.cgw.com*

Imaging Magazine, Telecom Library Inc., 12 West 21 Street, New York, NY 10010, WWW: *http://www.imagingmagazine.com*

Laser Focus World, Pennwell Publishing, 1421 S. Sheridan, P.O. Box 180, Tulsa, OK 74101-9847

NewMedia, Hypermedia Communications, Inc., 901 Mariner's Island Boulevard, Suite 365, San Mateo, CA 94404, email:*edit@newmedia.com*

Photonics Spectra, Laurin Publishing, Berkshire Common, P.O. Box 4949, Pittsfield, MA 01202-4949, email: *Photonics@MCIMail.com*

INDEX

D

G

H

I

J

K

L

M

N

O

P

Q

R

S

T

U

V

W

Y

Z

LICENSE AGREEMENT AND LIMITED WARRANTY

READ THE FOLLOWING TERMS AND CONDITIONS CAREFULLY BEFORE OPENING THIS SOFTWARE PACKAGE. THIS LEGAL DOCUMENT IS AN AGREEMENT BETWEEN YOU AND PRENTICE-HALL, INC. (THE "COMPANY"). BY OPENING THIS SEALED SOFTWARE PACKAGE, YOU ARE AGREEING TO BE BOUND BY THESE TERMS AND CONDITIONS. IF YOU DO NOT AGREE WITH THESE TERMS AND CONDITIONS, DO NOT OPEN THE SOFTWARE PACKAGE. PROMPTLY RETURN THE UNOPENED SOFTWARE PACKAGE AND ALL ACCOMPANYING ITEMS TO THE PLACE YOU OBTAINED THEM FOR A FULL REFUND OF ANY SUMS YOU HAVE PAID.

1. **GRANT OF LICENSE:** In consideration of your payment of the license fee, which is part of the price you paid for this product, and your agreement to abide by the terms and conditions of this Agreement, the Company grants to you a nonexclusive right to use and display the copy of the enclosed software program (hereinafter the "SOFTWARE") on a single computer (i.e., with a single CPU) at a single location so long as you comply with the terms of this Agreement. The Company reserves all rights not expressly granted to you under this Agreement.

2. **OWNERSHIP OF SOFTWARE:** You own only the magnetic or physical media (the enclosed disks) on which the SOFTWARE is recorded or fixed, but the Company retains all the rights, title, and ownership to the SOFTWARE recorded on the original disk copy(ies) and all subsequent copies of the SOFTWARE, regardless of the form or media on which the original or other copies may exist. This license is not a sale of the original SOFTWARE or any copy to you.

3. **COPY RESTRICTIONS:** This SOFTWARE and the accompanying printed materials and user manual (the "Documentation") are the subject of copyright. You may not copy the Documentation or the SOFTWARE, except that you may make a single copy of the SOFTWARE for backup or archival purposes only. You may be held legally responsible for any copying or copyright infringement which is caused or encouraged by your failure to abide by the terms of this restriction.

4. **USE RESTRICTIONS:** You may not network the SOFTWARE or otherwise use it on more than one computer or computer terminal at the same time. You may physically transfer the SOFTWARE from one computer to another provided that the SOFTWARE is used on only one computer at a time. You may not distribute copies of the SOFTWARE or Documentation to others. You may not reverse engineer, disassemble, decompile, modify, adapt, translate, or create derivative works based on the SOFTWARE or the Documentation without the prior written consent of the Company.

5. **TRANSFER RESTRICTIONS:** The enclosed SOFTWARE is licensed only to you and may not be transferred to any one else without the prior written consent of the Company. Any unauthorized transfer of the SOFTWARE shall result in the immediate termination of this Agreement.

6. **TERMINATION:** This license is effective until terminated. This license will terminate automatically without notice from the Company and become null and void if you fail to comply with any provisions or limitations of this license. Upon termination, you shall destroy the Documentation and all copies of the SOFTWARE. All provisions of this Agreement as to warranties, limitation of liability, remedies or damages, and our ownership rights shall survive termination.

7. **MISCELLANEOUS:** This Agreement shall be construed in accordance with the laws of the United States of America and the State of New York and shall benefit the Company, its affiliates, and assignees.

8. **LIMITED WARRANTY AND DISCLAIMER OF WARRANTY:** The Company warrants that the SOFTWARE, when properly used in accordance with the Documentation, will operate in substantial conformity with the description of the SOFTWARE set forth in the Documentation. The Company does not warrant that the SOFTWARE will meet your requirements or that the operation of the SOFTWARE will be uninterrupted or error-free. The Company warrants that the media on which the SOFTWARE is delivered shall be free from defects in materials and workmanship under normal use for a period of thirty (30) days from the date of your purchase. Your only remedy and the Company's only obligation under these limited warranties is, at the Company's option, return of the warranted item for a refund of any amounts paid by you or replacement of the item. Any replacement of SOFTWARE or media under the warranties shall not extend the original warranty period. The limited warranty set forth above shall not apply to any SOFTWARE which the Company determines in good faith has been subject to misuse, neglect, improper installation, repair, alteration, or damage by you. EXCEPT FOR THE EXPRESSED WARRANTIES SET FORTH ABOVE, THE COMPANY DISCLAIMS ALL WARRANTIES, EXPRESS OR IMPLIED, INCLUDING WITHOUT LIMITATION, THE IMPLIED WARRANTIES OF MERCHANTABILITY AND FITNESS FOR A PARTICULAR PURPOSE. EXCEPT FOR THE EXPRESS WARRANTY SET FORTH ABOVE, THE COMPANY DOES NOT WARRANT, GUARANTEE, OR MAKE ANY REPRESENTATION REGARDING THE USE OR THE RESULTS OF THE USE OF THE SOFTWARE IN TERMS OF ITS CORRECTNESS, ACCURACY, RELIABILITY, CURRENTNESS, OR OTHERWISE.

IN NO EVENT, SHALL THE COMPANY OR ITS EMPLOYEES, AGENTS, SUPPLIERS, OR CONTRACTORS BE LIABLE FOR ANY INCIDENTAL, INDIRECT, SPECIAL, OR CONSEQUENTIAL DAMAGES ARISING OUT OF OR IN CONNECTION WITH THE LICENSE GRANTED UNDER THIS AGREEMENT, OR FOR LOSS OF USE, LOSS OF DATA, LOSS OF INCOME OR PROFIT, OR OTHER LOSSES, SUSTAINED AS A RESULT OF INJURY TO ANY PERSON, OR LOSS OF OR DAMAGE TO PROPERTY, OR CLAIMS OF THIRD PARTIES, EVEN IF THE COMPANY OR AN AUTHORIZED REPRESENTATIVE OF THE COMPANY HAS BEEN ADVISED OF THE POSSIBILITY OF SUCH DAMAGES. IN NO EVENT SHALL LIABILITY OF THE COMPANY FOR DAMAGES WITH RESPECT TO THE SOFTWARE EXCEED THE AMOUNTS ACTUALLY PAID BY YOU, IF ANY, FOR THE SOFTWARE.

SOME JURISDICTIONS DO NOT ALLOW THE LIMITATION OF IMPLIED WARRANTIES OR LIABILITY FOR INCIDENTAL, INDIRECT, SPECIAL, OR CONSEQUENTIAL DAMAGES, SO THE ABOVE LIMITATIONS MAY NOT ALWAYS APPLY. THE WARRANTIES IN THIS AGREEMENT GIVE YOU SPECIFIC LEGAL RIGHTS AND YOU MAY ALSO HAVE OTHER RIGHTS WHICH VARY IN ACCORDANCE WITH LOCAL LAW.

ACKNOWLEDGMENT

YOU ACKNOWLEDGE THAT YOU HAVE READ THIS AGREEMENT, UNDERSTAND IT, AND AGREE TO BE BOUND BY ITS TERMS AND CONDITIONS. YOU ALSO AGREE THAT THIS AGREEMENT IS THE COMPLETE AND EXCLUSIVE STATEMENT OF THE AGREEMENT BETWEEN YOU AND THE COMPANY AND SUPERSEDES ALL PROPOSALS OR PRIOR AGREEMENTS, ORAL, OR WRITTEN, AND ANY OTHER COMMUNICATIONS BETWEEN YOU AND THE COMPANY OR ANY REPRESENTATIVE OF THE COMPANY RELATING TO THE SUBJECT MATTER OF THIS AGREEMENT.

Should you have any questions concerning this Agreement or if you wish to contact the Company for any reason, please contact in writing at the address below.

Robin Short
Prentice Hall PTR
One Lake Street
Upper Saddle River, New Jersey 07458